ECOLOGICAL INVESTIGATIONS OF STEPHEN ALFRED FORBES

ECOLOGICAL INVESTIGATIONS OF STEPHEN ALFRED FORBES

ARNO PRESS

A New York Times Company

New York / 1977

Editorial Supervision: LUCILLE MAIORCA

———◆———

Reprint Edition 1977 by Arno Press Inc.

Copyright © 1977 by Arno Press Inc.

HISTORY OF ECOLOGY
ISBN for complete set: 0-405-10369-7
See last pages of this volume for titles.

Manufactured in the United States of America

———◆——— # 3071665

Library of Congress Cataloging in Publication Data

Forbes, Stephen Alfred, 1844-1930.
  Ecological investigations of Stephen Alfred Forbes.

  (History of ecology)
  Reprinted from various sources.
  1. Animal ecology--Collected works. 2. Animals,
Food habits of--Collected works. I. Title. II. Series.
QH541.145.F65 1977 591.5'08 77-74217
ISBN 0-405-10387-5

CONTENTS

Howard, L[eland] O.
BIOGRAPHICAL MEMOIR OF STEPHEN ALFRED FORBES, 1844-1930.
With a Bibliography Compiled by H.C. Oesterling (Reprinted from *National Academy of Sciences of the United States of America Biographical Memoirs,* Volume XV), Washington, D.C., 1932

Forbes, Stephen Alfred
Papers from *Illinois Laboratory of Natural History Bulletin*

BIOGRAPHICAL MEMOIR
OF
STEPHEN ALFRED FORBES

L. O. Howard

Stephen A. Forbes

NATIONAL ACADEMY OF SCIENCES

OF THE UNITED STATES OF AMERICA

BIOGRAPHICAL MEMOIRS

VOLUME XV—FIRST MEMOIR

BIOGRAPHICAL MEMOIR

OF

STEPHEN ALFRED FORBES

1844-1930

BY

L. O. HOWARD

PRESENTED TO THE ACADEMY AT THE ANNUAL MEETING, 1931

STEPHEN ALFRED FORBES

1844-1930

BY L. O. HOWARD

Chronology

Born at Silver Creek, Stephenson County, Ill., May 29, 1844.
Attended the district school from school age to the age of 14.
1858-60, studied at home, under brother's instruction.
1860, attended Beloit Academy.
1861-5, soldier, U. S. Army.
1866-7, attended Rush Medical College.
1867, raised strawberries at Carbondale, Ill.
1867-9, studied and practiced medicine under a preceptor at Makanda, Ill.
1868-70, taught school and studied science at Makanda, Ill.
1870-1, taught school and studied science at Benton, Ill.
1871, studied at Illinois State Normal University.
1871-2, taught school and studied science at Mount Vernon, Ill.
1872-7, curator, museum of the Illinois State Natural History Society, at Normal, Ill.
1875-8, instructor in zoology, Illinois State Normal University.
1877, founded, and to 1917, Director, Illinois State Laboratory of Natural History.
1882-1917, State Entomologist of Illinois.
1884, Ph.D., University of Indiana.
1884-1909, professor of zoology and entomology, University of Illinois.
1886, awarded first-class medal of the Société d'Acclimatation de France, for scientific publications.
1888-1905, Dean, College of Science, University of Illinois.
1891-2, biological expeditions, Rocky Mountains, U. S. Fish Commission.
1893, special agent, U. S. Fish Commission; Director of the Aquarium, World's Columbian Exposition; Chairman, International Congress of Zoologists; President, American Association of Economic Entomologists; also pre-

3

pared natural history exhibit of Illinois for Columbian Exposition.

1894, established Illinois Biological Station.

1901-7, President, National Society of Horticultural Inspectors.

1905, LL.D., University of Illinois.

1908, President, Illinois Academy of Science; American Association of Economic Entomologists.

1909-21, professor of entomology, University of Illinois.

1912, President, Entomological Society of America.

1917-30, Chief, State Natural History Survey.

1918, elected member National Academy of Sciences.

Died at Urbana, Ill., March 13, 1930.

This chronology is taken from a pamphlet entitled "In Memoriam Stephen Alfred Forbes" published by the University of Illinois shortly after his death. To the chronology is added a list of the societies of which he was a member, which included among others the American Philosophical Society and the Entomological Society of France. It is also stated in the same paragraph that he was an honorary member of the American Ornithologists' Union; had been President of the Ecological Society of America, and that he was an honorary member of the International Congress of Entomology.

In the same pamphlet was printed a short article on Doctor Forbes' ancestry, education and character, written by his son Ernest Browning Forbes of the State College of Pennsylvania. He was the son of Isaac Forbes and Agnes Van Hoesen, and, racially speaking, was Scotch on his father's side and Dutch on his mother's, both theoretically and practically an admirable mixture. The first authentic record of his American ancestry is the marriage of Daniel Forbes in 1660 to Rebecca Perriman at Cambridge, Mass. There seem to be no authentic records of the origin of this Daniel or of his emigration from Scotland. From this Daniel, through Thomas, Aaron, Stephen, John and Isaac, came the subject of this sketch. Of his great grandfathers, three at least were soldiers in the Revolutionary War, namely, Stephen Forbes, who lived at Brookfield, Mass., and later at Hardwick in the same State, who served in a Hardwick company and was given land in Vermont; Garrett Van Hoesen (on the mother's side), a sergeant in a company of sharpshooters under General Schuyler; and Captain Isaac Sawyer, a Yale student who served with distinction in the Sullivan campaign against the Iroquois in 1779.

It is Professor Ernest Forbes' opinion that among the later ancestors of his father those who left the deepest impress were his mother, who was a high-strung woman, and his paternal grandmother whose name was Anna Sawyer and who is said to have been the strongest source of the intellect of all of her descendants.

At the time of Stephen's birth at Silver Creek, Stephenson County, Illinois, the family lived in a log house on a small farm and in a condition of privation general among the pioneer families of the Middle West. His father was obliged to sell the farm upon which this log house stood, as the result of having endorsed a note for a friend, and had to begin again on a new farm, in a house built of slabs.

There were six children, namely, Flavilla Anna, born 1824, married 1842; Francis Van Hoesen, born 1828, died 1848; Mary Elizabeth, born 1831, married, 1850; Henry, born in 1833; Stephen, 1844, and Agnes Verneth, known as "Nettie," born 1845.

When Stephen was ten years old his father died. The older brother Henry was then twenty-one, but had been independent since he was fourteen, and at the time was working his way towards a college education for which he had prepared at Freeport and Rockford and on which he had made about two years' progress by private study. Stephen, his mother and Nettie were living on a farm of 140 acres, poorly equipped and heavily mortgaged, living in a cabin of a single room in which was a cookstove at one end, two beds at the other, a trundle bed under one of them, and a dining table in the middle. At one side of the building was a little "lean-to" containing a single bed on which the father of the family died. The mother was broken in health, and the two younger children were practically helpless. Henry, however, stepped into the breach. Professor Forbes has written—

"Without an hour's hesitation he gave up his personal plans, abandoned his career for which he had already given brilliant promise, and took upon his own shoulders the burden of our support and education; and from that time on for the next seven years he was not our guardian merely, but he was our guardian angel.

"He saved enough on the farm to build us a comfortable home. He taught me French, sent me to Beloit to prepare for college, and helped my sister to the Woman's College at Northwestern University at Evanston; and when the Civil War broke out he sold the farm, paid off the still existing mortgage, gave my mother the residue of the family funds to live on with her eldest daughter Flavilla, borrowed the money with which to buy his horse and mine for the cavalry service, and to the war we went together, he as lieutenant and presently captain of our company, becoming later lieutenant colonel of the regiment and colonel by brevet." (From E. B. Forbes.)

As to the details of the Army service of S. A. Forbes, I am able to quote from an unpublished autobiograpical sketch in a letter that he wrote in 1923 and which I take from the *Scientific Monthly* of May, 1930:

"My army service, concerning which you ask particulars, was begun by my enlistment as a private in Company B, 7th Illinois Cavalry, in September, 1861, when I was seventeen years old. At eighteen I was made orderly sergeant; at nineteen, second lieutenant, and at twenty, captain of my company. Just after my eighteenth birthday, when sent to carry an important dispatch to a distant outpost near Corinth, Mississippi, I was put upon the wrong road and presently found myself inside the rebel lines as a prisoner of war. Telling my captors that I had a verbal message only, which I refused, and they did not compel me, to disclose, I availed myself of an opportunity to tear up my dispatch, secreting the fragments in the pistol holster on my saddle. At General Bragg's headquarters I was threatened with hanging if I did not produce my dispatch and was thoroughly searched for it, as was also my saddle, but nothing was found. Later in the day I had a brief interview with General Beauregard, in command of the rebel army after the battle of Shiloh, and a much longer one with a major of his staff, who ended by wishing me good luck, and telling me to appeal to him if I got into trouble. I was in prison four months, at Mobile, Alabama; Macon, Georgia; and Richmond, Virginia. Utilized my abundant leisure by studying Greek from books which I managed to buy at Mobile. When paroled and released, I was sent to a hospital for three months to recover from scurvy and malaria, acquired in prison. Rejoined my regiment, reenlisted for the war, and was mustered out in November, 1865."

This is a very short and very modest account of the boy's war adventures. He was only twenty-one when he was mustered out (he had the rank of Captain at twenty), and yet he had seen nearly five years of service. The experience, apparently, hurt neither his mind nor his physique. Perhaps in the long run it was physically good for him; and the boy who used his days in a Confederate prison studying Greek was surely one whose mind was not rusting. Writing of him in *Science,* April 11, 1930, Dr. Henry B. Ward says, "It was a source of great delight to hear the story of those days, when on rare occasions he could be persuaded to relate to younger friends some of his experiences in the field."

And now as to the matter of education. In the unpublished autobiographical sketch from which we have quoted the paragraph regarding his army service, he states, "My formal education . . . was incomplete and fragmentary, partly because of the poverty of my family, my father, a pioneer farmer, having died when I was ten years old, and partly because the Civil War took me away as I was getting ready to enter college."

In the chronological list beginning this account occur the items, "Attended the district school from school age to the age of fourteen; 1858-60 studied at home under brother's instruction;[1] 1860 attended Beloit Academy."

The attendance at Beloit Academy in 1860 was a difficult one to bring about, on account of lack of money. Ernest Forbes writes in his account, "When the young Stephen was obliged to give up his studies at Beloit Academy, because the last available dollar was gone, he had to take his trunk to the railroad station in a wheelbarrow." And he has since written me, "This year at Beloit was enjoyed at the expense of real hardship for the family at home. The letters written to Stephen contained repeated

---

[1] Henry taught Stephen the French language, and Professor Henry B. Ward states that he studied Italian and Spanish at this time.

suggestions that he try to find a cheaper boarding place, and serious debate of his expressed desire to spend one dollar and twenty cents for a glass lamp, since his eyes suffered as a result of study by candle light."

According to S. A. Forbes' manuscript sketch, when he returned from the war he began and nearly finished a course in medicine at Rush College in Chicago. We must not be satisfied with this bare statement. In his printed account, Professor Ernest Forbes says, "He discontinued his medical studies before their completion, however, because his savings from his career as a soldier became exhausted and because a series of incidents having to do mainly with surgical operations without the use of anesthetics convinced him that he was not temperamentally adapted to medical practice." And then he goes on to write, "His interest in natural science was determined by an academic tradition in the family, by an agricultural background, by four years' out-of-doors experience in the Army, by a naturally thoughtful habit, and by a continuing scientific interest after the cessation of his medical studies."

The chronology shows that he raised strawberries at Carbondale, Ill., and taught school and studied science at Makanda and Benton, and that he studied at the Illinois State Normal University in 1871. It was about this period (in 1870) that his first notes were published, under the heading "Botanical Notes (from Southern Illinois)" in the *American Entomologist and Botanist* for September and October of that year. This journal was then edited by C. V. Riley, and each number contained a botanical section under the editorship of George Vasey then living at Richview, Ill. The notes by Professor Forbes had evidently been sent in by request. They were confined to the counties of Union and Jackson, Illinois, gave a running account of the topography of the region, and discussed the plant growth in such a way as almost to foreshadow his future interest in the broad study of environment and association that became known as ecology. At all events they secured the warm approbation of Doctor Vasey, with whom he corresponded frequently on botanical subjects.

In his unpublished autobiographical sketch, Professor Forbes, after referring to his unfinished course in medicine, goes on to say, ". . . but becoming infatuated with a study of the botany of southern Illinois I changed my plans, taught school for a living, and studied natural history as an avocation, with the result that when Major J. W. Powell, afterwards U. S. Geologist, resigned in 1872 as Curator of the Museum of the State Natural History Society at Normal, Illinois, I was chosen to succeed him there."

The old Illinois State Natural History Society had been in existence since 1858. It was chartered in 1861. Its museum was housed in the building of the State Normal School at Normal. Its final business meetings were held in Bloomington, May 26th and June 22nd, 1871, and its museum was then formally passed over to the State of Illinois to be held by the State Board of Education "for the use and benefit of the State." The earliest curator was Dr. J. A. Sewall, instructor in chemistry in the State Normal School and afterwards President of the State University of Colorado. The second curator was Major J. W. Powell, and he held this post when he made his well known western explorations. Major Powell's deputy in the Museum, who, however, served as the actual curator, was Dr. George Vasey. In 1872 Doctor Vasey was appointed Botanist of the United States Department of Agriculture in Washington and Curator of the U. S. National Herbarium under the Smithsonian Institution. He took the oath of office April 13 of that year.

Doctor Vasey (who was then a man of fifty) had a high opinion of young Forbes. They both belonged to the Natural History Society; Forbes had studied at the State Normal University in 1871, and, moreover, Doctor Vasey must have been impressed by the character of the notes Forbes had sent in for publication in the *American Entomologist and Botanist*. I take it, and in fact I have been told, that it was largely Vasey's recommendation that secured Forbes the appointment as Curator of the Museum in 1872.

Most of the facts in the last two paragraphs are taken from a long, careful and admirable address given by Professor Forbes on the occasion of a meeting called to establish the Illinois Academy of Sciences. It is entitled "History of the Former State Natural History Societies of Illinois," and was published in *Science,* December 27, 1907, pages 892-898.

For the next two or three years he was diligently occupied with his duties as curator, but found time to publish several papers, mostly relating to the study of natural history and how natural history should be taught—really educational papers. Then came the first Summer School of Natural History and his appointment as Instructor in Zoology in the Illinois State Normal University. In his 1907 address on the natural history societies of Illinois, he mentions this summer school and its effect in bringing about the organization of the second natural history society of the State. It will be interesting to quote his words:

"Under the influence of Darwin and Agassiz and Huxley, a transforming wave of progress was sweeping through college and school, a wave whose strong upward swing was a joy to those fortunate enough to ride on its crest, but which smothered miserably many an unfortunate whose feet were mired in marsh mud."

Has this movement ever been so concisely and perfectly put into words? He goes on to say,

"This wave reached central Illinois in the early seventies, with the effect to bring about in 1875 a summer school of natural history at the State Normal School—only two years, it will be noticed, after the first session of the Agassiz School at Penikese. Wilder of Cornell and W. S. Barnard, just back from Europe with a doctor's degree, were members of its teaching staff, together with Burrill of the State University, Thomas, the State Entomologist, and the present writer, who was also director of the school. Besides an abundance of living plants and animals of our own environment, we had great boxes and barrels of marine material in large variety, some of it received alive, secured by a most active collector engaged for the purpose, who scoured the New England coast for us from Portland to Buzzard's Bay."

The school was a notable success. The Philadelphia Centennial Exposition of 1876 deranged plans for its immediate continuance, but a second summer session was held in 1878 and the students organized themselves into the second of the Illinois State natural history societies. Obviously this summer helped to turn Forbes' attention to aquatic things, for in December, '75, he published his first strictly zoological paper—on Amblystoma; and in the following year he published his "List of Illinois Crustacea with Descriptions of New Species"; and in 1877 was published the first of his papers on the food of birds, the first one on the food of fishes appearing in 1878. In 1878 also appeared the first of his strictly entomological publications, which happened to be on "Breeding Habits of Corixa" (an aquatic insect). It will thus be seen that much early work was done with aquatic forms. And during this early period he formulated plans for the starting of the State Natural History Survey which was a child of the Illinois State Laboratory of Natural History which was founded by Professor Forbes in 1877 and of which he continued to be the Director until 1917.

In his obituary notice in *Science* of April 11, 1930, Dr. Henry B. Ward says, "He will always be looked upon as the first and the leading worker in America on aquatic biology. When he started on his studies of fresh water organisms the inland waters of our country were practically unknown. He was the first man to write on the fauna of the Great Lakes, and to contribute to a knowledge of the food of fishes, a fundamental piece of work for the proper understanding of the factors concerned in solving questions that are involved in the preservation of our commercial fisheries."

It is safe to state further that he was the first man to start careful study of the food of birds, a subject that was taken up only a few years later by the old Section of Economic Ornithology started by C. Hart Merriam in the old Division of Entomology of the U. S. Department of Agriculture and that afterwards developed into the Bureau of Biological Survey.

In his work on the fauna and flora of the lakes and inland streams, he was a pioneer and was responsible for most important studies made by others. Professor C. A. Kofoid, now of the University of California, was associated with him for many years, and has sent me the following paragraph relating to this work:

"Professor Forbes early in his career took an interest in the microscopic life of fresh water. It bore fruit in his early papers on the food of fresh water fishes. His studies in this field led to his interest in fresh water Entomostraca, an interest which led ultimately to the studies by Dr. Sharp on the Ostracoda, by his own son Dr. E. B. Forbes on the Copepoda, and by Mr. Schacht on the genus Diaptomus. His brochure of 1887, read before the Peoria Scientific Association and later reprinted with emendations, on "The Lake as a Microcosm" is a classic not only in its interpretation of a variety of lakes which he had investigated but also because it is probably one of the first if not indeed the first biological paper which recognizes the phenomenon of animal associations, the foundation of the modern science of ecology. His later work in this field was limited, by the development of his entomological interests, to explorations of the aquatic life in the waters of Yellowstone Park, of Lake Geneva, and his activity in founding and maintaining the Biological Station on the Illinois River, housed in a floating laboratory at Havana, Ill., and engaged in 1894-1900 in a continuous survey of the plankton of that stream prior to the opening of the Chicago Drainage Canal. This work resulted in my "Studies on the Plankton of the Illinois River." His interest in this work never flagged and he followed its course intimately and later in conjunction with Mr. Richardson investigated and reported on the destructive effect which the influx of sewage later had upon the life in this river."

It will be noticed in studying the bibliographical list appended to this paper that he rarely published anything about insects until early in the 1880's, his first work having been in connection with aquatic forms and with the food of fishes and birds; but in 1882, on the retirement of Cyrus Thomas, he was made State Entomologist of Illinois and two years later Professor of Zoology and Entomology in the University of Illinois. From that time on his writings naturally came to relate preponderatingly to insects.

Before we go into his work with insects, however, more must be said about his earlier work. At the memorial exercises held at the University of Illinois on March 15, 1930, an address was made by Dr. H. J. Van Cleave, a distinguished zoologist of a younger generation, who has done much work in aquatic zoology, and, rather than to attempt myself to evaluate Forbes' work at this earlier period, I will quote the words of the vastly more competent Doctor Van Cleave:

"Very early in his program, he became impressed with the significance of interrelationships between organisms and their environment. Before ecology had been conceived as an offspring of the biological sciences, Professor Forbes had adopted the ecological point of view in his published writings. As early as 1887, in a spirit of prophetic anticipation of the coming ecological era in North America, he delivered a paper on *The Lake as a Microcosm,* wherein he set forth the themes of interdependence of organisms and community of interests in aggregations of living beings. This ecological point of view has dominated his entire program of research. His contributions on insects, as well as those on fishes and on birds have been conceived in light of relationship to the environment and have always acknowledged man and human interests as essential though by no means exclusive factors of importance in the environment of organisms. As a consequence, even his economic studies have been engendered in that breadth of biological interpretation that renders them distinctive in their field.

"In his pioneer work on the food of birds, of fishes, and of insects, he contributed a wholly new method of attack upon problems in the interpretation of the economic status of animals. Many and marked have been the tributes paid him by subsequent workers in recognition of the importance of these pioneer studies.

"His early interest in the food of birds in its relation to their economic importance, found an extension in later years to a numerical study of birds in different localities. Thereby he sought a more exact means of interpreting the true value of birds in relation to agriculture.

"The long series of papers on the biology of the Illinois River, stands as one of the greatest monuments to Professor Forbes. These investigations, starting in 1894, were carried on under his leadership with the collaboration of Frank Smith, C. A. Kofoid, and R. E. Richardson, whose names will be perpetually associated with that of S. A. Forbes. Intensive work on the life of the Illinois River and correlation of physical conditions with distribution of life of the stream were carried on for a long period of years. When changed conditions made a continuation of this project seem unprofitable the researches had extended to that point where it was widely acclaimed that the Illinois River had been more thoroughly investigated as to its biology than any other river in the world.

. . . . . .

"Much of his earlier work was upon the fresh water crustacea which at the time when he began his studies were practically

NATIONAL ACADEMY BIOGRAPHICAL MEMOIRS—VOL. XV

unknown in this country. A colleague, C. Dwight Marsh, an internationally recognized authority in this group, has evaluated Professor Forbes' earliest taxonomic studies with the statement: 'In North America articles were published regarding some forms in the early part of the century, but nothing recognizable appeared until S. A. Forbes commenced his series of papers. Although these papers were not extensive, they were exact and carefully worked out, and to Forbes may be given the credit of laying the foundation for all subsequent work in this country.'

"Some notion of the thoroughness of his studies may be gained from the fact that a paper which he published in 1876 is the most comprehensive work on the higher crustacea of Illinois which has appeared down to the present date. When Professor Forbes was accorded that signal honor of membership in the National Academy of Sciences, the significance of this early period of his work that had escaped the memory of many of his associates was again brought to light."

It is interesting to remember that the present writer was one of the proposers of Professor Forbes for the Academy, and that he drew especial attention to the significance of the early period of his work, largely for the purpose of emphasizing his great breadth of view as a zoologist and as an entomologist.

In 1882 Forbes was made State Entomologist of Illinois. At that time only three States had had officials with this title, namely New York, Illinois and Missouri; and with the appointment of C. V. Riley in Washington in 1878, Missouri abandoned work of this kind. The Illinois position was created by the State legislature in the winter of 1866-67. The first incumbent of the office had been a very extraordinary man, Benjamin D. Walsh, an Englishman by birth, of Cambridge training, who published only one report before his death. Walsh was succeeded by Dr. William LeBaron, who published four reports as appendices to the Transactions of the State Horticultural Society from 1871 to 1874. LeBaron died in harness, and was succeeded by Rev. Cyrus Thomas, who published a series of six reports extending over the years 1875 to 1880. After the publication of the last of these reports Thomas transferred his labors to the field of ethnology, and upon his withdrawal from office Forbes, then Director of the State Laboratory of Natural History, was appointed. He held the office from that time until 1917.

In 1894 the present writer gave an address entitled "A Brief Account of the Rise and Present Condition of Official Economic Entomology." At that time Forbes had published six reports, and I wrote of them in the following words:

"Prof. Forbes' reports are among the best which have been published. They are characterized by extreme care and by originality of treatment which has seldom been equaled. The practical end is the one which he has kept mainly in view. His experiments with the arsenites against the codling moth and the plum curculio were the first careful, scientific experiments in this direction which were made, and his investigations of the bacterial diseases of insects have placed him in the front rank of investigators in this line. His monographic treatment of the insects affecting the strawberry plant is a model of its kind, and the same may be said of his work upon the corn bill-bugs and of his studies of the chinch bug. In fact, whatever insect or group of insects has been the subject of his investigations, he has attacked the problem in a thoroughly original and eminently scientific and practical manner."

I feel sure, however, that, strong as is this praise, I did not do full justice to these reports nor to Professor Forbes' work. I was afraid of superlatives and hardly appreciated his breadth of view. From the start his work made a profound impression upon every one interested in the applications of science to agriculture, and he was chosen as the second President of the Association of Economic Entomologists, following C. V. Riley who, as the original proposer of the Association and as the official government entomologist, was naturally made its first President. It is interesting to note moreover that Forbes is the only man who was ever reëlected to that office. This reëlection occurred in the year 1908. It is perfectly safe to say that all through his active career he was looked upon as the leader among the economic entomologists of America—the man who commanded every one's respect and whose judgment was considered to be absolutely sound. And a man of this type was greatly needed among the men drawn from here, there and everywhere, in the opening up of this relatively new branch of applied science. He realized from the start and as a result of his former broad studies the necessity for the careful study of all of the ecological features connected with the undue increase of a given species. He appreciated this before the words *ecology* and *ecological* came into use, and in a way he may be said to have been the founder of the science of ecology in the United States.

In 1917 the position of State Entomologist was merged into the Illinois Natural History Survey, and he continued as Chief of the Survey until the time of his death.

In 1884 he was called to Urbana as the head of the Department of Zoology and Entomology in the University of Illinois. The State Laboratory of Natural History was transferred to the University, and in 1917 became the Illinois Natural History Survey. His professorship of zoology and entomology in the University was changed in 1909, and he remained Professor of Entomology until his retirement as Emeritus Professor in 1921. He was Dean of the College of Science from 1888 to 1905, and after that was the Chairman of the Committee on University Educational Policy of the University Senate.

In the list of his writings, appended to this paper, will be found more than four hundred titles, but this does not include any reference to articles in fugitive journals or to any small published notes. Neither does it include the titles of such of his writings as do not concern Science. Two of them, of transcendent merit have come to my attention. His "Gierson's Cavalry Raid" was published in the Transactions of the Illinois State Historical Society, 1907, and his wonderful essay "War as an Education" appeared in the Illinois Magazine for October, 1911. Of these titles, 19 refer to educational matters, 74 to zoology aside from entomology, 10 to botany, 20 to ecology, 52 to general natural history, and 224 (a little more than half) to entomology. But this list of titles in entomology is somewhat misleading from the fact that it includes only as individual titles his reports as State Entomologist. Of these he published eighteen, and all were lengthy and included many topics of which the majority might well have appeared as separate entries in his bibliography.

All of his writings on entomological topics stand out conspicuously from the mass of publications on this subject. Here for the first time in the history of economic entomology a man not only of striking originality but of very broad biological sympathies and experiences found himself engaged in the multi-farious problems of insect damage to human interests. None of his predecessors or contemporaries in this field had his broad experience in biology or his broad outlook on nature. Therefore, especially in his general addresses, he establishd beyond question his leadership among all of his fellow workers. No one who heard or read his address to the Economic Entomologists at the meeting of the American Association for the Advancement of Science in Baltimore in December, 1908 (*Journal of Economic Entomology,* vol. II, no. 1, February 1909, pp. 25-35), will ever forget it. It made a very deep impression which has resulted in no end of good to all of us in our work and in our views of things.

So much for Forbes' scientific work. As to the man himself, we must devote some attention. In a charming paper on his "Ancestry, Education and Character," his son, Ernest B. Forbes, gives a very intimate insight into his father's character. It is in fact too intimate in part to be quoted here, although it was quite appropriate as a contribution at the memorial services in Urbana. I have selected the paragraphs that follow as indicating the man very clearly aside from his scientific achievements.

During the years 1861-65, his four-years' experience in the army constituted as intensive a course of instruction as any obtainable in college, and it was all education to the mind athirst and prepared to receive it. Enlisting in Company B of the Seventh Illinois Cavalry at the age of 17, he pursued his military studies as diligently as he had applied himself to books, and was made an orderly sergeant at 18, a lieutenant at 19, and captain of his company at 20. His company was on active duty during almost the whole period of the war, and he was under fire on 22 occasions including engagements at Clinton, Port Hudson, Byhalia, Moscow, Somerville, Collinsville, Okolona, Memphis, Campbellsville, Lynville, Franklin, and Nashville.

While in the vicinity of Corinth, Miss., he was captured while carrying a despatch, and spent four months in prison at Mobile, Macon, and Richmond. During this time, under desperately discouraging conditions, he maintained his morale by studying Greek; and later, after exchange, during a period of inactivity, he bought Spanish books, which he learned to read. It was all school, to the born scholar.

Of the educational value of his military experience he writes: "In one respect particularly, our experience was a hopeful prophecy, if not itself a cause, of subsequent success. Any one who had kept the solitary flame of his separate intellectual life steadily burning through all the blasts and storms of war, might reasonably believe that nothing that should happen to him thereafter could possibly extinguish it; and this, as we all know, is more than can be inferred from the completion of an ordinary college course. The eager hunger with which the students among us attacked the full tables at home, after four years or more of semi-starvation on a few husks and scraps, and the enthusiastic appreciation with which we embraced such long deferred opportunities as still remained to us, made it certain that no artificial graduation-day would put an end to our studies, and this, after all, is the best outcome of an education. Those of us who survived the Civil War in good health and strength, with morals unstained and minds still alert, have had no final cause to regret what seemed at the time the complete wreckage of our plans of life. To us war was not hell, but at the worst a kind of purgatory, from whose flames we emerged with much of the dross burned out of our characters, and with a fair chance still left to each of us to win his proper place in the life of the world."

In politics my father was a Republican, from the day when as a 14-year-old boy he listened to the Lincoln and Douglas debate at Freeport, Ill., on August 26, 1858. He came away from this historic event "quite aflame with enthusiasm for the new Republican party and especially for Lincoln as its champion, and equally incensed against Douglas as the leader and champion of the Democrats."

Of this occasion he has written, in part:

"When Lincoln arose to open the debate, my first feeling was a genuine shock of surprise, of disappointment, of chagrin at his homeliness, his awkwardness, his plainness of attire—at the farthest remove from the bearing, look, and dress of a boy's ideal; but when he began his argument in his high, penetrating voice, calm, clear, connected, and so simple and lucid that even I could follow it without effort, I got the first impression of my life of a truly lofty character and a great mind in vigorous action.

"Lincoln, as you will remember, made the opening speech at Freeport, and although his most telling points were enthusiastically applauded, there was practically nothing in the responses of his audience to indicate that there were any Democrats among them. Douglas, in fact, opened his rejoiner with a compliment, not to the speaker for his calm and persuasive speech, but to the assembly for 'the kind and respectful attention which they yielded not only to political friends but to those opposed to them in politics.' He himself received at first the same kind of treatment, being even more frequently and vociferously applauded than Lincoln; but as he warmed up to his argument he began seemingly to try to irritate his opponents by calling them always 'black' Republicans, with an angry and contemptuous emphasis on the word 'black.' The taunt was received in silence for a few times, and then loud cries of 'white, white,' began to come from all directions, every time he used the offensive epithet, and the clamor presently became so great, after a peculiarly irritating application of it, that Douglas paused to remind his hearers 'that while Lincoln was speaking there was not a Democrat *vulgar* and *blackguard* enough to interrupt him.' It was at this point that my adhesion to republicanism became complete, and I shouted up to Douglas at the top of my boyish voice: 'Lincoln didn't use any such talk.' I was sharply reproved by those about me, and told that I must not 'talk back'; and so stood in mortified silence until Lincoln again took the stand, when he began by saying: 'The first thing I have to say to you is a word in regard to Judge Douglas' declaration about the "vulgarity and blackguardism" in the audience, . . . that no such thing as he says, was shown by any Democrat while I was speaking. Now, I only wish, by way of reply on this subject, to say that while *I* was speaking, I used no "vulgarity or blackguardism" toward any Democrat.' With this elaboration of my own sentiment I need not say that I was relieved and delighted, or that I joined in the hearty laughter and applause with which . . . his rejoinder was received."

As a Republican, however, my father was, as in most other ways, an independent, and he never hesitated to "scratch" his ticket in favor of the Democrat when he believed the Democrat to be the better man.

My father's religious status is made clear by the following, which was written in 1923: "I was, and still am, a rationalist and an agnostic, for whom what is known as faith is merely assumption, often practically necessary, since in active life one must very often act *as if* he believed what he does not and cannot really know, but unexcusable in purely theoretical matters,"— but he drew a line through the words "and still am," and wrote above them the words "as a younger man." I think that this change was dictated by a growing hope that there is more to life than a scientist knows, or can know. At this time he was much attracted by the beauty and the comfort of the orthodox beliefs of his childhood, which he had lost during his scientific career.

Throughout life my father was active in the support of liberal religion, and, with my mother, initiated the movement which resulted in the establishment of the Unitarian Church of Urbana.

My father was reticent concerning his inner and deeper life, and, in 1923, in writing of things he would do differently if he were to live again he said, "I would be open, frank, and free with respect to my theories of life, and especially of religion, instead of leaving my children, so far as I was concerned, to find their own way in these difficult and important matters, as best they individually could. It was not indifference or thoughtlessness that influenced me, but, for one thing, a belief in individual freedom in this field, unprejudiced by authority exercised upon the helpless and defenseless child, and for another a conviction that my personal views were so far out of harmony with those of the ordinary American community that my children might find themselves strangers and even outcasts among their fellows if they were to follow my lead."

Physically, my father was characteristically restless, active, and energetic. While at the height of his powers his course through the Natural History Building could be traced by the slamming of the doors behind him. The attitude of command, attained during his extensive military service, was habitual, and he expected action on the part of his subordinates.

Until the last few years he exercised with phenomenal persistence. In the early days he did a great deal of walking, running, rowing, wood-chopping, club swinging, roller-skating and horse-back riding. He rode a bicycle for many years, and averred that his use of this vehicle added ten years to his life.

Later he drove an automobile and became locally famous for a long series of minor mishaps,—resulting from the facts that the automobile came late in the period of his physical adaptability, and that he drove to the accompaniment of his intensely concentrated thinking, without knowing that he failed to give the job his whole attention.

He learned to skate, swim, and to play golf with the generation following his own, but did not become proficient in these exercises. "Common" labor was uncongenial to him, and he once said that, for him, it would be "like hitching a race-horse to a dump cart."

Mentally he was quick, acute and alert; he delighted to deal with nice distinctions, and especially with complicated situations; and had a decided predilection for statistical and graphic presentation of products of research.

He was a poor sleeper, about five hours being his normal amount. The remainder of the night he spent reading, in bed.

He was throughout life a prodigious reader, of a wide range of serious literature. When delving into treatises in the fields of philosophy, ethics, psychology, pedagogy, metaphysics, genetics, scientific theory, and the method of science, his mind was enjoying its choicest diet; and by way of lighter reading he devoured quantities of French fiction.

Temperamentally he was in all ways a man of refined sensibility. He had a developed appreciation of the dramatic; and greatly enjoyed the theatre. His soul was attuned to the poetic, and he was devoted to Browning. When deeply moved he occasionally wrote an exquisite poem—to express himself—and then, characteristically, destroyed it. He loved music, and he played the organ in the days before pianos were common. He was naturally imaginative, and by early training was deeply religious.

. . . . . .

Since he was of an intensely intellectual type, he was easily bored by boresome people, and made but few intimate friends. There was, however, almost always some small and choice group of kindred spirits with whom he met for feasts of philosophy, and these men he enjoyed immensely.

During his later years, as his professional interests became more perfectly organized, and more securely established, my father was able to relax a certain degree of the pressure under which he had formerly worked, and to enjoy knowing a greatly broadened group of acquaintances. These years, therefore, were increasingly rich in gratifying personal relations.

It is characteristic of the man that even among his closest friends he was fair and temperate in his expressions regarding those whom he disapproved. He was much too wise to let himself hate anyone. On occasion, however, he could demonstrate a power of description equivalent to vivisection.

The dominant motive in my father's life was overwhelmingly that of scientific research in the public service. Research was, with him, not a conscious passion, I think; it was life itself—as natural as breathing; and his public-mindedness was no less spontaneous. It is true that he managed the investment of his savings with great interest but to have been condemned to a life devoted to private gain would have been for him genuine imprisonment.

The final, supreme demonstration of his superb morale was when with health, strength, friends of his own age, and wife all gone, he was able, through sheer intellectuality, to continue at his work, stimulated by the doing of new things, in the true spirit of youth, until 9 days from the date of his death, at the age of 85 years, 9 months, and 13 days.

I have already quoted from a letter recently sent me by Prof. C. A. Kofoid that portion relating to Doctor Forbes' early work on aquatic life and animal associations, and I now introduce another passage from the same letter which refers to Professor Forbes in other ways:

Intellectually Professor Forbes was a man of rare ability marked by an acquisitive mind and an unusual perspective of values. He early mastered the languages needed in scientific research. He read widely, and no work of importance escaped his keen eye. It was his wont to glance over the incoming literature in the library of the State Laboratory of Natural History, and he often dropped a casual word to his associates on the matters which interested him or which might interest others. His library reflected very well the breadth and activity of his intellectual and scientific interests. In spite of a certain aloofness, due apparently to an inherent shyness, he enjoyed conversation on the highest planes. He sought out and cultivated the acquaintance of those with whom he found and to whom he gave intellectual stimulus. He was the leading spirit in the "Theory Club"— a group of six, consisting of Professors Kinley (later President), Palmer (Chemistry), Townsend (Mathematics), Daniels (Philosophy) and myself the youngest of the group. We met regularly in the home of some member, and digested and discussed in a heavily nicotinized atmosphere such mighty treatises as Karl Pearson's "Grammar of Science" and Ward's "Naturalism and Agnosticism." As a dialectician, Forbes was keen in analysis, alike in attack on and defense of the position of the author under discussion and of his critics among our numbers. We came to look upon him much as an arbiter. He was at his best when he was in the midst of the clarification of some knotty problem in the philosophical relations of science, when with uplifted cigar in hand he would discourse, wittily, fluently and pointedly, till his cigar went cold, on subjects far afield from Hessian fly and corn aphids but with a familiarity and readiness as masterful as though he were in a group of his entomological colleagues.

It would be inferred very naturally from this that Forbes wrote unusually well. He did more than that. He was one of the ablest writers among scientific men. I have never read anything more perfectly done than his article "War as an Education," published in the *Illinois Magazine* for October, 1911. Others have called attention to this especial ability. Henry B.

Ward says (loc. cit.) : "In all his writings he manifested a beauty of style that made them unusually appealing." Herbert Osborn called him "an adept in the use of language." C. L. Metcalf says, "His writings were characterized by their remarkable, simple, lucid expression." Dr. David Kinley, in his memorial address, refers especially to Forbes' "lucidity of expression," and goes on to say, "He was a master of literary style, writing with a simplicity and clarity that would serve as a model for many specialists in literature and rhetoric."

In the long quotation from Dr. E. B. Forbes a page or two back, the paragraph beginning with the words "Of the educational value of his military experience   .   .   ." quotes the last paragraph from "War as an Education," and it was well worth quoting. But simply for delightful reading, let me quote one more paragraph from that remarkable essay. It relates to the call of the Civil War to the youth of that time :

I had prepared for college, and was "studying ahead" on the college course of the time as well as I could on a farm and under the tuition of an older brother, when the echo of the cannon in Charleston harbor drove all such ideas out of my mind. To remain quietly at home, busy with books and teachers while my comrades were thronging to the front to fight for *my country,* was simply impossible, and the mere thought of it intolerable. Indeed, I think that most of us were secretly glad that we had been born in the time when it was possible for a boy to do anything so wildly and gloriously different from what had been planned for him as to go to war. It was not to us a dilemma, a sacrifice ; it was a privilege, an intoxicating opportunity ; we could not be made to stay at home. And this was not by any means the result of our training. It was because of something born in us ; it was in our blood—and it is in the blood of young Americans yet, peace gospels and Carnegie foundations to the contrary notwithstanding. The longing for adventure, the youthful spirit rising to the challenge of danger, the thrill of responsive feeling to the "call of the country," all merging in that irresistible swell of patriotic emotion which lifted a whole people on its mighty bosom—these were the influences within us and without us which made us feel for the time that war was the only thing in the world worth while.

Personally, Professor Forbes was tall (six feet or thereabouts), reasonably slender, muscular, active. People have spoken of his "military carriage," the result of early army training. Although I had known him for more than forty years, I never noticed anything especially military in his bearing. I have just looked up a snapshot taken of him by Paul Marchal in 1913, and which is published in that famous French writer's book "Les Sciences biologiques Appliquée à l'Agriculture dans les États Unis" (Figure 111), and I find that, while the figure is straight, the head is advanced, confirming my recollection of him at that time. The picture was taken on Doctor Forbes' birthday. Marchal and I happened to visit him on that day and stayed to the family birthday dinner. In earlier years he had a notably quick step and military carriage.

His son Ernest has referred to his father's physical activities and to his later interest in certain outdoor sports. Henry B. Ward, in his account in *Science,* referring to his automobile driving, says, ". . . he often chuckled at a comment on his arrest for speeding on his eightieth birthday." He was a handsome, well featured man, of, I should say, a scholarly type. A very interesting comment on his personal appearance was made by Sir William Ramsey in a letter to Doctor Kinley written after a visit to the University of Illinois. I quote two paragraphs:

When I was in Champaign, Professor Forbes, whom I saw frequently, struck me as being strangely familiar and I was always puzzled as if I ought to have recognized him. It was only after I left that the explanation occurred to me. He is the typical Aberdonian of one, and that the commonest, type. There are two distinct types of Aberdonian, one like him, and it is the most characteristic, a big, powerfully built frame of a certain characteristic build, and with a certain aspect of face. This is the type that is regularly associated with Aberdeen. If a painter were instructed to paint a picture of the battle of Harlaw, where the Aberdeen Burghers, led by Provost Davidson, defeated the whole force of the Highlands (so far as the Highlands ever could be united in anything) and were looking about for a characteristic set of five men from whom he should make up the type of Provost Davidson, Professor Forbes would certainly be one of the five.

. . . .

That is why Professor Forbes seemed to be such a familiar figure and personality. He was *the* typical Aberdonian of his own kind, who had preserved the type absolutely unchanged. The Aberdonian never yields to circumstances but dominates them and remains himself (except in the case of a few persons who are conceited and therefore affect an Anglified accent, which only makes them comic).

In 1873 Professor Forbes married Clara Shaw Gaston, and her death in 1930 preceded his own by a few weeks only. It was a long, happy and very loving companionship. Two of Mrs. Forbes' colonial ancestors came over on the *Mayflower*, namely, John Tilley and John Howland. In her were combined French, Scotch, English and Welsh blood.

The devoted couple had five children—Bertha Van Hoesen (Mrs. Burton R. Herring), Ernest Browning, Winifred, Ethel (Mrs. Frank W. Scott), and Richard Edwin. The last-named died at the age of six years, in 1903, but all of the others are still living.

I must not conclude this account of Forbes' long and distinguished career without especial reference to an article entitled "Stephen Alfred Forbes—An Appreciation," by Frank Smith, then just retiring from his professorship in the University of Illinois, and published in *The Audubon Bulletin*, 1926, No. 17. Doctor Smith, himself a broad zoologist, gives an especially careful and appreciative survey of Forbes' work other than in entomology.

It will undoubtedly appear from what I have written that although the subject of this memoir did most important work in botany and in a number of branches of zoology, he was far more than a specialist in any branch. He studied the birds and the fishes and the insects and the life of the rivers and lakes, all as elements of a great complex, and he studied them broadly in their relations to their surroundings. Man himself was his starting ecological factor. In fact, it will be difficult if not impossible to point out a naturalist of his generation who was more original or broader or sounder.

BIBLIOGRAPHY

COMPILED BY H. C. OESTERLING, FORMER EDITOR,
ILLINOIS STATE NATURAL HISTORY SURVEY

Key To Abbreviations of Titles of Illinois Publications

Bul. Sci. Assoc. Peoria, Ill....Bulletin of the Scientific Association of Peoria, Illinois.

Ill. Acad. Sci. Trans....Transactions of the Illinois State Academy of Science.

Ill. Adm. Rep....Administrative Report of the Directors of Departments of the State of Illinois.

Ill. Agr. Exp. Sta. Bul....Bulletin of the Agricultural Experiment Station of the University of Illinois.

Ill. Agr. Exp. Sta. Cir....Circular of the Agricultural Experiment Station of the University of Illinois.

Ill. Blue Book....Blue Book of the State of Illinois.

Ill. Crop Rep. Circ....Illinois Crop Report.

Ill. Dept. Agr. Trans....Transactions of the Department of Agriculture of the State of Illinois.

Ill. Ent. Exec. Rep....Executive Report of the Illinois State Entomologist.

Ill. Ent. Off. Bul....Bulletin of the Illinois State Entomologist's Office.

Ill. Ent. Off. Circ....Circular of the Illinois State Entomologist's Office.

Ill. Ent. Off. pamphlet...Pamphlet of the Illinois State Entomologist's Office.

Ill. Ent. Rep....Report of the State Entomologist on the Noxious and Beneficial Insects of the State of Illinois.

Ill. Farm. Inst. Rep....Report of the Illinois Farmers' Institute.

Ill. Fish Com. Rep....Report of the Illinois State Fish Commission.

Ill. Hort. Soc. Trans....Transactions of the Illinois State Horticultural Society.

Ill. Jour. Com....Illinois Journal of Commerce, Chicago.

Ill. Lab. Nat. Hist. Bul....Bulletin of the Illinois State Laboratory of Natural History.

Ill. Lab. Nat. Hist. Circ....Circular of the Illinois State Laboratory of Natural History.

Ill. Lab. Nat. Hist. pamphlet...Pamphlet of the Illinois State Laboratory of Natural History.

Ill. Nat. Hist. Surv. Bul....Bulletin of the Illinois State Natural History Survey.

Ill. Nat. Hist. Surv. Ent. Circ....Entomological Circular of the Illinois State Natural History Survey.

Ill. Nat. Hist. Surv. For. Circ....Forestry Circular of the Illinois State Natural History Survey.

Ill. Nat. Hist. Surv. pamphlet...Pamphlet of the Illinois State Natural History Survey.
Ill. Nat. Hist. Surv. Rep....Report on the Natural History Survey of the State of Illinois, issued by the State Laboratory of Natural History.
Ill. Sch. Jour....Illinois School Journal, Bloomington.
Ill. Sch. Rep....Report of the Superintendent of Public Instruction of the State of Illinois.
Ill. Schoolm....Illinois Schoolmaster.
Ill. Supt. Pub. Instr. Circ....Illinois Superintendent of Public Instruction Circular.
Ill. Univ. pamphlet...Pamphlet of the University of Illinois.
Ill. Univ. Quart....Quarterly and Fortnightly Notes of the Alumni Association of the University of Illinois.
Ill. Univ. Rep....Report of the Board of Trustees of the University of Illinois.
Illini (Univ. Ill.)....The Illini, student publication, University of Illinois.
Ill. Water Sup. Assoc. Proc....Proceedings [of the Fifth Meeting] of the Illinois Water Supply Association.

1870

Botanical notes (from Southern Illinois). Amer. Ent. & Bot., vol. 2, pp. 317-318, 352, Sept. & Oct. 1870.

1872

The independent study of natural history. Chicago (*later* Illinois) Schoolmaster, vol. 5, pp. 313-316, Nov. 1872.

1873

Natural history in the public schools. Ill. Schoolm., vol. 6, pp. 363-373, Nov. 1873.

1874

The states of matter. Ill. Scholm., vol. 7, pp. 317-319, Oct. 1874.
Suggestions to teachers of zoology (I). Ill. Schoolm., vol. 7, pp. 391-395, Dec. 1874.

1875

Suggestions to teachers of zoology (II). Ill. Schoolm., vol. 8, pp. 73-79, Mar. 1875.
The School and College Association of Natural History of the State of Illinois. Ill. Sch. Rep., vol. 10, pp. 143-146, 1875.
Amblystoma punctatum. Ill. Schoolm., vol. 8, pp. 402-406, Dec. 1875.

1876

List of zoological duplicates in the museum of the Illinois Natural History Society, for exchange with public schools. Ill. Lab. Nat. Hist. Circ., 5 pp., Jan. 1876.
The beginning point in zoology. Ill. Schoolm., vol. 9, pp. 207-211, June 1876.

The museum and the school. Ill. Schoolm., vol. 9, pp. 273-276, Aug. 1876.

Introduction (on the nature and purpose of the Illinois Museum of Natural History and its projected series of publications). Ill. Mus. (*later* Lab.) Nat. Hist. Bul., vol. 1, no. 1, pp. 1-2, Dec. 1876.

List of Illinois Crustacea, with descriptions of new species. Ill. Mus. (*later* Lab.) Nat. Hist. Bul., vol. 1, no. 1, pp. 3-16 (key to Illinois species, pp. 17-24, and appendix describing three extra-limital species, pp. 24-25), with 1 pl., Dec. 1876.

1877

The food of birds. Ill. Hort. Soc. Trans., vol. 10, pp. 37-44, 1877.

Report on the Museum of Natural History at Normal. Ill. Sch. Rep., vol. 11, pp. 324-331, 1877.

1878

Introduction (explaining change of title of bulletin series). Ill. Lab. Nat. Hist. Bul., vol. 1, no. 2, p. 1, June 1878.

Circular of information (on equipment, library, collections, investigations, etc.) Ill. Lab. Nat. Hist. Circ., 14 pp., June 1878.

The food of Illinois fishes. Ill. Lab. Nat. Hist. Bul., vol. 1, no. 2, pp. 71-89, June 1878.

Breeding habits of Corixa. Amer. Nat., vol. 12, p. 820, Dec. 1878.

Notes on the development of Amia. Amer. Assoc. Adv. Sci. Proc., vol. 27, pp. 296-298, 1878.

1879

On some sensory structures of young dog-fishes. Amer. Quart. Micr. Jour., vol. 1, pp. 257-260, pl. xix, July 1879.

The food of birds. Ill. Hort. Soc. Trans., vol. 12, pp. 140-145, 1879.

Report of the Director of the State Laboratory of Natural History (to the State Superintendent of Public Instruction), for 1877-1878. Ill. Sch. Rep., vol. 12, pp. 191-198, 1879. (*Based on* "Circular of Information" issued in June 1878).

1880

The food-habits of thrushes. Amer. Ent., vol. 1 (n. s.), pp. 12-13, Jan. 1880.

The food of birds : the thrush family. Ill. Hort. Soc. Trans., vol. 13, pp. 120-172, 1880. (*Extracts,* Amer. Nat., vol. 14, pp. 448-450, June 1880.)

The food of the bluebird (*Sialis sialis* L.). Amer. Ent., vol. 1 (n. s.), pp. 215-218, 231-234, Sept.-Oct. 1880. (*See supplementary note,* Amer. Nat., vol. 15, pp. 66-67, Jan. 1881.)

The food of the darters. Amer. Nat., vol. 14, pp. 693-703, Oct. 1880.

On some interactions of organisms. Ill. Lab. Nat. Hist. Bul., vol. 1, no. 3, pp. 3-17, Nov. 1880. (*2d. ed.,* pp. 3-18, 1903.)

The food of fishes (Acanthopteri). Ill. Lab. Nat. Hist. Bul., vol. 1, no. 3, pp. 18-65, Nov. 1880. (*2d. ed.,* pp. 19-70, 1903.) (*Also,* Ill. Fish Com. Rep. for 1884, pp. 90-127.)

On the food of young fishes. Ill. Lab. Nat. Hist. Bul., vol. 1, no. 3, pp.
66-79, Nov. 1880. (*2d. ed.*, pp. 71-85, 1903.) (*Also,* Proc. Central
Fishcult. Soc., 2d. meet., pp. 10-19, 1881.)

On the food of birds (the thrush family and the bluebird). Ill. Lab. Nat.
Hist. Bul., vol. 1, no. 3, pp. 80-148, Nov. 1880. (*2d. ed.*, pp. 86-161, 1903.)

Notes on insectivorous Coleoptera. Ill. Lab. Nat. Hist. Bul., vol. 1, no. 3,
pp. 153-160, Nov. 1880. (*2d. ed.*, pp. 167-176, 1903.)

1881

Supplementary report on the food of the thrush family. Ill. Hort. Soc.
Trans., vol. 14, pp. 106-126, 1881.

A few notes on the food of the meadow lark. Ill. Hort. Soc. Trans., vol.
14, pp. 234-237, 1881.

Supplementary note on the food of the bluebird. Amer. Nat., vol. 15, pp.
66-67, Jan. 1881.

A rare fish in Illinois. Amer. Nat., vol. 15, pp. 232-233, Mar. 1881.

The English sparrow in Illinois. Amer. Nat., vol. 15, pp. 392-393, May
1881.

How to study a common fish. Ill. Sch. Jour., vol. 1, no. 2, pp. 10-13, June
1881.

Guide to the morphology of the crawfish. Ill. Sch. Jour., vol. 1, nos. 3-4,
pp. 5-9, July-Aug. 1881.

Birds and canker worms. Amer. Agr., vol. 40, pp. 482-483, Nov. 1881;
vol. 41, pp. 18-19, Jan. 1882.

Report of the Director of the State Laboratory of Natural History (to
the State Superintendent of Public Instruction), for the year ending
June 30, 1880. Ill. Sch. Rep., vol. 13, pp. 127, 138-160, 1881. (*Also as
pamphlet,* 24 pp., 1881.)

1882

The blind cave-fishes and their allies. Amer. Nat., vol. 16, pp. 1-5, Jan.
1882.

On the first food of the white-fish. Amer. Field, vol. 17, p. 171, with 1
fig., Mar. 11, 1882. (*A preliminary report of investigations described
in* Ill. Lab. Nat. Hist. Bul., vol. 1, no. 6, pp. 95-109, May 1883).

(Letters to S. F. Baird on the food of young whitefish, *Coregonus clupeiformis.*) U. S. Fish Com. Bul., vol. 1, pp. 19-20, 269-270, 402-403, 1882.

On the lakes of Illinois. Ill. Sch. Jour., vol. 2, no. 1, pp. 17-18; no. 2,
pp. 14-15; May-June 1882.

A remarkable new rotifer (*Cupelopagis bucinedax*). Amer. Mo. Micro.
Jour., vol. 3, pp. 102-103, with 1 fig., June 1882. (*See note, ibid.,* p. 151,
Aug. 1882.)

On some entomostraca of Lake Michigan and adjacent waters. Amer.
Nat., vol. 16, pp. 537-542, 640-649, pl. ix, July-Aug. 1882.

Circular (inviting correspondence from agriculturists and horticulturists
on injurious insects). Ill. Crop Rep. Circ. 90, p. 46, July 1882.

The chinch-bug in 1882: field notes. Ill. Crop Rep. Circ. 92, p. 77, Aug. 1882.

Bacterium a parasite of the chinch bug. Amer. Nat., vol. 16, pp. 824-825, Oct. 1882. (*Cf.* Prairie Farmer, Dec. 9, 1882.)

List of localities and dates (of collections). Ill. Lab. Nat. Hist. Circ., 10 pp., 1882.

The regulative action of birds upon insect oscillations. Ill. Lab. Nat. Hist. Bul., vol. 1, no. 6, pp. 3-32, Dec. 1882. (*2d. ed.*, pp. 1-31, 1912.)

The ornithological balance-wheel. Ill. Hort. Soc. Trans., vol. 15, pp. 120-131, 1882.

The relations of birds to horticulture. Missouri Hort. Soc. Rep. for 1880-1881, pp. 170-184, 1882.

1883

Scolopendrella in Illinois. Amer. Nat., vol. 17, p. 91, Jan. 1883.

The food relations of the Carabidae and Coccinellidae. Ill. Lab. Nat. Hist. Bul., vol. 1, no. 6, pp. 33-64, Jan. 1883.

Partial list of the duplicate fishes and aquatic invertebrates of the Illinois State Laboratory of Natural History. Ill. Lab. Nat. Hist. Circ., 4 pp., Jan. 1883.

Insects affecting corn. Ill. Ent. Off. Circ., 21 pp., 1 pl., 1883.

(Letters to B. W. Thomas on resolving *Amphipleura pellucida* by central light.) Microscope, vol. 3, pp. 10-11, Apr. 1883.

A new wheat insect: the wheat bulb worm. Ill. Crop Rep. Circ. 96, pp. 16-17, with 2 fig., Apr. 1883.

The food of the smaller fresh-water fishes. Ill. Lab. Nat. Hist. Bul., vol. 1, no. 6, pp. 65-94, May 1883.

The first food of the common white-fish. Ill. Lab. Nat. Hist. Bul., vol. 1, no. 6, pp. 95-109, May 1883. (*Also,* U. S. Fish Com. Rep. for 1881, pp. 771-782, 1884.)

The corn-root worm. Ill. Crop Rep. Circ. 94, p. 122, 1883.

The strawberry crown-borer (*Tyloderma fragariae* Riley). Ill. Crop Rep. Circ. 98, pp. 42-43, May 1883.

Experiments on chinch-bugs: memoranda of experiments relating to use of kerosene emulsions. U. S. D. A., Div. Ent., Bul. no. 2, pp. 23-25, 1883.

An enemy of wheat—the wheat bulb-worm. Prairie Farmer, vol. 55, pp. 481-482, with 7 fig., Aug. 4, 1883.

Memoranda with regard to the contagious diseases of caterpillars and the possibility of using the virus of the same for economic purposes ("a full abstract"). Amer. Nat., vol. 17, pp. 1169-1170, Nov. 1883. (*Cf.* Can. Ent., vol. 15, pp. 171-172, Sept. 1883.)

(Proposed study of internal parasites of swine.) Ill. Lab. Nat. Hist. Circ., 1 p., Nov. 1883.

Notes on economic ornithology. Ill. Hort. Soc. Trans., vol. 16, pp. 58-71, 1883.

Contributions to horticultural entomology. I. The strawberry crown borer. II. The strawberry root worm. Ill. Hort. Soc. Trans., vol. 16, pp. 193-203, 1883.

Insects affecting the strawberry. Miss. Vall. Hort. Soc. Trans., vol. 1, pp. 50-85, with 21 fig., 1883.

(Letter on) the strawberry leaf roller. Minn. Hort. Soc. Rep. for 1883, pp. 323-324, 1883.

A new insect disease. Prairie Farmer, vol. 55, p. 630, Oct. 6, 1883.

Twelfth Report of the State Entomologist on the noxious and beneficial insects of the State of Illinois, (being the) first annual report of S. A. Forbes, for the year 1882. Ill. Dept. Agr. Trans., vol. 20, app., pp. 6-120, with 30 fig., 1883. (Also issued separately, Nov. 1883.) Contents as follows:

> Letter of transmittal (5-9); the corn root-worm (10-31); studies on the chinch-bug—I (32-63); the strawberry crown-borer (64-75); the strawberry crown miner (76-82); the melon plant-louse (83-91); experiments with the European cabbage worm (92-97); miscellaneous notes (98-104); the food relations of predaceous beetles (105-120). Appendix: the Lombardy poplar borer, by T. J. BURRILL (121-122); the Phytopti and other injurious plant mites, by H. GARMAN (123-143); observations on the Angoumois grain moth and its parasites, by F. M. WEBSTER (144-154).

Report of the Director of the State Laboratory of Natural History (to the State Superintendent of Public Instruction), for the two years ending June 30, 1882. Ill. Sch. Rep., vol. 14, pp. lx-lxxi, 1883. (Also as pamphlet, 12 pp., 1883.)

1884

On the life-histories and immature stages of three Eumolpini. Psyche, vol. 4, pp. 123-130, pl. i, Jan.-Feb. 1884. (See corrective note, ibid., vol. 4, pp. 167-168.)

Entomological notes of the season (1883). Ill. Crop Rep. Circ. 106, pp. 177-178, Mar. 1884.

Circular (on the corn plant louse and the Hessian fly). Ill. Crop Rep. Circ. 110, pp. 48-49, with 4 fig., May 1884.

The wheat-straw worm. Ill. Crop Rep. Circ. 112, pp. 14-15, with 1 fig., June 1884.

Supplementary report on insects affecting the strawberry. Miss. Vall. Hort. Soc. Trans., vol. 2, pp. 234-258, with 14 fig., 1884.

The tarnished plant bug. Minn. Hort. Soc. Rep. for 1884, pp. 339-342, 1884.

On a contagious disease of caterpillars. Ill. Hort. Soc. Trans., vol. 17, pp. 29-41, 1884.

Destruction of fish food by bladderwort (Utricularia). U. S. Fish Com. Bul., vol. 4, p. 443, 1884. (Reprinted from Forest and Stream for Sept. 4, 1884.)

Thirteenth Report of the State Entomologist on the noxious and beneficial insects of the State of Illinois, (*being the*) second annual report of S. A. Forbes, for the year 1883. Ill. Dept. Agr. Trans., vol. 21, *app.*, 183 + xxi pp., 15 pl., 1884. (*Also issued separately,* May 1884.) Contents as follows:

> Letter of transmittal, notes of the year, etc. (9-12) ; the wheat-bulb worm (13-29) ; the wheat-straw worm (30-38) ; on insects affecting sorghum and broom-corn (39-56) ; the black-headed grass maggot (57-59) ; insects injurious to the strawberry (60-180) ; insects injurious to the apple (181-183).

(Statement of the origin and function of the Illinois State Laboratory of Natural History.) Introduction to Ill. Lab. Nat. Hist. Bul., vol. 1, p. 2, 1884.

Report of the Director of the State Laboratory of Natural History (to the State Superintendent of Public Instruction), for the two years ending June 30, 1884. Ill. Sch. Rep., vol. 15, pp. lviii-lx, 1884.

1885

Aberration in the perch. Amer. Nat., vol. 19, p. 192, Feb. 1885.

Description of new Illinois fishes. Ill. Lab. Nat. Hist. Bul., vol. 2, pp. 135-139, March 1885.

Miscellaneous notes on farm insects. Ill. Crop Rep. Circ. 118, pp. 193-194, Mar. 1885.

The root web-worm (*Crambus zeelus* Fernald). Ill. Crop Rep. Circ. 122, pp. 38-39, with 1 fig., May 1885.

Further notes on the root web-worm. Ill. Crop Rep. Circ. 123, pp. 48-49, with 3 fig., June 1885.

On some Illinois locusts (grasshoppers). Ill. Crop Rep. Circ. 125, pp. 27-30, Aug. 1885.

[—with T. J. BURRILL] Report on agricultural experiment stations. Ill. Hort. Soc. Trans., vol. 18, pp. 52-71, 1885.

A season's work in horticultural entomology. Ill. Hort. Soc. Trans., vol. 18, pp. 117-127, 1885.

(Letter to William Jackson, Mar. 8, 1884, on remedies for the cabbage worm, etc.) Ill. Hort. Soc. Trans., vol. 18, pp. 258-259, 1885.

Fourteenth Report of the State Entomologist on the noxious and beneficial insects of the State of Illinois, (*being the*) third annual report of S. A. Forbes, for the year 1884. Ill. Dept. Agr. Trans., vol. 22, *app.*, vii + 136 + xi + 120 pp., 12 pl., 1885. (*Also issued separately,* Sept. 1885.) Contents as follows:

> Letter of transmittal (1) ; entomological calendar (3-8) ; contributions to agricultural entomology—I, on new and little-known corn insects (11-33), —II, notes on insects injurious to wheat (34-69), —III, brief notes on sorghum insects (70-71), —IV, on some clover insects (72-74) ; contributions to horticultural entomology

—I, on new and imperfectly known strawberry insects (77-82), —II, on a few grape insects (83-86), —III, on new insect enemies of the blackberry and raspberry (87-92), —IV, on the speckled cutworm as a cabbage worm (93-94), —V, notes on insects injurious to the apple and pear (95-102), —VI, on some insect enemies of the soft maple (103-111), —VII, insects injurious to the elm (112-115), —VIII, brief miscellaneous notes (116-118). Appendix: general indexes to the first twelve reports . . . (prepared by ANGE V. MILLER) —introduction (v-viii); systematic list of genera and species (ix-xix); general index to insects (1-68); general index to food plants (69-94); general index to remedies, natural and artificial (95-107). Index to the fourteenth report (108-120).

A catalogue of the native fishes of Illinois. Ill. Fish Com. Rep. for 1884, pp. 60-89, with ill., 1885. (*Also, ibid.* for 1900, pp. 59-80.)

1886

Miscellaneous essays on economic entomology. Ill. Dept. Agr. Trans., vol. 23, *app.,* pp. 1-126, 1886. (*Also issued separately,* 1886.) Contents as follows:

The entomological record for 1885 (5-25); experiments on the codling-moth and curculios (26-45) [*revised reprint from* Ill. Hort. Soc. Trans., vol. 19, pp. 103-124, 1886]; a second contribution to the life history of the corn plant-louse, by H. GARMAN (46-56); partial economic bibliography of Indian corn insects (with a systematic list of corn insects), by THOMAS F. HUNT (57-126).

(Letter) to nurserymen and fruit growers of Illinois (on insects injurious to young nursery stock). Ill. Ent. Off. Circ., 1 p., Feb. 1886.

Circular (on the Hessian fly). Ill. Ent. Off. Circ., 1 p., Mar. 1886.

Circular (on the plum curculio). Ill. Ent. Off. Circ., 1 p., Apr. 1886.

Studies on the contagious diseases of insects. Ill. Lab. Nat. Hist. Bul., vol. 2, pp. 257-321, with 1 pl., June 1886. (*Reprinted, in part,* Ill. Univ. Rep., vol. 13, pp. 294-301, 1887.)

Chinch bugs in Illinois. Prairie Farmer, vol. 58, p. 491, with 3 fig., July 31, 1886.

The chinch-bug in Illinois. Ill. Ent. Off. Circ., 8 pp., Sept. 1886. (*Reprinted,* Prairie Farmer, Sept. 25 and Oct. 2, 1886.)

Notes of the past year's work. Can. Ent., vol. 18, pp. 176-177, Sept. 1886.

Leptodora in America. Amer. Nat., vol. 20, pp. 1057-1058, Dec. 1886.

A contagious disease of the European cabbage worm, *Pieris rapae,* and its economic application. Soc. Prom. Agr. Sci. Proc., vol. 7, pp. 26-32, 1886. (*Also,* Ill. Univ. Rep., vol. 13, pp. 294-301, 1887.)

Experiments on the codling moth and curculios. Ill. Hort. Soc. Trans., vol. 19, pp. 103-124, 1886. (*Revised,* Ill. Dept. Agr. Trans., vol. 23, *app.,* 26-45, 1886.)

Report of the Director of the State Laboratory of Natural History (to the State Superintendent of Public Instruction), for the year ending June 30, 1885. Ill. Sch. Rep., vol. 16, lx-lxiii, 1886.

1887

The chinch-bug in southern Illinois. Ill. Ent. Off. Circ., 2 pp., Apr. 1887. (*Also*, Prairie Farmer, vol. 59, p. 279, Apr. 30, 1887.)

(Circular of inquiry on injury to grass lands by crane-fly larvae.) Ill. Ent. Off. Circ., 1 p., Apr. 1887.

The relations of ants and aphids. Amer. Nat., vol. 21, pp. 579-580, June 1887.

(Circular letter requesting information on chinch-bug infestations.) Ill. Ent. Off. Circ., 1 p., July 1887.

Arsenical poisons for the codling moth—record and discussion of experiments for 1885 and 1886. Ill. Ent. Off. Bul., no. 1, pp. 1-24, with 11 diagrams, 1887. (*Also*, Ill. Hort. Soc. Trans., vol. 20, pp. 109-118, 1887.)

On the chinch bug in Illinois: present condition and prospects for 1887 and 1888; remedial procedure recommended. Ill. Ent. Off. Bul., no. 2, pp. 27-43, 1887.

Second contribution to a knowledge of the life history of the Hessian fly. Ill. Ent. Off. Bul., no. 3, pp. 45-61, with 1 diagram, 1887. (Advance sheets from Ill. Ent. Rep. 15, 1889.)

The lake as a microcosm. Bul. Sci. Assoc. Peoria, Ill., pp. 77-87, 1887. (*Reprinted*, Ill. Nat. Hist. Surv. Bul., vol. 15, pp. 537-550, Nov. 1925.)

(Reports as professor of zoölogy and entomology, for 1885-1886.) Ill. Univ. Rep., vol. 13, pp. 47-48, 100-102, 160-163, 1887.

(Reports of the Director of the State Laboratory of Natural History, to the Regent of the University of Illinois, 1885-1886.) Ill. Univ. Rep., vol. 13, pp. 55, 93, 100-102, 1887.

1888

On the present state of our knowledge concerning contagious insect diseases. Psyche, vol. 5, pp. 3-12 (bibliog. pp. 15-22), Jan.-Feb. 1888.

The common apple-tree borers. Ill. Ent. Off. Circ., 4 pp., 7 fig., Feb. 1888.

A new parasite of the Hessian fly. Psyche, vol. 5, pp. 39-40, with 1 fig., Apr. 1888.

Studies of the food of fresh-water fishes. Ill. Lab. Nat. Hist. Bul., vol. 2, pp. 433-473, Apr. 1888.

On the food relations of fresh-water fishes: a summary and discussion (with detailed recapitulation of data). Ill. Lab. Nat. Hist. Bul., vol. 2, pp. 475-538, July 1888.

Note on chinch-bug diseases. Psyche, vol. 5, pp. 110-111, Sept.-Oct. 1888.

(Letter on epidemic diseases of the chinch bug.) Insect Life, vol. 1, p. 113, Oct. 1888.

The Western Society of Naturalists—a presidential address. Amer. Nat., vol. 22, pp. 988-996, Nov. 1888.

Relation of wheat culture to the chinch-bug. Soc. Prom. Agr. Sci. Proc., vol. 9, pp. 27-34, 1888. (*Abstract,* Insect Life, vol. 1, pp. 222-223, Jan. 1889.)

Unsolved problems in horticultural entomology. Ill. Hort. Soc. Trans., vol. 21, pp. 92-97, 1888.

Food of the fishes of the Mississippi Valley. Amer. Fish. Soc. Trans., vol. 17, pp. 37-66, 1888.

Report of the Director of the State Laboratory of Natural History (to the Trustees of the University of Illinois), as of June 6, 1887. Ill. Lab. Nat. Hist. pamphlet, 4 pp., 1888. (*See* Ill. Univ. Rep., vol. 14, p. 48, 1889.)

1889

Fifteenth Report of the State Entomologist on the noxious and beneficial insects of the State of Illinois, (*being the*) fourth report of S. A. Forbes, for the years 1885 and 1886. Ill. Dept. Agr. Trans., vol. 24, *app.,* vi + 115 pp., 4 fig., 1889. (*Also issued separately,* Apr. 1890.) Contents as follows:

Letter of transmittal (iv); the entomological record for 1885-86 (1-6); arsenical poisons for the codling moth (7-20); a second contribution to a knowledge of the life history of the Hessian fly (21-34) [*issued separately in advance,* 1887]; on the life history of the wheat bulb worm (35-39); on an outbreak of injurious locusts in central Illinois, by C. M. WEED (40-44); on some common insects injuriously affecting the foliage of young apple trees in the nursery and orchard, by C. M. WEED (45-85). Appendix: the present condition and prospects of the chinch-bug in Illinois (89-103) [*reprinted from* Ill. Ent. Off. Bul., no. 2, pp. 27-43, 1887].

Early occurrence of the chinch-bug in the Mississippi Valley. Insect Life, vol. 1, p. 249, Feb. 1889.

An early note on the periodical cicada. Insect Life, vol. 1, p. 313, Apr. 1889.

The grain plant louse in Illinois. Ill. Crop. Rep. Circ. 145, pp. 27-28, June 1889.

Arsenical poisons for the plum and peach curculio. Insect Life, vol. 2, pp. 3-7, July 1889.

Office and laboratory organization. Insect Life, vol. 2, pp. 185-187, Dec. 1889.

(Report as professor of zoology and entomology, for 1886-1887.) Ill. Univ. Rep., vol. 14, pp. 164-167, 1889.

Report of the Director of the State Laboratory of Natural History (to the Trustees of the University of Illinois), as of Oct. 31, 1888. Ill. Univ. Rep., vol. 14, pp. 185-193, 1889. (*Also as pamphlet,* 10 pp., 1889.)

1890

Note on the feeding habits of *Cermatia forceps* Raf. Amer. Nat., vol. 24, pp. 81-82, Jan. 1890.

The American plum borer, *Euzophera semifuneralis* Walk. Psyche, vol. 5, pp. 295-299, 3 fig., Jan. 1890.

Sixteenth Report of the State Entomologist on the noxious and beneficial insects of the State of Illinois, (*being the*) fifth report of S. A. Forbes, for the years 1887 and 1888. Ill. Dept. Agr. Trans., vol. 26, *app.*, xiii + 104 + 122 + ix pp., 6 graphs, 6 pl., 1890. (*Also issued separately,* Apr. 1890.) Contents as follows:

> Letter of transmittal (vii) ; general record for 1887 and 1888 (ix-xiii) ; studies on the chinch-bug II (1-57) (*issued separately in advance,* 1889; *pp. 7-32 reprinted,* 1914) ; the corn bill bugs (58-77) ; the meadow maggots, or leather-jackets (78-83) ; notes on cutworms (84-97) ; the burrowing web-worm (98-101). Appendix : economic bibliography of the chinch-bug, 1785-1888 (1-122).

Note on an American species of Phreoryctus. Amer. Nat., vol. 24, pp. 477-478, May 1890.

The Hessian fly. Ill. Crop. Rep. Circ. 149, pp. 32-34, with 1 pl., June 1890. (*Also,* Ill. Agr. Exp. Sta. Bul. 12, pp. 377-379, with 1 pl., Nov. 1890.)

An American terrestrial leech. Amer. Nat., vol. 24, pp. 646-649, July 1890. (*Also,* Ill. Lab. Nat. Hist. Bul., vol. 3, pp. 119-122, Sept. 1890.)

On an American earthworm of the family Phreoryctidae. Ill. Lab. Nat. Hist. Bul., vol. 3, pp. 107-117, with 3 pl., Sept. 1890.

New and old insects. Ill. Hort. Soc. Trans., vol. 23, pp. 242-251, 1890.

Synopsis of recent work with arsenical insecticides. Ill. Hort. Soc. Trans., vol. 23, pp. 310-324, 1890.

Preliminary report upon the invertebrate animals inhabiting Lakes Geneva and Mendota, Wisconsin, with an account of the fish epidemic in Lake Mendota in 1884. U. S. Fish Com. Bul., vol. 8, pp. 473-487, with 3 pl., 1890.

(Letter to the Board of Direction of the Agricultural Experiment Station, proposing field experiments for protection of crops against injury by chinch bugs.) Ill. Univ. Rep., vol. 15, pp. 17-18, 1890.

Biennial Report of the Director of the State Laboratory of Natural History (to the Trustees of the University of Illinois), for 1889-1890. Ill. Univ. Rep., vol. 15, pp. 243-245, 1890. (*Also as pamphlet,* 5 pp., 1891.)

1891

On some Lake Superior entomostraca. U. S. Fish Com. Rep. for 1887-1888, pp. 701-718, with 4 pl., 1891. (*Advance sheets issued* 1890.)

A summary history of the corn-root aphis. Insect Life, vol. 3, pp. 233-238, Jan. 1891.

On the life history of the white grubs. Insect Life, vol. 3, pp. 239-246, Jan. 1891.

The fruit bark beetle. Ill. Agr. Exp. Sta. Bul. 15, pp. 469-479, with 1 pl., Feb. 1891. (*Incorporated in* Ill. Ent. Rep. 17, pp. 1-20, 1891.)

Partial list of duplicate insects in the collection of the Illinois State Laboratory of Natural History. Ill. Lab. Nat. Hist. Circ., 19 pp., May 1891.

On a bacterial insect disease. N. Amer. Practit., pp. 401-405, Sept. 1891. (*Also,* Amer. Mo. Micr. Jour., vol. 12, pp. 246-249, Nov. 1891.)

The importation of a Hessian-fly parasite from Europe. Insect Life, vol. 4, pp. 179-181, Dec. 1891.

The fruit bark beetle. Ill. Ent. Off. Bul., no. 4, pp. 63-81, with 1 pl., 1891.

Seventeenth Report of the State Entomologist on the noxious and beneficial insects of the State of Illinois, (*being the*) sixth report of S. A. Forbes for the years 1889 and 1890. Ill. Dept. Agr. Trans., vol. 28, *app.,* xv + 90 + 36 + ix pp., 8 pl., *date on title-page* 1891. (*Also issued separately,* Feb. 1892.)   Contents as follows:

Letter of transmittal (vii) ; general record for 1889 and 1890 (ix-xv) ; the fruit bark beetle (1-20) [*issued as* Ill. Agr. Exp. Sta. Bul. 15, Feb. 1891] ; experiments with the arsenical poisons for the plum and peach curculio (21-25) ; the American plum borer (26-29) ; on the common white grubs (30-53) ; additional notes on the Hessian fly (54-63) ; a summary history of the corn-root aphis (64-70) ; on a bacterial disease of the larger corn-root worm (71-73) ; notes on the diseases of the chinch bug (74-87). Appendix : an analytical list of the entomological writings of William LeBaron, M. D., Second State Entomologist of Illinois, compiled with the assistance of C. M. WEED, JOHN MARTIN, MARY J. SNYDER, and C. A. HART. (1-36). Index to the seventeenth report (i-ix).

Horticultural entomology. Ill. Hort. Soc. Trans., vol. 24, pp. 88-95, 1891.

List of economic species of insects for public schools, with references to entomological literature. Ill. Lab. Nat. Hist. Circ., 13 pp., 1891.

The head of the English sparrow. Ill. Lab. Nat. Hist. Circ., 10 pp., 1891.

History and status of public school science work in Illinois. In "Educational Papers by Illinois Science Teachers," vol. 1, pp. 6-20, 1891.

Pedagogical contents of zoology. In "Educational Papers by Illinois Science Teachers," vol. 1, pp. 38-48, 1891.

Science in the country school. Interstate Sch. Rev. (Danville, Ill.), vol. 1, pp. 13 ff., 1891 (*not seen*).

1892

Bacteria normal to digestive organs of Hemiptera. Ill. Lab. Nat. Hist. Bul., vol. 4, pp. 1-7, Jan. 1892.

The chinch-bug in Illinois, 1891-92. Ill. Agr. Exp. Sta. Bul. 19, pp. 44-48, Feb. 1892.

Zoology in the public school : choice and arrangement of material. (Ill.) Public Sch. Jour., vol. 11, pp. 319-321, 375-377, 429-431, Mar.-May 1892.

The Illinois insectarium. Amer. Nat., vol. 26, pp. 353-354, Apr. 1892.

An all-around microscope. Amer. Mo. Micr. Jour., vol. 13, pp. 91-92, Apr. 1892.

Experiments with chinch-bugs. Psyche, vol. 6, p. 250, Apr. 1892.

Address of first vice-president (Assoc. Econ. Ent., Aug. 15, 1892, Rochester, N. Y.) on contagious diseases of insects. Insect Life, vol. 5, pp. 68-76, Nov. 1892.

(Food habits of the common yellow thrips.) Insect Life, vol. 5, pp. 126-127, Nov. 1892.

(Letter to James Fletcher, on the Natural History Hall of the University of Illinois.) Ottawa Naturalist, vol. 6, pp. 133-135, Dec. 1892.

The fruit-destroying insects of southern Illinois. Ill. Hort. Soc. Trans., vol. 25, pp. 116-123, 1892.

Answers to interrogations (on) horticultural entomology. Ill. Hort. Soc. Trans., vol. 25, pp. 258-259, 1892.

Report of the Illinois State Entomologist concerning operations under the Horticultural Inspection Act. U. S. D. A. Off. Exp. Sta. Bul. 12, pp. 1058-1060, 1892 (not seen).

1893

A preliminary report on the aquatic invertebrate fauna of the Yellowstone National Park, Wyoming, and of the Flathead region of Montana. U. S. Fish Com. Bul., vol. 11, pp. 207-258, with 6 pl., Apr. 1893.

(Provisional program for international congress of zoologists in connection with the Columbian Exposition at Chicago.) Amer. Nat., vol. 27, pp. 762-763, Aug. 1893.

Presidential address (Assoc. Econ. Ent., Aug. 14, 1893, Madison, Wisc.). Insect Life, vol. 6, pp. 61-70, Dec. 1893.

Injuries to fruits by twenty-five worst fruit insects. Ill. Hort. Soc. Trans., vol. 26, pp. 121-134 (see also pp. 199-204, with 25 fig.), 1893.

Biennial Report of the Director of the State Laboratory of Natural History (to the Trustees of the University of Illinois), for 1891-1892. Ill. Univ. Rep., vol. 16, pp. 284-288, 1893. (Also as pamphlet, 8 pp., 1893.)

1894

The chinch-bug in southern Illinois, 1894. Ill. Agr. Exp. Sta. Bul. 33, pp. 397-399, June 1894.

How to fight chinch bugs. Prairie Farmer, vol. 66, no. 31, p. 2, Aug. 4, 1894.

The biological station of the University of Illinois. Amer. Nat., vol. 28, pp. 723-724, 1894.

A program of practice in horticultural entomology. Ill. Hort. Soc. Trans., vol. 27, pp. 113-118, 1894.

The chinch bug in 1894. Prospect for 1895. Contagious disease experiments, etc. Ill. Ent. Off. Bul., no. 5, pp. 83-89, 1894. (A brief statement of results published in full as Ill. Agr. Exp. Sta. Bul. 38, 1895, and in Ill. Ent. Rep. 19, 1895.)

Damage to food-fish in Wabash River. U. S. 53d Congr., 2d Sess., H. R. Misc. Doc. no. 196, pp. 2-5, 1894. (*Also,* U. S. Fish Com. Rep. for 1893-1894, pp. 109-112, 1896.)

The Aquarium of the United States Fish Commission at the World's Columbian Exposition. Report of the director. U. S. Fish Com. Bul., vol. 13, pp. 143-158, 1894.

Biennial Report of the Director of the State Laboratory of Natural History (to the Trustees of the University of Illinois), for 1893-1894. Ill. Univ. Rep., vol. 17, pp. 304-323, 1894? (*Also as pamphlet,* 36 pp., 18 pl., 1894?)

1895

Eighteenth Report of the State Entomologist on the noxious and beneficial insects of the State of Illinois, (*being the*) seventh report of S. A. Forbes, for the years 1891 and 1892. Ill. Dept. Agr. Trans., vol. 31, *app.,* xi + 171 + xiii pp., 15 pl., *date on title-page 1894; probably not issued until 1895.* (*Also issued separately,* Mar. 1895.) (*2d. ed.,* x + 149 + xii pp., 15 pl., 1920.) Contents as follows:

Letter of transmittal (vi) ; general record for 1891 and 1892 (vii-xi) ; a monograph of insect injuries to Indian corn—Part I (3-171), and index (i-xiii) [*reissued, in part, as* Ill. Agr. Exp. Sta. Bul. 44, May 1896].

Experiments with the muscardine disease of the chinch-bug, and with the trap and barrier method for the destruction of that insect. Ill. Agr. Exp. Sta. Bul. 38, pp. 25-85, with 7 pl. Mar. 1895. (*Incorporated in* Ill. Ent. Rep. 19, pp. 5-176, 1896.)

Preliminary note on a contagious insect disease. Science, vol. 2, pp. 375-376, Sept. 20, 1895.

1896

Nineteenth Report of the State Entomologist on the noxious and beneficial insects of the State of Illinois, (*being the*) eighth report of S. A. Forbes, for the years 1893 and 1894. Ill. Dept. Agr. Trans., vol. 32, *app.,* 206 + 65 + v pp., 13 pl., *date on title-page 1895, not issued before Apr. 1896.* (*Also issued separately,* May 1896.) Contents as follows:

Letter of transmittal (3) ; experiments for the destruction of chinch-bugs as they emerge from fields of small grain at harvest [with an analytical list of American articles consulted, 1824-1894] (5-176) [*issued, in part, as* Ill. Agr. Exp. Sta. Bul. 38, Mar. 1895]; laboratory experiments with chinch-bugs (177-189) ; the white ant in Illinois [with a descriptive list of economic articles consulted] (190-204). Appendix: the Mediterranean flour moth (*Ephestia kuehniella* Zeller) in Europe and America, by W. G. JOHNSON (7-65). Index (i-v).

Insect injuries to the seed and root of Indian corn. Ill. Agr. Exp. Sta., Bul. 44, pp. 209-296, with 61 fig., May 1896. (*Extracted from* Ill. Ent. Rep. 18, pp. 3-171, 1894.)

Progress in economic entomology. Ill. Farm. Inst. Rep., vol. 1, pp. 103-112, with portrait, 1896.

Recent progress in horticultural entomology. Ill. Hort. Soc. Trans., vol. 29, pp. 137-148, 1896.

Summer opening of the biological experiment station of the University of Illinois. Ill. Univ. pamphlet, 24 pp., 11 pl., 1896.

Biennial Report of the Director of the State Laboratory of Natural History (to the Trustees of the University of Illinois), for 1895-1896, and special report of the Biological Experiment Station. Ill. Univ. Rep., vol. 18, pp. 298-326, with 20 pl., 1896? (*Also as pamphlet*, 31 pp., 20 pl., 1896?)

1897

The San Jose scale in Illinois. Ill. Agr. Exp. Sta. Bul. 48, pp. 413-428, with 2 fig., Apr. 1897. (*Cf*. Ill. Hort. Soc. Trans., vol. 30, pp. 109-124, 1897.) (*Incorporated in* Ill. Ent. Rep. 20, pp. 1-25, 1898.)

Circular notice concerning the San Jose scale and other fruit insects. Ill. Ent. Off. Circ., 4 pp., July 1897.

Agricultural methods with farm insects. Ill. Farm. Inst. Rep., vol. 2, pp. 22-25, with portrait, 1897.

Entomological work at the University (of Illinois). Ill. Farm. Inst. Rep., vol. 2, pp. 28-33, 1897.

Summer school of field biology and second summer opening of the University Biological Station: preliminary notice. Ill. Univ. pamphlet, 4 pp., 1897.

1898

Twentieth Report of the State Entomologist on the noxious and beneficial insects of the State of Illinois, (*being the*) ninth report of S. A. Forbes, for the years 1895 and 1896. Ill. Dept. Agr. Trans., vol. 34, *app.*, vi + 112 + xxxii pp., 12 pl., *date on the title-page 1897; probably issued in 1898.* (*Also issued separately*, Mar. 1898.) Contents as follows:

Introductory note (v-vi); the San Jose scale in Illinois (1-25) [*issued as* Ill. Agr. Exp. Sta. Bul. 48, Apr. 1897]; field observations on white grubs (26-34); midsummer measures against the chinch-bug (35-44); a study of the causes of the disappearance of a chinch-bug outbreak (45-74); the spontaneous occurrence of white muscardine among chinch-bugs in 1895 (75-78); miscellaneous chinch-bug experiments (79-102); an entomological train wrecker (103-105); note on a new disease of the army worm (106-109). Appendix: the white pine Chermes, by E. L. STORMENT (iii-xxvi).

The method of science. Sch. & Home Educ. (Bloomington, Ill.), vol. 18, pp. 113-119, Mar. 1898. (*Reprinted as pamphlet*, 23 pp., 1898.)

The season's campaign against the San Jose and other scale insects in Illinois. Ill. Hort. Soc. Trans., vol. 31, pp. 105-119, 1898.

State control of injurious insects. Ill. Farm. Inst. Rep., vol. 3, pp. 67-71, 1898.

Summer school of biology and third summer opening of the Illinois Biological Station. Ill. Univ. pamphlet, 7 pp., 5 fig., 1898.

Biennial Report of the Director of the State Laboratory of Natural History (to the Trustees of the University of Illinois), for 1897-1898, (including reports on the Biological Station and the Summer School). Ill. Univ. Rep., vol. 19, pp. 340-354, 1898? (*Also as pamphlet*, 31 pp., 10 pl., 1898?)

1899

The Museum of Natural History. Illini (Univ. Ill.), pp. 396-400, March 31, 1899.

The San Jose scale. Ill. Ent. Off. Circ. (*in poster form*), Apr. 1899.

Recent work on the San Jose scale in Illinois. Ill. Agr. Exp. Sta. Bul. 56, pp. 241-287, with 4 pl., July 1899. (*Incorporated in* Ill. Ent. Rep. 21, pp. 1-44, 1900.

Lessons from the year's work with the San Jose scale. Ill. Hort. Soc. Trans., vol. 32, pp. 50-61, 1899.

1900

[—with C. A. HART.] The economic entomology of the sugar beet. Ill. Agr. Exp. Sta. Bul. 60, pp. 397-532, with 97 fig. and 9 pl., Aug. 1900. (*Incorporated in* Ill. Ent. Rep. 21, pp. 49-184, 1900.)

The working of the San Jose scale law. Ill. Hort. Soc. Trans., vol. 33, pp. 150-159, 1900. (*Reprinted with new title:* annual statement of the Illinois State Entomologist, to the State Horticultural Society, concerning operations under the horticultural inspection act (as of Dec. 20, 1899). Ill. Ent. Off. pamphlet, 10 pp., 1900.)

Twenty-first Report of the State Entomologist on the noxious and beneficial insects of the State of Illinois, (*being the*) tenth report of S. A. Forbes. iv + 184 + xxviii pp., 9 pl., 97 fig., 1900. Contents as follows: Introductory note (iv); recent work on the San Jose scale in Illinois (1-44) [*issued as* Ill. Agr. Exp. Sta. Bul. 56, July 1899]; the economic entomology of the sugar beet (49-184) [*issued as* Ill. Agr. Exp. Sta. Bul. 60, Aug. 1900].

Report of the Illinois State Entomologist (to the Governor) concerning operations under the horticultural inspection act, (as of Oct. 31, 1900). Ill. Ent. Off. pamphlet, 30 pp., 1900. (*Cf.* Ill. Hort. Soc. Trans., vol. 34, pp. 213-215, 1901.)

1901

On the principal nursery pest likely to be distributed in trade. Ill. Agr. Exp. Sta. Circ. 36, 43 pp., 33 fig., July 1901. (*Incorporated in* Ill. Ent. Rep. 22, 1903.)

Nursery inspection and orchard insecticide work in Illinois. U. S. D. A. Off. Exp. Sta. Bul. 99, pp. 173-176, 1901.

The crown gall as a nursery pest. Ill. Hort. Soc. Trans., vol. 34, pp. 136-149, 4 pl., 1901.

Annual statement concerning operations under the horticultural inspection act. Ill. Hort. Soc. Trans., vol. 34, pp. 213-215, 1901.

Insects injurious to corn. Ill. Farm. Inst. Rep., vol. 6, pp. 80-82, 1901.

The method of science and the public school. N. Y. Science Teachers' Assoc. Proc. for 1900, pp. 768-788, 1901. (*Revised reprint,* 18 pp. 1901.)

Biennial Report of the Director of the State Laboratory of Natural History (to the Trustees of the University of Illinois), for 1899-1900. Ill. Univ. Rep., vol. 21, pp. 8-10, 1901. (*Also a revised version as pamphlet,* 12 pp., 1901.)

1902

Experiments with insecticides for the San Jose scale. Ill. Agr. Exp. Sta. Bul. 71, pp. 241-264, April 1902. (*Incorporated in* Ill. Ent. Rep. 22, pp. 67-90, 1903.)

Additional insecticide experiments for the San Jose scale. Ill. Agr. Exp. Sta. Bul. 72, pp. 265-268, May 1902. (*Incorporated in* Ill. Ent. Rep. 22, pp. 91-95, 1903.)

Insecticide experiments for the San Jose scale. Ill. Agr. Exp. Sta. Circ. 53, 2 pp., May 1902. (*An abstract of* Ill. Agr. Exp. Sta. Bul. 71 and 72.)

The corn bill-bugs in Illinois. Ill. Agr. Exp. Sta. Bul. 79, pp. 435-461, Oct. 1902. (*Incorporated in* Ill. Ent. Rep. 22, pp. 1-26, 1903.)

Methods and results of field insecticide work against the San Jose scale. Ill. Agr. Exp. Sta. Bul. 80, pp. 463-502, with 9 pl., Oct. 1902. (*Incorporated in* Ill. Ent. Rep. 22, pp. 27-66, 1903.)

The canker worm on apple trees and elms. Miss. Vall. Apple Growers' Assoc. Proc. for 1901-1902, pp. 38-41, 1902.

Entomological notes and inspection report for 1901. Ill. Hort. Soc. Trans., vol. 35, pp. 142-152, 1902.

Elementary biology in the public school. In "Superintendents' Round Table," (introductory discussion at DeKalb, Oct. 19, 1901, and synopsis of discussion at Chicago, Dec. 13, 1901) separate pamphlet, 20 pp., 1902.

1903

How to make the farm attractive to the educated. Unity, pp. 316-318, Jan. 15, 1903.

The scientific method in high school and college. Sch. Sci., vol. 3, pp. 53-66, May 1903.

Notes on the insecticide use of the gasoline blast lamp. Ill. Agr. Exp. Sta. Bul. 89, pp. 145-154, Nov. 1903.

Twenty-second Report of the State Entomologist on the noxious and beneficial insects of the State of Illinois, (*being the*) eleventh report of S. A. Forbes. viii + 149 + xx pp., 9 pl., 33 fig., 1903. Contents as follows:

Introductory note (viii); the corn bill-bugs in Illinois (1-26) [*issued as* Ill. Agr. Exp. Sta. Bul. 79, Oct. 1902]; methods and results of field insecticide work agaist the San Jose scale, 1899-1902

(27-66) [*issued as* Ill. Agr. Exp. Sta. Bul. 80, Oct. 1902]; experiments with lime and sulphur washes for the San Jose scale (67-90) [*issued as* Ill. Agr. Exp. Sta. Bul. 71, Apr. 1902]; experiments and observations on the use of crude petroleum and pure kerosene for the San Jose scale (91-95) [*issued as* Ill. Agr. Exp. Sta. Bul. 72, May 1902]; experiments with summer washes for the San Jose scale (96-97); on the principal nursery pests likely to be distributed in trade (98-138) [*issued as* Ill. Agr. Exp. Sta. Circ. 36, July 1901]; the canker-worm on shade and forest trees (139-144); the Colaspis root-worm (145-149).

Excellencies and defects of existing legislation for the control of insect and fungus pests. U. S. D. A., Off. Exp. Sta. Bul. 123, pp. 122-126, 1903.

Report of inspection and insecticide operations. Ill. Hort. Soc. Trans., vol. 36, pp. 120-121, 1903.

The San Jose and other scale insects. Ill. Hort. Soc. Trans., vol. 36, pp. 121-126, 1903.

Report of the Illinois State Entomologist (to the Governor) on the horticultural inspection law, (as of Feb. 1, 1903). Ill. Ent. Off. pamphlet, 30 pp., 1903. (*Cf.* Ill. Hort. Soc. Trans., vol. 35, pp. 142-152, 1902; vol. 36, pp. 120-121, 1903.)

(Report on) Entomology. Ill. Blue Book for 1903, pp. 474-475, 1903?

1904

Directions for the preparation of the Oregon wash. Ill. Hort. Soc. Trans., vol. 37, pp. 20-21, 1904.

Report of inspection and insecticide operations. Ill. Hort. Soc. Trans., vol. 37, pp. 229-238, 1904.

The more important insect injuries to Indian corn. Ill. Agr. Exp. Sta. Bul. 95, pp. 329-399, with 38 fig. and 5 pl., Nov. 1904. (*Incorporated in* Ill. Ent. Rep. 23, pp. 1-280, 1905.)

The economic entomology which the farmer should know. Ill. Farm. Inst. Rep., vol. 9, pp. 58-65, 1904. (*Also with new title:* The kind of economic entomology which the farmer ought to know. Ill. Ent. Off. pamphlet, 16 pp., 1904.)

The teaching of the scientific method. Nat. Educ. Assoc. Jour., vol. 43, pp. 879-888, 1904.

Report of the Director of the State Laboratory of Natural History (to the Trustees of the University of Illinois), as of June 6, 1903. Ill. Univ. Rep., vol. 22, pp. 92-93, 1904.

1905

Practical treatment of the San Jose scale. Ill. Agr. Exp. Sta. Circ. 85, 4 pp., Jan. 1905.

Illinois River plankton. Science, vol. 21, pp. 233-234, Feb. 10, 1905.

Economic and industrial aspects of secondary school biology. Sch. Sci. Math., vol. 5, pp. 173-183, Mar. 1905.

[—with R. E. RICHARDSON.] On a new shovelnose sturgeon from the Mississippi River. Ill. Lab. Nat. Hist. Bul., vol. 7, pp. 37-44, with 4 pl., May 1905.

Field experiments and observations on insects injurious to Indian corn. Ill. Agr. Exp. Sta. Bul. 104, pp. 95-152, with 2 fig., Oct. 1905. (*Incorporated in* Ill. Ent. Rep. 24, pp. 1-58, 1908.)

(Review of) American Insects, by Vernon L. Kellog. Sch. Sci. Math., vol. 5, pp. 777-778, Dec. 1905.

Plum curculio in the apple orchard: a spraying experiment. Ill. Hort. Soc. Trans., vol. 38, pp. 91-107, with 4 pl., 1905.

Report of nursery inspection and insecticide operations. Ill. Hort. Soc. Trans., vol. 38, pp. 108-114, 1905.

(Conference on) insects injurious to corn. Ill. Farm. Inst. Rep., vol. 10, pp. 35-55, with portrait and col. pl., 1905.

The principal insects injurious to the corn plant. Ill. Farm. Inst. Rep., vol. 10, pp. 220-308, with 69 fig., 1905. (*Composed of portions of* Ill. Ent. Rep. 18 and 23.)

Twenty-third Report of the State Entomologist on the noxious and beneficial insects of the State of Illinois, (*being the*) twelfth report of S. A. Forbes. 273 + xxxiii pp., 8 pl., 238 fig., 1905. (*2d. ed.,* 280 + xxxviii pp., 1920). Contents as follows:

Introductory note (viii); a monograph of insect injuries to Indian corn. Part II. The more important insect injuries to Indian corn (1-280) [*issued as* Ill. Agr. Exp. Sta. Bul. 95, Nov. 1904].

Executive report of the Illinois State Entomologist (to the Governor), including a report of operations under the horticultural inspection law, for 1903-1905. Ill. Ent. Off. pamphlet, 35 pp., 1905. (*Cf.* Ill. Hort. Soc. Trans., vol. 37, pp. 229-238, 1904; vol. 38, pp. 108-114, 1905.)

1906

Comparative experiments with various insecticides for the San Jose scale. Ill. Agr. Exp. Sta. Bul. 107, pp. 243-261, Apr. 1906. (*Incorporated in* Ill. Ent. Rep. 24, pp. 59-77, 1908.)

Spraying apples for the plum curculio. Ill. Agr. Exp. Sta. Bul. 108, pp. 265-286, with 3 diagrams and 4 pl., May 1906. (*Incorporated in* Ill. Ent. Rep. 24, pp. 78-99, 1908.)

The corn root-aphis and its attendant ant. U. S. D. A. Bur. Ent. Bul. 60, pp. 29-39, Sept. 1906.

Report of nursery inspection and insecticide operations. Ill. Hort. Soc. Trans., vol. 39, pp. 132-135, 1906.

Fifty years' progress in the practical control of the insect and fungus pests of Illinois horticulture. Ill. Hort. Soc. Trans., vol. 39, pp. 219-227, 1906.

(Report on) Entomology. Ill. Blue Book for 1905, pp. 528-529, 1906.

1907

The cottony maple scale in Illinois. Ill. Agr. Exp. Sta. Bul. 112, pp. 343-360, with 8 fig. and 3 pl., Jan. 1907. (*Incorporated in* Ill. Ent. Rep. 24, pp. 100-117, 1908.)

On the local distribution of certain Illinois fishes: an essay in statistical ecology. Ill. Lab. Nat. Hist. Bul., vol. 7, pp. 273-303, with 15 maps and 9 pl., April 1907.

An ornithological cross-section of Illinois in autumn. Ill. Lab. Nat. Hist. Bul., vol. 7, pp. 305-335, April 1907. (*2d. ed.,* pp. 305-332, 1914.)

On the life history, habits, and economic relations of the white-grubs and May-beetles. Ill. Agr. Exp. Sta. Bul. 116, pp. 447-480, with 1 fig., Aug. 1907. (*Incorporated in* Ill. Ent. Rep. 24, pp. 135-168, 1908.)

Report of nursery inspection and orchard insecticide operation. Ill. Hort. Soc. Trans., vol. 40, pp. 124-128, 1907.

Insects in relation to health. Ill. Farm. Inst. Rep., vol. 12, pp. 263-270, with portrait, 1907.

Executive report of the Illinois State Entomologist (to the Governor), including a report of operations under the horticultural inspection law, for 1905-1906. Ill. Ent. Off. pamphlet, 27 pp., 1907. (*Cf.* Ill. Hort. Soc. Trans., vol. 39, pp. 132-135, 1906; vol. 40, pp. 124-128, 1907.)

History of the former state natural history societies of Illinois. Science, vol. 26, pp. 892-898, Dec. 27, 1907. (*Also* Ill. Acad. Sci. Trans., vol. 1, pp. 18-30, 1908.)

1908

Practical treatment for the San Jose scale. Ill. Ent. Off. Circ., 2 pp., Feb. 1908.

Experiments with repellents against the corn root-aphis. Jour. Econ. Ent., vol. 1, pp. 81-83, Apr. 1908. (*Abstract of* Ill. Agr. Exp. Sta. Bul. 130, Dec. 1908.)

Twenty-fourth Report of the State Entomologist on the noxious and beneficial insects of the State of Illinois, (*being the*) thirteenth report of S. A. Forbes. 168 + xvi pp., 8 fig., 11 pl., 1908. Contents as follows:

Introductory note (vi), injury to corn by the timothy bill-bugs (1-7), field experiments on the corn root-aphis (8-29), field experiments for the protection of corn against chinch-bug injury (30-58) [*first 58 pp. issued as* Ill. Agr. Exp. Sta. Bul. 104, Oct. 1905] ; comparative experiments with various insecticides for the San Jose scale (59-77) [*issued as* Ill. Agr. Exp. Sta. Bul. 107, Apr. 1906]; spraying apples for the plum-curculio (78-99) [*issued as* Ill. Agr. Exp. Sta. Bul. 108, May 1906]; the cottony maple scale in Illinois (100-117) [*issued as* Ill. Agr. Exp. Bul. 112, Jan. 1907]; the elm twig-girdler (118-134) ; on the life history, habits, and economic relations of the white-grubs and May-beetles (135-168) [*issued as* Ill. Agr. Exp. Sta. Bul. 116, Aug. 1907].

Directions for the preparation and use of the lime and sulphur wash. Ill. Hort. Soc. Trans., vol. 41, pp. 21-22, 1908. (*Also, ibid.*, vol. 42, pp. 25-27, 1909; vol. 43, pp. 29-30, 1910.)

Report of nursery inspection and orchard insecticide operations. Ill. Hort. Soc. Trans., vol. 41, pp. 196-201, 1908.

(Abstract of) a statistical study of midsummer birds of Illinois. Science, vol. 27, pp. 918-920, June 12, 1908.

The midsummer bird life of Illinois: a statistical study. Amer. Nat., vol. 42, pp. 505-519, Aug. 1908. (*Reprinted,* Ill. Lab. Nat. Hist. Bul., vol. 9, pp. 373-385, Jan. 1913.)

Experiments with repellents against the corn root-aphis, 1905-1906. Ill. Agr. Exp. Sta. Bul. 130, pp. 3-28, Dec. 1908. (*Incorporated in* Ill. Ent. Rep. 25, pp. 1-26, 1909.)

Habits and behavior of the corn-field ant, *Lasius niger americanus.* Ill. Agr. Exp. Sta. Bul. 131, pp. 31-45, with 1 pl., Dec. 1908. (*Incorporated in* Ill. Ent. Rep. 25, pp. 27-40, 1909.)

Reports of the Director of the State Laboratory of Natural History (to the Trustees of the University of Illinois), for 1906 and 1907. Ill. Univ. Rep., vol. 24, pp. 29-30, 157-158, 393-394, 1908.

(Report on) Entomology. Ill. Blue Book for 1907, pp. 493-494, 1908.

1909

On the general and interior distribution of Illinois fishes. Ill. Lab. Nat. Hist. Bul., vol. 8, pp. 381-437, with 102 maps, Feb. 1909. (*Also in* "The Fishes of Illinois," Ill. Nat. Hist. Surv. Rep., vol. 3, pp. lxxii-cxxii, 1909.)

[—with R. E. Richardson.] The fishes of Illinois. Ill. Nat. Hist. Surv. Rep., vol. 3, cxxxi + 357 pp., 56 pl., 76 text fig.; 103 maps in a separate atlas; 1909. (*2d. ed.*, with 68 pl., 1920.)

[—with A. O. Gross.] The economic values of some common Illinois birds. Ill. Supt. Pub. Instr. Circ. 32, "Arbor and Bird Day," pp. 53-74, with 20 fig., 1909.

Aspects of progress in economic entomology. Jour. Econ. Ent., vol. 2, pp. 25-35, Feb. 1909.

Program of cooperative experiments by the farmer and the entomologist. Ill. Ent. Off. Circ., 4 pp., Mar. 1909.

The general entomological ecology of the Indian corn plant. Amer. Nat., vol. 43, pp. 286-301, May 1909. (*Reprinted,* Ill. Nat. Hist. Surv. Bul., vol. 16, pp. 447-457, Apr. 1927.)

Report on nursery inspection and orchard insecticide operations. Ill. Hort. Soc. Trans., vol. 42, pp. 63-66, 1909.

Cooperation between the farmer and the entomologist. Ill. Farm. Inst. Rep., vol. 14, pp. 128-135, 1909.

Twenty-fifth Report of the State Entomologist on the noxious and beneficial insects of the State of Illinois, (*being the*) fourteenth report of S. A. Forbes. 124 + xxiii pp., 35 fig., 3 pl., 1909. Contents as follows: Introduction (v-xi); experiments with repellents against the corn root-aphis, 1905 and 1906 (1-26) [*issued as* Ill. Agr. Exp. Sta. Bul. 130, Dec. 1908]; habits and behavior of the corn-field ant, *Lasius niger americanus* (27-40) [*issued as* Ill. Agr. Exp. Sta. Bul. 131, Dec. 1908]; the insect pests of clover and alfalfa, by J. W. Folsom (41-124) [*issued as* Ill. Agr. Exp. Sta. Bul. 134, Apr. 1909].

Introduction (to) contents and index of the reports of the State Entomologist of Illinois, XIII-XXIV, 1884-1908, pp. 5-7, 1909.

Executive report of the Illinois State Entomologist (to the Governor), including a report of operations under the horticultural inspection law, for 1907-1908. Ill. Ent. Off. pamphlet, 21 pp., 1909. (*Cf.* Ill. Hort. Soc, Trans., vol. 41, pp. 196-201, 1908; vol. 42, pp. 63-66, 1909.)

The Illinois State Laboratory of Natural History and the Illinois State Entomologist's Office. Ill. Acad. Sci. Trans., vol. 2, pp. 54-67, 1909.

1910

Report on nursery inspection and orchard insecticide operations. Ill. Hort. Soc. Trans., vol. 43, pp. 93-96, 1910.

The Brown-tail and gipsy moths. Ill. Hort. Soc. Trans., vol. 43, pp. 96-101, with 2 pl., 1910.

Recent work of the entomologist's office. Ill. Farm. Inst. Rep., vol. 15, pp. 140-150, with portrait, 1910.

The Hessian fly in Illinois. Ill. Agr. Exp. Sta. Circ. 146, 4 pp., Sept. 1910. (*Revised ed.,* 5 pp., June 1914.)

(Review) Catalogue of Canadian birds, by John Macoun. Science, vol. 32, pp. 343-344, Sept. 9, 1910.

Biological investigations on the Illinois River. Ill. Lab. Nat. Hist. pamphlet, 14 pp., 11 pl., 1910. *Includes:* I. The work of the Illinois Biological Station. (*Abstract,* Science, vol. 31, pp. 837-838, May 27, 1910.) II. The investigation of a river system in the interest of its fisheries. (*Cf.* Amer. Fish. Soc. Trans., vol. 40, pp. 179-193, with 2 pl., 1911).

Executive report of the Illinois State Entomologist (to the Governor), including a report of operations under the horticultural inspection law, for 1909-1910. Ill. Ent. Off. pamphlet, 11 pp., 1910. (*Cf.* Ill. Hort. Soc. Trans., vol. 43, pp. 93-96, 1910; vol. 44, pp. 90-93, 1911.)

Relations of the Illinois Academy of Science to the State. Ill. Acad. Sci. Trans., vol. 3, pp. 32-43, 1910.

[—with others.] Report of committee on an ecological survey (of Illinois). Ill. Acad. Sci. Trans., vol. 3, pp. 51-56, 1910.

1911

The chinch bug abroad again in Illinois. Ill. Ent. Off. Circ., 4 pp., May 1911.

Household pests and their relation to public health. (Chicago) City Club Bul., vol. 4, pp. 77-80, May 1911.

Economic entomology (abstract of an address). Ill. Farm. Inst. Rep., vol. 16, pp. 309-316, 1911.

Report on nursery inspection and orchard insecticide operations. Ill. Hort. Soc. Trans., vol. 44, pp. 90-93, 1911.

Forestry survey of Illinois. Ill. Hort. Soc. Trans., vol. 44, pp. 93-96, 1911.

Some important insects of Illinois shade trees and shrubs. Ill. Agr. Exp. Sta. Bul. 151, pp. 463-529, with 67 fig., Oct. 1911. (*Incorporated in* Ill. Ent. Rep. 26, pp. 1-67, 1911.) (*Reprinted,* Ill. Farm. Inst. Rep., vol. 21, pp. 244-294, 1916.)

The chinch-bug situation in Illinois: fall and winter measures to be taken. Ill. Agr. Exp. Sta. Circ. (*without serial number*), 4 pp., Nov. 1911. (*Reprinted,* Ill. Farm. Inst. Rep., vol. 17, pp. 300-302, 1912.)

[—with others.] (Report of the) committee on ecological survey (of Illinois). Ill. Acad. Sci. Trans., vol. 4, pp. 24-27, 1911.

J. A. West: in memoriam. Ill. Acad. Sci. Trans., vol. 4, p. 158, 1911.

Twenty-sixth Report of the State Entomologist on the noxious and beneficial insects of the State of Illinois, (*being the*) fifteenth report of S. A. Forbes. 160 + xix pp., 133 fig., 1911. Contents as follows:
> Some important insects of Illinois shade trees and shrubs (1-67) [*issued as* Ill. Agr. Exp. Sta. Bul. 151, Oct. 1911, *and reprinted in* Ill. Farm. Inst. Rep., vol. 21, pp. 244-294, 1916]; miscellaneous economic insects, by CHARLES A. HART (68-98); on the more important insects of the truck-farm and vegetable garden, by J. J. DAVIS (99-160) [*also in* Ill. Farm. Inst. Rep., vol. 16, pp. 216-263, 1911].

Report of the Director of the State Laboratory of Natural History (to the Trustees of the University of Illinois), as of June 3, 1909. Ill. Univ. Rep., vol. 25, pp. 200-201, 1911.

Illinois Biological Station. Internat. Rev. Hydrobiol., vol. 4, pp. 226-227, with 1 pl., 1911.

1912

Definite results of survey work on the Illinois River. Amer. Fish. Soc. Trans., vol. 41, pp. 75-89, with 1 map, 1912. (*Revised version, with new title:* Chemical and biological investigations on the Illinois River, midsummer of 1911. Ill. Lab. Nat. Hist. pamphlet, 9 pp., 1 map, 1912.)

What is the matter with the elms in Illinois? Ill. Agr. Exp. Sta. Bul. 154, pp. 3-22, with 4 fig. and 6 pl., Feb. 1912. (*Incorporated in* Ill. Ent. Rep. 27, pp. 1-20, 1912.)

What should the state require of a negligent owner of a dangerous orchard? Jour. Econ. Ent., vol. 5, pp. 205-207, Apr. 1912. (*Also,* Ill. Hort. Soc. Trans., vol. 45, pp. 77-81, 1912.)

Report on nursery inspection and orchard insecticide operations. Ill. Hort. Soc. Trans., vol. 45, pp. 74-75, 1912.

The chinch-bug situation in Illinois: plans for a cooperative campaign. Ill. Agr. Exp. Sta. Circ. (*without serial number*), 7 pp., May 1912. (*Also*, Ill. Farm. Inst. Rep., vol. 17, pp. 302-307, 1912.)

The 1912 chinch-bug campaign in Illinois. Ill. Ent. Off. Circ., 17 pp., 8 fig., Nov. 1912.

The native animal resources of the state. Ill. Acad. Sci. Trans., vol. 5, pp. 37-48, 1912.

Twenty-seventh Report of the State Entomologist on the noxious and beneficial insects of the State of Illinois, (*being the*) sixteenth report of S. A. Forbes. 143 + xvii pp., 6 pl., 93 fig., 1912. Contents as follows:

> What is the matter with the elms in Illinois? (1-20) [*issued as* Ill. Agr. Exp. Sta. Bul. 154, Feb. 1912]; on black-flies and buffalo-gnats (Simulium) as possible carriers of pellagra in Illinois (21-55) [*also in* Ill. Pellagra Com. Rep. for 1911, pp. 176-191, 1912, *and also issued separately*, 1912]; insects injurious to stored grains and their ground products, by A. A. GIRAULT (56-82) [*issued as* Ill. Agr. Exp. Sta. Bul. 156, July 1912]; report on insects injurious to flowering and ornamental greenhouse plants in Illinois, by J. J. DAVIS (83-143).

Report of the Director of the State Laboratory of Natural History (to the Trustees of the University of Illinois), for 1909-1910. Ill. Univ. Rep., vol. 26, pp. 68-70, 1912.

1913

The corn root-aphis in Illinois. Ill. Agr. Exp. Sta. Circ. (*without serial number*), 7 pp., Jan. 1913.

The Simulium-pellagra problem in Illinois. Science, vol. 37, pp. 86-91, Jan. 17, 1913. (*Also,* Ent. Congr. (Oxford) Trans., vol. 2, pp. 477-485, 1913.)

[—with R. E. RICHARDSON.] Studies on the biology of the upper Illinois River. Ill. Lab. Nat. Hist. Bul., vol. 9, pp. 481-574, with 21 pl., June 1913.

Biological and chemical conditions on the upper Illinois River. Ill. Water Sup. Assoc. Proc., vol. 5, pp. 161-170, with 1 map and 3 diagrams, 1913.

Needed legislation in the interests of horticulture. Ill. Hort. Soc. Trans., vol. 46, pp. 49-51, 1913.

The apple flea-weevil in Illinois. Ill. Hort. Soc. Trans., vol. 46, pp. 53-55, 1913.

Executive report of the Illinois State Entomologist (to the Governor), including a report of operations under the horticultural inspection law, for 1911-1912. Ill. Ent. Off. pamphlet, 24 pp., 1913. (*Cf.* Ill. Hort. Soc. Trans., vol. 45, pp. 74-81, 1912; vol. 46, pp. 49-51, 1913.)

Report of the committee on an ecological survey (of Illinois). Ill. Acad. Sci. Trans., vol. 6, pp. 18-23, 1913.

1914

The chinch-bug situation: present prospects and practical plans. Ill. Ent. Off. Circ., 3 pp., Feb. 1914.

The chinch-bug in Illinois in 1914: preparations for the summer campaign. Ill. Ent. Off. Circ., 11 pp., 3 pl., Apr. 1914.

Community operation and co-operation against fruit pests. Ill. Hort. Soc. Trans., vol. 47, pp. 171-176, 1914.

Fresh water fishes and their ecology. Ill. Lab. Nat. Hist., pamphlet, 19 pp., 10 pl., 21 maps, 1914.

Report of the Director of the State Laboratory of Natural History (to the Trustees of the University of Illinois), for 1911-1912. Ill. Univ. Rep., vol. 27, pp. 158-159, 1914.

[—with others.] Report of the committee on an ecological survey of the state. Ill. Acad. Sci. Trans., vol. 7, pp. 12-16, 1914.

1915

Recent Illinois work on the corn root-aphis and the control of its injuries. Ill. Agr. Exp. Sta. Bul. 178, pp. 405-466, with 18 fig., Jan. 1915. (*Incorporated in* Ill. Ent. Rep. 28, pp. 1-62, 1915.)

Observations and experiments on the San Jose scale. Ill. Agr. Exp. Sta. Bul. 180, pp. 545-561, with 3 fig., Mar. 1915. (*Incorporated in* Ill. Ent. Rep. 28, pp. 87-106, 1915.)

The chinch-bug in Illinois in 1915: preparations for the summer campaign. Ill. Agr. Exp. Sta. Circ. (*without serial number*), 16 pp., 1 map, 4 pl., Apr. 1915.

Preliminary note on the life history of the codling-moth in Illinois. Ill. Ent. Off. Circ., 15 pp., 6 pl., May 1915.

The danger to wheat from the Hessian fly. Ill. Agr. Exp. Sta. Circ. (*without serial number*), 1 p., Sept. 1915.

Twenty-eighth Report of the State Entomologist on the noxious and beneficial insects of the State of Illinois, (*being the*) seventeenth report of S. A. Forbes. 106 + ix pp., 24 fig., 4 pl. 1915. Contents as follows: Recent Illinois work on the corn root-aphis and the control of its injuries (1-62) [*issued as* Ill. Agr. Exp. Sta. Bul. 178, Jan. 1915]; observations and experiments on the San Jose scale (63-79) [*issued as* Ill. Agr. Exp. Sta. Bul. 180, Mar. 1915]; life history and habits of the northern corn root-worm (80-86); the San Jose scale, by P. A. GLENN (87-106) [*issued as* Ill. Agr. Sta. Circ. 180, Apr. 1915].

The ecological foundations of applied entomology. Ann. Ent. Soc. Amer., vol. 8, pp. 1-19, 1915.

Insect and fungous pests of the nursery that should be classed as dangerous. Ill. Hort. Soc. Trans., vol. 48, pp. 153-161, 1 pl., 1915.

Insect pests and their control. Ill. Farm. Inst. Rep., vol. 20, pp. 187-191, 1915.

The insect, the farmer, the teacher, the citizen, and the state. Ill. Lab. Nat. Hist. pamphlet, 14 pp., 1915.

Executive report of the Illinois State Entomologist (to the Governor), including a report of operations under the horticultural inspection law, for 1913-1914; and report of the Director of the State Laboratory of Natural History. Ill. Ent. Off. pamphlet, 10 pp., 1915. (*Cf.* Ill. Hort. Soc. Trans., vol. 47, pp. 178-184, 1914; vol. 48, pp. 67-78, 1915. Ill. Univ. Rep., vol. 28, pp. 158-159, 1916.)

Report of the committee on an ecological survey of the state. Ill. Acad. Sci. Trans., vol. 8, pp. 15-17, 1915.

1916

A general survey of the May-beetles (Phyllophaga) of Illinois. Ill. Agr. Exp. Sta. Bul. 186, pp. 215-257, with 1 map, Feb. 1916. (*Incorporated in* Ill. Ent. Rep. 29, pp. 23-65, 1916.

The influence of trees and crops on injury by white-grubs. Ill. Agr. Exp. Sta. Bul. 187, pp. 261-265, Feb. 1916. (*Incorporated in* Ill. Ent. Rep. 29, pp. 66-70, 1916.)

Obituary: Francis Marion Webster. Journ. Econ. Ent., vol. 9, pp. 239-241, with portrait, Feb. 1916.

Thomas Jonathan Burrill (obituary). Ill. Univ. Quart., vol. 1, pp. 409-417, July 1916.

The chinch-bug outbreak of 1910 to 1915. Ill. Agr. Exp. Sta. Circ. 189, pp. 3-59, with 7 fig., Aug. 1916. (*Incorporated in* Ill. Ent. Rep. 29, pp. 71-127, 1916.)

Report on nursery and orchard inspection. Ill. Hort. Soc. Trans., vol. 49, pp. 59-66, 1916.

[—with P. A. GLENN.] On the life history of the codling-moth. Ill. Hort. Soc. Trans., vol. 49, pp. 170-180, 12 pl., 1916.

Twenty-ninth Report of the State Entolomologist on the noxious and beneficial insects of the State of Illinois, (*being the*) eighteenth report of S. A. Forbes. 127 + ix pp., 2 diagrams, 6 maps, 12 pl., 1916. Contents as follows:

On the life history of the codling-moth (1-21); a general survey of the May-beetles (Phyllophaga) of Illinois (23-65) [*issued as* Ill.. Agr. Exp. Sta. Bul. 186, Feb. 1916]; the influence of trees and crops on injury by white-grubs (66-70) [*issued as* Ill. Agr. Exp. Sta. Bul. 187, Feb. 1916]; the chinch-bug outbreak of 1910 to 1915 (71-127) [*issued as* Ill. Agr. Exp. Sta. Circ. 189, Aug. 1916].

Report of the Director of the State Laboratory of Natural History (to the Trustees of the University of Illinois), for 1913-1914. Ill. Univ. Rep., Vol. 28, pp. 158-159, 1916. (*Also app.* to Ill. Ent. Exec. Rep. for 1913-1914.)

Report of the committee on an ecological survey of the state. Ill. Acad. Sci. Trans., vol. 9, pp. 14-16, 1916.

1917

Entomology in time of war. Ill. Ent. Off. Circ., 6 pp., 2 pl., 1917. (*Reprinted from the Chicago Herald for June 3, 1917.*)

Report of the committee on an ecological survey (of Illinois). Ill. Acad. Sci. Trans., vol. 10, pp. 30-34, 1917.

1918

How to reduce Hessian-fly injury to a minimum. Ill. Nat. Hist. Surv. Ent. Circ. (*without serial number*), 2 pp., Feb. 1918.

Report of ecological committee. Ill. Acad. Sci. Trans., vol. 11, pp. 15-18, 1918.

1919

[—with R. E. RICHARDSON.] Some recent changes in Illinois River biology. Ill. Nat. Hist. Surv. Bul., vol. 13, pp. 139-166, April 1919.

The danger from house-flies and how to control it. Ill. Nat. Hist. Surv. Ent. Circ. (*without serial number*), 4 pp., Nov. 1919.

Recent forestry survey of Illinois. Ill. Hort. Soc. Trans., vol. 52, pp. 103-110, 1919.

Forest and stream in Illinois. Ill. Nat. Hist. Surv. pamphlet, 15 pp., 1919.

(Annual report of the) State Natural History Survey Division (to the Director of the State Department of Registration and Education, as of June 30, 1918). Ill. Adm. Rep., vol. 1, pp. 606-608, 1919.

1920

Concerning a forestry survey and a forester for Illinois. Ill. Nat. Hist. Surv. For. Circ. no. 1, 7 pp., 2 fig., 1920.

The life cycle of insects; general discussion. Ann. Ent. Soc. Amer., vol. 13, pp. 193-201, June 1920.

The function of economic entomology. Ill. Farm. Inst. Rep., vol. 25, pp. 55-60, with portrait, 1920.

(Annual report of the) State Natural History Survey Division (to the Director of the State Department of Registration and Education, as of June 30, 1919). Ill. Adm. Rep., vol. 2, pp. 517-522, 1920.

1921

Streams pollution in the Illinois basin. Ill. Sportsman, vol. 6, no. 10, pp. 1-2, Mar. 1921.

[—with A. O. GROSS.] The orchard birds of an Illinois summer. Ill. Nat. Hist. Surv. Bul., vol. 14, pp. 1-8, with 1 map and 6 pl., June 1921.

(Annual report of the) State Natural History Survey Division (to the Director of the State Department of Registration and Education, as of June 30, 1920). Ill. Adm. Rep., vol. 3, pp. 784-789, 1921.

1922

The humanizing of ecology. Ecology, vol. 3, pp. 89-92, Apr. 1922.

[—with A. O. GROSS.] The numbers and local-distribution in summer of

Illinois land birds of the open country. Ill. Nat. Hist. Surv. Bul., vol.
14, pp. 187-218, with 36 pl., April 1922.
The State Natural History Survey. Ill. Blue Book for 1921-1922, pp. 311-
314, 1922.
(Annual report of the) State Natural History Survey Division (to the
Director of the State Department of Registration and Education, as
of June 30, 1921). Ill. Adm. Rep., vol. 4, pp. 804-807, 1922.

1923

[—with A. O. Gross.] On the numbers and local distribution of Illinois
land birds of the open country in winter, spring, and fall. Ill. Nat. Hist.
Surv. Bul., vol. 14, pp. 397-453, Oct. 1923.
(Annual report of the) State Natural History Survey Division (to the
Director of the State Department of Registration and Education, as of
June 30, 1922). Ill. Adm. Rep., vol. 5, pp. 909-914, 1923.

1924

Introduction (to second report on forest survey of Illinois). Ill. Nat. Hist.
Surv. Bul., vol. 15, art. 3, p. vii, Aug. 1924.
The State Natural History Survey. Ill. Blue Book for 1923-1924, pp. 384-
387, 1924.
(Annual report of the) State Natural History Survey Division (to the
Director of the State Department of Registration and Education, as of
June 30, 1923). Ill. Adm. Rep., vol. 6, pp. 1036-1039, 1924.
Sewage pollution of the Illinois River. Outdoor America, vol 3, no. 5, pp.
35-36, Dec. 1924.

1925

Introduction (to entomological survey of Salt Fork). Ill. Nat. Hist.
Surv. Bul., vol. 15, p. iii, Oct. 1925.
Introduction (to) contents and index of the reports of the State Entomol-
ogist of Illinois, xxv-xxix, 1909-1916, p. 5, 1925.
(Annual report of the) State Natural History Survey Division (to the
Director of the State Department of Registration and Education, as of
June 30, 1924). Ill. Adm. Rep., vol. 7, pp. 1251-1254, 1925.

1926

Aerial music in Yellowstone Park. Science, vol. 64, pp. 119-120, 451-452,
July 30, Nov. 5, 1926.
Recent progress in the forestry movement. Ill. Blue Book for 1925-1926,
pp. 468-472, with 2 pl., 1926.
Some dangers of pollution. Outdoor America, vol. 5, no. 2, pp. 30-39, 58-
59, Sept. 1926. (*Reprinted under new title:* The effects of stream pollu-
tion on fishes and their food. Ill. Nat. Hist. Surv. pamphlet, 13 pp.,
1928.)
(Annual report of the) State Natural History Survey Division (to the
Director of the State Department of Registration and Education, as of
June 30, 1925). Ill. Adm. Rep., vol. 8, pp. 956-958, 1926.

1927

Introduction (to an experimental investigation of the relations of the codling moth to weather and climate). Ill. Nat. Hist. Surv. Bul., vol. 16, pp. 311-313, Mar. 1927.

(Annual report of the) State Natural History Survey Division (to the Director of the State Department of Registration and Education, as of June 30, 1926). Ill. Adm. Rep., vol. 9, pp. 967-972, 1927.

1928

The biological survey of a river system—its objects, methods, and results. Ill. Nat. Hist. Surv. Bul., vol. 17, pp. 277-284, March 1928.

Forces line up to battle the corn borer. Ill. Jour. Com., vol. 10, no. 8, pp. 7, 14, 21, Aug. 1928.

Foreword (to report on bottom fauna of Illinois River 1913-1925). Ill. Nat. Hist. Surv. Bul., vol. 17, pp. 387-388, Dec. 1928.

(Annual report of the) State Natural History Survey Division (to the Director of the State Department of Registration and Education, as of June 30, 1927). Ill. Adm. Rep., vol. 10, pp. 1161-1167, 1928.

1929

Concerning certain ecological methods of the Illinois Natural History Survey. Ill. Acad. Sci. Trans., vol. 21, pp. 19-25, Feb. 1929.

[—with P. K. HOUDEK.] State Natural History Survey and the high school. Sch. Sci. Math., vol. 39, pp. 388-390, Apr. 1929.

What can the Natural History Survey do for the high school? In "Proceedings of the High School Conference . . . 1928." Ill. Univ. Bul., vol. 26, no. 17, pp. 41-44, 1929. (Also, Ill. Nat. Hist. Surv. pamphlet, 7 pp., 1930.)

Foreword (to report on trees of Illinois). Ill. Nat. Hist. Surv. Bul., vol. 18, p. 3, June 1929.

(Annual report of the) State Natural History Survey Division (to the Director of the State Department of Registration and Education, as of June 30, 1928). Ill. Adm. Rep., vol. 11, pp. 1287-1290, 1929.

1930

What the Natural History Survey is doing for the high school. Ill. Nat. Hist. Surv. pamphlet, 11 pp., 1930. (Also in "Proceedings of the High School Conference . . . 1929." Ill. Univ. Bul., vol. 27, no. 41, pp. 77-82, 1930.)

State Natural History Survey. Ill. Blue Book for 1929-1930, pp. 393-404, with 5 fig., 1930.

Forest resources. In "Illinois resources, development, possibilities," pub. by Ill. Chamber of Commerce, Chicago, pp. 87-90, 1930.

(Annual report of the) State Natural History Survey Division (to the Director of the State Department of Registration and Education, as of June 30, 1929). Ill. Adm. Rep., vol. 12, pp. 1040-1047, 1930.

THE FOOD OF ILLINOIS FISHES

Stephen Alfred Forbes

THE FOOD OF ILLINOIS FISHES.

By S. A. FORBES.

But little has been written on the food of the fresh water fishes of this country, and n othing whatever, so far as I can learn, on the food of the fishes of of this state. I have not found anything more elaborate than a short paper* by Prof. S. I. Smith, of Yale College, on the food of a few specimens of White Fish, Red Horse (*Myxostoma aureolum*), Yellow Perch and Sturgeon (*Acipenser rubicundus*), from Lakes Superior and Erie. An item † relating to the food of the White Fish was published by Dr. Stimpson, of the Chicago Academy of Sciences, in 1870, and a few scattered notes of single observations occur in various papers on classification. ‡

The importance of the subject, both to the scientific student and to the practical fish breeder, seems to warrant more systematic work; and a methodical investigation has therefore been begun at the State Laboratory, the first results of which are given in the following memoranda.

PURPOSES OF THE INVESTIGATION.

A thorough knowledge of this subject should contribute something to our theories of distribution, since the food of those forms having appetites at all discriminating must have much to do with their range. Light might even be thrown upon past distribution, and the causes be suggested of extensive migrations. The chosen *haunts* of different groups within their habitat, are probably determined largely by their gastronomic needs and preferences. Do the wide-spread species eat similar articles throughout their range, or are they wide-spread because they are omnivorous, or because their food habits are more *flexible* than those of other fishes? On the other hand, are the narrowly limited species ever restricted by the local character of their food?

---

* Report of U. S. Commissioner of Fish and Fisheries, Pt. II, p. 690.
† American Naturalist, Sept. 1870, p. 403.
‡ A paper by Dr. C. C. Abbot in the Report of the U. S. Fish Commissioner for 1875-6 will also repay examination.

We ought also to gain, by this means, some addition to our knowledge of the causes of variation, of the origin and increase, the decline and extinction of species, and of the remarkable persistence of such forms as the river gar. What groups crowd upon each other in the struggle for subsistence? Do closely allied species, living side by side, ever compete for food? What relation, if any, do specific and generic differences bear to differences of food? These, and many similar questions, may not improbably be helped toward a solution.

Several *structures* not now fully understood, ought to receive their explanation. The variously developed grinding surfaces on the pharyngeal teeth of some cyprinoids, the differences in the structure of the gill-rakers among sun-fishes and of the lips among suckers, are cases in point.

It seems likely, however, that the food habits of fishes will be found, like their structure, much less highly differentiated than those of birds. This is what we should expect *a priori*, and it is indicated by the observations I have made upon both classes.* Prominent peculiarities, having apparently an important bearing upon the taking of food, will probably be found merely to extend a little the capacities of the species, or to enable it to take those slight advantages of its competitors when the struggle for existence comes to the death grapple, which after all are sufficient to decide the contest. To bring out such facts as this, a great number of observations will be necessary, covering all varieties of circumstance, and made with reference to the relative proportions of the different elements in the food of each species. The Top Minnows, for example, will probably be found to take the surface-swimming insects *more frequently* than the *Cyprinidae* do, but not by any means to depend on them *chiefly*.

Really *intelligent fish-culture*, on any large scale, implies a full acquaintance with the food of the native species. It is a matter of especial importance that the predaceous fishes should all be known, as well as the kinds of fishes on which each chiefly preys. A knowledge of the food of all species worth saving is, of course, indispensable, in order that proper measures may be taken to preserve their food supplies. It will also be of interest to know what fishes there are at once worthless for human food and harmless in their habits, and therefore worth encouraging, or perhaps even hatching, as food for the more valuable "game fishes." The gizzard shad (*Dorosoma cepedianum*), seems to be a fish of this character, as it lives chiefly on vegetable food and minute crustacea, and contributes largely to the food of the marketable fishes. Apparently ignorant of this fact, the fishermen often leave long lines of this species to rot on the bank where the seines are hauled.

* See "The Food of Birds," in Trans. Ill. Hort. Soc., Vol. X, p. 37, 1876.

Some valuable fishes may be found dependent on food too liable to injury or destruction by man or nature, to make it worth while to cultivate them, while others, equally valuable, may be proven to subsist on food practically indestructible.

Such species as eat *mixed food*, so that, in case of scarcity of one kind, another may be drawn upon, are evidently more promising, other things being equal, than those of a more limited diet.

That a full understanding of the *competitions* among the fishes of a stream or lake is necessary to anything better than guess-work in fish-culture, or an expensive and improvident trusting to luck, is evident at once. *

The *scavenger* fishes, which, by devouring the filth of streams, help to purify them, are doubtless worthy of recognition. Whether a filth-eating fish is better or healthier food than a bird or a mammal of similar habits, may, perhaps, be profitably discussed.

An acquaintance with the subject sufficient for the purposes above mentioned must, of course, include the whole life of the fish, at all ages and in all seasons. It is not impossible, for example, that the draining of stagnant waters connected with a stream may unfavorably affect some of its fishes, by lessening the supply of *Entomostraca*, especially *Cladocera*, for the food of the fry.

So much may properly be said concerning the purpose and promise of the research, to justify the labor given to it,—especially since the general neglect it has received may seem to indicate that it is not worth elaborate study.

METHODS.

The stomachs and intestines were taken out of the fishes just as these came from the seine; were labeled with specific name, place and date, and preserved in strong alcohol. They were afterward opened and the contents examined (usually with the microscope). Notes were made upon the objects found in each, as far as they were recognizable—the species being determined, if possible, otherwise the genus, family, order, or even only the class. The contents of each stomach were then bottled separately in alcohol, labeled and preserved for future verification and further study. The emptied stomachs have also been kept for anatomical purposes, and as a means of verifying the species. It was found unnecessary to remove the stomachs of the minnows, as these were well enough preserved in the bodies of the fishes themselves.

---

* That fishes and land birds should ever come into competition, seems at first sight remarkable; nevertheless some of the former eat large numbers of land insects which fall into the water. The supply of these would, of course, be limited by the depredations of birds.

In summing up, all the notes on the food of each species were collated, and an attempt was made to arrange the essential facts in a compact and simple form. The classification of fishes used is that of the preceding paper on the fishes of Illinois.

RESULTS.

Only a mere beginning has as yet been made. One hundred and forty-nine specimens have been examined, representing fifty-four species—taken chiefly (except the minnows) from the Illinois River, near Peoria and Henry, in June and November, 1877, and April and May, 1878. The specimens were all of a fair average size. In this preliminary report upon so small a number of specimens, it has not been deemed worth while to specify dates and places.

When the facts relating to any species are numerous and varied enough to make systematic condensation desirable, the articles of food have been arranged according to the natural classification of plants and animals, in such a way that one wishing to know only the general conclusions reached can readily learn them, without being embarrassed by unessential details.

The importance of a knowledge of the *proportions* of the different elements of the food has been kept in mind, and an attempt made to indicate these rudely by placing after each the number of specimens of the species in which the given element was found. Thus, under *Lepiopomus pallidus* (No. 18), of which two specimens were examined, "Chrysomelidæ 2" indicates that *one or more beetles* of this family were found in the stomachs of each of *two specimens* of that species. The figures in parentheses placed after the family and specific names of *fishes* indicate the number of specimens examined.

DETAILS OF FOOD.

DARTERS. ETHEOSTOMATIDAE. (9.)

Entomostraca and larvae of diptera and neuroptera.

1. *Sand Darter.* Pleurolepis pellucidus, Ag. (2.) Larvae of small diptera.

2. *Black-sided Darter.* Alvordius maculatus, Grd. (1.) Small diptera (gnats), larvae of May-flies (Ephemeridae), and many unknown minute eggs?, oval, tuberculated, with tubercles in longitudinal rows.

3. *Johnny Darter.* Boleosoma maculata, Ag. (1.) Several Cyclops and many larvae of gnats.

4. *Banded Darter.* Nanostoma zonalis, Put. (1.) Larvae of gnats, including some with antennae similar to those of Corethra pictipennis.

5. *Rough-cheeked Darter.* Poecilichthys asprigenis, Forbes, (2.) Larvae of Chironymus (diptera), and other aquatic larvae : also pupae of a small Ephemerid approaching Baetisca.

6. *Striped Darter.* Etheostoma lineolata, Ag. (1.) Many larvae of Chironymous (diptera).

7. ——— ——— Boleichthys elegans, Gir. (1.) Larvae of gnats and of May-flies, with a few copepoda.

PERCHES. PERCIDAE. (7.)

Chiefly fishes, including perch, bass, sun-fish, gizzard shad, minnows, (Cyprinidae) and cat-fish; also water-bugs (Corixa).

8. *Ringed Perch.* Perca americana, Schrank, (1). A cyprinoid fish.

9. *Black "Salmon." Wall-eyed Pike.* Stizostethium vitreum, Mitch. (1). A bony fish with sub-globular stomach ; probably one of the suckers.

10. *White "Salmon." Wall-eyed Pike.* Stizostethium canadense, Smith, (4). A common perch (Perca), a sun-fish (Ichthelidae), a black bass (Micropterus salmoides), a gizzard shad (Dorosoma), a cat-fish (Siluridae), and an undetermined bony fish with cycloid scales.

11. *Wall-eyed Pike.* Stizostethium, (species undetermined) (1). An unrecognizable bony fish and several water-bugs (Corixa alternata, Say).

WHITE BASS. LABRACIDAE. (4.)

Sun-fish, larvae of neuroptera and diptera, and other insects.

12. *White Bass.* Roccus chrysops, Raf. (3). Chiefly larvae of May-flies (Ephemeridae) ; also a sun-fish, (Centrarchidae) and another spiny-finned fish, and a few larvae of Chironymous and other diptera.

13. *Yellow Bass.* Morone interrupta, Gill. (1). Chiefly larvae of Ephemeridae (May-flies). An amphipod crustacean (Allorchestes dentata, Sm.), some larvae of dragon flies (Agrion) and a few young grasshoppers.

BLACK BASS AND SUN FISHES. CENTRARCHIDAE. (31.)

Food mixed, animal and vegetable, the former largely predominating. A few fishes (a darter, another percoid fish, and two or three cyprinoid minnows), a multitude of insects, land and water, representing all orders but hymenoptera; arachnida (spiders and water mites), amphipod and isopod crustacea (Allorchestes and Asellus), hosts of entomostraca (cladocera and copepoda, chiefly the former), a few mollusks, bivalve and univalve, an earth worm, and masses of Plumatella-like polyzoa ; also a good deal of Potamogeton, Ceratophyllum and other water weeds, and algae, together with miscellaneous floating vegetable trash.

14. *Black Bass.* Micropterus pallidus, Raf. (3). A large mouse, a percoid fish, a small stone roller (Campostoma anomalum), pupae of dragon flies, a water bug (Zaitha fluminea) and a few confervoid algae.

15. *Black Croppie.* Pomoxys nigro-maculatus Le S.(10). Chiefly larvae of May-flies (Ephemeridae). Many larvae of small diptera (various species of gnats), and occasionally a small percoid fish. The following is a detailed exhibit of the food of these ten specimens:

FISHES (Ctenoid.) 2.

INSECTS 10.

COLEOPTERA 3 (Larvae of Gyrinidae 2.)
DIPTERA 6 (Gnats and their larvae.)
HEMIPTERA 2 (Corixa alternata, Say.)
NEUROPTERA 9 (Larva of Ephemeridae 8, pupae of Agrioninae 1).

CRUSTACEA 3 (Entomostraca.)

CLADOCERA 2 (* Ceriodaphnia angulata (Say) Forbes 2). * Bosmina (Sp. ? 1).
COPEPODA 1 (Diaptomus).
POLYZOA Sp ? 1.

A few seeds and blossoms of trees in 2.

16. *White Croppie.* Pomoxys annularis, Raf. (9). Specimens taken in midsummer were feeding chiefly on the larvae of May flies. A number collected in March were distended with Ceriodaphnia angulata and larvae of may flies and dragon flies (Agrioninae). A small fish was found occasionally, and a few water bugs.

FISHES 4.

Etheostomatidae 1, Cyprinidae 1 (Luxilus analostanus), undetermined cycloid fish 1.

INSECTS 8.

COLEOPTERA 1 (larvae of Gyrindae).
HEMIPTERA 1 (Corixa alternata, Say).
NEUROPTERA 8 (Larvae of Ephemeridae 5, of Agrioninae 5).

CRUSTACEA 5 (Entomostraca).

Ceriodaphnia angulata, Say, 5.

17. *Croppies.* Pomoxys, Sp. undet. (2). A cyprinoid minnow, a few spiders, a hemipter, and larvae of May flies.

18. *Common Sun Fish. Blue Sun Fish.* Lepiopomus pallidus, Mit. (5). Almost wholly insects (many of them terrestrial.) A few mollusks and a little pond weed (Potamogeton).

---

* See appendix.

INSECTS 4.

LEPIDOPTERA 2 (Caterpillars).

COLEOPTERA 3,—Carabidae 1 (Agonoderus pallipes), Gyrinidae 1 (larva), Scarabaeidae 1 (Aphodius inquinatus), Chrysomelidae 2 (Diabrotica 12—guttata and a Haltica ?).

DIPTERA 2 (larvae of gnats).

ORTHOPTERA 1 (Phaneroptera curvicauda, a Tettix and a cricket.)

HEMIPTERA 3 (Corixa alternata 2, Arma ? 1).

NEUROPTERA 2 (Larvae of Ephemeridae.)

ARACHNIDA 3.

Spiders 2, Hydrachnidae (water mites) 1.

MOLLUSKS, 2.

Gasteropoda 2 (Physa, Planorbis).

Also an earthworm, some Potamogeton, and a number of unrecognized small seeds.

19. *Blue-cheeked Sun-fish.* Lepiopomus ischyrus, Nels. (1). Full of hornwort (Ceratophyllum demersum) and a polyzoan (Plumatella ?); also fragments of small bivalve shells, some small crustacea (Asellus, * Allorchestes dentata, Sm., and Cypris, sp.) and a little mixed vegetable matter.

20. *Bream. Pumpkin Seed.* Eupomotis aureus, Walb. (1). Several Aselli, univalve mollusks, and some unrecognized vegetable matter.

PIRATE PERCHES. APHODODERIDAE.

21. *Western Pirate Perch.* †Aphododerus isolepis, Nels. (3). The largest specimen (3 in. long) had eaten several Aselli, some larvae of diptera, a Corixa and another water-bug—apparently a Galgulus. The second in size (2¼ in.) contained only a small cycloid fish and several larvae of neuroptera. In the stomach of the smallest were several ostracoda (Cypris,) a larval Corixa and a few gnats.

MAIGRES. SCIAENIDAE.

22. *Sheepshead.* Haploidonotus grunniens, Raf. (7). Mollusks and larvae of May flies, with a few larvae of gnats.

Unios 2, Planorbis 2, Limnea 1, Ephemeridae 6, diptera 2.

---

* See appendix.

† An observation of the intestines shows that one effect of the remarkable change in the position of the vent in this species is the lengthening of the alimentary canal and a consequent increase of the digestive surface. The intestine passes from its origin at the stomach first upward, then backward, then downward, reaching the ventral wall at a point about half way from the bases to the tips of the ventrals. In the smallest specimens, it opens at this point. In the others, it turns forward along the middle line of the belly, and opens at a point more or less to the front, according to the size of the fish, leaving a seam of naked skin behind.

STICKLEBACKS. GASTEROSTEIDAE.

23. *Brook Stickleback.* Eucalia inconstans, Kirt. (1). Entomostraca (Cyclopidae and Eurycercus), and many mulberry-like masses of eggs (mollusks?)

SILVERSIDES. ATHERINIDAE.

24. *River Silverside.* Labidesthes sicculus; Cope. (5). Chiefly minute crustacea (copepoda and cladocera). A few small diptera and larvae of small dragon flies.
Cyclopidae 3, Daphniadae 3, (Bosmina,* Daphnia *), Lynceidae 1, (Eurycercus *), small diptera 3, larvae of Agrioninae 1.

TOOTHED MINNOWS. CYPRINODONTIDAE. (7).

Chiefly insects, aquatic and terrestrial; crustacea (amphipoda and cladocera), and gasteropod mollusks.

25. *Barred Killifish.* Fundulus diaphanus, Le S. (2). Planorbis and Pisidium, larvae of small diptera, Allorchestes, and cladocera of the family Lynceidae.

26. *Top Minnows.* Zygonectes notatus, Raf. (3). Bones of a small fish. Several small winged hymenoptera, small wingless ants, a spring beetle (Elateridae), a few hemiptera, several diptera and diptera larvae and pupae, a small spider, Crangonyx gracilis, and a number of Daphniidae and Lynceidae (Eurycercus, &c.)

27. *Striped Top Minnow.* Zygonectes dispar, Ag. (2). Physa and Planorbis, hemiptera and small diptera, and a few Lynceidae.

MUD MINNOWS. UMBRIDAE.

28. *Mud Minnow.* Melanura limi, Kirt. (2). Many water mites (Hydrachnidae), diptera larvae, Cypris, Planorbis, and fragments of unrecognizable insects.

PIKES. ESOCIDAE. (8.)

Fishes of several families, and tadpoles.

29. *Northern Pike.* Esox lucius, L. (8.) Only fishes. The eight specimens contained remains of 12 fishes, among which were a black bass (Ambloplites rupestris), and another ctenoid fish, a gizzard shad (Dorosoma), a toothed herring (Hyodon), 3 cyprinoids (1 Campostoma anomalum and 1 Alburnops?), an Ichthyobus and another large cycloid fish.

* See appendix.

30. *Little Pickerel.* Esox salmoneus, Raf. (2.) A small fish and two tadpoles of frogs.

MOON EYES. HYODONTIDAE.

31. *Toothed Herring.* Hyodon tergisus, Le S. (3). Numerous insects, including a bee, carabid beetle, some aquatic hemiptera, numerous larvae of Ephemeridae; a remarkable crustacean of which but a single specimen has hitherto been found in this country (Leptodora hyalina, * Lillj.); some rotten wood, elm seeds and other vegetable trash, evidently gathered from the drift-wood where one of the specimens had found its food.

HERRINGS. CLUPEIDAE. (6.)

32. *Ohio Golden Shad.* Pomolobus chrysochloris, Raf. (1.) A small bony fish and a few fragments of insects.

33. *Gizzard Shad. Hickory Shad.* Dorosoma cepedianum, Le S., var. heterurum, Raf. (5.) Extremely dirty. Every stomach examined was at least half full of mud. The food was chiefly vegetable, consisting largely of algae and a few crustacea, such as would naturally be entangled in the vegetation eaten. The objects recognized were a few diptera larvae, Leptodora hyalina, Cypris, Cyclops, masses of Chara, confervoid algae, desmids, and vast numbers of diatoms.

MINNOWS. CYPRINIDAE. (15.)

Time has failed for the examination of any sufficient number of this family, some of the most important genera having been omitted entirely. Enough has been learned, however, to show that these fishes live, some chiefly on aquatic vegetation (especially algae) and others on insects, of which small diptera and their larvae seem to constitute the greater part. The section whose teeth are not hooked probably eat vegetable food more largely than those with raptatorial teeth, no insects at all having been found in the intestines of the five specimens of that section examined.—I found a surprising amount of dirt in the intestines of these herbivorous minnows,— too much, I think, to have been taken incidentally.

34. *Stone Roller.* Campostoma anomalum, Raf. (1.) Full of dirt and confervoid algae.

35. *Blunt-nosed Minnow.* Hyborhynchus notatus, Raf. (3.) Full of dirt, with fragments of endogenous vegetation, confervoid algae, and many diatoms.

36. *Silvery Minnow.* Hybognathus argyritis, Grd. (1.) Full of sand and an immense number of diatoms, with a few filaments of confervoid algae and fragments of other vegetable matter.

---

* See appendix.

37. *Red-finned Shiner*. Luxilus cornutus, Mit. (2.) Chiefly vegetable. Fragments of unrecognizable insects, a mass of confervoid algae and parts of a netted-veined leaf.

38. *Silver Fin*. Luxilus analostanus, Gir. (3.) Insects of several orders and a few algae.

39. *Red Fin*. Lythrurus diplaemius, Raf. (1.) A few small diptera.

40. *Emerald Minnow*. Notropis atherinoides, Raf. (2.) Several small gnats, an unknown hemipter and a little vegetable matter.

41. *Silver-mouthed Dace*. Ericymba buccata, Cope. (1.) A few small larvae of diptera, much sand and indeterminable matter, partly vegetable.

42. *Common Chub*. Semotilus corporalis, Mit. (1.) Many Ephemera larvae and a larva apparently belonging to the Dytiscidae.

SUCKERS. CATOSTOMIDAE. (24.)

My observations on this family indicate that its food is chiefly animal, consisting principally of mollusks, insect larvae and entomostraca. It is not impossible that in this, as in some other cases, the proportion of vegetable food is under-estimated, owing to the rapidity with which it is digested, as compared with the chitinous and calcareous coverings of arthropods and mollusks. The intestines of the family contain usually large quantities of mud.

There seems to be a well defined difference between the food of the Catostominae (Red Horse) and that of the Bubalichthyinae (Buffaloes), the former group feeding much more freely on mollusks than the latter, and less generally on entomostraca. Of the eight Myxostomas examined, the principal food of each was small, thin-shelled Unionidae (Anodonta), while no entomostraca at all were observed in them, and annulate worms (Naididae) were found in but two specimens. On the other hand, mollusks were found in only four out of fourteen carp and buffaloes (gasteropods 2, bivalves 2), and in these insignificant in quantity, while large numbers of entomostraca were noticed in twelve of the specimens. The intestines of many of the buffaloes were filled with a yellowish, shreddy, corpuscular fluid which I could only interpret as altered intestinal mucus and broken down membrane. The fishermen at Peoria report, however, that these fishes frequent the mouths of the gutters from the still-houses of which the river front is redolent, apparently feeding upon the distillery slops.

43. *Red Horse*. Myxostoma macrolepidotum, Le S. (3.) Chiefly mollusks, (Unionidae, Physa, Planorbis.) A number of ringed worms (Naididae), fragments of Chara and endogenous vegetation, and much mud.

44. *Golden Red Horse.* Myxostoma aureolum, Le S. (2.) Chiefly mollusks, (Anodonta, Paludina), a few slender ringed worms (Naididae), much dirt and a little unrecognized vegetation.

45. Myzostoma, sp. undet. (3.) Only mud and Unionidae.

46. *Chub Sucker.* Eremyzon sucetta, Lac. (2.) Confervoid algae, diatoms, mud.

47. *Carp Suckers.* Carpiodes, sp. (6.) I have found it quite impossible to recognize the species of this genus with certainty by the descriptions extant, and until they have been revised prefer not to attempt to separate them.

Mollusca 1 (gasteropoda).

Insecta 3 (Chironomus larvae).

Crustacea 4; Cladocera 3 (Bosmina 1, Ceriodaphnia 1, Lynceidae 1), Ostracoda 2 (Cypris), Copepoda 3 (Cyclops 2, Canthocamptus 2.) Also nematoid worms 1, and various algae 2.

48. *Red-mouth Buffalo.* Ichthyobus bubalus, Raf. (6.) Chiefly entomostraca.

Mollusca 1 (Limnea ?).

Insecta 4 (larvae of diptera.)

Crustacea 6, Cladocera 5, (Bosmina 4, Daphnia 2, Ceriodaphnia angulata 4, Lynceidae 2) Ostracoda 2 (Cypris), Copepoda 5 (Cyclops 5, Canthocamptus 1). Also diatoms 1, and seeds and glumes of grasses 1.

49. *Black Buffalo.* Bubalichthys niger, Raf, (2.) Larvae of gnats and May flies, Unionidae, Ceriodaphnia, Cyclops, a Lumbricus, diatoms, mud.

CAT FISHES. SILURIDAE, (8.)

Very miscellaneous feeders, eating both animal and vegetable food, the former the more freely. Bones and pieces of fish were found, evidently from too large animals to have been swallowed alive. Aquatic larvae of all kinds, worms, water bugs, mollusks (rarely), algae, stems and leaves of both exogens and endogens, masses of fine roots, with an occasional craw-fish make up the bill of fare.

50. *Common Channel Cat.* Ichthaelurus punctatus, Raf. (6.) Bones and pieces of large fishes 3, fragments of bivalve mollusks (no shells) 1, land insects 1 (bees, plant bugs), aquatic larvae 6 (Dytiscidae, Agrioninae

and other dragon flies, May flies, caddis flies), water bugs, 1 (Corixa and Notonecta), vegetable matter 4, (algae, Naiadaceae, roots stems and leaves of various plants).

51. *Black Cat Fish.* Amiurus melas, Raf. (2). Taken in small prairie creeks, McLean Co. Stomach of one was full of purely vegetable food, consisting chiefly of a mass of confervoid algae ; that of the other contained no vegetation, but exhibited fragments of various insects, some of them terrestrial, and remains of young craw-fishes and aquatic larvae.

DOG FISHES. AMIIDAE.

52. *Dog Fish. Grinnel.* Amia calva, L. (1.) A single small specimen, 5 in. long, from S. Ill., had eaten some Ephemera larvae, a few ostracoda (Cypris) and some confervoid algae, with numerous diatoms.

GAR PIKES. LEPIDOSTEIDAE.

53. *Broad-nosed Gar.* Lepidosteus platystomus, Raf. Seven or eight specimens were opened, but the stomachs of all but one were entirely empty. This one contained a common river craw-fish, (Cambarus immunis, Hagen.) Is the gar a nocturnal feeder?

SPOON-BILLED CATS. POLYODONTIDAE.

54. *Shovel Fish. Bill Fish.* Polyodon folium, Lac. (5.) This is by far the most remarkable fish in our rivers, and is not less remarkable in its food than its structure. By the fishermen it is supposed to live on the slime and mud of the river bottom. The alimentary canal of each of the five specimens examined was found full of a brownish, half fluid mass, which, when placed under the microscope, was seen to be made up chiefly (in one case almost wholly) of countless myriads of entomostraca, of nearly every form known to occur in our waters, including many that have been seen as yet nowhere but in the stomachs of these fishes. Mixed with these, in varying proportion, were several undetermined and probably undescribed species of water worms (Annulata), most of them belonging to the family Naididae. Sometimes as much as a fourth of the mass was composed of vegetable matter,—largely algae, but including fragments of all the aquatic plants known by me to occur in the waters of the Illinois, except Ceratophyllum. Occasional leeches (Clepsine), water beetles (Coptotomus interrogatus &c.), a few larvae of diptera and Ephemerae and water bugs (Corixa) were noticed. Among the crustacea several specimens of the remarkable Leptodora hyalina already referred to were found.

I have not had time for anything more than a general examination of the mass of matter presented,—sometimes more than a pint from a single fish, —and cannot, therefore, give a list of the species. Curiously, very little mud was mixed with the food.

The remarkably developed gill-rakers of this species thus receive their explanation. These are very numerous and fine, arrangèd in a double row on each gill arch, and are twice as long as the filaments of the gill. By their interlacing they form a strainer scarcely less effective than the fringes of the baleen plates of the whale, and probably allow the passage of the fine silt of the river bed when this is thrown into the water by the shovel of the fish, but arrest everything as large as a Cylops. The fish is said by the fish-ermen to plow up the mud in feeding, with its spatula-like snout, and then to swim slowly backward through the muddy water. Its mouth, it may be noticed, is very large, even for a fish.

It is possible that this wholesale destruction of entomostraca may affect the food supply of other and more valuable fishes, especially of the very young of the predaceous species. We cannot yet say, however, where the *stress* of the struggle comes in the life of any given species, and conse-quently are unable either to relieve or heighten it at will, or to perceive the full effect of the forces already at work. Fuller knowledge must pre-cede any but the most cautious and conservative recommendations.

RECAPITULATION.

A summary of the leading facts presented in this paper is given in the following table. The figures taken from left to right give, in a general way, the food of each family and species, and taken vertically show the fishes which eat any given kind of food. The line of totals will show some-thing of the relative importance to fishes as a class of the different food ele-ments. The figures in the table have the same use as those in the preceding list, showing the *number of specimens* of the given species found to eat the food mentioned at the head of the column.

It will be seen that, estimated in this way, the most important kinds of food are insects, crustacea, plants, fishes and mollusks, in the order named. These data apply entirely to adult fishes, however; if the young were also taken into account, crustacean food would doubtless be found more important.

The best food fishes ("fine fish" of the markets—Percidae, Labracidae, Centrarchidae, Esocidae) are chiefly piscivorous, except the Centrarchidae, which are nearly omnivorous, but prefer insects. Scarcely any fishes ex-amined, except some Cyprinidae, can be called strictly herbivorous, al-though the gizzard shad (Dorosoma) is chiefly so, the animal food taken be-ing probably incidental.

That the sheepshead, with its enormous crushing pharyngeals, should eat fewer mollusks than the red-horse, was scarcely to be expected, and may yet prove untrue.

Cat-fishes are the only ones shown to be scavengers. The fishermen, however, attribute similar habits to sheepshead and buffalo-fish.

Cyprinidae seem to be divided into two sections, corresponding to the shapes of the pharyngeal teeth, those without raptatorial hooks being her-bivorous.

All these general statements ought, perhaps, to be put in the form of questions for future solution, since the number of specimens is too small and the space and time represented too limited to justify settled conclusions.

No. on List	Names of Fishes	No. examined	Fishes	Ctenoid	Cycloid	Siluroid	Mollusks	Univalve	Bivalve	Insects	Terrestrial	Hymenoptera	Lepidoptera	Diptera	Coleoptera	Orthoptera	Hemiptera	Neuroptera	Aquatic	Diptera	Coleoptera	Hemiptera	Neuroptera	Arachnida	Araneidae	Hydrachnidae	Myriapoda	Crustacea	Decapoda	Tetradecapoda	Entomostraca	Worms	Polyzoa	Plants	Phaenogam	Cryptogam	Garbage
	Etheostomatidae																																				
1	Pleurolepis pellucidus	9								9								4	9	7			4					2			2						
2	Alvordius maculatus	2								2								1	2	2			1					1			1						
3	Boleosoma maculatum	1								1									1	1								1			1						
4	Nanostoma zonale	1								1									1	1																	
5	Poecilichthys asprigenis	1								2				1				1	2	1			1					1			1						
6	Etheostoma lineolatum	2								2							1		2	1			1														
7	Boleichthys elegans	1								1									1				1														
	Percidae																																				
8	Perca americana	7	3	3						4	1				1			3	4				3	3	3			1			1						
9	Stizostethium vitreum	1	1	3						3	1			1				3	3				3														
10	S. canadense	1			1																																
11	Stizostethium, sp.	4		3	2																																
	Labracidae																																				
12	Roccus chrysops	4	2	2																																	
13	Morone interrupta	3	2	2																									1								
	Centrarchidae																																				
14	Micropterus pallidus	31	8	4	5	1	4	3	1	25	4	1		3	1	1		21	24	3	1	1	21	4	3	1		10		2	9	1	2	4	4	1	
15	Pomoxys nigro-maculatus	3	2	2	1																							3			3		1	1	2		
16	P. annularis	10	2	2	2		2		2	10	1			1	3		1	9	10	2		1	9					3			5	1	1	1	2	1	
17	Pomoxys, sp.	9	3	2	1					8								8	8			2	8					5			1	1	1	1			
18	Lepiopomus pallidus	2	1	2	1		2		1	2	1		1	2			2		2	1			2			2					1	1	1	1			
19	L. ischyrus	5	1		1		1		1	4	3	1						3	3				2	3	2	1						1	1	1	1		
20	Eupomotis aureus	1			1																																
	Aphododeridae																																				
21	Aphododerus sayanus*	3	1	1						3	1			1				2	3	1			2					2			1						
	Sciaenidae																																				
22	Haploidonotus grunniens	7					3	2	2	6				2				6	6	2			6					2			1	1	1	1	1		

* While these pages are being printed, I learn from Dr. Jordan that he finds no constant difference between the eastern and western forms of Aphododerus.

Table of fish food contents (rotated 90° on the page).

No. on List.	Names of Fishes.	No. examined.	FISHES.	Ctenoid.	Cycloid.	Siluroid.	MOLLUSKS.	Univalve.	Bivalve.	INSECTS.	Terrestrial.	Hymenoptera.	Lepidoptera.	Diptera.	Coleoptera.	Orthoptera.	Hemiptera.	Neuroptera.	Aquatic.	Diptera.	Coleoptera.	Hemiptera.	Neuroptera.	ARACHNIDA.	Araneidae.	Hydrachnidae.	MYRIAPODA.	CRUSTACEA.	Decapoda.	Tetradecapoda.	Entomostraca.	Worms.	POLYZOA.	PLANTS.	Phaenogams.	Cryptogams.	GARBAGE.
	Gasterostidae																																					
23	Eucalia inconstans	1																1										1			1						
	Atherinidae																																				
24	Labidesthes sicculus	5						3	3	3	1			3				1	4	4	2							5			5						
	Cyprinodontidae																																				
25	Fundulus diaphanus	7				3	3	1	7	4	1			2	1		2		2	2	1	2			1			2	2		5						
26	Zygonectes notatus	3				1	1	1	3	3	1			1		1			1	1	1		1			1		2		2	2						
27	Z. dispar	2							2	2						1			1	1		1			1			2		1	2						
	Umbridae																																				
28	Melanura limi	2															1		1	1		1						1		1	1						
	Esocidae																																				
29	Esox lucius	8	8		7				2	2																											
30	E. salmoneus, (tadpoles 1)	2	1	3					2																												
	Hyodontidae																																				
31	Hyodon tergisus	3	1						3	2	1			1		1			1		1					2											
	Clupeidae																																				
32	Pomolobus chrysochloris	1							1						1		1		3	3	2						1				2			4	4	1	
33	Dorosoma cepedianum	5							3	5									3	3	3					1								8	2	6	
	Cyprinidae																																				
34	Campostoma anomalum	15																	5	5	2													1		1	
35	Hyborynchus notatus	1							1	1																								3	1	2	
36	Hybognathus argyritis	3							3										3	3	2																
37	Luxilus cornutus	1																																1	1	1	
38	L. analostanus	2							2	1	1			2		2			3	3	2			1													
39	Lythrurus diplaemius	3							3	2	1			1			1		3	2	1						1				1			1	1	1	
40	Notropis atherinoides	2							1	1				2																				1	1	1	
41	Ericymba buccata	1							2	2					2				1	1			1					1									
42	Semotilus corporalis	1							1	1	1				1			1	1	1							1?							1			

No. on List	Names of Fishes	No. examined	FISHES	Ctenoid	Cycloid	Siluroid	MOLLUSKS	Univalve	Bivalve	INSECTS	Terrestrial	Hymenoptera	Lepidoptera	Diptera	Coleoptera	Orthoptera	Hemiptera	Neuroptera	Aquatic	Diptera	Coleoptera	Hemiptera	Neuroptera	ARACHNIDA	Araneidae	Hydrachnidae	Myriapoda	CRUSTACEA	Decapoda	Tetradecapoda	Entomostraca	Worms	Polyzoa	PLANTS	Phaenogams	Cryptogams	GARBAGE
	Catostomidae	24					11	5	9	11	1			9					9	9								11		5	10			10		7	
43	Myxostoma aureolum	2						1		2	1			1					1	1											1			1	1		
44	M. macrolepidotum	3							1	1	1			1						1																	
45	Myxostoma, sp.	3						2	1	1	1			1						1														1	1		
46	Erimyzon sucetta	2					3		2	3				4					4	4									4		4				2	1	2	
47	Carpiodes, sp.	6					2		3	4				4	1	1			6	4								6			6			3	3	2	
48	Ichthyobus bubalus	6					3			2				2	1					2	1							1		1	1			3	1	1	
49	Bubalichthys niger	2						2	1															6										1		1	
50	Siluridae						1		1						1																						
51	Ichthaelurus punctatus	6					1		1	6	1						1	1	1			1	1				1									
52	Amiurus melas	2					2	2		1					1													1			1	1		4	4	3	3
	Amiidae									1	1							1		1															1		
53	Amia calva	1																															1				
	Lepidosteidae									1	1						1					1							1	1							
54	Lepidosteus platystomus	1																											1	1							
	Polyodontidae																																				
	Polyodon folium	5												5					5	5									5				5		5	3	5	3
	TOTAL	149	30	17	12	1	22	14	12	95	22	5	2	9	8	2	5	.	75	41	10	13	3	7	4	3	1	46	2	16	37	7	2	35	14	25	3

NOTE.—I have just succeeded in obtaining, too late for previous notice, a copy of an elaborate paper on the Fisheries of the Great Lakes, by Mr. J. W. Milner, Asst. U. S. Fish Commissioner, published in the first report of the commissioner, 1872-3. It contains full notes on the food of the White Fish, Lake Trout, Lake Herring and Sturgeon. An article by Prof. A. E. Verrill in the same report contains interesting matter for a comparison of the food of allied marine and fresh water species.

APPENDIX.

ON THE CRUSTACEA EATEN BY FISHES.

I have recognized the following genera and species of crustacea in the stomachs of the fishes of the preceding list, several of them being new to the state. The material afforded has been by no means exhaustively studied, and the list of species could probably be quadrupled. I have refrained from formal description of some species which are evidently new, preferring to wait for specimens in more perfect condition.

Cambarus immunis, Hagen.

This is the only craw-fish I have yet noticed in the stomachs of fishes, and this I have seen but once (in the short-nosed gar), unless young individuals eaten by a small cat-fish (*Amiurus melas*) also belonged to this commonest of our species.

Allorchestes dentata, (Smith) Faxon.

Specimens of *Lepiopomus pallidus,* taken in Crystal Lake, McHenry Co., in June, were feeding chiefly on this crustacean. It has also been found in the same species, in *L. ischyrus* and in *Morone interrupta* from the Illinois at Peoria.

Crangonyx gracilis, Smith.

The western form of this species (see Bull. No. 1, p. 6) occurs abundantly throughout central and southern Illinois. It is a very agile and voracious creature, behaving in a jar of entomostraca like a tiger in a sheep-fold. I have noticed that ponds in which it is at all common are nearly or quite destitute of *Eubranchipus.* The "handiness" with which it uses its anterior feet in feeding is quite amusing. I have found it eaten only by the Top Minnow (*Zygonectes notatus.*)

Asellus intermedius, Forbes.

Eaten by *Aphododerus* from Union Co. A species of *Asellus* described by Mr. O. P. Hay, in the paper following this, as *A. militaris,* has recently been collected in the Illinois River, and has been noted in the stomachs of

Lepiopomus ischyrus and *Eupomotis aureus.* Another form which, from its variability, I have not yet ventured to describe as distinct from *intermedius*, is very common in slow streams and fresh pools in McLean Co., especially in early spring, and has reached me also from La Salle Co., and from Wisconsin. Its size is equal to that of *communis*, and it differs from typical *intermedius* also in the much more robust development of all its appendages, and in the large size of the second joint of the outer ramus of the second genital plates of the male. The form and proportions of these genital plates must be used with caution, however, in describing species, as they evidently vary greatly.

Leptodora hyalina, Lilljeborg.

This extremely curious crustacean, which may be known by its peculiar, slender form (that of a true cross, the arms of which are the swimming appendages), by its extreme transparency and by the single eye in the front end of its cylindrical head, has hitherto been observed in this country only by Prof. S. I. Smith, by whom a single specimen was dredged in L. Superior in 1871. * It evidently stands between the other Cladocera and the Phyllopoda in many respects, having no slight resemblance to a larval *Eubranchipus.*

It occurs in considerable numbers in Peoria Lake, a mere expansion of the Illinois River, the depth of which does not exceed eighty feet. Specimens taken in a small surface net, in June, 1877, were lost in transit, and it was not again seen until found in the stomachs of *Polyodon, Dorosoma* and *Hyodon.* It is not at all certain that this is identical with the European species, all the specimens yet studied being too imperfect to decide this point.

Eurycercus lamellatus, Muell. ?

Specimens apparently of this species appear in the stomachs of fishes from Crystal Lake, McHenry Co., (*Apeltes, Labidesthes, Fundulus*) and also in shovel fishes from Peoria Lake. It is likewise common in ponds in McLean Co.

Bosmina, sp. ?

This genus belongs to a section of *Dapniadae* (*Lyncodaphnia*) distinguished by the long and strong anterior antennae and by the reduced importance of the posterior pair. The former are tapering, curved and cylindrical, (containing in our species about 14 slightly spinulose joints, with a tuft of bristles on the front of the third) and project from the front of the head like a bifid beak. Occurs in myriads in food of shovel fish, in carp, buffalo, &c., and in *Labidesthes* from Crystal Lake.

Ceriodaphnia angulata, (Say) Forbes.

Very abundant in central Illinois, (McLean and Rock I. counties), but

* Invertebrate Fauna of L. Superior, p. 696.

not hitherto reported. The following is Say's description, in Jour. Acad. Nat. Science, Phil., Vol. I, p. 440, 1818 :

"*D. angulata.* Body viewed laterally, sub-oval, contracted before, gibbous above near the posterior edge, beneath ventricose in the middle ; back sub-ovate, acute behind and contracted before : sides striate with numerous minute, parallel, obliqe lines. Hind edge of the body with a prominent angle in the middle, which is obtuse at tip; above the angle it is ciliated. *Antennae,* 4 filaments on the superior branch, and 5 on the inferior branch ; color white or red. Length $\frac{1}{10}$ of an inch. Cabinet of the Academy. Very common in the stagnant marsh water of the forests of the Southern States."

In the Illinois specimens the head is marked off from the body by a dorsal indentation. The color is usually white. Found in the stomachs of carp, buffalo, sun-fish, &c. It constituted the principal part of the food of a number of croppies taken in April, from the Illinois R. The eggs beneath the carapace were so numerous as to give an orange color to the whole mass of the food at this time.

Daphnia pulex? L.

The species referred to by Prof. Smith, under this name,[*] is our commonest *Daphnia,* occurring everywhere in immense numbers. It is eaten by *Polyodon* and by many small fishes.

Daphnia galeata, Sars.

A species probably the same as that figured by Prof Smith in the paper already cited, was found in Crystal Lake,—a shallow sheet of water about 2 miles long—and was eaten in numbers by the abundant little silversides (*Labidesthes.*)

Canthocamptus illinoisensis, Forbes.

This minute crustacean was frequently found in carp, buffalo and shovel fishes from the Illinois R.

Diaptomus sanguineus, Forbes.

In *Pomoxys nigro-maculatus.* Numbers of the genus unrecognizable as to species were observed in a variety of fishes.

Many *Cyclops* and *Cypris,* the species of which I have not attempted to discriminate, occurred in fishes from all waters and of a dozen families.

---

[*] Loc. cit.

ON SOME INTERACTIONS OF
ORGANISMS

Stephen Alfred Forbes

INTRODUCTION.

THE State Legislature of Illinois authorized, at its last session, an investigation of the food of the birds of the state, with especial reference to agriculture and horticulture, and a similar investigation of the food of fishes, with especial reference to fish culture, making to the State Laboratory of Natural History, at Normal, an appropriation of $350 *per annum* for the expenses of these researches, and the papers of this series are the first results of this work.

The appropriation for the publication of the bulletins of the Laboratory has been used, first for the printing of the reports thus implicitly called for, and afterwards for such original papers on the natural history of the state as were offered by their authors.

S. A. FORBES,
Director Ill. State Lab. of Nat. History.

Normal, Ill., Nov. 1, 1880.

ON SOME INTERACTIONS OF ORGANISMS.*

By S. A. FORBES.

While the structural relations of living organisms, as expressed in a classification, can best be figured by a tree,— the various groups, past and present, being related to each other either as twigs to twigs, as twigs to branches, or as branches to the main stem,— yet this illustration does not at all express their *functional* relations. While the anatomical characters of the various groups may show that they are all branches of a common stock, from which they have arisen by repeated divisions and continued divergences, the history of their lives will show that they are now much more intimately and variously bound together by mutual interactions than are twigs of the same branch,— that with respect to their vital activities they occupy rather the relation of organs of the same animal body. If for a type of their classification we look to the vegetable world, for an illustration of their mutual actions and reäctions we must look to the animal world. The serious modification of any group, either in numbers, habits, or distribution, must modify, considerably, various other groups; and each of these must transmit the change in turn, or initiate some other form of change, the disturbance thus propagating itself in a far extending circle.

While the whole organic world, viewed as a living unit, thus differs from the single plant by the much greater interdependence

---

* As details accumulated relating to the food of animals and similar subjects, it was found that a proper discussion of them would necessarily lead, step by step, to a full review of certain parts of the general subject of the reäctions between groups of organisms and their surroundings, organic and inorganic. Without such a review, the facts can not be safely generalized, nor the conclusions clearly apprehended to which they point. It has therefore seemed best to prepare the way for the discussion of special subjects by this general discussion of the subject at large.

The practical importance of this larger view is illustrated by the fact that, if the current ideas of the value of parasitic and predaceous insects are accepted, we must condemn the bluebird to extermination as a pest; while, if the conclusions of this paper are essentially sound, this bird is a very useful species and should be carefully preserved.

of its parts, on the other hand, it differs from the single animal in the fact that, notwithstanding this intimate and instant sympathy of part with part, it has an immense vitality. To cut off the leg of an animal is often sufficient to destroy its life, but one might cut off the *head* of the animal world, so to speak, without seriously impairing its energy. Suddenly to annihilate every living vertebrate would doubtless set on foot some tremendous revolutions in the life of the earth, but it is certain that in time the wound would heal,—that Nature would finish by readjusting her machinery, and would then go on much as before. In fact, any subkingdom of animals or any class of plants might thus be struck out, without the slightest danger that terrestrial life would perish as a consequence. The functions of the missing member would be taken on in part by other members, and in part be rendered needless by new adjustments.

We see many present illustrations of this fact, as in Australia, where there is but one native carnivorous animal, and that probably not indigenous; in several Pacific islands where mammals are unknown, and in New Zealand and the Galapagos, where insects are extremely few and the flowers, therefore, chiefly colorless and odorless. We see, likewise, illustrations of the same truth in the conditions of vegetable and animal life in earlier geological periods. Plants and insects, for example, existed together through vast periods of time when there were neither mammals nor birds on earth to supervise or regulate their relations.

If this is true of such immense and revolutionary disturbances, it is all the more certain that this same spontaneous action of natural forces must in time reduce the smaller disturbances of the primitive order caused every-where by civilized man, and must end by adjusting the whole scheme of organic relations to his interests as completely as to the interests of any other species. It is also plain that if man understands clearly the disorders which arise in the system of Nature as a result of the rapid progressive changes in his own condition and activities, and understands also the processes of Nature which tend to lessen and remove these disorders, he may, by his own intelligent interference, often avoid or greatly mitigate the evils of his situation, as well as hasten their remedy and removal.

Some general notion of the original order of Nature, which

obtains where civilization has not penetrated, will be needful for an understanding of the most important consequences of the modifications of that order which man brings to pass,— for an understanding of the relations of our own industrial operations and interests to the general laws and activities of the organic world under whose constant influence we must live and work.

There is a general consent that primeval nature, as in the uninhabited forest or the untilled plain, presents a settled harmony of interaction among organic groups which is in strong contrast with the many serious mal-adjustments of plants and animals found in countries occupied by man. This is so familiar a fact that I need not dwell upon it, but will cite the reader to the generally accessible "Introduction to Entomology," by Kirby & Spence, for a sufficient statement of it. It will be more to my purpose to discuss the subject from a different standpoint. To determine the primitive order of Nature by induction alone requires such a vast number of observations in all parts of the world, for so long a period of time, that more positive and satisfactory conclusions may perhaps be reached if we call in the aid of first principles,— traveling to our end by the *a priori* road.

For the purposes of this inquiry I shall assume as established laws of life, the reality of the struggle for existence, the appearance of variations, and the frequent inheritance of such as conduce to the good of the individual and the species,— in short, the evolution of species and higher groups under the influence of natural selection. I shall also postulate, as an accepted law of Nature, the generalization that the species is maintained at the cost of the individual; — that, as a general rule, the rate of reproduction is in inverse ratio to the grade of individual development and activity; or, as Spencer tersely states this law, "Individuation and Genesis are antagonistic." Evidently a species can not long maintain itself in numbers greater than can find sufficient food, year after year. If it is a phytophagous insect, for example, it will soon dwindle if it seriously lessens the numbers of the plants upon which it feeds, either directly, by eating them up, or indirectly, by so weakening them that they labor under a marked disadvantage in the struggle with other plants for foothold, light, air and food. The interest of the insect is therefore identical with the interest of the plant it feeds upon. Whatever injuriously affects the latter, equally injures the former; and whatever favors the latter, equally favors the former. This must, therefore, be regarded as the extreme normal limit of the numbers of a phytophagous species,— a limit such that its depredations shall do no especial harm to the plants upon which it depends for food, but shall remove only the excess of foliage or fruit, or else superfluous individuals which must either perish otherwise, if not eaten, or, surviving, must injure their species by over-crowding. If the plant-feeder multiply beyond the above limit, evidently the diminution of its food supply will soon reäct to diminish its own numbers; a counter reäction will then take place in favor of the plant, and so on through an oscillation of indefinite continuance.

On the other hand, the reduction of the phytophagous insect below the normal number, will evidently injure the food plant by preventing a reduction of its excess of growth or numbers, and will also set up an oscillation like the preceding, except that the steps will be taken in reverse order.*

I next point out the fact that precisely the same reasoning applies to predaceous and parasitic insects. Their interests, also, are identical with the interests of the species they parasitize or prey upon. A diminution of their food reäcts to decrease their own numbers. They are thus vitally interested in confining their depredations to the excess of individuals produced, or to redundant or otherwise unessential structures. It is only by a sort of unlucky accident that a destructive species really injures the species preyed upon.

The discussion has thus far affected only such organisms as are confined to a single species. It remains to see how it applies to such as have several sources of support open to them,— such, for instance, as feed indifferently upon several plants or upon a variety of animals, or both. Let us take, first, the case of a predaceous beetle feeding upon a variety of other insects,— either indifferently, upon whatever species is most numerous or most accessible, or preferably upon certain species, resorting to others only in case of an insufficiency of its favorite food.

It is at once evident that, taking the group of its food-insects as a unit, the same reasoning applies as if it were restricted to a sin-

---

* See Principles of Biology, by Herbert Spencer, Vol. II, pp. 397–478.

gle species for food: that is, it is interested in the maintenance of
these food-species at the highest number consistent with the general conditions of the environment,— interested to confine its own
depredations to that surplus of its food which would otherwise
perish if not eaten,— interested, therefore, in establishing a rate
of reproduction for itself which will not unduly lessen its food supply. Its interest in the numbers of each species of the group it
eats will evidently be the same as its interest in the group as a
whole, since the group as a whole can be kept at the highest number possible only by keeping each species at the highest number
possible.

If the predatory insect prefer some species of the group to
others, we need only say that whatever interest it has in any
species of the group, will be an interest in keeping up its numbers
to the highest limit; and any failure in this respect will injure it
in precisely the ratio of the value of that species as an element of
its food. It would be most injured by any thing injuriously
affecting the species it most preferred; — the *preferences* of animals being, according to the doctrine of evolution, like their
instincts, inherited tendencies toward the things which have proved
beneficial to their progenitors.

This argument holds for birds as well as for insects, for animals
of all kinds, in fact, whether their food be simple or mixed, animal
or vegetable, or both. It also applies to parasitic plants. The
ideal adjustment is one in which the reproductive rate of each
species should be so exactly adapted to its food supply and to the
various drains upon it that the species preyed upon should normally produce an excess sufficient for the species it supports.
And this statement evidently applies throughout the entire scale
of being. Among all orders of plants and animals, the ideal balance of Nature is one promotive of the highest good of all the
species. In this ideal state, towards which Nature seems continually striving, every food-producing species of plant or animal
would grow and multiply at a rate sufficient to furnish the required
amount of food, and every depredating species would reproduce at
a rate no higher than just sufficient to appropriate the food thus
furnished.

We must now point out how this common interest is naturally
subserved,— how the mutually beneficial balance between animals
and their food is ordinarily maintained.

Exact adjustment is doubtless never reached any where, even for a single year. It is usually closely approached in primitive nature, but the chances are practically infinite against its becoming really complete, and mal-adjustment in some degree is therefore the general rule. All species must oscillate more or less. Even the most stable features of the organic environment are too unstable to allow the establishment of any perfectly uniform habit of growth and increase in any species. The most unvarying species will at one time crowd its boundaries vigorously, and, at another, sensibly recede from them. That such an oscillation is injurious to a species may be briefly shown. The most favorable condition of a species is that in which its numbers are maintained at the highest possible average limit; and this, as already demonstrated, requires that its food supplies should likewise be maintained at the highest possible limit,— that the species should, in fact, confine its appropriations to the unessential surplus of its food. But when the numbers of an oscillating species are above this average limit, it will devour more than this surplus of its food,—its food supplies will be directly lessened. On the other hand, when the oscillating species falls below this limit, its food supplies, reäcting, of course can not increase *beyond* the highest possible limit, but will reach it and there stop. The average amount of food will therefore be less than it might be if the species dependent upon it did not oscillate,—and, the food being less, the average number of the species itself must be smaller. Our problem is, therefore, to determine how these innumerable small oscillations, due to imperfect adjustment, are usually kept within bounds,—to discover the forces and laws which tend to prevent either inordinate increase or decrease of any species, and also those by which widely oscillating species are brought into subjection and reduced to a condition of prosperous uniformity. We may know in general that such laws and forces are constantly at work, and that the tendency of things is towards this healthful equilibrium, because we see substantially such an equilibrium widely established and steadily maintained through long periods of time, notwithstanding the great number and kaleidoscopic variability of the forces by which each species is impressed. But this idea will repay more detailed elucidation. We will notice, first, some of the checks upon injurious oscillations arising out of the laws of the individual organism, and afterwards those which are brought to bear upon it from without.

It will at once be seen that, in any case, the mal-adjustments possible are of only two kinds,—the rate of reproduction in the species must be either relatively too small or relatively too great. If it be relatively too small,—if the species bring forth fewer young than could mature, on the average, under existing circumstances, whatever may be the oscillations arising, they will tend to disappear with the disappearance of the species. The average numbers of such a species being, in the most favorable event, less than they might be, it will be at a certain disadvantage in the general struggle for existence—it will eventually yield to some more prolific species with which it comes in competition. If, for any reason, its rate of multiplication be or become too high, the law of the antagonism between individuation and genesis will constantly tend to bring it within the proper limit. Reproduction being more active than is necessary, the individual force and activity will be less than it might be,—the species will be at a disadvantage in the search for food, and in all its other activities, as compared with other species more exactly adjusted, or, as compared with members of its own species which tend to a better adjustment. As soon as a better adjusted competitor appears, the other must begin to suffer, and in the long course of evolution will almost certainly disappear. The fact of survival is therefore usually sufficient evidence of a fairly complete adjustment of the rate of reproduction to the drains upon the species.

For the sake of illustration, let us take an instance—and the most difficult we can find, for the application of these ideas,—the case of a caterpillar and its hymenopterous parasite.

If the rate of increase of the parasite be relatively too great, that is, if more parasites are produced than can find places of deposit for their eggs in the bodies of the mere *excess* of caterpillars, some of them will deposit their eggs in caterpillars which would otherwise come to maturity,—that is, the number of caterpillars will be gradually diminished. With this diminution of their hosts the parasites will find it more and more difficult favorably to bestow all their eggs, and many of them will fail of development. The multiplication of the parasites will thus be checked, and their numbers will finally become so far reduced that less than the then excess of caterpillars will be infested by them, in which case the caterpillars will commence to increase in numbers, and so

2

on indefinitely. Briefly, the excessive rate of increase of the parasite will keep up an oscillation of numbers in both parasite and host which will cross and recross a certain average line.

Let us now look at the method by which Nature may check this injurious fluctuation.

Let us suppose two groups of a parasitic species at work on the same species of caterpillar, of which one (A) is distinguished by a tendency to an excessive reproductive rate, while the other (B) multiplies no faster than is consistent with the best interest of its host. A, producing more eggs than B, must either parasitize more caterpillars than B, or must deposit a greater number of eggs in each. It can not parasitize more caterpillars than B, because this would require a greater activity,— a higher individuation,— and this is contrary to the law that individuation and genesis are antagonistic. Instead of being more active than B, it will then be less active, and will, therefore, deposit more eggs in each caterpillar. B, however, can not have acquired the habit of depositing too few eggs in each caterpillar, as that would compel it to search habitually for a greater number of larvæ than necessary,— to have acquired, that is, a habit of wasting energy,—which is, as already said, contrary to evolution. A will, therefore, sometimes deposit too many eggs in a caterpillar, and will then either lose the whole deposit, or bring forth a weakened offspring, which will, in the long run, give way to the more vigorous progeny of B. This regular production of a wasted excess will constitute an uncompensated drain upon variety A, which will end, like any other radical defect, in its yielding to its better adjusted rival.

Or if, notwithstanding the foregoing, we suppose this excessive reproductive rate to have become fully established, then the parasite-ridden species will evidently labor under such a disadvantage in the struggle for existence that it will probably be crowded out, in time, by some more fortunate rival. If the pair are permanently ill-adjusted, so that permanent loss of numbers follows, they will be treated by the laws of natural selection as a single imperfect animal,— they will be pushed to the wall by some better adjusted caterpillar and parasite, or by some insect free from troublesome companions. We may be sure, therefore, that, as a general rule, in the course of evolution, only those species have been able to survive whose parasites, if any, were not prolific enough sensibly to limit the numbers of their hosts for any length of time.

We notice incidentally that it is thus made unlikely that an injurious species can be exterminated, can even be permanently lessened in numbers, by a parasite strictly dependent upon it,— a conclusion which remarkably diminishes the economical rôle of parasitism. The same line of argument will, of course, apply, with slight modifications, to any animal, or even to any plant dependent upon any other animal or any other plant for existence.

From the foregoing argument we conclude that, since the interest of a species of plant or animal and the interest of its " enemies " are identical, and since the operations of natural selection tend constantly to bring about an adjustment of the species and its enemies which shall best promote this common interest; therefore *the annihilation of all the established " enemies " of a species would, as a rule, have no effect to increase its final average numbers.* This being a general law, applying to all organisms, it is plain that the real and final limits of a species are the *inorganic* features of its environment,— soil, climate, seasonal peculiarities, and the like.

In treating of the external forces brought to bear upon an oscillating species to restrain its disastrous fluctuations, I shall mention only a part of the organic checks to which it is subject.

It is a general truth, that those animals and plants are least likely to oscillate widely which are preyed upon by the greatest number of species, of the most varied habit. Then the occasional diminution of a single enemy will not greatly affect them, as any consequent excess of their own numbers will be largely cut down by their other enemies, and especially as, in most cases, the backward oscillations of one set of enemies will be neutralized by the forward oscillations of another set. But by the operations of natural selection, most animals are compelled to maintain a varied food habit,— so that if one element fails, others may be available. Thus each species preyed upon is likely to have a number of enemies, which will assist each other in keeping it properly in check.

Against the uprising of inordinate numbers of insects, commonly harmless but capable of becoming temporarily injurious, the most valuable and reliable protection is undoubtedly afforded by those predaceous birds and insects which eat a *mixed food*, so that in the absence or diminution of any one element of their food, their own numbers are not seriously affected. Resorting, then, to other food supplies, they are found ready, on occasion, for immediate and overwhelming attack against any threatening foe. Especially does the wonderful locomotive power of birds, enabling them to escape scarcity in one region which might otherwise decimate them, by simply passing to another more favorable one, without the loss of a life, fit them, above all other animals and agencies, to arrest disorder at the start,— to head off aspiring and destructive rebellion before it has had time fairly to make head. But we should not therefrom derive the general, but false and mischievous notion, that the indefinite multiplication of either birds or predaceous insects is good. Too many of either is nearly or quite as harmful as too few.

And this brings us to the application of these principles to the interests of civilized man. We must note how the new forces which he brings into the field expend themselves among those we have been studying, and to what reäctions they are in turn subjected. We must first see how far the primitive natural order of life lends itself to the supply of man's needs, to the accomplishment of his purposes; and must determine, in a general way, where he may be content to leave it undisturbed, where he should address himself to its improvement, and where he is compelled to attempt wholly to set it aside, substituting artificial arrangements of his own, devised solely in his own interest.

Some of Nature's arrangements man finds himself unable to improve upon for his own benefit. No one thinks of cultivating the forest to hasten the growth of the wood, or of trimming the wild oak or the maple, or of planting artificially the nuts and acorns in the woods to increase the number of the trees.

We are content to leave things there to go on essentially in the old way, merely anticipating the processes of natural death and decay by removing the trees before they spontaneously perish, and glad if the revolutions of organic life which we set up in the country around do not penetrate to the forest, visiting the leaves and trunks of the trees with the scourge of excessive insect depredations

Usually, however, we find the ready-made system of Nature less to our liking, and all our cultures are attempts to set it aside more

or less completely. In the pasture and the meadow, it answers
our purpose to substitute other species for the grasses growing
there spontaneously, and these adapt themselves easily to the circumstances which have proved favorable to their native predecessors. But in the grain-field and fruit-garden the case is different.
Not only do we bring in species often very unlike any aboriginal
vegetation and still further altered by long cultivation, but we
propose an end quite different from that for whose accomplishment
all the arrangements of Nature have been made.

According to the settled order, the whole economy of every
fully-established plant and animal is directed to the production of
one more plant or animal to take the place of the first one when it
perishes. All the excess of growth and reproduction is a reward
to friends or a tribute to powerful enemies, intended to make only
this one end secure. But man is not content with this. He does
not raise apple-trees for the sake of raising more apple-trees. He
would cut off all excess not useful to himself, and all that is useful
he would stimulate to the utmost, and appropriate to his own
benefit. In carrying out this purpose he finds himself opposed
and harassed at every step by rules and customs of the natural
world established long ages before he was seen upon the earth; —
laws certainly too powerful for him wholly to defy, customs too
deeply rooted for him to overturn without the most complicated
consequences And yet even here, we see that the primitive
order is not an evil, it is simply insufficient. It is good as far as
it goes, and must be carefully respected in its essence, however
far it may be modified in detail. We find abundant reason for a
belief in its usual beneficence and for a reluctance to disturb it
without urgent necessity.

At the best the disturbances we must originate will be tremendous. Old combinations will necessarily be broken up and new
ones entered into. As in a country undergoing a radical change
in its form of government, disorders will almost certainly break
out,— some of them fearfully destructive and temporarily uncon-
trollable; but the general tendency towards a just equilibrium
will make itself felt, and intelligent effort will mitigate some evils
and avoid others. Without attempting to go into details,— which
would be quite unnecessary for my purpose,— I will endeavor

briefly to show the bearing of some of these ideas upon practical conduct.

To man, as to nature at large, the question of adjustment is of vast importance, since the eminently destructive species are the widely oscillating ones. Those insects which are well adjusted to their environments, organic and inorganic, are either harmless or inflict but moderate injury (our ordinary crickets and grasshoppers are examples); while those that are imperfectly adjusted, whose numbers are, therefore, subject to wide fluctuations, like the Colorado grasshopper, the chinch-bug and the army-worm, are the enemies which we have reason to dread. Man should then especially address his efforts, first, to prevent any unnecessary disturbance of the settled order of the life of his region which will convert relatively stationary species into widely oscillating ones; second, to destroy or render stationary all the oscillating species injurious to him; or, failing in this, to restrict their oscillations within the narrowest limits possible.

For example, remembering that every species oscillates to some extent, and is held to relatively constant numbers by the joint action of several restraining forces, we see that the removal or weakening of any check or barrier is sufficient to widen and intensify this dangerous oscillation; may even convert a perfectly harmless species into a frightful pest. Witness the maple bark louse, which is so rare in natural forests as scarcely ever to be seen, limited there as it is by its feeble locomotive power and the scattered situation of the trees it infests. With the multiplication and concentration of its food in towns, it has increased enormously, and, if it has not done the gravest injury, it is because the trees attacked by it are of comparatively slight economical value, and because it has finally reached new limits which hem it in once more.

We are therefore sure that the destruction of any species of insectivorous bird or predaceous insect, is a thing to be done, if at all only after the fullest acquaintance with the facts. The natural presumptions are nearly all in their favor. It is also certain that the species best worth preserving are the mixed feeders and not those of narrowly restricted dietary (parasites, for instance),—that while the destruction of the latter would cause injurious oscillations in the species affected by them, they afford a very uncertain safeguard against the *rise* of such oscillations. In fact, their undue increase would be finally as dangerous as their diminution.

Notwithstanding the strong presumption in favor of the natural system, when we remember that the purposes of man and what, for convenience' sake, we may call the purposes of Nature do not fully harmonize, we find it incredible that, acting intelligently, we should not be able to modify existing arrangements to our advantage,— especially since much of the progress of the race is due to such modifications made in the past.

We should observe, in passing, that the principal general problem of economical biology is that of the discovery of the laws of oscillation in plants and animals, and of the methods of Nature for its prevention and control.

For all this, evidently, the first, indispensable requisite is a *thorough knowledge of the natural order,*— an *intelligently conducted natural history survey.* Without the general knowledge which such a survey would give us, all our measures must be empirical, temporary, uncertain, and often dangerous.

Next we must know the nature, extent and most important consequences of the disturbances of this order necessarily resulting from human interference,— we must study the methods by which Nature reduces these disturbances, and learn how to second her efforts to our own best advantage.

But far the most important general conclusion we have reached is a conviction of the general beneficence of Nature, a profound respect for the natural order, a belief that the part of wisdom is essentially that of practical conservatism in dealing with the system of things by which we are surrounded.

Summary.

The argument and conclusions of this paper may be thus briefly recapitulated: —

We find a mutual interdependence of organic groups and a modifiability of their habits, numbers and distribution, which brings them under the control of man. We also see that, after the most violent disturbances of their internal relations, a favorable readjustment eventually occurs. Starting with the general laws of multiplication and natural selection, it is first observed that every species of plant or animal dependent upon living organic food is interested to establish such a rate of reproduction as will, first, meet all the drains to which it is itself subjected, and still leave a sufficient progeny to maintain its own numbers, and, second, leave a sufficient supply of its own food-species to keep them undiminished, year after year. That is, we find that the interests of any destructive plant or animal are identical with the interests of its food-supply.

This common interest of the organism and its organic food is continually promoted by natural selection, by which those that unduly weaken the sources of their own support are eventually crowded out by others with a better adjusted rate of increase; but, because of the immense number, variability and complexity of the forces involved, a complete adjustment is never reached. Whether the rate of multiplication of the food-producing species be relatively too great or relatively too small, the result is to cause an oscillation of numbers of both depredating species and its food. These oscillations of a species are both directly and indirectly injurious to it, and tend, in various ways, to diminish the average of its numbers, especially by lessening the general average amount of the food available for it. By the operations of natural selection, therefore, widely oscillating species, thus placed at a marked disadvantage as compared with more stable ones, are either eliminated, or else reduced to order more or less completely; — they tend to become so adjusted to their food supplies as to appropriate only their surplus and excess.

Hence, as a general thing, the real limits of a species are not set by its organic environment, but by the inorganic; and the removal of the organic checks upon a species would not finally diminish its average numbers.

Among the external checks upon the oscillations of species of insects, the most important are those predaceous insects and insectivorous birds which eat a varied food, using most freely those elements of their dietary which are, for the time being, most abundant.

When we compare the results of the primitive natural order with the interests of man, we see that, with much coincidence, there is also considerable conflict. While the natural order is directed to the mere maintenance of the species, the necessities of man usually require much more. They require that the plant or animal should be urged to excessive and superfluous growth and increase, and that all the surplus, variously and widely distributed in nature, should now be appropriated to the supply of human wants. From the consequent human interferences with the established system of things, numerous disturbances arise,— many of them full of danger, others fruitful of positive evil. Oscillations of species appear, not less injurious to man than to the plants and animals more directly involved. Indeed, most of the serious insect injuries, for example, are due to species whose injurious oscillations have resulted from changes of the organic balance initiated by man.

To avoid or mitigate the evils likely to arise, and to adapt the life of his region more exactly to his purposes, man must study the natural order as a whole, and must understand the disturbances to which it has been subject. Especially he must know the forces which tend to the reduction of these disturbances and those which tend to perpetuate or aggravate them, in order that he may reinforce the first and weaken or divert the second.

The main lesson of conduct taught us by these facts and reasonings is that of conservative action and exhaustive inquiry. Reasoning unwarranted by facts, and facts not correctly and sufficiently reasoned out, are equally worthless and dangerous for practical use.

3

THE FOOD OF FISHES

Stephen Alfred Forbes

THE FOOD OF FISHES.

By S. A. FORBES.

For a clear conception of the general and intricate interdependence of the different forms of organic life upon the earth, one can not do better than to study thoroughly the life of a permanent body of fresh water,—a river or smaller stream, or, better than these, a lake. The animals of such a body of water are, as a whole, curiously *isolated*,—closely related among themselves in all their interests, but so far independent of the life of the land about them that, if every terrestrial plant and animal were annihilated, it would doubtless be long before the general multitude of the inhabitants of the lake or stream would feel the effects of this event in any very important way.

Further, the greater difficulty of communication between the different parts of a water system as compared with the different regions of the land, is such that the former are much the more sharply limited. There is very much less interchange of all kinds between two branches of the same stream, for example, than between the tracts of land which they separate. Consequently, one finds in a single body of water a far more complete and independent equilibrium of organic life and activity than in any equal body of land. It forms a little world within itself,—a microcosm within which all the elemental forces are at work and the play of life goes on in full, but on so small a scale as to bring it easily within the mental grasp.

Nowhere can one see more clearly illustrated what may be called the *sensibility* of such an organic complex,— expressed by the fact that whatever affects any species belonging to it, must speedily have its influence of some sort upon the whole assemblage. He will thus be made to see the impossibility of studying any form successfully out of relation to the other forms,—the necessity for taking a comprehensive survey of the whole as a condition to a satisfactory understanding of any part. If one wishes to become acquainted with the black bass, for example, he will learn but little if he limits himself to that species. He must evidently study also the species upon which it depends for its existence, and the various conditions upon which *these* depend. He must likewise study the species with which it comes in competition, and the entire system of conditions affecting their prosperity. Leaving out any of these, he is like one who undertakes to make out the construction of a watch, but overlooks one wheel; and by the time he has studied all these sufficiently, he will find that he has run through the whole complicated mechanism of the aquatic life of the locality, both animal and vegetable, of which his species forms but a single element.*

In such a general survey of the plants and animals of a region, the study of their food relations will be found to afford an admirable objective point. Doubtless, of all the features of the environment of an individual, none affect it at the same time so powerfully, so variously and so intimately as the elements of its food. Even climate, season, soil and the inorganic circumstances generally, influence an animal through its food quite as much as by their direct action. It is through the food relation that animals touch each other and the surrounding world at the greatest number of points, here they crowd upon each other the most closely, at this point the struggle for existence becomes sharpest and most deadly; and, finally, it is through the food relation almost entirely that animals are brought in contact with the material interests of man. Both for the student of science and for the economist, therefore, we find this subject of peculiar interest and value. It includes many of the most important relations of a species, and may properly be made the nucleus about which all the facts of its natural history are gathered.

In a paper on the food of Illinois fishes published in the second bulletin of this Laboratory, the subject was treated in a general and cursory way, the amount of material upon which that paper was based being insufficient for exact or detailed description. The favor with which that preliminary notice was received, has

---

* I can not too strongly emphasize the fact—frequently illustrated, I venture to hope, by the papers of this series—that a comprehensive survey of our entire natural history is absolutely essential to a good *working knowledge* of those parts of it which chiefly attract popular attention,— that is, its edible fishes, its injurious and beneficial insects, and its parasitic plants. Such a survey, however, should not stop with a study of the dead forms of Nature, ending in mere lists and descriptions. To have an *applicable* value, it must treat the life of the region as an organic unit, must study it *in action*, and direct principal attention to the laws of its activity.

made it possible to undertake a more serious investigation; and
this paper contains an account of the food of the Acanthopteri of
the state which I believe to be nearly or quite sufficient for the
student of science and for the practical fish culturist. It is still
necessary only to study the food of specimens under a half-inch
in length, and to test the value of the general conclusions here
reached, by occasional examinations of fishes taken from other
waters at other seasons of the year. Among the results of this
study, those relating to the food of the young are especially
worthy of attention, and these have therefore been summed up
separately.

The explanation of certain structural conditions about the
mouth, throat and gills, has proceeded so far as to make it very
likely that a number of definite general correspondences between
structure and food will be made out, which will enable us to tell
with considerable accuracy and detail what the food of an unknown
fish must be, by a mere inspection of the fish itself; provided, of
course, that we know what food is accessible to it in its habitat.
It seems likely to prove to be a general rule that a fish makes
scarcely more than a *mechanical* selection from the articles of food
accessible to it, taking almost indifferently whatever edible things
the water contains which its habitual range and its peculiar alimentary apparatus enable it to appropriate, and eating of these
in about the ratio of their relative abundance and the ease with
which they can be appropriated at any time and place. If this is
so, knowing the structure of a fish and the contents of a body of
water, we shall be able to tell, *a priori*, what the fish will eat if
placed therein.

This is, in fact, the objective point of the present investigation,—
to arrive at a knowledge of the correlations of structure and food
habits sufficiently detailed and exact to make the tedious and difficult labor of examining the contents of stomachs unnecessary
hereafter. Some generalizations of this sort are given in the following pages, and others relate to genera not included in this
report.

The method of this paper differs from that of the previous one
referred to by the calculation of the *ratios* of the different kinds
of food for each species or group of individuals. These ratios
were obtained by averaging careful estimates of the relative
amounts of the different food elements found in each stomach.

It is proposed to follow a similar method hereafter down through the remaining orders of the class. Most of the material has been collected for this purpose, and much of it has been already studied.

Order TELEOCEPHALI.
Suborder ACANTHOPTERI.

This suborder includes all Illinois fishes which have the anterior dorsal fin (where there are two) or the first rays of the dorsal (where there is but one) stiff, spinous and sharp, and united by an evident membrane; excepting only the remarkable "brook silversides," which is placed by Drs. Gill and Jordan in another group. It embraces all our game fishes except those belonging to the pickerel family (*Esocidæ*) and the salmon family (*Salmonidæ*). Its principal members are the darters, the various species of perch and bass, the sunfishes and the sheepshead. Forty-six species of the order have been collected in the state, but only thirty-four of these are common enough to form features of any importance in our fish fauna.

The most numerous family of the group is the *Centrarchidæ* (sunfishes); the most important species are the two kinds of black bass, the pike-perch or "wall-eyed pike,"* the common perch, the white bass, and the croppie or silver bass.

The following account of the food of this suborder is based upon the careful microscopic study of the contents of four hundred and twenty-five stomachs, representing six families, twenty genera † and thirty-three species.

These were all collected by myself or one of my assistants (Mr. W. H. Garman), and labeled at the time with name of species, locality and date. While the northern half of the state is most fully represented, several trips to southern Illinois contributed to the material studied, and it is believed that the results arrived at are substantially true for our whole area.

---

* It is generally to be desired that the absurd names of "Salmon" and "Jack Salmon" for these species should be suppressed. They might as well be called suckers or catfishes or minnows, as far as accuracy is concerned. Common names are many times harder to kill than the cat of the proverb, however; and it is probable that unnumbered generations will continue to call the pike-perch "salmon"; the sunfishes, "perch"; and the black bass, "trout."

† The classification of this paper is substantially that of Jordan's Manual of the Vertebrates of North America, etc., Ed. 2, 1878.

Family ETHEOSTOMATIDÆ.   The Darters.

What the humming-birds are in our avifauna, the "darters" are among our fresh-water fishes. Minute, agile, beautiful, delighting in the clear, swift waters of rocky streams, no group of fishes is more interesting to the collector; and in the present state of their classification, none will better repay his study. Notwithstanding their trivial size, they do not seem to be *dwarfed* so much as *concentrated* fishes — each carrying in its little body all the activity, spirit, grace, complexity of detail and perfection of finish to be found in a perch or a "wall-eyed pike."

They are generally distributed, in suitable streams throughout the state; but we have found them much the most abundant in northern Illinois,— in the upper Galena R., in Yellow creek near Freeport, and in tributaries of the Kishwaukee at Belvidere.

A short and strong minnow-seine of very fine mesh is needed in collecting them. Rapid hauls, made almost on the run, down stream, in swift and shallow water, will be found the most successful. Two or three species, of wider range, will be taken in ordinary situations, in collecting for minnows generally; but the brightest and most characteristic forms can only be got by special effort.*

I shall give here a description of the food of the family, based upon a study of the contents of seventy stomachs representing fifteen species, collected in all parts of Illinois, in several months of four successive years. These indicate much more than their number would imply, since from those collected at each time and place, as many were commonly studied as were necessary to give a full idea of the food of the species then and there. The different individuals from the same date and locality usually agreed so closely in food, that the study of from two to five gave all the facts obtainable from several times as many. The data here given, therefore, really exhibit the food of the family at different seasons in twenty-nine localities within the state.

The genus *Pleurolepis* is comparatively rare in Illinois, as there are few of the sandy streams in the state, which it inhabits. Seven individuals were examined — four of *P. pellucidus* and three of *P. asprellus*. The food of these specimens was remarkably

---

* For a very entertaining and instructive account of these fishes, the reader is referred to papers in the American Naturalist, by Messrs. Jordan and Copeland, Vol. X, pp. 335–341, and Vol. XI, pp. 86–88.

uniform — the only elements found being the larvæ of small diptera and Ephemerids. Eighty-one per cent. of the food of all consisted of the larvæ of Chironomus,*— a small, gnat-like insect,— twelve per cent. of the larvæ of other small diptera, and the remaining seven per cent. of Ephemerid larvæ (May flies).

Twelve specimens of the genus *Alvordius* were studied — seven of *maculatus* and five of *phoxocephalus.* These represented five different localities and dates. This is a larger species than the preceding, and to this fact is probably due the predominance (seventy-five per cent.) in its food of the larvæ and pupæ of May flies (Ephemeridæ). These included four per cent. of the larvæ of *Palingenia bilineata,* Say, one of the largest Ephemerids in our streams. The remaining kinds were larvæ of dragon flies (Agrionidæ), four per cent.; larvæ of Chironomus, seven per cent., *Corixa tumida,* Uhl., thirteen per cent., and Cyclops, one per cent.

The genus *Boleosoma,* regarded by Dr. Jordan as the typical darter, was represented by twelve specimens from eight localities — nine of *maculatum,* two of *olmstedi* and one of *camurum.†* These specimens show but slight food differences from other darters of similar size; the only notable variation being the appearance of fifteen per cent. of case-worms (larvæ of Phryganeidæ). Sixty-six per cent. of the food was Chironomus larvæ, seven per cent. larvæ of other minute diptera, and the remaining twelve per cent. was larvæ of small Ephemerids, and a few Cyclops.

I studied the food of two specimens of *Pœcilichthys variatus,* four of *P. spectabilis* and two of *P. asprigenis* — making eight of the genus, representing six localities. Fifty-eight per cent. of small larvæ of diptera (forty-nine per cent. of Chironomus), thirty-two per cent. of larvæ and pupæ of small Ephemerids, and ten per cent. of case-worms made up the entire bill of fare.

Percina caprodes, the largest of the group, departs from all the foregoing species by the prominence given to crustacean food —

---

* The larvæ of Chironomus are among the most important elements of fish food in our waters, appearing in abundance in the stomachs of the young of a great variety of species. They have been too little studied in this country to allow specific determination.

† *Boleosoma maculatum* and *B. olmstedi* should undoubtedly be united. Specimens in the laboratory collection present the extremes of both forms, together with numerous intermediate stages of each character used to distinguish them.

This whole group exhibits a surprising variability, perhaps due to its comparatively recent origin.

thirty per cent. of Entomostraca and three per cent. the smallest of our Amphipoda, *Allorchestes dentata* (Smith) Faxon. Most of the Entomostraca were *Cladocera*, including Daphnia, Eurycercus and Daphnella.*

Here occurred the only instance of molluscan food in the group. One specimen had taken a few individuals of *Ancylus rivularis*, Say. Reduced ratios of Chironomus and Ephemerid larvæ, and a few *Corisa tumida* complete the list.

Of *Nanostoma zonale*, less common than the others, but two individuals were examined, and these had eaten nothing but larvæ of small diptera, including sixty-five per cent. of Chironomus.

Six specimens of *Etheostoma flabellare* var. *lineolata*, from four localities, had eaten sixty-one per cent. of Chironomus larvæ, twenty-seven per cent. larvæ of small Ephemerids, and twelve per cent. of Copepoda (Cyclops).

Boleichthys elegans, found only in the southern part of the state (three specimens examined), had eaten only dipterous larvæ (thirty-seven per cent.) and Ephemerid larvæ (sixty-three per cent.). This is a larger, heavier species than most of the others, and, therefore, like Alvordius, prefers Ephemerids to gnats.

Last and least comes *Microperca punctulata*, represented by nine specimens from four localities in Northern Illinois. This smallest of the darters shares with Percina, the largest, the peculiarity of a large ratio of crustacean food, which made up sixty-four per cent. of the total. The principal kinds were Cyclops, Chydorus, young *Gammarus fasciatus*, Say, and young *Crangonyx gracilis*, Smith. The remaining elements were Chironomus larvæ (thirty-four per cent.) and a trace of Ephemerids (two per cent.).

It will be seen that the family, taken as a whole, divides into two sections, distinguished by the abundance or deficiency of crustacean food. This is easily explained by the fact that Percina and Microperca range much more freely than the other genera — being frequently found among weeds and Algæ in comparatively slow water with muddy bottom, while the others are rather closely confined to swift and rocky shallows.

In discussing the food of the whole group, taken as a unit, it may best be compared with the food of the young of other percoids. It is thus seen to be remarkable for the predominance of

---

* Daphnella was found in a Percina from the Calumet river, at South Chicago, but not in condition to permit the determination of the species.

the larvæ of Chironomus and small Ephemeridæ — the former of these comprising forty-four per cent. and the latter twenty-three per cent. of the whole food of the seventy specimens. In young black bass (*Micropterus pallidus*), on the other hand, the averages of nine specimens, ranging from five-eighths inch to one and a half inches in length, were, in general terms, as follows: Cladocera forty-two per cent., Copepoda seven per cent., young fishes twenty per cent., Corixa and young Notonecta twenty-nine per cent., and larval Chironomus only two per cent. The search for the cause of this difference leads naturally to an examination of the whole economy of these little fishes, and opens up the question of their origin as a group.

The close relation of the Etheostomatidæ to the Percidæ requires us to believe that the two groups have but recently diverged, if, indeed, they are yet distinctly separate.

We must inquire, therefore, into the causes which have operated upon a group of percoids to limit their range to such apparently unfavorable situations, to diminish their size, to develop unduly the paired fins and reduce the air-bladder, to remove the scales of several species more or less completely from the head, breast, neck and ventral region, and to restrict their food chiefly to the few forms mentioned above.

No species can long maintain itself anywhere which can not, in some way, find a sufficient supply of food, and also protect itself against its enemies. In the contest with its enemies it may acquire defensive structures or powers of escape sufficient for its protection, or a reproductive capacity which will compensate for large losses, or it may become adapted to some place of refuge where other fishes will not follow. What better refuge could a harassed fish desire than the hiding-places among stones in the shallows of a stream, where the water dashes ceaselessly by with a swiftness few fish can stem? And if, at the same time, the refugee develops a swimming power which enables it to dart like a flash against the strongest current, its safety would seem to be insured. But what food could it find in such a place? Let us turn over the stones in such a stream, sweeping the roiled water at the same time with a small cloth net, and we shall find — larvæ of Chironomus and small Ephemerids and other such prey, and little else; food too minute and difficult of access to support a large fish, but answering very well if our immigrant *can keep down his size.* Here the principles of natural selection assert their power. The limited supply of food early arrests the growth of the young; while every fish which passes the allowable maximum is forced for food to brave the dangers of the deeper waters, where the chances are that it falls a prey. On the other hand, the smaller the size of those which escape this alternative, the less likely will they be to attract the appetite of the small gar or other guerilla which may occasionally raid their retreat, and the more easily will they slip about under stones in search of their microscopic game.*

Like other fishes, the darters must have their periods of repose, all the more urgent because of the constant struggle with the swift current which their habitat imposes. Shut out from the deep still pools and slow eddies where the larger species lurk, they are forced to spend their leisure on or beneath the bottom of the stream, resting on their extended pectorals and anal, or wholly buried in the sand. Possibly this fact is correlated with the absence or rudimentary condition of the air-bladder; as it is a rule with many exceptions — but still, probably, a rule — that this organ is wanting in fishes which live chiefly at the bottom.

Doubtless the search for food has much to do with this selection of a habitat. I have found that the young of nearly all species of our fresh-water fishes are competitors for food, feeding almost entirely on entomostraca and the larvæ of minute diptera.† As a tree sends out its roots in all directions in search of nourishment, so each of the larger divisions of animals extends its various groups into every place where available food occurs, each group becoming adapted to the special features of its situation. Given this supply of certain kinds of food, nearly inaccessible to the ordinary fish, it is to be expected that some fishes would become especially fitted to its utilization. Thus the Etheostomatidæ as a group are explained, in a word, by the hypothesis of the progressive adaptation of the young of certain Percidæ to a peculiar place of refuge and a peculiarly situated food supply.

Perhaps we may, without violence, call these the mountaineers among fishes. Forced from the populous and fertile valleys of the river beds and lake bottoms, they have taken refuge from their enemies in the rocky highlands where the free waters play in ceaseless torrents, and there they have wrested from stubborn nature a meagre living. Although diminished in size by their

---

* In Boleosoma, which is normally scaled in front of the dorsal fin. we often find the skin of this region bare in large specimens, and showing evident signs of rubbing.

† Several of the Catostomidæ (suckers) are an exception to this rule, feeding when young chiefly on Algæ and Protozoa.

continual struggle with the elements, they have developed an activity and hardihood, a vigor of life and glow of high color almost unknown among the easier livers of the lower lands.

The appended table will facilitate a comparison of the records of the different genera. The percentages were obtained by estimating carefully the ratios of each element of the food of each individual, and averaging these ratios for all the individuals of a species:

DETAILS OF THE FOOD OF THE ETHEOSTOMATIDÆ.

	Pleurolepis.	Alvordius.	Boleosoma.	Pœcilichthys.	Percina.	Nanostoma.	Etheostoma.	Boleichthys.	Microperca.
Number of specimens	7	12	12	8	11	2	6	3	9
I. Mollusca					01				
Ancylus rivularis Say					01				
II. Insecta	100	99	96	100	65	100	88	100	36
1. Diptera	93	07	73	58	43	100	61	37	34
Undetermined larvæ	12	01	07	09	02	35		10	
Chironomus larvæ	81	06	66	49	41	65	61	27	34
2. Hemiptera		13			05				
Corixa		13			05				
Undetermined					03				
Larvæ		07			02				
C. tumida Uhl		06							
3. Neuroptera	07	79	23	42	17		27	63	02
Ephemeridæ	07	75	08	32	09		27	63	02
Pupæ			08	14					
Larvæ	07	63	08	18	09		27	63	02
Palingenia		04							
Agrionidæ (pupæ)		04							
Phryganeidæ (larvæ)			15	10	08				
III. Crustacea		01	04		33		12		64
1. Amphipoda					03				12
Gammarus, yg									06
Crangonyx, "									06
Allorchestes dentata Sm.					03				
2. Cladocera					24				27
Undetermined					05				
Daphniidæ					06				
Daphnia					07				
Sididæ					05				
Daphnella					05				
Lynceidæ					01				03
Chydorus									24
Eurycercus					01				
3. Ostracoda					01				
Cyprididæ					01				06
Undetermined					01				06
Cypris									
4. Copepoda		01	04		05		12		19
Cyclops		01	04		05		12		19
Confervoid Algæ					01				

Family PERCIDÆ.   The Perches.

This family consists, in this state, of three species,—the common yellow perch and the two species of pike-perch or "wall-eyed pike." I have examined the food of seventy-five specimens of this family, so distributed in time and space as to give a satisfactory idea of the usual food

PERCA AMERICANA, Schrank.   THE COMMON PERCH.   RINGED PERCH.

This exceedingly well-known species is most abundant along the shores of Lake Michigan and in the small streams and lakes of the northeastern part of the state, becoming less common to to the south and west. In the Illinois river at Peoria and Henry it occurs in limited numbers, but in southern Illinois disappears so completely that even its name (there generally pronounced "pearch") is transferred to a different family, the sun-fishes (Centrarchidæ).

My knowledge of the food of this species is derived from the study of the contents of forty-nine stomachs, of which thirty were from adults and the remaining nineteen from fishes ranging $1\frac{3}{16}$ inch to four inches in length. Ten localities and as many dates are represented by these specimens. Some were taken in the Illinois river, others in Lake Michigan and its southern tributaries, and still others in Fox R. at McHenry, and in the lakes connected with that stream. One lot included in these notes was bought in the Chicago market. They were evidently of the river form of the species, and, judging from the contents of their stomachs, which included a crustacean* not known to occur in Illinois but found abundantly in Michigan, I conclude that they were from that state or from Wisconsin.

Food of the Young

Finding that the food of most fishes differs with age, I have grouped the young according to size, and averaged the food for each group separately,—the first group consisting usually of those under an inch in length, the second of those from one to two, etc.

Two perch under an inch in length had eaten nothing but Entomostraca,—about equal quantities of Cyclops and Daphnias. It

Mancasellus tenax **Harger.**

was not until the specimens reached an inch and a half in length
that insects of any considerable size appeared in the food. A
single smaller fish had eaten a few minute larvæ of Chironomus,
but otherwise the food at this age consisted wholly of Entomostraca.

About thirty-four per cent. of the food of nine specimens ranging from 1⅛ to 2 inches in length consisted of insects, and sixty-
six per cent. of crustaceans. The only insects recognized were the
larvæ and pupæ of Chironomus (eleven per cent.), small water
bugs — *Corixa tumida*, Uhl, *C. alternata*, Say, etc. (twenty-three
per cent.),—and a trace of larvæ of May-flies (Ephemeridæ).
The Crustacea were chiefly Cladocera and Copepoda — thirty-six
per cent. and twenty-four per cent. respectively. Four of the
nine had eaten small quantities of a small amphipod crustacean,
Allorchestes dentata, which is very abundant north, and has, in
fact, about the same distribution in the state as the perch itself.
The Cladocera were chiefly Daphniidæ (twenty-seven per cent.)
including *Daphnia pulex*, L., *Simocephalus americanus*, Birge,
and *Bosmina longirostris*. Specimens of Chydorus and Pleuroxus
made up the principal part of the nine per cent. of Lynceidæ
eaten. The Copepoda were all Cyclops and Diaptomus.

Four specimens two and a half inches long, all taken at Peoria,
in November, 1878, had eaten nothing but Hemiptera (twelve per
cent.) and Neuroptera (eighty-eight per cent.). The Hemiptera
were all *Corixa alternata*, and the Neuroptera were nearly all
the extremely common larva of one of our most abundant May-
flies (*Palingenia bilineata*, Say). Larvæ of small dragon-flies
(Agrionini) made five per cent. of the food. The simplicity of the
food of these specimens is probably due partly to the fact that
they were all caught at the same time and place, and partly to
the wintry weather when they were taken.

Four specimens, from three and a half to four inches long, representing two localities and dates, had eaten a greater variety of
articles; the food, in fact, now closely approaching that of the
adult. Forty-five per cent. of the food was insects,— chiefly larvæ
of May-flies — and fifty-five per cent. Crustacea,— chiefly Amphipoda and Cladocera. Other insect elements were larvæ of
Chironomus six per cent., and four per cent. Corixas. The Cladocera were all Daphnia, and the Amphipoda were *Allorchestes den-*

tata. A single specimen from Long L., near Pekin, Ill., had eaten an Isopod crustacean (Asellus). Cyprididæ, another family of minute crustaceans, formed eight per cent. of the whole food of these specimens.

Food of the Adult.

The thirty mature individuals may best be treated in two groups, the first from streams and the second from Lake Michigan.

Four of the first group were bought in the Chicago market, in March, 1880; six were taken from the upper Fox, in May; four were from Calumet R. at South Chicago, taken in August, 1878, and four were caught in October of that year, from the Illinois at Peoria.

We notice, first, the entire disappearance of Entomostraca, which are thus seen to be food proper to the young. We next observe the appearance of mollusca (nineteen per cent.), which are evidently no insignificant food resource of the species. Unio, Cyclas, Succinea, *Physa heterostropha,* Say, and *Valvata tricari-nata,* Say, are the mollusks recognized. Notwithstanding the lack of Entomostraca, Crustacea are the most important resource of these river specimens, constituting forty-eight per cent. of their food. Crawfishes (Cambarus) and our common little fresh-water shrimp (*Palæmonetes exilipes,* St.) compose ten per cent. of the whole; the previously noticed Allorchestes amounts to fifteen per cent., and species of Asellus, and *Mancasellus tenax* to twenty-three per cent. The Mancaselli were all from the specimens from the Chicago market Insects are also an important item,— amounting to twenty-four per cent., nearly all being the larvæ of Neuroptera,— May-flies (Ephemeridæ), dragon-flies and case-flies (Phryganeidæ). A single specimen from Peoria Lake had eaten one small fish — a "darter" of the genus Pœcilichthys.

The second group, of twelve specimens from Lake Michigan, presents a curious and instructive contrast in food to the foregoing. Mollusks and insects wholly disappear, and Crustacea are limited to the commonest crawfish of the lakes (*Cambarus virilis,* Hagen), which forms fourteen per cent. of the food. The remaining eighty-six per cent. consisted wholly of fishes, all minnows (Cyprinidæ) as far as recognized except one, and that was some undetermined percoid,— probably itself a perch.

It will thus be seen that the common perch has a food history of three periods,—the periods of infancy, youth, and mature age. In the first it lives wholly on Entomostraca and the minutest larvæ of Diptera; in the second, commencing when the fish is about an inch and a half in length, it takes up first the smaller and then the larger kinds of aquatic insects in gradually increasing ratio, the entomostracan food at the same time diminishing in importance; and in the third it appropriates, in addition, mollusks, crawfishes and fishes,—in the lake specimens depending almost wholly on the last two elements.

We have here the first instance of a fact which we shall see again and again illustrated,—that the young, having at first an alimentary apparatus too small and delicate to dispose of any insects but the minutest larvæ, live almost wholly on minute crustaceans.

It is proper to note that the lake and river perch are by some good authorities regarded as separate species,—the latter being much more highly colored than the former. I have not found so strict a separation of the two forms as that described by Mr. E. W. Nelson, but have frequently taken both in the same haul of the seine in different parts of Calumet R. and in Lake George, Ind.,—a body of water communicating with Lake Michigan by an outlet three or four miles long. Occasional pale specimens are also taken far from the lakes, in the Fox and Illinois rivers. The difference of color is probably due partly to the smaller amount of light to which those inhabiting the deeper waters of the lake are exposed, and partly to their piscivorous habit combined with the comparatively few lurking-places afforded them. There is some evidence that fish food bleaches a fish directly, and a good deal that it does so indirectly, by increasing the importance of an inconspicuous appearance.

STIZOSTETHIUM CANADENSE, Smith. GRAY PIKE-PERCH. SAUGER. "JACK SALMON."

Fourteen specimens of this excellent fish were examined, all of which were from the Illinois R., ten taken in October, 1878, one in June, 1877, and three in November, 1877. It is evidently a very destructive species. These specimens had eaten nothing but fishes. In three cases these were unrecognizable, and in two others I could only tell that they were Acanthopteri,. Four of the remaining "Pike" had eaten hickory shad (*Dorysoma cepedianum*), two had eaten catfish (Siluridæ) of which one was an Amiurus, two had eaten sheepshead (*Haploidonotus grunniens*), and one had taken a black bass and some sunfish (Centrarchidæ). The presence in the stomach of one of these fishes, of a catfish of medium size, with its poisonous pectoral and dorsal spines unbroken, was a striking illustration of the gastric energy of this species.

STIZOSTETHIUM VITREUM, Mitch.   PIKE-PERCH.   WALL-EYED PIKE.   "SALMON."

This is far the finest of our river-fishes,—second to no fresh-water species except, possibly, some of the salmon family. It occurs in the great lakes, and throughout the state generally in the larger streams. It is a much larger fish than the preceding, not unfrequently reaching a weight of twenty pounds. Certainly no fish of our waters is better deserving of attention than this. The only drawback to its increase is in its voracity; but, although it devours an immense number of other fishes, there is no evidence that it is wantonly destructive or that it eats more in proportion to its weight than the black bass.

Twelve of this species were examined, two of which were under three inches in length, and the others adult.

Food of the Young.

A specimen two inches long, taken in the Illinois R., at Pekin, June 2, 1880, had eaten only a minute fish. One two and a half inches long, taken at the same place in June, 1878, had also eaten a small fish and a few Entomostraca (Cyprididæ and Daphniidæ). The appearance of these Entomostraca in the food of a fish of this size, makes it altogether probable that Stizostethium, like Perca, wholly depends on these minute Crustacea, when very young.

Food of the Adult.

The remaining specimens, taken from three localities, had eaten nothing but fishes, one half of them only the hickory shad or skipjack (*Dorysoma cepedianum*). In one other specimen, this species was associated with a minnow (Cyprinidæ), and in still another with a small sunfish with three anal spines (Centrarchidæ). One of the remaining stomachs contained only an unrecognizable fish, and the other two contained Cyprinidæ, including the creek chub, *Semotilus corporalis.*

The two species of this genus agree so closely in food that they may well be discussed together. Apart from their exclusively piscivorous habit, the most interesting fact shown is the importance of the hickory shad as food for this fish. We shall find accumulating evidence that this shad, utterly useless for human food, is, notwithstanding, one of the most valuable fishes in our streams. Nevertheless, not the slightest attention is paid to its preservation, much less to its encouragement. The fishermen commonly regard these fishes as a mere nuisance, and leave them to die on the bank by hundreds, rather than take the trouble to return them to the water. They are a very delicate species, and are easily killed by rough handling in the seine, but the majority of those captured might be saved with a little care.

The abundance of these fishes as compared with some other species in the river might seem to indicate that they are common enough as it is. Few realize, however, the number of fishes needed to feed a pike-perch to maturity. Two or three items from my notes will furnish the basis for an intelligent estimate of this number.

From the stomach of a *Stizostethium canadense* caught in Peoria Lake, October 27, 1878, I took ten well-preserved specimens of *Dorysoma*, each from three to four inches long; and from a *Stizostethium vitreum* I took seven of the same species, none under four inches in length. As the Dorysoma is a very thin, high fish, with a serrate belly, these were as large as a pike-perch can well swallow; and we may safely suppose that not less than five of this species would make a full meal for the pike-perch. The species is a very active hunter, and it is not at all probable that one can live and thrive on less than three such meals a week. The specimens above mentioned were taken in cold autumn weather, when most other fishes were eating but little; but, since fishes generally take relatively little food in winter, we will suppose that the pike-perch eats, during the year, on an average, at this rate per week for forty weeks, giving us a total *per annum* of six hundred Dorysomas destroyed by one pike-

5

perch. We can not reckon the average life of a Stizostethium at
less than three years, and it is probably nearer five. The smallest
estimate we can reasonably make as the food of each pike-perch
would therefore be somewhere between eighteen hundred and
three thousand fishes like Dorysoma. A hundred pike-perch,
such as should be taken each year along a few miles of a river
like the Illinois, would therefore require one hundred and eighty
thousand to three hundred thousand fishes for their food. Finally,
when we take into account that a number of other species also
prey upon Dorysoma, and that the whole number destroyed in all
ways must not exceed the mere *surplus* reproduced—otherwise
the species would be extinguished,—we can form some approximate idea of the multitudes in which the food species must abound
if we would support any great number of predaceous fishes. Dorysoma, being a mud-eater and a vegetarian, taking animal food
only during the Entmostracan period, can probably be more readily maintained in large numbers in our muddy streams than any
other fish.

It is evident that the increase of edible fishes without a corresponding supply of food will be largely time and labor thrown
away. Probably if protected from wanton and ignorant destruction, the Dorysoma would abound sufficiently, as it is enormously
prolific.

The following table is similar to that given for the preceding
family. The mark † is used to indicate the occurrence of an element in too small an amount to figure in the ratios:

TABLE OF THE FOOD OF THE PERCIDÆ.

	PERCA				River specimens.	Lake specimens.	STIZOSTETHIUM.
	One inch and under.	One to two inches.	Two to three inches.	Three to four inches.			
Number of specimens examined.......	2	9	4	4	18	12	26
I. FISHES					06	86	100
Undetermined....						50	23
Acanthopteri....						08	21
Undetermined....						08	08
Pœcilichthys					06		
Centrarchidæ....							05
Undetermined....							03
Micropterus....							02
Haploidonotus....							08
Dorysoma....							41
Cyprinidæ....						28	09
Undetermined....						28	05
Semotilus....							04
Siluridæ....							08
Undetermined....							04
Amiurus....							04
II. MOLLUSCA....					19		
Physa heterostropha				†	05		
Succinea....					04		
Valvata 3-carinata					01		
Cyclas					05		
Unio....					04		
III. INSECTA....		34	100	45	24		
Pupæ....					†		
1. *Diptera* (larvæ)....		10		06	01		
Undetermined....		01					
Chironomus		09		06	01		
2. *Hemiptera*....		23	12	04			
Corixa....		23		04			
Undetermined....		11		04			
C. alternata....			12				
C. tumida....		12					
3. *Neuroptera* (larvæ)....		01	88	35	23		
Ephemeridæ....		01	83	35	08		
Undetermined....				35	03		
Palingenia			83		05		
Agrionidæ			05		04		
Libellulidæ....					08		
Phryganeidæ					03		

TABLE OF THE FOOD OF PERCIDÆ—Continued.

	One inch and under.	One to two inches.	Two to three inches.	Three to four inches.	River specimens.	Lake specimens.	STIZOSTETHIUM.
			PERCA.				
Number of specimens examined.......	2	9	4	4	18	12	26
IV. CRUSTACEA....		66		55	48	14	†
1. *Decapoda*					10	14	.....
Cambarus.......					04	14	
Palæmonetes.......					06		......
2. *Amphipoda.*		06		24	15		......
Undetermined		02					
Allorchestes.......		04		24	15		......
3. *Isopoda*				01	23		......
Asellus.......				01	11		
Mancasellus					12		
4. *Entomostraca*	100	60		30			†
Cladocera....	55	36		22			......
Daphniidæ	55	27		22			†
Undetermined.......	55	13					†
Simocephalus.......		†					
Daphnia.......		12		22			
Bosmina.......		02					
Lynceidæ.......		09					
Undetermined.......		07					
Pleuroxus.......		†					
Chydorus.......	†	02					
Copepoda.......	45	24					
Cyclops.......	45	24					
Diaptomus.......		†					
Ostracoda (Cypris).......		†		08			†
V. VEGETATION.......					03		......

Family LABRACIDÆ.　The Bass.

We have but two species of this family, the white bass and the brassy bass (*Roccus chrysops* and *Morone interrupta*).　As far as their food is concerned, these are evidently equivalent species, agreeing closely in their general relations, and differing only in their distribution.

ROCCUS CHRYSOPS, Raf. WHITE BASS.

This species is of medium abundance throughout the northern half of the state,—most common in Lake Michigan. A curious fact of its distribution is its rarity in Fox River and the lakes connected with that stream. Indeed, during several days' active collecting in this region we did not see a single specimen, neither could we hear of the occurrence of the species in those waters, although we made careful inquiry for it among experienced fishermen.

My notes on its food relate only to eleven specimens, of which three, taken at South Chicago, in August, were young, but of unknown size. Two of these had eaten only Chironomus larvæ and the larvæ of a remarkable Ephmerid? not yet determined, and the stomach of the third contained only a minute fish. The remaining eight individuals had depended chiefly on the larvæ of May-flies (sixty-nine per cent.). The other important articles of their food were twenty per cent. fishes (including one sun-fish—Centrarchidæ) and eight per cent. Isopod Crustacea (Asellus). Several attempts to secure food from Lake Michigan specimens were unsuccessful, as, being taken in pound-nets, their stomachs were always empty. Those studied were from various interior situations in the northern third of the state.

MORONE INTERRUPTA, Gill. STRIPED BASS. BRASSY BASS.

This fish replaces the preceding in the southern half of the state, the Illinois River forming a neutral zone between the respective territories of the two species.

The food of six specimens of this species was studied, all taken from the Illinois River from May to October.

Four of these were young. The smallest, one and a fourth inches long, taken at Peoria, in June, 1878, had eaten about equally of small *Dorysoma cepedianum* and Entomostraca,—forty per cent. Leptodora and ten per cent. Cyclops. One, an inch and a half in length, taken at the same time and place, had eaten only Dorysoma, with a trace of Cyclops. The next, one and five-eighths inches in length, had eaten a small undetermined fish and a few Daphnias. The fourth, one and seven-eighths inches long, caught at Peoria, in October, had eaten only larvæ and pupæ of Chironomus.

The two adult specimens were feeding chiefly upon the larvæ of Neuroptera,—especially May-flies. An *Allorchestes dentata* and a few small grasshoppers also appeared in the food.

It will be seen that this species apparently agrees closely with the preceding in its food. The large amount of crustacean food in the smallest specimen shows that we should probably find still smaller Labracidæ depending upon these as strictly as the Percidæ.

Family CENTRARCHIDÆ. The Sun-fishes.

This interesting group, known, in some of its members, to every one who has ever seen a dozen fishes, is represented, in Illinois, by sixteen species, as the species of this family are now understood. The two black bass, included in this family for technical reasons, are, of course, the most important species. The rock bass, the croppie and the common sun-fish (*Lepiopomus pallidus*), although not fishes of the first class, would be seriously missed if we were to lose them; and boyhood in the country would be quite another thing if it were not for the "pumpkin seed" in the mill-pond, whose barbaric splendor thrills the heart of the youthful fisherman as the more delicate beauties of the trout or salmon do those of tougher fibre.

I have studied the food of thirteen species of this group, as indicated by two hundred and thirty-seven specimens, well distributed in time and area.

Decided differences in food made out in the various genera, have been found to coincide with differences in a few structures about the mouth in such a way that one may predict, from an examination of these structures, what the leading peculiarities of the average food of any genus will be.

MICROPTERUS PALLIDUS, Raf. LARGE-MOUTHED BLACK BASS.

This famous species is too well known to require extended comment. The ordinary fishermen rarely distinguish it from the following; and, indeed, sportsmen do not always recognize the difference.

I have examined the food of thirty-one specimens of this species, fourteen of which were adults, and the remainder young, of different ages.

Food of the Young.

The first group, consisting of five specimens under one inch in length (ranging from ⅝ to ¾ in.), represent three localities,—Crystal Lake, in McHenry county, the Illinois River at Pekin, Tazewell county, and the same stream at Starved Rock, in La Salle county. They were taken in June, July and August of three different years. It is evident, therefore, that the common features of their food can not well be attributed to any other fact than their similar size.

The entire food of these fishes consisted of small Crustacea,—all Entomostraca except seven per cent., eaten by a single fish, which consisted of the very young of some undetermined amphipod,— probably Allorchestes. Eighty-seven per cent. of the food was Cladocera, principally *Bosmina longirostris*, Müll. *Simocephalus americanus*, Birge, was also an important element; and traces appear of Chydorus, Pleuroxus and *Eurycercus lamellatus*. About six per cent. of Cyclops had been eaten.

In the food of the next group—six specimens, from 1¼ to 1½ inches long—minute fishes and insects appear. The fishes (twenty-nine per cent.) were not large enough to determine. The insects (forty-six per cent.) were mostly young water bugs (Corixa), the principal part of which were about half grown. The adults were all *Corixa tumida*, Uhl. The Entomostraca drop to twenty-five per cent., about equally Cladocera and Cyclops. Among the former were many specimens of *Simocephalus americanus*, and a few of the rare and curious Leptodora mentioned in a previous paper.* The specimen in which this was found was taken at Peoria, in June, 1878. All of this group were taken from the Illinois River, but at different places and dates. Some, taken at the same place and time as others of the preceding group, differed from them in the smaller number of Entomostraca eaten, and the larger number of insects,—differences evidently only to be explained as due to the different sizes of the fishes.

The next two specimens, between two and three inches long, had eaten only insects, chiefly *Corixa tumida*.

Four specimens, ranging from three to three and a half inches in length, all taken from a lake in the Illinois river bottom, in Oc-

---

* See Bull. No. 2, Ill. State Lab. Nat. Hist., p. 88.

tober, 1879, had eaten nothing but insects,—almost wholly Corixas
and the larvæ of May-flies (Ephemeridæ). The Corixas were
C. alternata, Say, and *C. tumida*, Uhl.

Food of the Adult.

Turning to the food of the fourteen adults, we note the total
disappearance of Entomostraca, the merely accidental occurrence
of insects, the appearance of crawfishes (*Cambarus immunis*), which
amount to seven per cent. of the whole food, and the great predominance of fishes (eighty-six per cent.). These were of sufficient variety to show that no group is safe from the appetite of
the bass unless it be the gar.

Perch, minnows, catfish and hickory shad were recognizable.
The last were much the most abundant, occurring in eight of the
specimens, and constituting fifty-eight per cent. of the food of the
whole number. They ranged from three to six in each stomach,
and were from three to four inches long. It should be noted,
however, that these were all eaten by fishes taken at the same
place and time. A large mouse was found in the stomach of one
bass from the Illinois River.

We may generalize these data by saying that this black bass
lives, at first, wholly on Entomostraca; that it commences to take
the smallest aquatic insects when about an inch in length, and
that minute fishes appear in its diet almost as early. From this
forward, the Entomostraca diminish in importance, and the insects
and fishes become larger and more abundant in the food. The
adults eat voraciously of a great variety of fishes—especially the
hickory shad (Dorysoma),—and feed upon crawfishes also to some
extent.

MICROPTERUS SALMOIDES, Lac.   SMALL-MOUTHED BLACK BASS.

This species, called also tiger bass, river bass, etc., is the black
bass, *par excellence*. It ranges, usually, in deeper and clearer
water than the preceding; but both are often taken together.

I have made full notes of the food of twenty-seven specimens,—
three adult and the others young. I had none of this species
under an inch in length; but, judging from the general resemblance
of the food of this and the preceding bass at later ages, I do not
doubt that this will also be found to feed at first on Entomostraca,

although insect food is possibly more important to it from the beginning.

Seven individuals, from one to two inches in length, were all taken in July from rocky ripples in the Fox River, at Dayton, Ill., a few miles above the mouth of the stream. These had eaten only five per cent. of Entomostraca,—the whole remainder of the food consisting of insects, of which *Corixa tumida*, young and adult, and larvæ of May-flies and darning-needles (Agrionidæ) were the most important kinds. Four per cent. of the larvæ of Chironomus are worthy of notice. The scarcity of Entomostraca in the food of fishes as small as these is probably due to the situation in which these specimens occurred, as few Entomostraca are to be found in swift water. The same fact will account for the presence of Chironomus larvæ,—found abundantly under stones in rapid streams.

The next ten specimens, between two and three inches long, were taken in July, partly at the same place as the preceding, and partly from the Illinois River, a few miles below the mouth of the Fox. These differed from the smaller specimens chiefly in the appearance of fishes in the food (five per cent.) and in the absence of Neuroptera. Probably the last of these differences, at least, was accidental. A few larvæ of aquatic Coleoptera (Hydrophilidæ and Dytiscidæ) were noticed. Corixas, including *C. tumida*, Uhl, and *C. signata*, Fieb.,* amounted to eighty-two per cent. of the food.

In those ranging from three to four inches in length (seven individuals), the fishes eaten rise to fourteen per cent., but the insects drop away to seven per cent., and the Crustacea rise to seventy-nine. Here, however, difference of locality interferes to prevent any satisfactory comparison with other ages,—as these specimens were all taken in August, from Calumet River, at South Chicago. This slow stream, clogged with Algæ and a great variety of other aquatic plants in midsummer, also swarms with Crustacea,—especially the little *Allorchestes dentata*. This species made sixty-three per cent. of the food of these specimens; and an undetermined species of Asellus, fourteen per cent. A few *Gammarus fasciatus* were also found. The insects were Corixa and larvæ of Agrionidæ.

---

* Determined by Mr. Uhler.

6

It will be seen that, excepting the gradual increase of the number of fishes eaten, these data show no especial difference in the young of different ages. Smaller specimens and a larger number from a greater variety of situations, would be necessary to exhibit this difference.

The food of the young as a whole, apparently, does not differ essentially from that of the large-mouthed species, except in the probably greater importance of the insect element,– especially Corixas, which in these twenty-four specimens amounted to fifty per cent. of the food—and the inferior importance of fishes.

This peculiarity is expressed in a slightly different manner, in the food of the adult. The three specimens examined had eaten only fishes (*Noturus flavus* and *Percina caprodes*) and crawfishes (*Cambarus propinquus*),—thirty-eight per cent. of the former and sixty-two per cent. of the latter.

This is the first of several instances in which the ratio of fishes in the food of allied species and genera was found to correspond to the size of the mouth, being largest in those with the largest oral opening.*

---

* The frequency with which these two species of black bass are confounded makes it desirable that a single reliable character should be selected by which they can be invariably distinguished, whatever the age of the specimen. This character is afforded by the size of the scales, the small-mouthed species having the smaller scales. In this species there are eleven longitudinal rows of scales between the dorsal fin and the row of perforated scales running along the middle of the side, called the lateral line. In the large-mouthed species, there are never more than nine such rows. The young are easily distinguished by the longitudinal black stripe along the side of the large-mouthed bass, which is wanting in the young of the other species.

TABLE OF THE FOOD OF MICROPTERUS.

	M. PALLIDUS.					M. SALMOIDES.			
	Under one inch.	One inch to two.	Two to three inches.	Three to four inches.	Adults.	One inch to two.	Two to three inches.	Three to four inches.	Adults.
Number of specimens	5	6	2	4	14	7	10	7	3
I. FISHES		29			86		05	14	38
Acanthopteri					15				
Percina					08				
Perca									13
Dorysoma					58				
Cyprinidæ					06		05		
Campostoma					06				
Siluridæ					07				
Noturus flavus									25
II. INSECTS		46	100	100	†	95	89	07	
Undetermined larvæ			50						
1. Diptera (larvæ)		02				04	05		
Culicidæ							01		
Chironomus		02				04			
Muscidæ							04		
2. Coleoptera (larvæ)							02		
Dytiscidæ							02		
Hydrophilidæ							†		
3. Hemiptera		44	50	57	†	51	82	03	
Terrestrial									
Zaitha					†				
Corixa		42	50	57		51	82	03	
Notonecta		02							
4. Neuroptera (larvæ)				43	†	39	†	04	
Ephemeridæ				43		28	†		
Agrionidæ					†	11		04	
III. CRUSTACEA	100	25		†	07	05	04	79	62
1. Decapoda					07				62
Cambarus					07				62
2. Amphipoda	07						65		
Gammarus							02		
Allorchestes							63		
3. Isopoda (Asellus)								14	
4. Entomostraca	93	25				05			
Cladocera	87	14		†		05	04		
Daphniidæ	26	12		†		05	04		
Simocephalus	18	06							
Bosmina	60								
Leptodora		02							
Copepoda (Cyclops)	06	11				†	†		
IV. VEGETATION							02		
Endogenous							02		
Algæ					†	†			
Miscellaneous					07				

AMBLOPLITES RUPESTRIS, Raf.   ROCK BASS.

This favorite and widely distributed species does not differ from the other fishes mentioned in respect to the food of the young. The smallest specimen examined, five-eighths of an inch long, contained only a few Cladocera (Pleuroxus). Another, three-fourths of an inch long, had eaten Daphnids (seventy-five per cent.), Cyclops (ten per cent.), and larvæ of Chironomus A third, seven-eighths of an inch long, contained only minute fragments of a few larvæ of Neuroptera. These specimens were all taken from Fox River, in July, 1879. The remaining young of the year were living chiefly on Corixa (eighty-three per cent.), as were also the young of the year preceding (ninety per cent.), as far as could be judged from the food of two specimens, from three to four inches in length. Some land insects, Ephemerids, water beetles, and a few Allorchestes were also found in the food.

Four adult specimens, taken at Ottawa, on the 8th of July, had eaten some minute fishes (fifteen per cent.), a few water beetles, including *Tropisternus limbatus*, over forty per cent. of Neuroptera larvæ, and about thirty per cent. of small crawfishes. The Neuroptera included Baëtis and other Ephemerids (twenty per cent.), Agrionidæ and large Libellulidæ, and fifteen per cent. of case-flies (Phryganeidæ.) Pond-weed (Potamogeton) found in two stomachs, had probably been taken accidentally.

CHÆNOBRYTTUS GULOSUS, C. & V.   WIDE-MOUTHED SUNFISH.

This fine species is among the commonest of the family in the lakes and ponds of southern Illinois, where it is commonly known as the " Goggle Eye."

The northern limit of its range, as far as known, is the Illinois River valley. In numbers and habitat it replaces in the south the *Eupomotis aureus* of the north; but this equivalence is only apparent, as the two species differ widely in food. From its size and abundance, it is no insignificant food resource.

Food of the Young.

My smallest specimens were from lakes in the Mississippi bottom, near Bird's Point, Missouri. Two of these, one inch long and under, taken in September, 1879, had eaten only *Bosmina longirostris* and Cyclops. Insect food first appears in specimens one and a half inches long. Eight specimens, between one and three inches long, six of which were taken from a lake in the Illinois bottoms, near Pekin, in October, 1879, and two from a lake in Kentucky, near Cairo, Illinois, had eaten about forty per cent. Entomostraca, thirty per cent. Neuroptera larvæ, and thirty per cent. Corixas and Diptera larvæ. *Daphnia pulex, Simocephalus americanus, Bosmina longirostris,* Chydorus, Pleuroxus and Cyclops, were among the Entomostraca. *Corixa alternata* was found among the Hemiptera. Most of the Diptera (i. e., fifteen per cent.) were larval Chironomus.

Food of the Adult.

Six adults, from rivers, streams and lakes in central and southern Illinois, show the usual change in food, carried farther than in the preceding species. Entomostraca disappear—except a few Chydorus in a single specimen—and fishes become the principal reliance, amounting to forty-seven per cent of the food. Corixas, larvæ of *Palingenia bilineata,* and some terrestrial Coleoptera—*Anomala binotata* which made half the food of one specimen, are the remaining items.

The especially piscivorous habit of this species is probably related to the size of its mouth, which is much the largest among the sunfishes proper. A similar relation has already been noticed between the two black bass.

TABLE OF FOOD OF AMBLOPLITES AND CHÆNOBRYTTUS.

	AMBLOPLITES.				CHÆNOBRYTTUS.			
	Under one inch.	One inch to two.	Three to four inches.	Adults.	One inch and under.	One inch to two.	Two to three inches.	Adults.
Number of specimens examined.	3	3	2	4	2	4	4	6
I. FISHES				15				47
II. INSECTS	39	99	97	52		57	64	53
Undetermined larvæ		16		01				
Caterpillars		17						
1. *Diptera* (larvæ)	05	03		06		32	04	
Chironomus	05	03				26	04	

The Food of Fishes.

Table of Food of Ambloplites and Chænobryttus—Continued.

	AMBLOPLITES.				CHÆNOBRYTTUS.			
	Under one inch.	One inch to two.	Three to four inches.	Adults.	One inch and under.	One inch to two.	Two to three inches.	Adults.
2. Coleoptera...			02	03				10
Terrestrial...			01					10
Aquatic...								
Dytiscidæ			01	01				
Hydrophilidæ				02				
3. Hemiptera...		63	95				20	18
Corixa...		63	90				20	18
Hygrotrechus (yg.).			05					
4. Neuroptera (làrvæ)	34			42		25	40	25
Ephemeridæ				21		03	12	25
Palingenia								25
Baëtis				01				
Agrionidæ				01		10	28	
Libellulidæ				05				
Phryganeidæ				15				
III. Arachnida (Hydrachna)				†				
IV. Crustacea...	61		03	31	100	43	34	†
Decapoda (Cambarus)				31				
Amphipoda			03					
Entomostraca	61	01			100	43	34	†
Cladocera	58				70	24	34	†
Daphnia							21	
Bosmina					70			
Pleuroxus	33					†		
Chydorus							†	†
Copepoda	0.3	01			30	19		
V. Vegetation...				02			02	
Potamogeton..				02			02	
Algæ								

Apomotis cyanellus, Raf. Blue-spotted Sunfish.

This species, distributed throughout the state, is especially abundant in Central Illinois, where it is the common fish of the ponds and smaller streams,—" the sunfish " of the country school-boy and the picnic party. It is the constant companion of the "bull-head" (Amiurus) and "shiner" (Notemigonus) in the small stagnant ponds of the prairie regions, and of the " chub minnow " (Semotilus) in muddy creeks. It was found abundant with Centrarchus, Aphredoderus and *Amiurus catus*, in the rapidly drying mud-holes,* only a few feet across, left by the retreating overflow of the Mississippi bottoms, in Union county.

Food of the Young.

The smallest of nineteen specimens studied, was one inch in length- -taken in July, in a prairie pond near Normal. Ninety-five per cent. of its food was Cyclops and three per cent. Daphnids. The trifling remainder consisted of a Corixa just hatched, and a Chironomus larva.

Nine specimens, ranging from one to two and a fourth inches in length, vary so little in food that it is not worth while to treat them separately. These were taken from various ponds, streams and lakes in Central Illinois. Their food was distributed quite generally through the various orders of insects and crustaceans accessible to them, showing the indifferent appetite of this fish and the general effectiveness of its collecting apparatus.

Larvæ of Chironomus, Dytiscidæ, Staphylinidæ, Corixas, Ephemerid larvæ, Decapoda, Isopoda, Cladocera, Cyprids and Copepoda were all found in considerable quantities in the food of these specimens. As usual, the most important insects were Corixas and May-flies,—sixteen per cent. of the former and twenty-nine per cent. of the latter. About eight per cent. of the food was Cladocera (Daphnia, Simocephalus, Pleuroxus, Chydorus).

Food of the Adult.

The eight adults, from Northern and Southern Illinois, differed from the young in the disappearance of Entomostraca from the food, the larger size of the insects taken, and the appearance of fishes and crawfishes.

Among the insects were a large Hydrophilus unknown to me, but nearly as large as *H. triangularis*, the larva of *Corydalis cornutus*, of Libellula and of some Ephemerid. The fishes composed about thirty-six per cent. of the food. The only recognizable specimens were a small Cyprinoid and a young buffalo-fish

---

* All the specimens taken from these holes, so muddy that the water was almost opaque, were of a peculiarly bleached appearance,—many of them almost colorless,—a fact of interest relative to the laws of coloration among fishes.

(*Ichthyobus bubalus*). Crawfishes and the river shrimp (Palæ-monetes) had been eaten by two of the specimens.

LEPIOPOMUS PALLIDUS, Mit.   COMMON SUNFISH.

This abundant, hardy and voracious species, is found through-out the state, and may be regarded as the typical sunfish.   It is most plentiful in the larger rivers in Central Illinois, being replaced in ponds by *Apomotis cyanellus.*

Consistently with its wide range and varied habitat, it is a gen-eral feeder for a sunfish,—peculiar only in the fact of its strictly non-predaceous character.   Of forty-five specimens examined, only one had eaten a fish, and that one only a single small darter.

Undifferentiated Centrarchidæ.—I introduce here the food of six specimens of this family which were too small for determina-tion.   They were too deep for Micropterus, and, as they had but three anal spines, could not have been Ambloplites or Pomoxys. They were probably *Lepiopomus pallidus.*   All were taken from the Illinois River,—a part of them near La Salle, in July, 1879,— the others from Peoria, in June, 1878.

The smallest (seven-sixteenths of an inch long) had eaten only Daphniidæ.   The next in size (one-half inch) contained Cyclops (ninety-eight per cent.) and Chydorus.   Nearly the whole of the food of the remaining four was Daphniidæ (ninety-four per cent.), including *Daphnia pulex.*

Food of the Young.

My smallest specimens, five in number, ranging from three-fourths of an inch to one inch, were taken in August, September and October, at Pekin, Peoria and Mackinaw Creek, Woodford county.   Neither locality nor date seems to have made any marked difference in their food, the principal elements of which were En-tomostraca and Chironomus larvæ,—fifty-seven per cent. and thirty-seven per cent. respectively.

A few water-spiders (Hydrachnidæ) and undetermined Amphi-poda were the other items.   The Entomostraca were all Cyclops (twenty per cent.) and Cladocera (*Simocephalus vetulus* and *amer-icanus, Bosmina longirostris* and *Pleuroxus dentatus.*)

Nine specimens, between two and three inches long, were caught at the same times and places as the preceding, except that one specimen from Mackinaw Creek was taken in June, and one taken in September was from Clear Lake, Kentucky. The greater size of these specimens was indicated by the appearance of a few Neuroptera larvæ in the food—eight per cent. In other essential respects, the food was like that of the foregoing group. One specimen had eaten largely of water-mites and another of Cyprids (fifty per cent.), and these elements have therefore greater prominence in the averages. Chironomus larvæ and Entomostraca now sum up eighty-one per cent.

In the third group of the young, consisting of seven fishes, between two and three inches long, the Chironomus larvæ remain about as before (thirty per cent.), Corixas appear (twenty-five per cent.) and Neuroptera larvæ rise to fourteen per cent. Entomostraca now fall away to a trifle, and larger percentages of Amphipoda appear. Single fishes had eaten the larvæ of a Gyrinid beetle, portions of the Polyzoan, *Pectinatella magnifica,** Leidy, and an earthworm,—the latter probably nibbled from some fisherman's hook.

These specimens were all from the Illinois River, in June, July, October and November.

Food of the Adult.

The twenty-four adults examined were from various parts of the state north of the center; and, as the food has been found to differ so widely according to the local situation, I have treated them in three groups,—the first including those taken in the clear, inland, northern lakes; the second those from Calumet River, at South Chicago, and the shallow, muddy lakes of that vicinity, and the third those from the Illinois River, from Ottawa to Peoria.

The specimens from the northern lakes were taken in May and June. Sixty-two per cent. of the food consisted of Neuroptera,— eight per cent. being a black caddis-fly (*Sialis infumata*) and the remainder the larvæ of large dragon-flies (Libellulidæ), Agrions (eleven per cent.) and Baëtis (two per cent.). *Allorchestes dentata* was the next most important element (twenty-seven per cent.). A number of terrestrial insects besides Sialis appeared in the food.

---

* This animal forms the large, translucent masses found in midsummer in the slow water along the margins of the Illinois River and elsewhere throughout the state, usually collected about a stick or a stem of a waterweed. They vary from the size of a walnut to that of half a bushel. The fragments were easily recognized by the peculiar form and armature of the winter eggs (statoblasts), which are discoidal and bordered with a row of slender double hooks, shaped something like an anchor.

These included a Harpalid beetle, an *Aphodius fimetarius*, and some grasshoppers (Tettigidæ, etc.).

The second group, of four, from Calumet River, and from Lake George, Indiana, was peculiar in the number of tetradecapod Crustacea and case-worms taken, and especially in the amount of vegetation eaten.

The crustacea were Allorchestes (thirty-two per cent.) and Asellus (twenty per cent.). The vegetation was present in such quantities as to make it evident that it had been taken as food. It amounted to about a fourth of the contents of these stomachs. The stomach of one fish was packed with a piece of the stem of a plant (apparently a Scirpus) a third of an inch in diameter and six inches long. Three others contained smaller amounts of confervoid Algæ.

The fifteen specimens remaining were taken from the Illinois in May, July, August, October and November. Their food was especially noticeable for the presence of mollusks (sixteen per cent.), for the number and variety of land insects (fifteen per cent.), and for the large amount of vegetation it contained (thirty-one per cent.). A single small fish,—the only one taken by these forty-five specimens—was also noted.

The mollusks included Planorbis, Physa, Amnicola and Vivipara. Among the insects, were ants, caterpillars, flies, *Anisodactylus discoideus* and other Harpalids, *Aphodius inquinatus*, wireworms, minute Curculios, *Cryptocephalus 4-maculatus*, *Diabrotica 12-guttata*, Colorado potato beetles, Flea Beetles, Plant Bugs (Cydnidæ), Crickets (Nemobius), Locusts and Katy-dids (*Phaneroptera curvicauda*), grasshoppers and caseflies.

The vegetable food, as far as determined, consisted of Ceratophyllum, *Nais flexilis* and confervoid Algæ. Fragments of Polyzoa were noticed. *Coptotomus interrogatus*, Gyrinid larvæ,* *Tropisternus limbatus* and other Hydrophilidæ, larval and adult, a large Nepa, larvæ of *Palingenia bilineata* and other May-flies, of Agrions and dragon-flies were among the aquatic insects taken.

The crustacea were limited to small crawfishes (two per cent.), a trace of Allorchestes, and a few Aselli (four per cent.).

On comparing specimens from northern Illinois with those taken from the Illinois River in the same month, I find that there are no common seasonal food characters, and that the differences of

* Several of these little-known larvæ were found in the stomachs of this species,—some of them in suitable condition for description.

food are therefore due to difference of locality and not to difference of time represented by the groups. Concerning the entire number of adults, we can therefore say that their food ranges through the whole list of the smaller mollusks, terrestrial and aquatic insects, and smaller crustaceans (above Entomostraca) accessible in their localities, and that they feed largely on aquatic vegetation. A striking negative feature is the almost total absence of fishes in the food,—a fact which corresponds with the relatively small size of the mouth.

TABLE OF FOOD OF APOMOTIS AND LEPIOPOMUS.

	APOMOTIS.			LEPIOPOMUS.						
	One inch long.	One to four inches.	Adults.	One inch and under.	One inch to two.	Two to three inches.	Northern lakes.	Calumet river and lakes.	Illinois river.	Total adults.
Number of specimens..	1	10	8	5	9	7	5	4	15	24
I. FISHES.			36						01	01
Undetermined			10							
Etheostomatidæ.									01	
Cyprinidæ.			13							
Ichthyobus			13							
II. MOLLUSKS.									16	10
1. *Gasteropoda*									16	10
Undetermined									04	
Planorbis.									†	
Physa.									05	
Vivipara.								†	07	
Amnicola.									†	†
2. *Acephala.*									†	
III. INSECTA	02	79	42	37	34	72	70	23	43	45
Undetermined		12	01			02				
Terrestrial.		07					05	01	15	12
Aquatic.	02	60	41	37	34	70	65	22	28	33
1. *Diptera*(Chironomus)	01	09		37	26	30		†	01	01
2. *Coleoptera*		02	05			01	01	01	09	06
Undetermined								†	04	
Dytiscidæ		02							02	
Undetermined		02								
Larvæ.									01	
Coptotomus.									01	
Gyrinidæ (larvæ).						01		01	01	
Hydrophilidæ.							01		02	
Undetermined		05							01	
Tropisternus									01	

TABLE OF FOOD OF APOMOTIS AND LEPIOPOMUS—Continued.

	APOMOTIS			LEPIOPOMUS						
	One inch long.	One to four inches.	Adults.	One inch and under.	One to two inches.	Two to three inches.	Northern lakes.	Calumet river and lakes.	Illinois river.	Total adults.
Number of specimens..	1	10	8	5	9	7	5	4	15	24
3. *Hemiptera*	01	16	03	†	......	25	02	01	04	02
Undetermined.............				†						
Corixa....................	01	16	03	......	......	25	01	......	04	......
Nepa..................							01			
Ranatra								01		
4. *Neuroptera*	......	33	33	......	08	14	62	20	14	24
Larvæ (undetermined)					01					
Ephemeridæ (larvæ)...	......	29	27	......	07		02	......	05	03
Baëtis...................							02			
Palingenia			01						04	02
Agrionidæ (larvæ).......			06			14	11	......	†	02
Libellulidæ (larvæ)......	......	04	02				41	......	02	10
Sialidæ..................			04				08	......	07	06
Sialis.................							03	......	07	
Corydalis (larvæ)										
Phryganeidæ							......	20	......	03
IV. ARACHNIDA......				04	11	01	01	......		
Spiders.................							01	†		
Hydrachnidæ........ •				04	11	01			†	
V. CRUSTACEA.........	98	21	20	59	55	06	27	52	06	18
Decapoda...............		01	20					......	02	01
Undetermined.........		01								
Cambarus			08					......	02	
Palæmonetes..........			12							
Amphipoda............				02		05	27	32	......	11
Undetermined.........				02						
Allorchestes...........						05	27	32	†	......
Isopoda (*Asellus*).......		04						20	04	06
Entomostraca...........	98	16	......	57	55	01		†		
Cladocera.............		08	......	37	14			†		
Daphniidæ	03	07	......	33	12					
Lynceidæ		01	......	04	02	01		†	......	
Ostracoda.............		03	......		06			†	†	
Copepoda.............	95	05	......	20	35	†				
VI. VERMES......						20		......	03	02
Undetermined.............						†				
Polyzoa..............						11		......	03	02
Lumbricus						09				
VII. VEGETATION....			02			01	02	25	31	24
Phænogamous.............			02			01	02	21	19	16
Algæ.................								04	12	08

XENOTIS MEGALOTIS, Raf. LONG-EARED SUN-FISH.

This little species is not at all common in the state, but has been taken by us from the middle course of Fox R., from tributaries of the Illinois R., and from ponds in Union county, in southern Illinois.

Unfortunately, the three specimens examined had not lately taken food, and only a very imperfect notion of their usual aliment can be given. Corixa, Ephemerid larvæ, Chironomus larvæ, the tube of a case-worm, a few fish-scales and an undeterminable aquatic beetle were the only objects found.

XENOTIS PELTASTES.* Cope.

This beautiful little fish, hitherto taken in this state only in very small number from Fox R., was found quite abundant in the "slip" at South Chicago, in June, 1880. The three opened had eaten more larvæ of Chironomus than any thing else (sixty per cent.). Next came sixteen per cent. of mollusks, then Allorchestes and Asellus, Corixa, Gyrinid larvæ, and a few terrestrial larvæ (Chrysomelidæ). The large percentage of Chironomus was probably owing to the situation, a foul and muddy little bay, serving as a harbor for fishing-boats.

EUPOMOTIS AUREUS, Wahl. PUMPKIN SEED. BREAM.

This species swarms in the lakes and ponds of northeastern Illinois, but is much less abundant in the Illinois R.; and in the southern part of the state is almost unknown. The cause of this limitation of its range is apparently climatic; as there is certainly nothing in its food, nor, apparently, in any of its habits, to exclude it from our southern waters. Indeed, I do not see that its place is taken by any other fish to the southward. No other, unless *Eupomotis pallidus*, resembles it in food, and this is too infrequent to replace it. My knowledge of its food is based upon the study of twenty-five specimens, ranging from one and one-half inches upward, taken from the Illinois, Fox and Calumet rivers, and from Long, Crystal and Nipisink lakes and Lake George, in Central and Northern Illinois and Indiana. The months of May, June, July, August and October are represented by these specimens.

Food of the Young.

The nine smaller specimens, from one and one-half to two inches long, show at once two prominent peculiarities of the food. The larvæ of Chironomus compose fifty-one per cent. of the food,

---

* It is considered doubtful, by Dr. Jordan, if this species and the preceding are distinct.

and Entomostraca of the order Ostracoda (Cyprids), twenty-six. As both these are found most abundantly in muddy bottoms, it is evident that the fish is, at least at first, a bottom feeder. Traces of mollusks appear thus early, as well as a few Ephemerid larvæ (five per cent.). The remainder of the food was insects' eggs and Daphnids,—chiefly *Simocephalus americanus*—(twelve per cent.). Chydorus was found in five specimens, but in too small quantity to figure in the averages.

Five specimens were studied between two and three inches long. In these the same food characters continue, modified somewhat by the introduction of larger objects. The Chironomus larvæ stand at forty-four per cent., and the Cyprids at eighteen per cent. Fourteen per cent. of Allorchestes and eleven per cent. of Neuroptera larvæ are the only important elements remaining. Two per cent. of young Unios were noticed. Nearly half of the food of two larger specimens, between two and three inches long, consisted of mollusks,—chiefly Physa. A few Chironomi and about equal quantities of Ephemerid larvæ and Allorchestes were all the remaining food. Entomostraca therefore disappear at this point.

Food of the Adult.

Forty-six per cent. of the food of the nine adults consisted of Mollusca, including Planorbis, Amnicola and *Valvata tricarinata*, and six per cent. of undetermined bivalves.

The insect food was twenty per cent. of the whole, Crustacea twenty-two per cent., and vegetation twelve per cent. Half of the last was Chara, and the remainder chiefly Myriophyllum and Algæ. The Crustacea were all Allorchestes and Asellus. The insects included a trace of Chironomus larvæ and few water beetles (Hydrophilidæ), and the usual Neuroptera larvæ, among which case-flies of the genus Leptocerus were noticed.

Not a trace of fishes was found in the stomachs of these specimens; and this fact, together with the large percentage of molluscan food, constituted the leading alimentary peculiarities of the species.

The first of these is doubtless related to the small mouth,—the second to the stout, blunt pharyngeal teeth,—a character used in defining the genus. In all the preceding species the pharyngeals are set with more slender, pointed teeth.

EUPOMOTIS PALLIDUS, Ag. PALE SUN-FISH.

Having but few specimens of this rather uncommon species, I have examined the food of but one,—enough to indicate that it probably agrees closely with the preceding species.

This fish, taken in Clear Lake, Ky., had eaten largely of small Mollusca,—young Unionidæ, Planorbis, Amnicola, etc. These amounted to seventy-five per cent. of the food. The remaining elements were Chironomus larvæ, several small water beetles, (*Hydroporus hybridus, Cnemidotus* 12-*punctatus,* and *Haliplus,* sp.), an unknown aquatic pupa and a little pond weed.

TABLE OF FOOD OF EUPOMOTIS AND CENTRARCHUS.

	EUPOMOTIS.				CENTRARCHUS		
	One inch to two.	Two to three inches.	Three to four inches.	Adults.	One inch and under.	One inch to two.	Adults.
Number of specimens examined	9	5	2	9	5	1	2
I. MOLLUSCA	01	04	45	46			
Undetermined				07			
Gasteropoda	01	02	45	33			
Amnicola				04			
Vivipara				11			
Planorbis				01			
Physa			40				
Acephala		02		06			
II. INSECTA	60	61	30	20	24	20	91
Undetermined	03	03					07
1. Diptera (larvæ)	52	45	05	01	21		06
Chironomus	51	44	05	01	21		06
2. Coleoptera		02		02			
Hydrophilidæ		02		02			
3. Hemiptera						20	23
Corixa						20	23
4. Neuroptera (larvæ)	05	11	25	17	03		55
Ephemeridæ	05	05	25	03	03		55
Palingenia				03			
Agrionidæ		06					
Libellulidæ				08			
Phryganeidæ				02			
III. ARACHNIDA					01		
Hydrachnidæ					01		
IV. CRUSTACEA	39	35	25	22	75	80	
Amphipoda (Allorchestes)		14	25	13	04		
Isopoda (Asellus)				09			
Entomostraca	39	21			71	80	
Cladocera	13	03			12		
Simocephalus	12	01			12		
Ostracoda	26	18			21	35	
Copepoda					38	45	09
V. VEGETATION		†		12			
Myriophyllum				04			
Chara				06			
Algæ			†	02			

CENTRARCHUS IRIDEUS, Lac.

This little species is found in considerable numbers in ponds and streams in the southern hill-country of Illinois. My specimens, all taken in July, are from ponds and streams in the Mississippi bottoms in Union and Jackson counties, and from Cache R. and its tributaries in Johnson county.

Five of the young, from three-fourths of an inch to an inch in length, had eaten seventy-one per cent. of Entomostraca and twenty-one per cent. of larvæ of Chironomus; and, for the rest, about equal quantities of Ephemerid larvæ and young Allorchestes, with a trace of water mites (Hydrachnidæ).

Thirty-eight per cent. of the food was Cyclops; Cyprids amounted to twenty-one per cent.; and twelve per cent. of Simocephalus completed the ratio of Entomostraca. The smallest specimen, three-fourths of an inch long, had eaten sixty per cent. Simocephalus and forty per cent. Cyclops.

About a fifth of the food of one specimen, an inch and an eighth in length, consisted of minute young Corixas, the remainder being about equally Cyclops and Cyprids.

Only two specimens were examined which could be classed as adults,— one three and a fourth inches long, the other smaller. These indicate that the food of full-grown individuals differs from that of the young chiefly in the addition of considerable quantities of terrestrial and aquatic insects.

The gill-rakers of this species are numerous, long and slender,— a fact reflected in the food. Fifteen per cent. of the contents of the stomach of the largest specimen consisted of Cyclops and five per cent. of Chironomus larvæ. Consistently with the small mouth and pointed pharyngeal teeth, no traces of fishes or mollusks were found in the food.

POMOXYS NIGROMACULATUS, Lac. BLACK CROPPIE. LAKE CROPPIE. SILVER BASS. BUTTER BASS.

POMOXYS ANNULARIS, Raf. WHITE CROPPIE. TIMBER CROPPIE. SILVER BASS.

These two species, often not distinguished even by experienced fishermen, agree so closely in food that I have not thought it worth while to treat them separately. In the Illinois and Mississippi rivers they are much the most valuable and important of the family, excepting the black bass. They are nowhere else so abundant in the state, although occurring in the larger rivers generally and in the great lakes. The first species is commonest to the north, and the second southward, as far as my observation goes. In the Illinois, they are about equally abundant. These fishes are everywhere great favorites, and rank among the most important and promising of our smaller species. They are rarely found in creeks or small ponds, but seem to require deeper water for their maintenance.

The gill-rakers of this species are numerous, long, and finely-toothed, constituting the most efficient straining apparatus to be found among the sun-fishes. The pharyngeal teeth are sharp, and the mouth is rather wide and considerably enlarged by the lengthening of the lower jaw.

Consistently with the hypothesis concerning the meaning of the gill-rakers which I had already formed from a study of the preceding species, before I came to this, I found that the young continued to feed almost exclusively upon Entomostraca much longer than the other sun-fishes. Six specimens between three and four inches long, had eaten little else than Entomostraca and the larvæ of minute Diptera (Chironomus and Corethra). Even full-grown specimens were found eating Cladocera more freely than any other food. As might be inferred from the pharyngeals, not a trace of molluscan food was found in the forty-two specimens examined, while fishes formed nine per cent. of the food of the twenty-seven adults. Most of these were eaten late in the season, when Entomostraca and insect larvæ became less abundant.

Food of the Young.

The smallest specimen, three-fourths of an inch long, had eaten about equal quantities of Cyclops and Simocephalus, with only a few Pleuroxus beside. Three, an inch long and under, had confined their food entirely to Entomostraca and Chironomus larvæ,—the latter forming about a fourth of the whole. A third of the Entomostraca were Cyclops, the remainder chiefly Simocephalus.

Six specimens between one and three inches long, differed especially in the introduction of about eighteen per cent. of Corixas and three per cent. of small Ephemerid larvæ. Chironomus larvæ were reduced to seven per cent. The Entomostraca were about

8

equally divided between Cyclops and Cladocera. One specimen taken in July, 1879, from the canal near Ottawa, had taken a large number of Daphnella.

Six specimens between three and four inches long were examined. Eighty-three per cent. of their food was Entomostraca, about three-fourths of this amount being Cyclops, and the remainder nearly all Simocephalus. Twelve per cent. of larvæ of Chironomus and Corethra, three per cent. Corixas and two per cent. larvæ of small Ephemerids were the insect elements. Chydorus, Pleuroxus and Cypris were present in small numbers.

These fifteen young, agreeing so closely in food, irrespective of size, were nevertheless from a variety of situations and dates. All were from the Illinois river, its lakes and tributaries, from Ottawa to Pekin, but ranged in time from June to October of three different years.

Six were *P. nigromaculatus*, seven were *P. annularis*, and two were not identified specifically.

Food of the Adults.

An examination of the notes on the twenty-seven adults shows material differences of food at different parts of the year. As all but one were taken from the Illinois river, I have not the means of noting the correspondence of food with locality.

Five specimens taken at Peoria, in March, were found feeding most freely upon Cladocera, which composed fifty-five per cent. of their food. These were chiefly of the two species *Simocephalus retulus* and *S. americanus*. These little Entomostraca were taken at that time in such quantity as visibly to distend the stomach when seen from the outside, and the immense numbers of their eggs gave a reddish color to the contents of the alimentary canal. The larvæ of Neuroptera, both "darning-needles" and May-flies (Palingenia), were also eaten in considerable numbers (thirty-nine per cent.). A small Hybopsis, a little darter (*Boleosoma maculata*) and an unrecognizable fish were found in these stomachs, making about six per cent. of the food. Only trivial numbers of Entomostraca appear after this time.

Nine specimens, taken in April, likewise at Peoria, were feeding chiefly upon Neuroptera larvæ (eighty-six per cent.), especially upon that almost invaluable element of fish food, the larvæ of

Palingenia bilineata (sixty-six per cent.). A few larvæ of Gyrinidæ and Dytiscidæ were noted (three per cent.), and a few Corixas also. A *Gammarus fasciatus* and a little Ceratophyllum, etc., were noticed; and also the flower of an elm and the feather of a bird.

A single specimen from Pistakee lake, in McHenry county, taken in May, gave evidence of a similar reliance upon Neuropterous larvæ (eighty-five per cent.). Here, however, in the absence of Palingenia, Agrions and the larger dragon-flies were resorted to. A little vegetation had been taken with these (*Ceratophyllum demersum* and *Lemna trisulca* ten per cent.), probably by accident, as this lake was full of aquatic plants, and it would hardly have been possible for a fish to catch living food from the water without getting more or less vegetation at the same time. A single Hymenopter,—the only land insect found eaten by this species,—was taken from this stomach. A specimen taken in June at Peoria had eaten about equally of minute unrecognized fish-fry and Palingenia larvæ. One caught at Ottawa, in July, had eaten only insects,—Corixa twenty-five per cent., Palingenia larvæ seventy-five per cent.

Five croppies from Peoria, in October, 1878, and five from Henry, thirty miles above, in November, 1877, indicate that the autumnal food of the species is again different. These had eaten, respectively, thirty-nine per cent. and twenty-eight per cent. of small fishes,—partly Cyprinidae and partly undetermined Acanthopteri. The remainder of their food was composed chiefly of Palingenia larvæ. One October specimen had eaten two larvæ of the large "Helgramite," *Corydalis cornutus*. Although these fishes were taken directly from the seine, and opened upon the spot, the food in their stomachs did not average more than a fourth of the quantity in those taken in early spring. The weather during both these months was uncomfortably cold, with falling snow, and the food of these specimens probably gives a correct hint of the winter food of the species.

Fourteen of the above were *Pomoxys nigromaculatus* and twelve *P. annularis*,—one not having been determined.

TABLE OF FOOD OF POMOXYS.

	One inch and under.	One inch to three.	Three to four inches.	March, Peoria.	April, Peoria.	May, Pistakee L.	June, Peoria.	July, Ottowa.	October, Peoria.	November, Henry.	Total adults.
No. of specimens......	3	6	6	5	9	1	1	1	5	5	27
I. FISHES.				06			50		39	28	15
Acanthopteri									10	08	03
Boleosoma				01							
Cycloid...............				04					11		
Cyprinidæ...				04					11		03
Hybopsis..				04							01
II. INSECTS...........	28	28	17	39	90	90	50	100	61	72	73
1. *Hymenoptera*						05					
2. *Diptera* (larvæ)....	28	07	12						01	02	01
Corethra		†	03						01		
Chironomus	28	07	09		†	†				02	
3. *Coleoptera* (larvæ)..					03	†				03	01
Gyrinidæ					02					03	
Dytiscidæ.............					01						
4. *Hemiptera*		18	03		01			25			01
Corixa		18	03		01	†		25			01
5. *Neuroptera* (larvæ).		03	02	39	86	85	50	75	60	67	68
Ephemeridæ...........		03	02	15	72				44		54
Palingenia.............				15	66		50	75	44	67	52
Agrionidæ.............				24	14	55					12
Libellulidæ						30					01
Sialidæ (Corydalis)									16		01
III. CRUSTACEA. ...	72	72	83	57	10						12
Gammarus.............					†						
Entomostraca	72	72	83	55	10						12
Cladocera.............	49	33	17	55	10						12
Daphniidæ.- ..	46	18	17	55	†						12
Lynceidæ.........	03	0!	†								
Sididæ.............		14									
Ostracoda.............			01	†	†						
Copepoda...:.......	23	39	65		†		†				
IV. Vegetation					†	†	10	†	†		

Summary of the Family.

For the purpose of a comparative recapitulation of the above data respecting the food of the sun-fishes, I have prepared three condensed tables, showing, upon the same page, the food of the different genera in parallel columns. The first table exhibits the food of the youngest specimens, the second, of those of intermediate size, and the third, of those which may properly be regarded as mature.

By an inspection of the first table, it will be seen that the thirty specimens, one inch long and under, representing eight genera, which appear thereon, have eaten little else than Entomostraca and larvæ of Chironomus,—these two elements amounting to ninety-three per cent of the food. The only exception to this rule (that of the rock bass) is apparent rather than real. The large percentage of Neuropterous larvæ appearing under the name of that species is a technical ratio, inserted only for the sake of consistency, being based upon the fact that one of the specimens examined contained no food except a few traces of some indeterminable minute larva of that order. The minor differences in the food of the generic groups are doubtless due to differences of locality, and the like. That Ostracoda, for example, were found only in the stomachs of Centrarchus, is accounted for by the fact that the youngest specimens of this genus were taken from small mud-holes, favorable to the occurrence of Entomostraca of that order. The uniformity of food at this time implies that the selective apparatus of these fishes, whatever its construction, has not yet grown beyond the size of these minute animal forms.

From the second table of one hundred and six specimens we learn that with a general change of food from Entomostraca and Chironomus to larger crustacea and insects, there appear certain differences,—notably the continuance of Entomostraca as the most important element in Pomoxys, and the occurrence of mollusks in Eupomotis and of fishes in Micropterus. It is important to recall, at this point, that Pomoxys has the largest, finest and most numerous gill-rakers of the group,—the best *straining* apparatus, in short,—that Eupomotis has stout and pharyngeal teeth, and that the black bass have relatively the widest mouths of all. It is also to be noted that the large-mouthed bass commenced to take fish when an inch and a quarter long, and the small-mouthed species not until it reached a length of two and a half inches.

It will also be observed that Entomostraca are least abundant in the food of the small-mouthed black bass and the rock bass,—species found usually in swift and shallow water, when of this size. The importance of water-bugs (Corixa) to the first three species of this table is evident.

From the table of adult food we find that these commencing peculiarities of the preceding table become here more prominent. All the Entomostraca of this table, except insignificant traces, now appear in the food of Pomoxys; the molluscan food of Eupomotis is nearly five times that of any other genus; and the ratios of fish food, running from eighty-six per cent. down to nothing, when arranged in a series, are seen to correspond, with curious exactness, to a series of the species themselves arranged according to the relative sizes of their mouths.

I was disappointed in being unable to find any food characteristics corresponding to such minor differences in the lengths of the gill-rakers of the anterior arch as appear in Lepiopomus, Apomotis, etc., on the one hand, and Xenotis and Eupomotis on the other. If such peculiarities exist, they can probably be determined only by taking at one time and place a number of specimens of unlike character in this particular.

While I believe that the generalizations made above will hold good, at least for fishes of similar form and internal structure among the Acanthopteri, I do not wish to be understood as extending them at present beyond this order. Doubtless, while the characters mentioned must assist greatly in determining the food of a species *a priori*, they are not by any means sufficient for this purpose when taken by themselves. The discussion of other features, external and internal, bearing upon this subject, must be postponed to a later period of the investigation.

TABLE OF FOOD OF YOUNG CENTRARCHIDÆ. (*One inch and under.*)

	Micropterus pallidus.	Micropterus salmoides.	Ambloplites.	Chænobryttus.	Apomotis.	Lepiopomus.	Xenotis.	Eupomotis.	Centrarchus.	Pomoxys.	Undetermined.	Total.
Number of specimens	5	...	3	2	1	5	...	...	5	3	6	30
I. INSECTS		...	39	...	...	02	37	...	...	24	28	...	17
1. *Diptera* (larvæ)		...	05	...	...	01	37	...	...	21	28	...	13
Chironomus		...	05	...	...	01	37	...	...	21	28	...	13
2. *Hemiptera*		...	...	...	...	01	...	...	...	...	...	...
Corixa (young)		...	...	...	...	01	...	...	...	...	...	...
3. *Neuroptera* (larvæ)		...	34	...	...	...	...	...	...	03	...	...	04
II. ARACHNIDA (Hydrachna)		...	...	...	...	...	04	...	...	01	...	...	01
III. CRUSTACEA	100	...	61	100	98	59	...	...	75	72	100	82
Tetradecapoda	07	...	...	...	...	02	...	...	04	...	...	02
Entomostraca	93	...	61	100	98	57	...	...	71	72	100	80
Cladocera	87	...	58	70	03	37	...	...	12	49	88	53
Ostracoda		...	...	...	...	...	...	...	21	...	†	04
Copepoda	06	...	03	30	95	20	...	...	38	23	22	23

TABLE OF FOOD OF YOUNG CENTRARCHIDÆ. (*One to four inches.*)

	Micropterus pallidus.	Micropterus salmoides.	Ambloplites.	Chænobryttus.	Apomotis.	Lepiopomus.	Xenotis.	Eupomotis.	Centrarchus.	Pomoxys.
Number of specimens...	12	24	5	8	10	16	......	16	3	12
I. FISHES	15	06	......	......	......	......	......	......	......	......
II. MOLLUSCA	......	......	......	......	......	......	......	08	......	......
III. INSECTA	72	67	98	61	78	51	......	56	67	22
1. *Diptera* (larvæ)	01	03	02	18	09	28	......	44	04	09
Chironomus	01	01	02	15	07	28	......	43	04	08
2. *Coléoptera*	......	02	01	......	09	†	......	01	......	......
Terrestrial	......	......	......	......	07	......	......	......	......	......
Aquatic	......	02	01	......	02	†	......	01	......	......
3. *Hemiptera* (Aquatic)	50	50	78	10	16	11	......	......	23	11
Corixa	49	50	76	10	16	11	......	......	23	11
4. *Neuroptera*	14	12	......	33	33	11	......	10	37	02
Ephemeridæ	14	08	......	08	29	05	......	08	37	02
Odonata	......	04	......	19	04	06	......	02	......	......
IV. ARACHNIDA	......	......	......	......	......	07	......	......	......	......
V. CRUSTACEA	13	26	02	38	22	32	......	36	......	78
Decapoda	......	......	......	......	01	......	......	......	......	......
Tretadecapoda	......	23	01	......	04	......	......	08	......	......
Entomostraca	13	03	01	38	17	32	......	28	33	78
Cladocera	07	03	......	29	09	08	......	08	......	25
Ostracoda	......	......	......	......	03	03	......	20	12	01
Copepoda	06	†	01	09	05	20	......	......	21	52
VI. POLYZOA	......	......	......	......	......	05	......	......	......	......
VII. LUMBRICUS	......	......	......	......	......	04	......	......	......	......
VIII. VEGETATION	......	01	......	01	......	01	......	......	......	......

TABLE OF FOOD OF ADULT CENTRARCHIDÆ.

	Micropterus pallidus.	Micropterus salmoides.	Ambloplites.	Chænobryttus.	Apomotis.	Lepiopomus.	Xenotis.	Eupomotis.	Pomoxys.
No. of specimens examined..	14	3	4	6	8	24	6	9	27
I. FISHES	86	38	15	46	36	01	07	......	15
Acanthopteri	15	......	......	......	......	01	......	......	03
Dorysoma	58	......	......	......	......	......	......	......	......
Cyprinidæ	06	......	......	......	12	......	......	......	03
Siluridæ	07	......	......	......	......	......	......	......	......
II. MOLLUSCA	......	......	......	......	......	10	08	46	......
Gasteropoda	......	......	......	......	......	10	04	33	......

TABLE OF FOOD OF ADULT CENTRARCHIDÆ — Continued.

	Micropterus pallidus.	Micropterus salmoides.	Ambloplites.	Chænobryttus.	Apomotis.	Lepiopomus.	Xenotis.	Eupomotis.	Pomoxys.
No. of specimens examined...	14	3	4	6	8	24	6	9	27
Acephala						†	04	06	
III. INSECTA	†		52	54	42	45	82	20	73
1. *Diptera* (larvæ)			06			01	37	01	01
Chironomus							37	01	
2. *Coleoptera*			03	10	05	13	06	02	01
Terrestrial				10		06	04		
Aquatic			03		05			02	01
3. *Hemiptera* (Corixa)	†			18	03	02	17		01
4. *Orthoptera*						02			
5. *Neuroptera*	†		42	25	33	24	16	17	68
Ephemeridæ			21	25	27	03		04	52
Palingenia				25	01	02		03	52
Odonata			06		02	12		08	13
Sialidæ					04	06			01
Phryganeidæ			15			03	08	01	
IV. ARACHINDA			†						
V. CRUSTACEA	07	62	31	†	20	18	03	22	12
Decapoda	07	62	31		20	01			
Tetradecapoda						17	03	22	
Entomostraca				†		†			12
Cladocera				†		†			12
Ostracoda						†			†
Copepoda									†
VI. POLYZOA						02			
VII. VEGETATION	†		02		02	24		12	†
Miscellaneous	07								

HAPLOIDONOTUS GRUNNIENS, Raf. SHEEPSHEAD. GRUNTING PERCH.

This species is abundant in Lake Michigan and the larger rivers, occurring in the smaller streams rarely, at periods of exceptionally high water. It is sometimes eaten, but is regarded usually as unfit for food.

But six of the twenty-five specimens studied were young, and the smallest of these, from the Ohio R., in September, was an inch and an eighth in length. Seventy-five per cent. of its food was larvæ of Chironomus and twenty-five per cent. larvæ of *Palingenia bilineata*. Besides the usual indications that the food of the very young is made up of minute animals, we see here evidence that this species seeks its food from the first upon the bottom. In a specimen two inches long, the Chironomus larvæ fell to fifteen per cent. while the Palingenia larvæ rose to eighty per cent., and other Ephemerids and Cyclops made up the remainder of the food.

Four specimens, also from the Ohio, at Cairo, from two to four inches long, were found to have recently fed upon Ephemerid larvæ and larvæ of aquatic beetles, Gyrinidæ and Hydrophilidæ, in about equal quantities. Only five per cent. of their food was Chironomus.

Sixteen individuals of medium size were taken from the Illinois and Ohio rivers, in April, June, September and October of four different years. There was nothing in the contents of these stomachs to indicate any difference in food resulting from these differences of date and situation. The food, on the contrary, was remarkably simple and uniform, consisting chiefly of the larvæ of Neuroptera (eighty-four per cent.), of which *Palingenia bilineata* formed altogether the most important part (seventy-six per cent.),— the remaining eight per cent. being dragon-flies. A single small sucker (Catostomidæ), a few mollusks (Planorbis, young Unios and thin-shelled Anodontas), and some Aselli complete the brief dietary of this group.

It is not until we examine the food of full-grown specimens that we wholly appreciate the utility of the enormous crushing pharyngeal jaws with their pavement teeth, found in this species. The entire food of the three large specimens examined, taken at Peoria, in April and October, proved to consist of mollusks only, including forty-six per cent. of the thick and heavy water snail, *Melantho decisa*, whose shell probably no other fish in our rivers could break. Cyclas, Anodonta and undeterminable Gasteropoda composed the remainder of the food.

9

ON THE FOOD OF YOUNG FISHES

Stephen Alfred Forbes

ON THE FOOD OF YOUNG FISHES.

BY S. A. FORBES.

I cannot learn that anything has been recorded respecting the food of young fishes in this country,* nor have I been able to find anything upon this subject in such part of the ichthyological literature of Europe as is accessible to me. From the lack of all mention of the use of Entomostraca as the food of young fishes in the general review of the relations of these Crustacea to organic nature given by Gerstaecker in Bronn's Thier-Reich† I infer that whatever systematic investigation the subject may have received, the results have not attracted any general attention.

This seems a surprising fact when one considers the vast amount of labor which has been expended upon this class of animals, and reflects for a moment upon the interest to science and to practical fish-culture of a knowledge of the food-resources of fishes and of the competitions of the various species in the search for subsistence.

Although I cannot yet treat this subject as fully as it deserves, the results of such study as I have been able to make, during the past season, of the contents of the stomachs and intestines of small specimens, seem to justify this preliminary notice.

It was early apparent, in the course of the investigation, that the food of many fishes differs greatly according to age; and it was soon found that the life of most of our fishes divides into at least two periods, and of many into three, with respect to the kinds of food chiefly taken. Further, in the first of these periods, a remarkable similarity of food was noticed among species and families whose later food-habits are widely different.

The full-grown black bass, for example, feeds principally on

---

*Perhaps exception should be made of a note relating to the occurrence of diatoms in the stomachs of two young whitefishes, published in the appendix to the Report of the U. S. Fish Commissioner for 1872-3, p. 57.

†Classen and Ordnungen des Thier-Reichs, Band V, Abtheilung 1, ss. 750 u. 1057.

fishes and craw-fishes, the sheepshead on mollusks, the gizzard shad on mud and Algæ, while the catfishes are nearly omnivorous; yet these are all found to agree so closely in food when very small that one could not possibly tell from the contents of the stomachs which group he was dealing with.

It is my purpose in this paper to give what facts I have relating to the food of our fresh-water species during this first period of the fish's life. These facts were derived from the examination of one hundred and twenty-six specimens, ranging from three-eighths of an inch in length up to an inch and half, and in a few cases to two and three inches. These specimens belong to twenty-four genera and represent eleven families. In two or three genera none were obtained small enough to be regarded as belonging strictly to this first food-period, but the earliest food is nevertheless plainly inferrible; and the general distribution and variety of the species studied is such that I think the main conclusions will be found to stand the test of full investigation. As the first period is evidently much shorter with some species than with others, and doubtless varies in the same species according to situation and circumstances generally, of course no common limit of size could be set up, but the smallest specimens of each species were selected until a size was reached where a marked difference of food appeared.

ACANTHOPTERI.

Although the young Acanthopteri have already been discussed in the preceding paper on the food of that group, it will be convenient to review the facts concerning these young fishes for the purpose of comparing their food with that of the other orders.

The food of six *common perch* (*Perca americana*), from an inch to an inch and a quarter long, consisted wholly of Entomostraca and larvæ of Chironomus,—eight per cent. Chironomus, fifty-two per cent. Cladocera and forty per cent. Copepoda.

No very small *Labracidæ* were found, the youngest being a Morone an inch and a quarter long. Half of the food of this consisted of Entomostraca (chiefly Cladocera), and the other half was minute gizzard shad.

A group of forty-three sun-fishes (Centrarchidæ), from five-eighths of an inch to two inches long, was made up as follows:—of five specimens of Micropterus under three-fourths of an inch long, two of Ambloplites of the same size, two of Chænobryttus from seven-eighths of an inch to one inch, one of Apomotis an inch in length, nine of Lepiopomus from an inch to an inch and a fourth, nine of Eupomotis from one and a half to two inches, five of Centrachus one inch and under, four of Pomoxys from three-fourths of an inch to an inch and a half, and six indeterminable specimens, probably Lepiopomus, from seven-sixteenths to five-eighths of an inch long. Ninety-six per cent. of the food of these forty-three specimens consisted of Entomostraca and larvæ of Chironomus,— seventy of the first and twenty-six of the second—the trivial remainder consisting of Neuroptera larvæ and young Amphipoda with traces of water mites, Corixas and mollusks (the last in Eupomotis). The Entomostraca were forty-two per cent. Cladocera, nineteen per cent. Copepoda and nine per cent. Ostracoda.

A single Haploidonotus an inch an eighth in length, had eaten Chironomus larvæ (seventy-five per cent.) and larvæ of *Palingenia bilineata.*

Esocidæ.

I did not have the good fortune to obtain any young of the common pike, and can only report on the food of a single *Esox salmoneus* an inch and a fourth in length. This specimen, taken at Pekin, Ill., on the 2d of June, had already begun its life labor of the elimination of little fishes, these making about two-fifths of its food. The remainder consisted of Crustacea, composed about equally of young Amphipoda, Daphniidæ and Lynceidæ. The presence of so large a quantity of these minute Entomostraca in the stomach of a pickerel of this size, is sufficient evidence that they form the principal part of its food at an earlier age.

Clupeidæ.

We come next to twelve specimens of the *gizzard shad* (Dorysoma), whose minute fry swarm in countless numbers in the waters of our larger rivers in mid-summer. These were taken in June and July, from the Illinois R., from Ottawa to Peoria. The smallest of the group were twenty mm. long by two mm. wide,—as slender as cyprinoids and nearly cylindrical, although the adult is a high, thin fish. I was greatly interested by the discovery that the maxillaries of these smallest specimens are provided with teeth,—a single row of nine or ten on the lower edge,—although the mouth of the adult is entirely toothless and smooth. The internal structure also differs remarkably from that of the adult, especially in the much greater simplicity of the digestive apparatus. In a young gizzard shad seven-tenths of an inch long by one-tenth high, the intestine was found to pass from the anterior end of the stomach to the vent with only one short forward turn of about a fourth the length of the body cavity, made a little way behind the stomach. Although the mucous surface of the intestine was at this time very rugose, showing a commencing complication of the digestive system, there was no trace of pyloric cœca. The intestine was filled with Cypris, Chydorus, Alona, Cyclops, etc.

On the other hand, in a fish three and three-fourths inches long, showing the general characters of the adult, the intestine passed upward and backward from its origin, running without flexure the whole length of the body cavity (this part being covered with an immense number of pyloric cœca), then turned forward to the stomach, ran back from there about one-third of the way to the vent, then turned forward and ran a tortuous course beneath the stomach to the pericardial membrane and back again, also tortuously, two-thirds of the way to the vent. From this point it ran forward again to the stomach, and crossing to the left side, ran repeatedly backward and forward in the posterior part of the body cavity, making seven turns between the stomach and vent before opening; thus extending, in all, about eight times the length of the periviseral cavity. This intestine was well filled with mud with only a slight sprinkling of unicellular Algæ.

Much as these young resemble young Cyprinidæ, they can be easily distinguished from them by the very long anal fin; and from the brook silversides (Labidesthes), to which they bear some superficial resemblance, by the absence of a spinous dorsal.

These twelve fishes, all under two inches in length, had eaten about ninety per cent. of Entomostraca, two per cent. of Chironomus larvæ, and for the remainder, Algæ. The Crustacea were about equally Cladocera and Copepoda. Among the former were *Daphnia pulex, Simocephalus americanus, Ceriodaphnia dentata* Bosmina, Chydorus and Alona. In a specimen three-quarters of an inch long which I took from the stomach of a *Morone interrupta*, I found a few specimens of *Leptodora hya-*

lina (?) Lillj. The Copepoda were all Cyclops, as far as recognizable.

CYPRINIDÆ.

A single minute minnow, three-eighths of an inch long, which I could not determine specifically, had eaten Daphnids (twenty-five per cent.) and Chironomus larvæ.

The specimens of the common *club minnow* (*Semotilus corporalis*), ranging from five-eighths inch to one inch, indicate somewhat doubtfully an exception to the general rule respecting the early food of fishes. Only seven per cent. of their food was Entomostraca, and the whole remainder consisted of filamentous Algæ. It should be noted, however, that twenty per cent. of the food of the smallest specimen, which was five-eighths of an inch long, was Cyclops, and it may be that Semotilus lives wholly on Entomostraca at first, merely changing its habit earlier than most of its allies.

Two specimens of *Notropis*, an inch and a half in length, had eaten nothing but Daphnids.

CATOSTOMIDÆ.

Thirty specimens, representing five genera of this peculiar family, were studied. A very curious feature of the food of the young is the frequent dependence of suckers of considerable size—six inches long or more—upon food still more trivial than Copepoda or Lynceidæ; viz., upon rotifers, Protozoa and unicellular Algæ. While only such Protozoa were found as are furnished with firm tests or carapaces, yet the abundance of Difflugia and Arcella in the intestines of these fishes leaves little doubt that the more perishable Protozoa must also be taken in considerable quantity. It is an interesting fact that even here the smallest specimens were found feeding on Entomostraca only, and it is therefore possible that these form the first food of the family.

Ten specimens of the *stone-roller* (*Hypentelium nigricans*), ranging from one and three-eighths to three inches, represent two dates and localities. The four smallest, none longer than an inch and three-fourths, were taken from the lower Fox, July 9, 1879. The others were obtained from Mackinaw Cr., in Woodford co., Ill., in the latter part of August. The situations were similar, both streams being swift and rocky where these fishes were caught.

Their food was chiefly the larvæ of Chironomus (ninety per cent.), the remaining tenth being principally made up of Alona (six per cent.). Ostracoda, Copepoda and Algæ each made about one per cent. of the food. The Algæ were mostly diatoms and desmids, Closterium being especially common. Many Difflugia and Arcella were also found in these fishes.

We trace in this a remarkable resemblance to the food of the darters, which, it will be remembered, frequent similar situations. Lacking the sucking mouth of Hypentelium, they do not take Protozoa or unicellular Algæ, but in other particulars agree closely with this species. This curious fish is peculiar among the suckers in the unusual development of the pectoral fins,—a distinguishing feature of the darters likewise,—doubtless related, in both cases, to the constant struggle with a swift current. We may also remark the darter-like glow of color in the young of this species,—a very peculiar distinction among the Catostomidæ. This is one among many facts which indicate that exposure to light has great primary effect on the color of fishes,—an effect often suppressed, through natural selection, by secondary influences, but manifesting itself where these are not brought into play.

This species is in marked *contrast* with the darters, not only in the rapidity of its growth and the ultimate size attained, but in the form and size of the head, which in the darters is small and pointed, but in these fishes is unusually large, square and strong.

The principle of adaptation has here resulted in a different line of development. While the little Etheostomatidæ have become fitted to slip and pry about beneath the stones for their food, Hypentelium has acquired the power of rolling the stones before it. As it grows larger, it resorts, of course, to deeper water, but always prefers the rocky reaches of the stream. The moulding power of natural selection could scarcely have a better illustration than that afforded by the adaptive characters, both similar and dissimilar, of these two widely separated groups of fishes.

A single small specimen of *black sucker* (*Minytrema melan-ops*) was too large properly to come within this group; but, although six inches long, most of its food was Cyclops (eighty per cent.). Other items were Alona, Difflugia, Closterium and very young Uniones.

72   The Food of Young Fishes.

Four chub-suckers (*Eremyzon sucetta*), two of which were three-fourths of an inch and two an inch and a quarter long, differed greatly in food from the foregoing. The two smaller specimens, from Long L., near Pekin, taken June 2, 1880, had eaten only Cladocera, with a trace of water mites. Chydorus was the principal element of their food (eighty per cent.), but Pleuroxus, Alona and *Scapholeberis mucronatus* were also present. In the two larger specimens, locality and date unknown, a surprising number and variety of the minutest animal and vegetable forms were found. Squamella, Anurea of several species, *Rotifer vulgaris* and other Rotifera; Difflugia and Arcella* among the Protozoa; Chroöcoccus, Closterium, Cosmarium, Staurastrum and various diatoms among the Algæ, were the principal genera. A minute Agrion larva, a very young Amphipod, and larval Copepoda (Nauplius), were the only other kinds recognized. It was obviously impossible to · make any estimate of the ratios of such minute and varied objects occurring in such great quantity, and I have contented myself with a simple enumeration.

A specimen three inches long, from Peoria Lake, in October, had eaten only Copepoda (Canthocamptus) with a trace of Chironomus larvæ.

Ten specimens of *red-horse* (Myxostoma), varying in length from an inch to two and three-fourths, taken in July and August, from the Fox and Illinois rivers and from Mackinaw Cr., show no important differences of food.

In the smaller specimens, taken from the Fox and Illinois, Entomostraca, especially Cyprids, were relatively more important, sometimes constituting nearly the whole food; but no attempt was made to fix precise ratios. In the four larger specimens from Woodford Co., tests of Difflugia were estimated to form eighty-five per cent. of the contents of the intestines. These specimens were taken one at a time, several miles apart, along a rocky part of the stream. Besides the species of Difflugia and Arcella given in the foot-note, various desmids and diatoms were abundant, with fila-

* Slides of the food of this genus and Myxostoma were submitted to Dr. Jos. Leidy, of Philadelphia, and Prof. W. S. Barnard, of Cornell University, N. Y., and these gentlemen kindly sent me the following names of Rhizopoda as occurring therein:—From Prof. Barnard, *Difflugia acuminata, pyriformis, constricta* and *globosa;* from Dr. Leidy, *D. pyriformis, acuminata, globulosa, lobistoma* and *Arcella vulgaris* and *discoides.*

mentous Algæ, rotifers (Squamella and *Rotifer vulgaris*), Cyclops, Alona, Pleuroxus and water mites, Chironomus and other Diptera larvæ, some undeterminable vegetable matter and a single Thrips (Hemiptera). The small percentage of Chironomus larvæ shows that this species has not the habit of the stone-roller.

Two specimens of the *common sucker (Catostomus commersonii)*, six inches and six and three-fourths in length, taken from Mackinaw Cr., in August and June, had eaten food so similar to that of the preceding genus that detailed description is unnecessary.

Two specimens of the commonest *buffalo fish* (Ichthyobus), seven-eighths of an inch long, had eaten most freely of unicellular Algæ (sixty-three per cent.), of which only Protococcus and Closterium were recognized. Specimens of Anurea were reckoned at twenty-seven per cent., and the remainder of the food consisted of Copepoda and Cladocera. These specimens were taken from the Illinois R., in early June.

Four *carp-suckers* (Carpiodes), seven-eighths inch to two inches long, taken from the Illinois and from Clear L., in Kentucky, had fed like the preceding genus, except that the Entomostraca were in larger quantity (forty-eight per cent.), and included a number of Ostracoda, while the rotifers were comparatively few. The Daphniidæ of the Illinois R. specimens were nearly all *Scapholeberis mucronatus*. Canthocamptus in trivial numbers was also found in a single specimen.

Reviewing the food of these thirty young suckers, we see that they differ from the other families studied in the larger food-resources open to them; for, while the structure of their mouths does not prohibit their taking Entomostraca, it enables them to draw upon the multitudes of minute organisms found upon the bottom. Evidently they have no means of selecting such microscopic structures from the mud in which these most frequently rest and considerable quantities of dirt are consequently often found in the intestines; but from the "richness" of the contents I infer that they doubtless have the power of distinguishing mud containing a large percentage of organic matter from relatively barren portions.

10

SILURIDÆ.

Numerous specimens of the young of this family show that, notwithstanding its many peculiarities of structure and habit, it is no exception to the general rule respecting the food of the young. The smallest of these specimens were from a little school of minute fry, taken in June from the friendly protection of an old oyster-can in the Illinois R. These little creatures were colorless and seemingly almost helpless, and only three-eighths of an inch in length. They had already begun to eat, however, and their stomachs were well filled with Cyclops and a few Daphnids and Chironomus larvæ. These were certainly Amiurus, but it was of course impossible to tell the species.

Other specimens of this genus, making thirteen in all, none longer than an inch and five-eighths, were obtained from various places on the Illinois, and from mud-holes in the Mississippi bottoms, in Union Co. These thirteen individuals were feeding almost wholly on Entomostraca and larvæ of Chironomus, the latter composing seventy-four per cent. and the former eighteen per cent. of their food. Twenty-two per cent. of Cladocera include *Simocephalus americanus* and *S. vetulus*, Ceriodaphnia, and *Macrothrix laticornis*,* Jur., a species not hitherto reported from this country. Among the Lynceidæ (ten per cent.) I recognized Chydorus, *Pleuroxus dentatus*, Alona and *Eurycercus lamellatus*, and among the Ostracoda a species of Candona answering precisely to the description of *Candona bifasciata*, Say. A few young Amphipoda and a few unknown insects' eggs account for the remainder of the food.

Six specimens of *Noturus sialis*, varying in length from seven-eighths of an inch to an inch and a quarter, differed from the foregoing in the much larger proportion of Chironomus larvæ (forty-one per cent.) and in the twenty-six per cent. of young *Allorchestes dentata*,—eaten by the larger specimens. These had also taken seven per cent. larvæ of Ephemeridæ. Those under an inch in length were peculiar only in the large ratio of Chironomus larvæ (sixty-five per cent.), a fact probably indicating that this species seeks its food chiefly on the muddy bottoms.

---

* Possibly this is not the species cited, but a careful comparison with the description and figures in Lilljeborg's "Crustacea ex Ordinibus Tribus," etc., failed to show any difference.

No specimens of the other genera of cat-fishes were taken small enough to show their earliest food, but as far as can be judged from the food of four specimens of Ictalurus, from two and a half to three and a half inches long, the other genera will not be found to differ especially from the foregoing.

AMIIDÆ.

A single dog-fish (Amia), one and three-fourths inches long, taken in June, had eaten seventy per cent. of Entomostraca,— about equally Copepoda and Cladocera,—and two per cent. of larvæ and pupæ of Chironomus. A few young Allorchestes and some Corixas complete the brief list.

Several specimens of Amia under one inch in length, whose anatomy I studied three years ago, I remember to have had their intestines packed with Entomostraca.

LEPIDOSTEIDÆ.

Here also I shall have to content myself with such hints of the food of the young as are given by two or three specimens, as the youngest are not yet common enough in our collections to supply more material for a study of their food. One of the two smallest gars examined, an inch and a fourth in length, taken in June, near Peoria, had filled itself with *Scapholeberis mucronata*, and the other had taken only a minute fish. A specimen two inches long and only an eighth of an inch in depth, furnished a striking illustration of the voracity of this terror of our streams, as its stomach contained sixteen minute Cyprinoids.

Summary.

A sufficient recapitulation of the foregoing data is afforded by the appended table of the food of the different genera. It may be worth while to say that all the material upon which the foregoing statements rest, as well as all that used in the preceding paper, has been carefully preserved, and may be seen at any time by those interested, at the State Laboratory of Natural History.

The general conclusion from these observations is the supreme importance of Entomostraca and the minute aquatic larvæ of Diptera as food for nearly or quite all of our fresh-water fishes,—a conclusion that gives these trivial and neglected creatures, of whose very existence the majority of the people are scarcely aware, a prominent place among the most valuable animals of the state, for without them all our waters would be virtually depopulated. Other facts of eminent interest thus brought to view are the magnitude and intensity of the competition for food among the young of all orders of fishes, where a stream is fully stocked, and the injurious character of such a species as the shovel-fish, which feeds on Entomostraca throughout its life. It is probable that all fishes which are not especially adapted to the food requirements of the more valuable fishes, are hurtful to them, because they limit the food available for the young. The sun-fishes, whose shape protects them from many enemies, and the cat-fishes, with their armor of poisoned spines, are instances in point. While their young compete with the young bass and wall-eyed pike for food, they do not furnish the latter any important food resource in later years. On the other hand, such species as the herbivorous minnows and the cylindrical suckers, which depend upon Entomostraca to a less extent when young, or take up other food at a relatively early period, are those which seem to promise best as food for the higher fishes.

It is a curious corollary from the above reasoning that a prolific species having an abundant food supply, and itself the most important food of predaceous fishes, may, by extraordinary multiplication, so diminish the food of the young of the latter as to cause, through its own abundance, a serious diminution of the numbers of the very species which prey upon it. To put this statement into more concrete form, it is not certain that the excessive increase of the gizzard shad, for instance, would be a benefit to the black bass and pike-perch which feed so largely upon it. In fact, it is clear that the great overstocking of a stream with gizzard shad would, by eventually reducing the supply of Entomostraca, cause a corresponding reduction in the numbers of all the species of that stream by starvation of the young; and this decimation, applying to all in the same ratio, would take effect upon the *ordinary* number of the other species, but upon the *extraordinary* number of the gizzard shad,—would reduce the other species below the usual limit, but might not even cut off the *excess* of the shad above that limit. Consequently, important as is the supply of food fishes for the predaceous species, it is not less important that the predaceous species should be supplied to eat up the food. Here, as elsewhere, only harm can come from an imperfect balance of the forces of organic nature, whether the excess be upon one side or the other.

In the effort to increase the valuable fishes of a lake or stream, it is not sufficient that the food of these species should be increased alone, but at the same time special measures must be taken to secure a corresponding multiplication of the predaceous fishes themselves, otherwise precisely the reverse result may be produced from that intended.

As a further illustration of some of the practical bearings of these facts, it may be noticed that the free access of fishes to the ponds, lakes and marshes connected with a stream is a matter of the highest importance. Running water is relatively destitute of Entomostraca, and hence fishes denied access while breeding to slow or stagnant water in which Entomostraca abound, have no chance to multiply. The condition of fish life in the lower Fox R. will illustrate this point. This stream takes its rise in the numerous lakes of northwestern Illinois and southern Wisconsin, but in its lower course has few branches and no stagnant waters draining into it. Its own current is swift and much of its bed is rocky, while the vast expanse of water of which it forms the outlet prevents any great oscillations of its level with the consequent flooding of adjacent lands. This part of the stream is therefore peculiarly unfit for breeding purposes, and we should expect few fish to maintain themselves in it if denied access to the immense and teeming breeding grounds of the upper part of the river. Such access is effectually cut off by several dams, unprovided with fish-ways, which have been thrown across the stream. A fish which enters the river from above therefore cannot get back to breed,—a fact which must unfavorably affect the number of fishes in both river and lakes, and is apparently one cause of an unusual scarcity of game fishes in that stream.

The Food of Young Fishes.

TABLE OF FOOD OF YOUNG FISHES.

	Perca.	Morone.	Centrarchidæ.*	Haploidonotus.	Esox.	Dorysoma.	Cyprinidæ, sp.	Semotilus.	Notropis.
Number of specimens	6	1	43	1	1	12	1	3	2
Size in inches	1@1¼	1¼	⅝@2	1⅛	1½	½@1¼	¾	⅝@1	1½

KINDS OF FOOD.	Ratios in which each element of food was found.								
I. FISHES		50			40				
Dorysoma		50							
II. MOLLUSKS			†						
III. INSECTS	08		28	100		02	75		
1. Diptera (larvæ)	08		26	75		02	75		
Chironomus	08		26	75		02	75		
Corethra						†			
2. Hemiptera (young)			†						
Corixa			†						
3. Neuroptera (larvæ)			02	25					
Ephemeridæ				25					
Palingenia				25					
IV. HYDRACHNIDÆ			†						
V. CRUSTACEA	92	50	72			60	90	25	07	100
Amphipoda (young)			02			20			
Entomostraca	92	50	70			40	90	25	07	100
Cladocera	52	40	42			40	42	25		100
Sididæ			02						
Daphniidæ	50		36			20	34	25		100
Lynceidæ	02		04			20	04		
Leptodoridæ							02		
Ostracoda			09			†			
Copepoda	40	10	19			48		07	
VI. ALGÆ						08		93	

* For detailed tables of the food of the young of this family see the preceding paper on the food of the Acanthopteri.

TABLE OF FOOD OF YOUNG FISHES — Continued.

	Hypnetelium.	Eremyzon.	Myxostoma.	Ichthyobus.	Carpiodes.	Amiurus.	Noturus.	Amia.	Lepidosteus.	TOTAL.
Number of specimens...	10	4	10	2	4	13	6	1	2	126
Sizes in inches............	$1\frac{3}{8}$@ 3	$\frac{3}{4}$@ $1\frac{1}{4}$	1@ $2\frac{3}{4}$	$\frac{7}{8}$	$\frac{7}{8}$@2	$\frac{3}{8}$@ $1\frac{5}{8}$	$\frac{7}{8}$@ $1\frac{1}{4}$	$1\frac{3}{4}$	$1\frac{1}{4}$	
KINDS OF FOOD.	Ratios in which each element of food was found.									
I. FISHES....................	......	......	......	......	......	......	......	......	50	....
II. INSECTS................	90	11	†	......	......	22	59	25	......	....
Eggs	......	04	......	......	......	......	......	......	......	....
1. *Diptera* (larvæ).........	90	......	†	......	......	18	42	10	......	....
Chironomus..................	90	......	†	......	......	16	41	10	......	....
2. *Coleoptera* (larvæ)......	......	......	......	......	......	......	10	......	......	....
3. *Hemiptera*................	†	†	†	......	......	......	......	15	......	....
Corixa.........................	......	......	......	......	......	......	07	15	......	....
4. *Neuroptera* (larvæ)......	†	07	......	......	......	......	07	......	......	....
Ephemeridæ.................	†	04	......	......	......	......	......	......	......	....
III. HYDRACHNIDÆ ...	†	03	†	......	......	......	......	......	......	....
IV. CRUSTACEA.........	09	51	†	10	48	78	41	75	50	....
Amphipoda (young).......	......	†	......	......	......	04	20	05	......	....
Entomostraca	09	50	†	10	48	74	15	70	50	....
Cladocera....................	07	49	†	05	23	22	03	35	50	....
Daphniidæ	01	02	......	......	23	11	......	35	50	....
Lynceidæ...............	06	47	†	......	......	10	06	......	......	....
Ostracoda..................	01	01	†	......	15	12	......	......	......	....
Copepoda..................	01	†	†	05	10	40	09	35	......	....
V. ROTIFERA	......	......	†	27	†	......	......	......	......	....
VI. PROTOZOA..........	†	15	†	......	†	......	......	......	......	....
VII. ALGÆ.............	01	20	†	63	63	......	02	......	......	....

THE FOOD OF BIRDS

Stephen Alfred Forbes

THE FOOD OF BIRDS.

BY S. A. FORBES.

Excluding the inhabitants of the great seas, birds are the most abundant of the Vertebrata, occupying in this great sub-kingdom the same prominent position that insects do among invertebrate animals. These two classes thus constitute exceptions to the general rule that the higher and more active animals of each group are the less abundant,—a fact doubtless largely due to the immense advantage given them by their power of flight. It is this which, by making migration possible, enables birds to choose their climates and their seasons, thus avoiding, in a great measure, one of the most destructive checks upon the multiplication of animals. Their disproportionate number, their universal distribution, the remarkable locomotive power which enables them readily to escape unfavorable conditions, and their immense activity and higher rate of life, requiring for their maintenance an amount of food relatively enormous, give to birds in their relation to the pursuits and interests of man a significance which only here and there one seems ever fully to have realized. A few figures will illustrate and enforce this proposition.

The careful estimates of three ornithologists and experienced collectors give, as an average of the whole bird-life of Illinois, three birds per acre during the six summer months. That is to say, if all the birds of the year, except the swimmers, were concentrated in these six months, equally distributed throughout them and equally scattered over the state, we should have three birds on every acre of land. It is my own opinion that about two-thirds of the food of birds consists of insects, and that this insect food will average, at the lowest reasonable estimate, twenty insects or insects' eggs per day for each individual of these two-thirds, giving a total for the year of seven thousand two hundred per acre, or two hundred and fifty billions for the state a number which, placed one to each square inch of surface, would cover an area of forty thousand acres.

Estimates of the average number of insects per square yard in this state give us, at farthest, ten thousand per acre for our whole area. On this basis, if the operations of the birds were to be suspended, the rate of increase of these insect hosts would be accelerated about seventy per cent., and their numbers, instead of remaining year by year at the present average figure, would be increased over two-thirds each year. Any one familiar with geometrical ratios will understand the inevitable result. In the second year we should find insects nearly three times as numerous as now, and, in about twelve years, if this increase were not otherwise checked, we should have the entire state carpeted with insects, one to the square inch over our whole territory. I have so arranged this computation as to exclude the insoluble question of the relative value of birds and predaceous or parasitic insects, unless we suppose that birds eat an undue *proportion* of beneficial species.

This is intended only as an illustration of the great power of birds for good or evil, and not as a prediction of the consequences of their total destruction. These consequences would not be by any means so simple, but would apparently be fully as grave.

Let us take another view of this matter. According to the computation of our first State Entomologist, Mr. Walsh, the average damage done by insects in Illinois amounts to twenty million dollars a year. These are large figures, certainly; but when we find that this means only about fifty-six cents an acre, we begin to see their probability. At any rate, few intelligent farmers or gardeners would refuse an offer to insure complete protection, year after year, against insects of all sorts for *twenty-five* cents an acre per annum; and we will, therefore, place the damage at one-half of the above amount—ten million dollars per annum.

Supposing that, as a consequence of this investigation, we are able to take measures which shall result in the increase, by so much as one per cent., of the efficiency of birds as an insect police, the effect would be a diminution of the above injury to the amount of sixty-six thousand dollars per annum, equivalent to the addition of over one and one-half million dollars to the permanent value of our property; or if, as is in fact a most moderate estimate, we should succeed in increasing the efficiency of birds five per cent., we should thereby add eight and one-fourth million dollars to the permanent wealth of the state, provided, as before, that birds do not eat unduly of beneficial species.

11

These figures will be at once rejected by most naturalists as absurdly low. The young robin of Prof. Treadwell (a bird whose fame has extended over both hemispheres) required not less than sixty earth-worms a day to keep it alive. A pair of European jays have been found, Dr. Brewer informs us, to feed their brood half a million caterpillars in a season, and to eat a million of the eggs in a winter. I have myself taken one hundred and seventy-five larvæ of Bibio from the stomach of a single robin, and the intestine probably contained as many more.

Compared with these numbers, my two thousand four hundred insects a year for each bird, seem certainly many times too few; and similar criticisms might very probably be made on other items of the estimate. I prefer, however, to put these matters with a moderation which will command general assent, especially as we see that the importance of the subject does not require exaggeration. Of course the individual farmer or gardener could, by intelligent and careful management, if he knew just what to do, increase the value of his own birds far beyond his individual share of the above-mentioned aggregate.

The subject has, also, a considerable scientific interest. Since the struggle for existence is chiefly a struggle for subsistence, a careful comparative account of the food of various competing species and genera, at different places and seasons and at all ages of the individual, such as has not heretofore been made for any class of animals, cannot fail to throw much light upon the details, causes and effects of this struggle. The flexibility of the food-habits of the widely ranging species, the direct effects of normal departures from the usual average of food elements upon the origin of variations, and the general reäctions of birds upon their organic environment, are examples of subjects upon which light should be thrown by this investigation.

That an element of such transcendent importance to all agricultural pursuits, and, through these, to the general welfare, ranking evidently among the larger forces of nature which affect powerfully and continuously the most essential interests of the country, should never have been made the subject of continuous, systematic and accurate study, seems, at first, a surprising phenomenon. It is a subject, however, presenting few attractions to the scientific student, requiring a great amount of time, a good knowledge of ornithology, a minute acquaintance with considerable parts of entomology and botany, and a good degree of skill with the microscope, while it profits the student but slightly relatively to the work done, by way of an increase of his knowledge. What little he learns is gained àt every disadvantage. His material is in the worst possible condition for study; and the personal result of his labor is a continual discouragement to him. That whatever individual impulse should have been turned in this direction should have been exhausted long before definite or conclusive results were reached, was, therefore, inevitable. The student soon turned his attention to matters more attractive and more fruitful in knowledge and reputation. In short, this is emphatically one of those questions which, if studied exhaustively at all, must be studied chiefly in the public interest.

The primary purpose of this investigation is the determination of the exact relation of the different species of birds, and of the class in general, to agriculture and horticulture; it would be disgraceful to those in charge of this investigation if the opportunity were to be thrown away which it offers for an increase of that knowledge of the habits and relations of birds whose interest is strictly scientific rather than practical, and this has therefore been held in mind throughout as a legitimate secondary purpose. We need a full knowledge of the direct and indirect benefits and injuries attributable to each species,—the ratio of benefit to injury, where both are apparent, the numbers, distribution and migrations of all, and, in fact, a full acquaintance with their entire natural history.

The direct injuries due to birds commonly take the form of depredations upon the fruits of the garden and orchard, and upon the grain in the fields. It is, of course, necessary to know the species chargeable with these, and the ratio which such injuries bear to the benefit likewise attributable to them. The good done by birds is almost wholly indirect, consisting chiefly in the destruction of insects which would become directly or indirectly injurious if allowed to live. Much of the apparent evil for which they are held responsible is also indirect; viz., the destruction of parasitic and predaceous insects which, if not destroyed, would help to diminish the numbers of injurious species. I wish, however, to call especial attention to the fact that *the regular and continuous destruction of parasitic and predaceous insects by birds is not necessarily an evil.* Paradoxical as this statement may seem, it is fully borne out by the following facts:—

The most serious losses of the farmer and gardener due to in-

sects are not consequent upon the ordinary and uniform depredations of those species whose numbers remain nearly constant, year
after year, but upon excessive and extraordinary depredations of
those whose numbers are subject to wide fluctuations. Vegetation has become so far adjusted to our crickets and ordinary grasshoppers, etc., that the foliage they eat can be spared without injury to the plant, and the damage done by them is commonly
imperceptible.* It is far otherwise, however, with the vast hordes
of the Rocky Mountain locust, of the Colorado potato-beetle, of the
chinch-bug and of the army-worm, and many other species which
occasionally swarm prodigiously and then almost disappear from
view. The injurious species are chiefly the oscillating ones, and
the dangerous species are those which show a tendency to oscillate.
Anything which tends to limit the fluctuations of an oscillating
species, or to prevent the oscillation of a stable species is, therefore, highly useful, while anything which tends to intensify an
oscillation, or to convert a stable species into an oscillating one, is
as highly pernicious.

Now a species is stable because the rate of its reproduction is
uniform, because the checks upon its increase are substantially unvarying, and because these two forces balance each other. To set
up any vibration in any one of these checks, will necessarily cause
a corresponding vibration in the numbers of the species limited by
it. More explicitly, to set up an oscillation in a predaceous or
parasitic species must produce a reverse oscillation in the species
parasitized or preyed upon. As the former increases, the latter
must diminish, and *vice versa.* But either a marked decrease or a
marked increase of a species will cause it to oscillate, unless made
with extreme slowness,—a slowness so extreme as to allow progressive adjustments of all kinds to keep pace with it.

Taking a predaceous beetle as an example, we see that a rapid
decrease of its numbers, partly relieving the species which it preys
upon from one of the usual checks upon its multiplication, will
effect an increase in those species,—will thus render the food of
the predatory insect more abundant. This will, in turn, facilitate
individual maintenance of the predatory insect and thus stimulate
reproduction, initiating a forward movement, which, proceeding at
a geometrical ratio, must continue until the predaceous species
becomes too numerous for its food, or reaches other limitations;

*See Kirby and Spence's Introduction to Entomology, 4th Ed., 1822,
Vol. I, pp. 247–258.

when destruction of the excess produced will send it back below the average line again. An oscillation will thus necessarily arise which must be reproduced in the food species connected with it.

On the other hand, if the predaceous species be suddenly increased in number by a diminished power or stringency in one of its accustomed checks, the process will simply be reversed, but the resulting oscillation will be the same. The predaceous species will increase geometrically until its food supply becomes insufficient for it, then by starvation and diminished reproduction it will be again reduced, and so on indefinitely. *Any* marked disturbance of a *fixed adjustment* between the rate of reproduction and the death rate, whether it result in increase or decrease, whether it affects a beneficial or an injurious species, is, therefore, in itself, an *immediate* evil; only to be incurred where the ultimate good is a certain and liberal compensation.

Again, it is becoming evident that carnivorous insects and insectivorous birds all have their food-preferences. Probably no one species—certainly no one family—of birds or insects would quite take the place of another. Supposing, then, that some birds eat predaceous insects, in part, as well as phytophagous ones,—eat the former, perhaps, in undue ratio,—still as the chances are practically infinite that the predaceous insects it eats would not, if allowed to live, eat precisely the same amount and kind of injurious insects as the bird itself, by destroying the bird we should merely liberate a second cause of numerous oscillations. Those species neglected by the carnivorous insects would increase beyond their bounds, and those eaten by them would be unduly diminished. It follows from the foregoing reasoning that, as a general rule, *a bird should not be discredited for the regular and established habit of destroying predaceous or parasitic insects*, unless it can be shown that those insects would, if left to themselves, check the fluctuations of some injurious species, or afford a better safeguard against the possible fluctuations of others. It must also be shown that this prospective good will not be overbalanced by some greater evil. In short, the whole burden of proof is on the side of those who would disturb the fixed order of Nature.*

The most important question respecting the relations of birds to insects is, therefore, the determination of those species of birds

---

*For a discussion of the general subject, see Herbert Spencer's Principles of Biology, Vol. 2, Pt. VI, chap. II, p. 397; and the preceding paper, "On Some Interactions of Organisms."

which serve the most useful purpose as a *constant* check upon those insects which are either injurious or capable of becoming so if they appear in largely increased numbers. Fortunately, whatever oscillations or irregularities may arise, and whatever may be their cause, the general tendency of things is towards their correction. In course of time, if new disturbances do not continually unsettle even the newest arrangements, they will usually right themselves more or less completely. The methods of this spontaneous restoration of the unsettled balance of natural forces, are, of course, worthy of the most careful study. It is only by working in harmony with them that we ourselves can help to readjust the disturbed order. A fuller treatment of this matter may best be postponed until the general discussion of results obtained by the investigation. Enough has been said to show that the subject, although complicated and difficult, will richly repay the study necessary to its mastery. A full and accurate knowledge of the mutual relations of the various forms of organic life of a region, both normal and abnormal, is certainly quite as essential to the general welfare as a knowledge of the chemistry and geology of its soils, the peculiarities of its meteorology, or any other part of the inorganic environment.

Concerning the special subject of this paper, the knowledge we need is such that we shall be able to afford for every species a tolerably correct answer to the questions, What would be the main consequences if this species were exterminated? if it were reduced to half its present numbers? What if it were doubled in number? if it were quadrupled? When this is is known, we shall evidently be able to act wisely and with the best results. That these questions are not unanswerable, I shall undertake to prove by answering them, in substance, for several species, in this paper, and by demonstrating the sufficient accuracy of the answers.

Methods.

Three methods are possible in determining the food of birds. The birds may be fed in confinement, and the kinds of food apparently preferred and the amount eaten may be noted. This evidently shows only what the bird *will* eat when restrained of its liberty, of such food as may be placed before it, and furnishes few data which we can use with safety in making up an account of its food in freedom, when foraging for itself. The state of confinement is so abnormal for a bird that on this account, also, we can rarely reason from its habits in that state to its ordinary habits. This method is, therefore, available only for the solution of a few separate questions. A far more useful method, and, in fact, the usual one, is that of watching birds while taking their natural food in the free state. Now and then a fact may be learned in this way which would escape detection in any other,—such asthe perforation of the cocoons of Cecropia by the downy woodpecker reported by F. M. Webster,*—but usually this method is of wholly secondary usefulness. The difficulty is very great of telling with certainty, in the great majority of cases, just what a bird is eating, even if one watches it with a glass. The notion of the food resulting must be distorted, as the species will be seen much more frequently and clearly in some of its haunts than in others. It is impossible by the use of this method, even to *guess* intelligently at the *ratios* of the different elements of the food,—a matter of the first importance to an understanding of the subject. It yields very few facts for the time expended, and these, in nearly every instance, could have been learned in much less time, with far greater certainty, and in far greater detail, by the following method. Finally, it affords no means of reviewing observations, but the impressions received from the hasty and imperfect glance of a moment must either be rejected wholly or must stand as verified observations.

By the third method, however, that of examining the contents of the stomachs after death, each bird usually affords a large number of objects which can be studied critically, and in detail, and can be indefinitely preserved for reference. These objects give a nearly or quite complete and impartial record of the food for some hours past,—those elements taken in a thicket or a tree-top being as evident as those taken on open ground. They are usually identifiable by the skilled student. Even very minute fragments will tell as much as the out-of-door observer can learn under the most favorable circumstances. In the great majority of cases it is possible so far to fix the kinds of food as to bring every element clearly into one of the three classes, beneficial, injurious or neutral. And here opportunity is afforded for careful and trustworthy estimates of the ratios each element bears to the other, so that the average significance of the food can be discovered. Practically, this is indispensable. Whatever method fails of this, while its results may

---

*In an unpublished paper read at the meeting of the Illinois State Nat. Hist. Soc., at Bloomington, Feb., 1880.

be interesting, and may have a certain general value, can never afford a basis for anything better than indefinite opinion. It can never settle the case for or against the birds.

This method, while by far the best of the three, has its slight disadvantages. Some things eaten by birds leave no appreciable trace in the stomach. For example, it is difficult, by this method, to determine with certainty those birds which greatly injure grapes by breaking the skin of the fruit and sipping the juice. This difficulty applies only to liquid food. Other errors may arise from the shorter or longer periods for which different kinds of food will last in the stomach; but of this we have no proof. I have depended almost wholly on this third method of investigation, because it is evidently the most profitable and reliable, and because the method of cursory observation having been resorted to heretofore, most of the recorded facts are due to it. So far as one method could correct the deficiencies of the other, it was desirable that this more tedious and laborious but more fruitful one should be given greater prominence.

The stomachs of birds shot at all times of the year and in all parts of the state, have been preserved in alcohol, each labeled with name, date and locality. The contents of these stomachs were afterwards transferred, for permanent preservation, to separate vials bearing copies of the original labels. They were then examined, bit by bit, with the microscope, with whatever powers were necessary to the fullest possible understanding of each fragment. It has been no uncommon thing to spend half a day over a single bird. Full notes of the materials found in each stomach were made on separate slips, and after this careful examination an estimate was made and recorded of the ratios of the different elements to the whole mass of the food of each individual. Objects which I was not able to identify have usually been sent to some more experienced specialist, except where determination was evidently impossible.*

These memoranda were afterwards classified and the data arranged in tabular form, so as to give a complete recapitulation and summary of the food of each species for each month. The tables thus constructed have furnished the basis for the discussion of the

*For assistance of this sort, I am indebted above all others to Prof. C. V. Riley, chief of the U. S. Entomological Commission at Washington, D. C. I have called upon him especially for the identification of larvæ, and my drafts have never been dishonored.

food of the species; and a similar tabular summary of the food of the family has been used in a similar way. Thus every fact observed appears in the final conclusion, and receives, there, its due weight.

Family TURDIDÆ. The Thrushes.*

This family consists, in Illinois, of nine species of birds; the robin, the cat-bird, the brown thrush, the wood thrush, the hermit thrush, Swainson's thrush, the Alice thrush, the mocking-bird and Wilson's thrush or the Veery. The first four of these stay with us in this latitude during the summer; the others emigrate beyond our borders, except the mocking-bird, and that only reaches the southern third of the state in any considerable numbers. I have now carefully studied the food of three hundred and fifteen specimens of this family, shot in various parts of Illinois, and in all months from February to October.

TURDUS MIGRATORIUS, L. THE ROBIN.

This bird, as familiar to every one as the domestic cat, is the most abundant of the thrushes and plays so large a part in the economy of the farm and garden as to make the question of its food one of unusual importance. The species ranges from the Atlantic to the Pacific and from the Mexican plateau to the Arctic circle, at home in all the latitudes and longitudes of this vast and varied country. I cannot, of course, attempt to determine, at present, the food of the species throughout this immense area, but shall endeavor to show only what it eats under ordinary circumstances within the limits of Illinois. The species is not strictly migratory, but is reported as wintering, sometimes in considerable numbers, as far north as the White Mountains, in New Hampshire. It occurs but very rarely in winter in central or northern Illinois, as there is at that season not sufficient food to tempt it to brave our prairie winds. On the other hand, it is comparatively rare in southern Illinois in summer, but usually abundant there in autumn and winter, so that as far as this state is concerned, it is practically a migrant within our limits. In the latitude of Bloomington its advent depends on the forwardness of the season, but it usually

---

*The general reader is referred to the "recapitulations" and the discussions of the "economical relations" of each species for the most important facts of these papers.

appears not far from the first of March, and the last of the species are gone by October 15th or November 1st.

The nesting habit of this species is so varied that no special provision need be made by those wishing to encourage its multiplication. The lower branches of orchard trees are probably its favorite situation, but it selects the most various places and uses little art or caution in the concealment of its nest.

February.

The robin appeared at Bloomington, this year, in considerable numbers, about the middle of February, the spring being an unusually early and open one.

Eleven specimens were shot at Normal, on the 27th and 28th, and their stomachs carefully searched for food. We first note that ninety-nine per cent. of the food of these birds was insects, the remaining one per cent. being spiders. About fourteen per cent. of the food of these early birds consisted of caterpillars, all of them eaten by three birds, while seventy-six per cent. taken by every bird, was the larvæ of a slow, torpid fly, abundant in early summer, closely related to the Tipulids or crane flies (*Bibio albipennis*, Say). Prof. J. W. P. Jenks, now of Brown University, found this same larva to constitute about nine-tenths of the food of the robins examined by him in Massachusetts in February and March, 1858,— a fact which indicates a remarkable fixity of food habits, unaffected by twenty years of time and a distance of a thousand miles. The caterpillars were partly cut-worms, about one-third of them being recognized as the "speckled cut-worm" (*Mamestra subjuncta*, G. & R.), a species supposed to be injurious to cabbages.*
Coleoptera occurred in the stomachs of these birds only in small numbers, comprising about four per cent. of the food. Half of these were Carabidæ, eaten by six of the eleven birds, a fourth were scavenger beetles (*Aphodius inquinatus*) and a fourth were larvæ of Lampyridæ, including one of Chauliognathus. A few fragments of curculios were also found.

Grasshoppers were present in about the same quantity as beetles, but only two birds had eaten them. One had taken *Tragocephala infuscata* and another a Tettigidea.

The Hemiptera (one per cent.) were chiefly soldier-bugs (Cyd-

*Prof. Riley, by whom my specimens were determined, says that he reared the larva on cabbage, which it ate voraciously.

nidæ), eaten by five of the birds. The spiders had been taken by two birds, and one had eaten a small thousand-legs (Iulus).

The striking feature of the month is the great predominance of the larva of Bibio in the food, a fact which will seem of small or great importance according to our views of the habits of this larva. By Dr. Fitch, former state entomologist of New York, as quoted by Prof. Jenks,* it was believed to be especially injurious to grass lands, and the robin was therefore credited with an indispensable service to the farmer. Dr. Fitch gave no actual observations, however, and his opinion was apparently speculative. Mr. Walsh † and Prof. Riley have since reported that the larva feeds only on decaying vegetation and is therefore harmless, if not indeed useful. Prof. Riley has, in fact, reared it in rotten leaves where no living vegetation was accessible. Finding the robin feeding on it so excessively in spring, I took some specimens from among the roots of grass and weeds in a raspberry garden and others from the stomach of a robin, examined the contents of the intestine with a microscope, and mounted the material for permanent preservation. These larvæ were filled with vegetation, some of which was recognized as the leaves and rootlets of the grass-like weeds of the vicinity, while the remainder evidently consisted of the leaves of net-veined plants, probably trees, by which the ground was over-shadowed. The frequency with which these tissues were found penetrated by fungi showed that this vegetation was in a decaying condition. I next looked through my notes of the contents of the stomachs of meadow larks shot at the very time when the robins were stuffing themselves with this Bibio larva, and found that the meadow larks had not eaten so much as one. As they search the ground more closely than the robin, relying almost as fully on insect food, this seemed good evidence that the larva occurs here chiefly in situations frequented by the robin and not by the meadow lark,—that is, in gardens, groves and the like. It was only in such situations that I was able to find it myself. There is, therefore, no present evidence that this larva is now injurious even in the slightest degree, and the robin is not entitled to any very positive credit for its destruction. There is some probability, however, that if the insect were allowed to in-

---

*Journal of the Massachusetts Horticultural Society, Boston, March, 1859, p. 152.

† The Practical Entomologist, Vol. 2, No. 4, p. 45, January, 1867.

crease indefinitely, it would become injurious to living vegetation; and if so, the high rate of its multiplication would make it a seriously destructive pest. The immense numbers annually destroyed by the robin may be inferred from the fact that I have counted as many as one hundred and seventy-five from the stomach of a single bird; and as fully half of the food of the robin for a month consists only of this insect, fifty larvæ a day for each robin, or one thousand five hundred for the month, will be a very moderate estimate.

About five per cent. of the food of February consisted of beneficial insects.

March.

Nine birds were shot on four different days of March, between the 9th and 31st, six of them in McLean county and three at Galena. Four of these had eaten Bibio larvæ again, which amounted to thirty-seven per cent. of the food of the month. Four birds are to be credited with the thirty per cent. of caterpillars destroyed. About two-thirds of these were cut-worms, among which *Agrotis messoria** was recognized. A few were the larvæ of Arctiidæ, probably Callimorphã. Eighteen per cent. of the food, eaten by seven of the birds, was made up of Coleoptera, two-thirds of which were scavenger beetles (*Aphodius fimetarius* and *A. inquinatus*). Carabidæ and their larvæ made but two per cent. of the food. Harpalus was the only genus distinguished. A few Histeridæ, a few wire-worms (larval Elateridæ), a soldier beetle (*Telephorus bilineatus*), and traces of long-snouted curculios† were the remaining beetles. Hemiptera were found in somewhat larger number and variety than in the preceding month. Among these were the raptatorial species, *Coriscus ferus*, and also *Phytocoris lineolaris*, *Cœnus delia* and *Euschistus servus*. The soldier bugs (Cydnidæ) made about two-thirds of the three per cent. of Hemiptera taken in this month. Grasshoppers were present in about the same amount as before, and the same species appeared in the

---

*All the cut-worms but one mentioned in this paper were determined by Prof. Riley.

†I have used throughout this paper the somewhat artificial divisions of Longirostres and Brevirostres as applied to the Rhynchophora, because nearly all the especially injurious species belong to the former section. In fact, I have not hesitated to use an obsolete classification wherever the groups thus formed correspond better to the differences of food habit or of economical value than those made by the highest modern authorities.

food. A few spiders and thousand-legs and berries of sumach (*Rhus glabra*) complete the list. The large percentages of cut-worms, Bibio larvæ and dung beetles are thus seen to be the principal features of the food of these birds. Excluding the Bibionidæ, about thirty-seven per cent. of the food was composed of injurious insects and six per cent. of beneficial species.

April.

The robin is represented in my notes of this month by seventeen birds shot at Normal, Warsaw, Elizabeth and Hanover (Jo Daviess county), Waukegan and Evanston, at various dates between the 2d and 27th. The high insect averages are still maintained. Caterpillars are nearly as abundant as before and make about a fourth of the food. Arctiidæ and Phalænidæ (measuring worms) appear in some quantity, but of unrecognized species. The larvæ of Bibio fall to eight per cent. and do not again appear in the food during the year.

A strong upward jump in the ratios of Coleoptera, which rise in this month to forty-two per cent., is doubtless due to the greater activity of beetles during this season of their amours. The effect is clearly seen by running along the line of averages for Coleoptera from February to October, viz.: 4, 18, 42, 44, 15, 9, 7, 5, 3. The upward swell which commences in March and dies away in June, corresponds to the time when the procreative impulse overcomes the usual discretion of these insects, and draws them out more freely into the open air. It is in this month that the bird makes its principal attack on the predaceous beetles, which are represented by an average of seventeen per cent., eaten by eleven of the birds. Thirteen heads of *Harpalus herbivagus*, for example, were taken from the stomach of a single robin. Other species of Harpalus, *Brachylobus lithophilus, Anisodactylus baltimorensis, Geopinus incrassatus, Pterostichus* and *Amara* were observed. Scarabæidæ also occur in unusual abundance at this time (fifteen per cent.), as might be anticipated by one who recalls the numbers in which they are now seen flying in the air. May beetles (Lachnosterna) make about half of these, and Aphodii the other half. A single bird had happened upon an interesting store of water beetles (Hydrophilidæ) which included a specimen of *Hydrocharis obtusatus*, several of *Philhydrus cinctus*, and a number of Helophori unknown to me. Rhyncophora amount to about three per cent. of the food.   Only Centrinus and *Graphor-hinus vadosus* were recognized.   Minor items were the traces no-ticed of Elateridæ, Lampyridæ and Chrysomelidæ.

Hemiptera stand at about the ordinary average (three per cent.), as usual chiefly Cydnidæ.   *Coriscus ferus*, some indeterminable Reduvid, *Podisus modestus* and *Hymenarcys nervosa* were the principal forms.   The Orthoptera (five per cent.) call for no espe-cial remark; neither do the Arachnida (one per cent.).   One bird had eaten a predaceous thousand-legs (Geophilus), and two had eaten earthworms (five per cent.).   The infrequent occurrence of the last in the stomachs of robins surprised me.   It is probably due partly to the greater digestibility of these soft worms as com-pared with the chitinized skins of insect larvæ, and partly to the fact that the greater part of those taken by the robin are fed to the young.   A few sumach berries eaten by the woodland robins shot in northern Illinois complete the dietary of the month.

The April. food of the robin is, therefore, especially noticeable for the greatly diminishing number of Bibio larvæ and the excess-ive number of beetles eaten, especially of the Carabidæ and Scarabæidæ.

May.

Fourteen birds were studied for this month, all but two of them from various parts of northern Illinois.   The record of May is sub-stantially a duplicate of the April list, except in a few particulars. The Bibio larvæ are replaced by seven per cent. of adult crane-flies (Tipulidæ) and the Carabidæ drop to four per cent., the balance being almost exactly replaced by the scavenger beetles and leaf-chafers added.   Chlænius and *Agonoderus partiarius* are among the captures of these birds.   Lachnosterna rises to its highest point in May, and is represented by seventeen per cent. of the food.   Wire-worms (Elateridæ) are likewise unusually abundant, for some unexplained reason, amounting to eight per cent.   A single robin had eaten a single potato beetle (*Chrysomela 10-lineata*), and one had taken a specimen of *Prometopia 6-maculata*. *Cænus delia* appears among the Cydnidæ and Polydesmus among the thousand-legs; and sumach berries again occur.

June.

With June the robin revolutionizes his commissariat.   The in-sect ratios, which have averaged ninety-five per cent. during the

preceding months; now drop to forty-two, and remain at or below
this point for the rest of the year; and this lack is compensated
by the appearance of fifty-five per cent. of cherries and raspberries.
The loss falls chiefly upon the Diptera and Coleoptera, the former
dropping from eleven per cent. to less than one, and the latter
from forty-four per cent. to fifteen. Among the families of Coleoptera we see from the table that it is the Scarabæidæ which benefit chiefly by this diversion of the robin's activities; for, while the
other families remain about as before, this family drops from twenty-two per cent. in the preceding month to one in this.

Taking up the details of the food of the thirteen June robins,
ranging from the 10th to the 29th, all shot at Normal, we first
notice the larger percentage of ants. These have hitherto occurred in but trifling numbers,—(three per cent. in the preceding
month),—but are now more than twice as common in the food.
This fact is doubtless due to the same cause as the still greater
relative abundance of the ants in June in the food of the bluebird,—to the abundance of the winged perfect forms of some
species at this time. Caterpillars stand at seventeen per cent.,
seven per cent being cut-worms. Carabidæ form six per cent. of
the food. Among the adults were *Callida punctata, Cratacun 'us
dubius,* Agonoderus and Anisodactylus. Wire-worms were again
numerous, four per cent being eaten by seven of the birds. Forty-seven per cent. of the food of these birds was cherries and eight
per cent. raspberries.

July.

The fourteen July birds were evidently reveling in the fruit garden, raspberries, blackberries, and currants forming seventy-nine
per cent. of the food.*

On the other hand, but twenty per cent. of the food was insects
and one per cent was spiders. The caterpillars furnish only four
parts of the food, and beetles but nine parts, of which two-thirds
were Carabidæ. Evarthrus, Pterostichus and Amara were noticed
among these. Scarabæidæ, Elateridæ, and Rhyncophora each one
per cent., a mere trace of Hemiptera, four per cent. of Orthoptera
(chiefly crickets), eaten by two of the birds, and one per cent. each
of Arachnida and Myriapoda are the remaining trivial details.

*I have not ordinarily attempted to distinguish raspberries from black-
berries in the stomachs of birds, but have set down either one of the other,
according to the advancement of the season.

August.

This month is represented by twenty birds, all shot at Normal,* at repeated intervals from the fourth to the thirtieth. With the disappearance of blackberries, the food of this bird returns substantially to the status of June. Insects increase again to forty-three per cent. and fruits fall to fifty-six. Ants remain at the usual point of insignificance, caterpillars rise again to seventeen per cent., about two-thirds of them Noctuidæ. Coleoptera figure at seven per cent., only two per cent. being Carabidæ. *Rhyncophora* rise to four per cent., eaten by nine of the birds; and, except a stray Nepa picked up by one robin, Hemiptera appear in trifling quantity. Crickets and grasshoppers are more abundant, amounting to ten per cent. of the food.

The cherries made forty-four parts of the food of the month, eaten by fourteen of the birds, *but two-thirds of these cherries were wild.* Tame grapes make three per cent. of the food, berries of the mountain ash about four per cent., and blackberries from the woods not far from five per cent.

September.

Twelve birds, all but one shot at Normal, September 25th, and that one at Aurora on the 13th, show no more remarkable peculiarity than the substitution of ants for most of the caterpillars, the former composing now fifteen per cent. of the food, and the latter but five. The ants were largely winged, but of different species from those taken most freely in June.† The Carabidæ of this month were chiefly larvæ. Among the Hemiptera (three per cent.) were found *Mormidea ligens* and *Cœnus delia.* No trace of spiders or myriapods was found, and only two per cent. of grasshoppers. The fruits stand at seventy per cent., fifty-two per cent. being grapes and the remainder berries of the mountain ash and moonseed (Menispermum).

---

* The general cessation of taxidermist's field work in mid-summer has prevented the supply of any material for this month and the preceding, except that obtained by ourselves in McLean county.

† Examining the tables of food of the blue-bird, brown thrush and robin, I find throughout a curious inverse relation between the ratios of ants and caterpillars, the latter falling away in June to about the same degree that ants increase during the time of their most conspicuous activity. I cannot even guess why ants should thus replace caterpillars in the food.

October and December.

The robin commences to withdraw to the south in October, and his operations in central Illinois have little interest during this month. At Normal the species became rare earlier than usual this year, and but three specimens were secured. These were feeding largely on wild grapes (fifty-three per cent.) and ants (thirty-five per cent). Six per cent. of the food was caterpillars and two per cent. wire-worms (Elateridæ). I have seen the bird eating apples in all the autumn months, but have never found the remains of this fruit in the stomach, and doubt if any especial harm is done in this way.

A single bird shot at Cairo in December, piping loudly from a tree-top for company, the only one of the entire family seen during a week's winter shooting in southern Illinois, had evidently been feeding on the berries of the mistletoe. By the inhabitants of that region, troops of robins which commonly winter there were said to have gone south in November, a fact attributed by them to the failure of the wild grapes in the woods that year.

Recapitulation.

The food of the robin, as indicated by the stomachs of one hundred and fourteen specimens, consists almost entirely of insects from February to May inclusive, but from that time forward these make but little over a third of its food, the remainder (sixty-four per cent.) being composed of fruits, tame and wild, in varying proportions, according to the local situation and surroundings. Insects make almost precisely two-thirds of the food of the year, taken as a whole.

In early spring the bird depends chiefly for food upon the larvæ of a single species of fly (*Bibio albipennis*, Say), which it picks from among the leaves and roots of grass and weeds in gardens, and similar situations. In February this made three-fourths of the food of eleven specimens, and in March more than a third of the food of nine. While this larva is not at present injurious, but feeds ordinarily on decaying vegetation, it might possibly do injury to meadows and pastures if allowed to multiply without restraint.

But few ants are eaten by this bird until late in the fall, when

13

the swarming of the sexual forms of some of the species seems to attract its appetite, in the relative dearth of other insects.

Caterpillars make up, in March, April and May, fully a fourth of its food, about half of these being cut-worms and other similar forms. Later, these are largely given up for fruit, and in the latter half of the season make only about one-tenth of the food. The average of caterpillars for the year is seventeen per cent.

Beetles, commencing at four per cent. in February, when but few specimens have yet been aroused from their cold winter's sleep, rise to forty-four per cent. in April and May, when their procreative energies are most active and urge them out into the air in swarms. With the appearance of the small fruits, these, also, are neglected by the robin, and the average for the last four months of the season falls away to six per cent., eighteen being that for the year.

This discrimination affects chiefly the scavenger beetles and the "June beetles," the other families maintaining about their original numbers throughout, with only an upward wave in April. The predaceous beetles average six per cent. of the food of the year, the leaf-chafers three per cent., the wire-worms two per cent., and the snout-beetles one per cent.

The robin's depredations upon the true bugs (Hemiptera) are but trivial, amounting only to three per cent. of the food, but nearly all of these belong to species regarded more or less positively as beneficial.

The ratio of grasshoppers and crickets (four per cent.) seems trivial, at first sight. We note, however, that these were eaten by twenty-six of the birds, and that, consequently, at least twenty-six of the insects must have been destroyed. Remembering that these figures are based upon a single day's food, or even less, for each bird, we see that these robins were eating at an average rate of at least twenty-six grasshoppers or crickets a day, for seven months, giving us a minimum total of 5,500 Orthoptera for the year.

Only one per cent. of the food was spiders. Thousand-legs were eaten by eight of the birds, and by these in merely trivial quantity.

Coming now to the fruits, we find that tame cherries, blackberries, raspberries, currants and grapes, excluding wild fruit of all descriptions, make about one-fourth of the food of the species for the year, the wild fruits making another tenth. In the absence of the latter, the robins would doubtless attack the garden fruits more vigorously.*

Concerning these general statements, the all-important question is, of course, the sufficiency of their basis.

Granting that the observations have been exactly made and correctly generalized, how far may the conclusions reached be expected to hold good in the future? These conclusions actually rest upon the food of a hundred and fourteen birds for probably about half a day each. Can we safely reason from these to the food of the thousands and hundreds of thousands of robins of the state, day after day, the whole season through?

In a paper published last winter in the Transactions of the Illinois Horticultural Society, I made the following reply to substantially the same question:—

" If the same species will eat substantially the same food, year after year, in the same situation, then, of course, a good deal may properly be inferred from comparatively few data; but if the food varies widely, either arbitrarily or under slight changes of condition, then we can infer but little. Upon this fundamental question I have two suggestions to make.

" First, if several species allied in structure, occupying the same territory at the same time, living side by side, with the same sources of food supply open to them, are found, on the examination of a limited number of stomachs, to present several characteristic differences of food, so that the investigator can point out definite peculiarities of the food of each species, and finds these peculiarities reasonably constant, year after year, then we may say unquestionably, without going farther, that there is a fixity of food-habits in this group of birds which will allow us to reason from the data observed. . . . . .

" Second, if there are any other habits of the species in which there does not seem to be any greater reason for invariableness

---

* No man should needlessly sacrifice a wild cherry-tree or a fruiting vine or shrub of any kind. Ordinary common sense would teach the preservation of as much of the worthless natural food of frugivorous birds as possible, as a diversion from the cultivated fruits of the orchard and garden.

than in those relating to the food, which are nevertheless found to be substantially unvarying, then we may, with considerable force, argue the probability of a like unvarying character in the habits of alimentation.

"Respecting the first of these tests, you will see, when I sum up the food of the family now under consideration and bring the data respecting the various species into comparison with each other, that I have made out certain very well-marked specific differences of food, even among those eating at the same table; that the different species of this group, while agreeing in many particulars of food as they do in structure, present also certain peculiarities, so marked that I can usually determine the species by the contents of three or four stomachs.

"For the second test we may properly use the nesting habit. There seems to be no more cogent reason why one species should select from the same store-house different materials for its nest from those used by another closely allied species of nearly the same size and similar general habits, and building in the same locality, than why each should use a similar fixed discrimination in selecting its food. Yet no expert, scarcely a school-boy even, will hesitate a moment between the nest of a robin and that of a cat-bird; and the descriptions of the two given in the books are so different as to enable any novice to distinguish between them at a glance. In fact, a friend mentions, as I write, two birds whose nests are much more easily distinguished than the birds themselves."

I have now to add what we may regard as a decisive crucial test of the conclusion implied above. In the paper quoted from, I gave the details and a summary of the food of forty-one robins in a table similar to those presented in this paper, and a comparison of the averages of that table with those of the table on pages 104, 105, 106, 107, may be easily made. While any serious differences in the averages of these two tables would not necessarily condemn the later one, but, at the worst, would leave its sufficiency in doubt, a substantial agreement of the two would be conclusive proof of the correctness of both. It is incredible that the averages of a hundred and fourteen specimens should agree essentially with those of forty-one, unless both were framed upon identical principles and were sufficiently true to the facts for all practical purposes. I will, therefore, place the principal averages of these tables side by side, premising that the later table not only includes nearly three times as many specimens as the earlier, but covers two months' more time.

The figures for the first and second tables, taken alternately, are as follows:—Insects, seventy per cent. and sixty-five per cent.; caterpillars, eighteen per cent. and seventeen per cent.; Diptera, eighteen per cent. and seventeen per cent.; Coleoptera, nineteen per cent. and eighteen per cent.; Carabidæ, seven per cent. and five per cent.; Scarabæidæ, four per cent. and seven per cent.; Lachnosterna, two per cent. and three per cent.; Elateridæ, three per cent. and two per cent.; Rhyncophora, three per cent. and two per cent.; Chrysomelidæ, one per cent. and a trace; Hemiptera, four per cent. and three per cent.; Orthoptera, eight per per cent. and four per cent.; Arachnida, a trace and one per cent.; Myriapoda, two per cent. and a trace; garden fruits, twenty-eight per cent. and twenty-nine per cent.

As I did not discriminate, in the former table, between tame and wild edible fruits, I have included the latter in both, and excluded the inedible fruits. I believe that the agreement in these figures, taking into account the earlier and later months covered by the second table, is quite remarkable, and can be explained only on the supposition that the fuller table presents a reasonably accurate summary of the food of the robin as a species in at least the northern half of the state, and under the ordinary conditions of the last five or six years. Of course, I had no idea how these averages were coming out until my notes were finished and the ratios were calculated for the whole.

ECONOMICAL RELATIONS.

We come now to the intricate, delicate and difficult question of the economical relations of this species,— a question rendered less important by the general considerations urged elsewhere, but, nevertheless, deserving careful attention. While it is true that every insectivorous bird must be respected, whatever its other habits, at least until we clearly understand its function in the general order and are certain that its removal will do no harm which we can not remedy or endure better than we can support its injuries, yet an idea of the relative importance of edible fruits and insects of both the beneficial and injurious classes in the diet of the bird is necessary as a step to this clear and complete understanding of the matter.

Glancing at the bottom of the table of the food of the species, on page 107, the reader will see three lines of figures running across the page, showing for each month the percentages of beneficial, injurious and neutral species of insects and fruits eaten by these birds. The figures at the right give similar percentages for all the birds for the entire year. Following the upper line, we note the small percentages of injury done in the early spring, the marked increase of injury in April, due to the excessive destruction of predaceous beetles, and the heavy percentages of the fruiting months. The general average of beneficial elements destroyed for the year is thirty-six per cent. On the second line we notice an inverse variation. Commencing with a ratio of ninety-four per cent. of injurious elements eaten in February (if we include the larva of Bibio in these), the record runs down to seven per cent. in September, the general average for the year being forty-three per cent.

This comparison, however, is merely a quantitative one. Injurious or beneficial elements are balanced against each other according to their bulk and not their quality. A quart of caterpillars counts as the equivalent of a quart of blackberries, and, on the other hand, as the equivalent, also, of a quart of predaceous beetles. It is evident, therefore, that we cannot get at any close estimate of the economical values of this species in this indiscriminate way.

A nearer approximation to the truth may be made by critically comparing the general averages for the year found in the vertical column at the right of the table. Here we have the following totals of injurious and beneficial species:—Of the first, caterpillars, seventeen parts (including eight parts cut-worms); Bibio larvæ, fifteen parts; leaf-chafers, three parts; wire-worms, two parts; snout-beetles, two parts; crickets and grasshoppers, four parts. Of the second, predaceous beetles, six parts; predaceous bugs, three parts; garden fruits, twenty-four parts. Now, the opinions of entomologists would probably be found to differ somewhat widely on the question of the relative values of these various elements, and each must form his own opinion from the data given.* My own judgment is that, taking into consideration only the immediate present effect of the robin upon the fruits and insects of the state, ignoring for the moment the important secondary disturbances likely to arise if the number of the species were greatly lessened, and balancing these elements carefully against each other (applying to them, in fact, the operation of cancellation in arithmetic), we can reduce the question finally to about this form:—Will the destruction of seventeen quarts of average caterpillars, including at least eight quarts of cut-worms, pay for twenty-four quarts of cherries, blackberries, currants and grapes?

To this question I, for my own part, can only reply that I do not believe that the horticulturist can sell his small fruits anywhere in the ordinary markets of the world at so high a price as to the robin, provided that he uses proper diligence that the little huckster doesn't overreach him in the bargain. In other words, while the bird is far too valuable to exterminate, at least until we are sure we can replace him by some cheaper assistant, yet he is not so precious that we need hesitate to protect our fruits from outrageous injury. Indeed, it seems likely that the ordinary destruction of robins by gardeners does not more than compensate for the destruction of birds of prey in the interests of the poultry yard,— removing that excess of robins which, in the more natural order, would fall victims to the hawks and owls.

---

* Concerning the value of predaceous beetles, the reader is especially requested to examine the papers on that subject in the present bulletin. It is probable that their services have been greatly over-estimated.

TABLE OF THE FOOD OF THE ROBIN. (*Turdus migratorius*, L.)

KINDS OF FOOD.	Jan.	Feb.	March.	April.	May.	June.	July.	August.	Sept.	Oct.	Nov.	Dec.	TOTAL.	Ratio of each element to whole of food.
Number of specimens	11	9	17	14	13	14	20	12	3	...			1,114	
I. MOLLUSKS				1	1									
				01	†									
II. INSECTS	11	9	17	14	13	13	18	9	3				107	.65
	.99	.97	91	94	42	20	43	30	44					
1. *Hymenoptera*		1	5	7	6	3	10	7	2				41	.04
		†	02	03	07	01	03	15	35					
Apidæ			1				1						2	
			†				†							
Formicidæ			4	7	6	3	7	7	2				36	.04
			02	03	07	.01	02	15	35					
Ichneumonidæ						1	†						1	
Chalcididæ		1											1	
		†												
2. *Lepidoptera* (larvæ)	3	4	12	11	8	3	11	2	2				56	.17
	.14	.30	.24	23	17	04	.17	05	06					
Arctiidæ		1	3										4	
		.01	.02											
Noctuidæ	1	3	1	6	4			7	1				23	08
	.05	.19	.04	15	07		12	01						
Phalænidæ			1										1	
			05											
3. *Diptera*	11	4	6	4	1	1	1						28	17
	.76	.38	12	.11	†	01	†							
Tipulidæ				2									2	01
				07										
Bibionidæ	11	4	3										18	15
	.76	.37	08											
4. *Coleoptera*	8	7	16	11	11	8	14	3	3				81	18
	.04	.18	.42	44	15	09	07	05	03					
Carabidæ	6	5	11	6	6	3	7	2	1				47	05
	.02	.02	.17	04	06	06	02	03	01					
·Harpalidæ	4	4	11	6	5	2	5	2	1				40	05
	.01	.02	16	04	01	06	02	03	01					
Larvæ		1					1	2	1				5	
		.01				†	01	02						
Dytiscidæ							1						1	
							†							
Hydrophilidæ			1										1	
			01											
Staphylinidæ			1										1	
			01											
Histeridæ		1	8	1		1							11	01
		01	05	†		†								

TABLE OF THE FOOD OF THE ROBIN. (*Turdus migratorius, L.*)—Continued.

	Jan.	Feb.	March	April	May	June	July	August	Sept.	Oct.	Nov.	Dec.	Total	Ratio of each element to whole of food.
Number of specimens	11	9	17	14	13	14	20	12	3	...			1,114	

KINDS OF FOOD. — Number of specimens, and ratios in which each element of food was found.

Kinds of Food.	Jan.	Feb.	March	April	May	June	July	August	Sept.	Oct.	Nov.	Dec.	Total	Ratio
Nitidulidæ							1	1					2	
							†	†						
Scarabæidæ	3	4	13	12	4	3	1						40	07
	.01	12	.15	22	01	01	01							
Lachnosterna			2	6	1	1							10	03
			07	17	†	†								
Elateridæ		2	2	4	7	2	1	1	1				20	02
		01	†	08	04	01	†	†	02					
Lampyridæ	2	1	1		1								5	
	.01	01	1		02									
Rhyncophora	2	2	5	7	4	1	9						30	02
	†	01	03	02	01	01	04							
Brevirostres			3	4									7	
			02	01										
Longirostres	1	2	3	4			2						12	
	†	01	01	01			02							
Chrysomelidæ				1	2	2		1		1			7	
				†	01	01		†		†				
Doryphora					1								1	
					01									
5. Hemiptera	5	3	5	6	3	2	6	5					35	03
	01	03	03	05	01	†	06	03						
Nepa								1					1	
								5						
Coriscus		1	1	1				1					4	
		†	†	†				†						
Reduviidæ		1	1	1	1			2					6	
		†	†	†	†			†						
Phytocoreidæ	1							1					2	
	†							†						
Lygæidæ (Blissus)								1					1	
								†						
Coreidæ	1	1											2	
	†	†												
Cydnidæ	5	3	5	4	1	1	3	4					26	02
	01	02	.03	04	†	†	†	02						
6. Orthoptera	2	2	6	5	1	2	6	2					26	04
	04	05	05	04	01	04	10	02						
Gryllidæ				1			1	3					5	01
				01			03	06						
Acrididæ	2	2	3	5	1	2	3	2					20	03
	.04	05	03	04	01	01	04	02						
III. Arachnida	2	1	3	1			4	3					13	01
	01	01	01	†			01	01						

14

TABLE OF THE FOOD OF THE ROBIN.   (*Turdus migratorius, L.*)—Continued.)

	Jan.	Feb.	March	April	May	June	July	August	Sept.	Oct.	Nov.	Dec.	TOTAL	Ratio of each element to whole of food.
Number of specimens	11	9	17	14	13	14	20	12	3	...		1	114	

KINDS OF FOOD.	Jan.	Feb.	March	April	May	June	July	August	Sept.	Oct.	Nov.	Dec.	TOTAL	Ratio
IV. MYRIAPODA		1/†	3/01	3/01	1/02								8	
Geophilus			1/†	1/01									2	
Polydesmus					1/†								1	
Iulidæ		1/†	2/01	2/—/2	1/01								6	
V. EARTHWORMS (*Lumbricus*)				2/05									2	
VI. FRUITS AND SEEDS	1/†	2/01	1/01	3/04	13/58	14/79	17/56	11/70	3/56				65	34
Blackberries							12/56						12	07
Raspberries						3/08	1/05						4	02
Cherries						10/47	*14/44						24	11
Currants						6/17							6	02
Grapes							1/03	7/52	b2/53				10	07
Mistletoe (Phoradendron)												1/100	1	
Mountain Ash								1/04	1/08				2	01
Sumach (Rhus)			2/01	1/01	2/04								5	01
Hackberry (Celtis)								1/05					1	
Moonseed (Menispermum)									2/04				2	
Polygonum			1/†										1	
Grass	1/†				1/†	1/02							3	
Corn						1/01							1	

* 28 per cent. wild.   b All wild.

TABLE OF THE FOOD OF THE ROBIN. (*Turdus migratorius, L.*)—Concluded.

	Jan.	Feb.	March.	April.	May.	June.	July.	August.	Sept.	Oct.	Nov.	Dec.	TOTAL.	Ratios.
						Percentages for each month.								
Beneficial species...............	...	05	06	21	09	°64	°85	°50	°57	.....	.....	.....		36
Injurious species	...	*94	*74	*47	55	24	10	31	07	.....	.....	.....		43
Neutral species................	...	01	20	‡82	‡36	‡12	‡05	‡19	‡36	.....	.....	.....		21

* Includes Bibio.   ° Includes fruits.   ‡ Includes ants.

[NOTE.—In the foregoing tables, the integers indicate the number of birds found to eat the element against which they are placed, and the decimals express the ratio of this element to the whole food of the month.   October and December were omitted in computing the general averages for the year, on account of the small number of birds examined for those months.]

MIMUS CAROLINENSIS, L.   THE CATBIRD.

This bird, scarcely less abundant than the robin, arrives later and makes a shorter stay, coming late in April or early in May, and disappearing from this latitude usually in September.   It also occupies a larger territory in the state in mid-summer than the robin, being not at all rare in extreme southern Illinois in July and August.   I do not know that it ever winters northward.   Its habits and favorite haunts are so similar to those of the robin that one might not unreasonably anticipate that, respecting their food, both could be considered as one species; but we shall see proof that there are specific food characteristics to separate them.

How indefinite and uncertain is the present knowledge of the food of this especially notorious species, may be seen by comparing my notes with the statement made in the recent and elaborate work of Baird, Brewer and Ridgeway.

" The food of the cat-bird is almost exclusively the larvæ of the larger insects.   For these it searches both among the bushes and the fallen leaves, as well as the furrows of newly-plowed fields and cultivated gardens.   The benefit it thus confers upon the farmer and upon the horticulturist is very great, and can hardly be overestimated."

My observations of this bird cover the five months from May to September, inclusive.

May.

The specimens of this month range from the 1st to the 31st, and from Warsaw and Normal, in central Illinois, to Savanna, McHenry and Waukegan in the northern part of the state. Five of the birds of the month were taken in northern Illinois and seventeen in the central part of the state. All of these birds had eaten insects, which amounted to eighty-three per cent. of the food, the remainder consisting of spiders, three per cent.; thousand-legs (*Myriapoda*), seven per cent.; and seven per cent. of the dry berries of the sumach (*Rhus glabra*). Among the insects were about equal ratios of ants, crane-flies and beetles, the first composing eighteen per cent. of the food, the second nineteen and the third twenty-three. Caterpillars formed twelve per cent. of the food, and about one-sixth of these were distinctly recognizable as cut-worms (*Noctuidæ*). More than one-third of the beetles were Carabidæ, including specimens of Platynus and *Harpalus pennsylvanicus*. Only one per cent. of the food consisted of Scarabæidæ, and five per cent. of snout-beetles (*Rhyncophora*). Nearly all of the latter belonged to the section Brevirostres, in which are found few of the injurious species of the group. Those recognized were *Graphorhinus vadosus* and *Ithycerus noveboracensis.* Among the one per cent. of plant-beetles (*Chrysomelidæ*) only *Gastrophysa polygoni* was specifically determinable. Minor items among the Coleoptera are the water-beetles, including *Colymbetes biguttatus* and an undetermined species of *Hydrobius.* The *Hemiptera* amounted to only one per cent. of the food, and all of these were *Cydnidæ.* The *Orthoptera*, including a few specimens of the white cricket (*Œcanthus*) and of the common spring grasshoppers, amounted in all to four per cent. of the food. A single specimen of the young of the walking-stick (*Diapheromera femorata*) had been eaten by one of the birds. Spiders amounted to three per cent. of the food. The *Myriapoda* included several specimens of *Lithobius* and three species of Polydesmus, viz: *P. serratus, P. virginiensis* and *P. canadensis.*

It will be seen at once that the striking feature of the food of this bird in May, as compared with that of the robin, is the abundance of ants and crane-flies, a characteristic which we shall find persistent until the opening of the fruit season revolutionizes the food of both species.

June.

The food of June undergoes so complete a change when the small fruits begin to ripen that the record may best be given in two divisions, the first of which agrees closely with that of May, while the second approaches more nearly to that of July. In the first part of the month, ants were eaten by the nineteen birds examined in about the same ratio as in May. Crane-flies appear in the food only in the early days of the month. Among the Coleoptera the principal peculiarity is the greater importance of the May-beetles (Lachnosterna). A few strawberries and cherries were eaten by this bird previous to the fifteenth of the month, but these fruits were not taken in sufficient amount materially to influence the averages. After the seventeenth, however, only one per cent. of the food consisted of ants, and only about three per cent. of caterpillars. The May-beetles disappear almost entirely, and the other insect elements are reduced to equal insignificance, while the same fruits constitute by far the larger part of the food. These include currants and cherries in about equal parts, and about twice as many raspberries as of both the others taken together. Treating the food of the month as a whole, we find that forty-nine per cent. of it consists of insects, three per cent. of spiders and three per cent. of thousand-legs, while forty-five per cent. consists of fruits,—twenty-one per cent. being raspberries, twelve per cent. cherries, three per cent. strawberries and eight per cent. currants. The ants of the month amounted to but eleven per cent. and the crane flies to seven per cent. The Lepidoptera stand at ten per cent. and the Coleoptera at seventeen,—nearly one-third of the latter being Carabidæ. The Hemiptera made about one per cent. of the food and the Orthoptera two per cent. A single bird louse (*Mallophaga*) was found in the stomach of one of these birds.

July.

The record of this month rests upon eleven specimens, all from central Illinois, taken from the first to the twenty-third of the month. These indicate most clearly an eminent preference of the species for the small fruits, which composed three-fourths of their food; sixty-four per cent. being blackberries alone. Spiders and myriapods, are found in about the same ratio as in June. The latter are all Iulidæ, a part of them, at least, belonging to the genus Iulus. The only Orthoptera noted were specimens of the large black cricket of the fields (*Gryllus abbreviatus*) eaten by a single bird. The *Hemiptera* almost disappear, a single Thrips being the only representative of the order. The Coleoptera amounted only to nine per cent. of the food, and more than two-thirds of these were predaceous beetles, eaten by eight birds; among these were noted *Cicindela lecontei*, Pterostichus, Evarthrus, *Cratacanthus dubius, Anisodactylus baltimorensis* and Harpalus. Only a single bird had taken caterpillars, which constituted three per cent. of the food of the month. No trace of Diptera was found in the stomachs of these birds, and only four had eaten ants, which made two per cent. of the total food. Insects proper thus amounted to eighteen per cent. of the whole.

It is clear, from the foregoing, that the cat-bird in mid-summer eats only such insects as come in its way while regaling itself on the smaller fruits.

August.

Twelve birds were obtained in this month, the first on the 7th and the last on the 30th, all from McLean and adjoining counties. Three of these were young, but as no difference of food was noticed corresponding to age, these are not treated separately.

The food record of August resembles that of June, owing, doubtless, to the diminution of the smaller garden fruits at this time and to the fact that the wild fruits have not yet generally come into bearing. The insect percentages are, therefore, much larger than in July, and it is instructive to notice that this increase is first apparent and most evident in the ratios of ants, an indication of the positive preference of the cat-bird for this food. Nearly one-half of the forty-six per cent. of insects eaten in this month were ants. A bee, a gall-fly and an ichneumon were noticed among the other Hymenoptera. Forty per cent. of the food was caterpillars, a considerable proportion of which were cut-worms. Only six per cent. of the food was Coleoptera, and the only predaceous beetle taken by these birds was one specimen of *Cratacanthus dubius*. Three per cent. of the food was scavenger beetles, including Geotrupes and *Bolbocerus farctus*. It is in this month that the Meloidæ appear abundantly on golden-rods and other Compositæ; but only a single Epicauta was found in the food of one of these birds. The few plant beetles noticed included a single *Diabrotica vittata*. Seven per cent. of Hemiptera were eaten; largely chinch bugs, taken by one of the birds. This fearful pest of the grain-fields was sufficiently abundant in the vicinity of Normal this year sensibly to injure the crops of grain. Nearly all the species of birds examined were found to eat them to some extent, but in quantities so trifling as probably to have little or no effect upon their multiplication. It is evident, however, that the birds have no especial prejudice against them. The remainder of the Hemiptera were the ordinary "soldier-bugs," belonging to the genus Euschistus.

Orthoptera appear in somewhat larger ratio, amounting to seven per cent. of the food, an indication, doubtless, of the commencement of the autumnal multiplication of this order which will be found reflected to a very notable degree in the food of the blue-bird further on. Only traces of spiders and thousand-legs were discovered. Fifty-four parts of fruit were eaten, sixteen of which were wild. Nearly all of the garden fruits were blackberries,— cherries constituting but three per cent. of the food for the month.

September.

The cat-bird leaves our latitude in September, and only six specimens were secured,— all of them on or before the 17th, in the vicinity of Normal and Bloomington. The chief peculiarity of the food of the month is the substitution of cherries and wild fruits for blackberries. Seventy-six per cent. of the food at this time consisted of fruits, all wild but the grapes, which amounted to fourteen per cent. Elderberries, wild cherries and the fruit of the Virginia creeper were the most important elements. Carnivorous thousand-legs amounted to three per cent. of the food and insects proper to twenty-one per cent., nearly half of which were ants. But few caterpillars had been eaten by these birds, and only seven per cent. of Coleoptera,— five per cent. being Harpalidæ. The lower orders of insects were conspicuous only by their absence.

We are now prepared for the review of the general averages of the season, and the indications which these afford of the economical value of the cat-bird. Taking the record of the year together as found in the vertical column at the right of the table on pages 116, 117, 118, the seventy birds of the species examined are found to have eaten forty-three parts of insects, two parts of spiders and harvest-men, three parts of thousand-legs and fifty-two parts of fruits. Only thirty-three per cent. of the food consisted of tame fruits, four per cent. being raspberries, twenty per cent. blackberries, one per cent. currants, four per cent. tame cherries, one per cent. strawberries and three per cent. grapes. Scrutinizing more closely the details of the insect food, we find that ants form twelve per cent. of the total for the season; Diptera, chiefly crane-flies, about five per cent.; Lepidoptera six per cent., and beetles twelve per cent., one-third of which are Carabidæ. The scavenger beetles and leaf-chafers are three per cent. of the food; plant-beetles, one per cent., and snout-beetles, belonging chiefly to the leaf-eating Brevirostres, likewise one per cent. Two parts of Hemiptera and three of Orthoptera are the only other items that we need notice. It will be seen that ants and beetles occur in about equal ratios, and-that these are the most important insect elements in the food. Diptera and Lepidoptera taken together about equal one of the former elements.

Recapitulation.

In the cat-bird as in the robin the insect averages are highest in the early months, and fall rapidly away from May to July,—rising again in August and declining in September. The ratios of insects taken for the five months covered by this table are as follows:- 83, 49, 18, 46, 21. The same double curve is especially apparent in the averages of ants, the corresponding ratios for which are 18, 11, 2, 20, 9. Beetles gradually diminish to July and then remain tolerably constant for the season. The predaceous ground beetles maintain themselves at nearly uniform figures throughout. The Scarabæidæ are, of course, most abundant in May and June, when the leaf-chafers are abroad. The snout-beetles observed were all taken in the months of May and June, and belonged chiefly to species whose injuries are confined to the leaves of trees. Only trifling ratios of plant beetles were eaten by these birds. Hemiptera also occur in insignificant quantity, the only notable fact being the presence of chinch-bugs in the food of one bird. Orthoptera seemed to be most abundant in the late and early months, diminishing in June and July. Considerable numbers of Arachnida and Myriapoda are eaten by the cat-bird,—a point in which it contrasts notably with the robin. No earth-worms were detected in the food. With respect to the fruits taken by this bird, we find that the general ratios for the corresponding months agree closely with those of the robin. Berries of the sumach are eaten in May, but raspberries and blackberries are the most prominent elements of June, July and August. Wild cherries take the place of these fruits in September, and grapes are then eaten to some slight extent.

A comparison of the statements of this paper with the report published in the Transactions of the Illinois Horticultural Society for 1879, will give some interesting results. The former paper relates to thirty-seven specimens, obtained during the three months of May, June and July; and the present paper relates to seventy birds, taken during five months from May to September. As both the additional months extend the fruit season, we should expect the insect averages would now be smaller than before and that the averages of fruit would show a corresponding increase. This I find to be the principal difference between these tables. The various insect elements stand in about the same ratio to each other as before, except the ants (whose swarming in autumn accounts for their greater prominence in the food), and the Hemiptera and Orthoptera. The first of these orders figures more largely in the general averages for 1880 because this was a "chinch-bug year" in central Illinois; and the second because grasshoppers, locusts and crickets greatly increase in numbers during the later months. In the earlier table, insects amount to fifty-six per cent. of the food; in the later, only to forty-three; ants are respectively ten and twelve, Diptera thirteen and five, Lepidoptera ten and seven, Coleoptera nineteen and twelve, Carabidæ eight and five, leaf-chafers four and three, snout-beetles three and one, Hemiptera one and two, Orthoptera two and three, Arachnida three and two, Myriapoda six and three and the edible fruits twenty-seven and forty-one.

The Cat-bird and the Robin.

In order to a more exact comparison of the food-habits of the cat-bird and the robin, I have computed the averages of the principal elements of the robin's food for the period of five months covered by the cat-bird's record, and give these here alternately with the corresponding averages of, the cat-bird. The ants eaten by the robin during these months amounted to five per cent. of the food, and those by the cat-bird to twelve per cent. Diptera were two per cent. and five per cent., Lepidoptera thirteen per cent. and seven per cent., Coleoptera thirteen and twelve, Carabidæ four and five, leaf-chafers three and two, wire-worms three and a trace, snout-beetles two and one, Hemiptera three and two, Orthoptera four and three, Arachnida a trace and two, Myriapoda a a trace and three; raspberries and blackberries fourteen and twenty-four, cherries eighteen and twelve, currants three and one, grapes eleven and three, and strawberries, none by the robin and one per cent. by the cat-bird. From this it will be seen that the notable differences in the food-habits of these birds are the much larger ratios of ants, Diptera and berries eaten by the cat-bird; and of Lepidoptera, wire-worms, cherries and grapes eaten by the robin. It also appears that the cat-bird has a much more hearty appetite for spiders and thousand-legs than the robin.

It is not likely that there is any such active competition for food between these two species as this close agreement in the kinds taken at the same place and season would imply. The stress of the robin's struggle for subsistence evidently comes in early spring, before the advent of the cat-bird; and by the time the latter appears there is probably an abundance of food for both species. The earlier departure of the cat-bird likewise prevents any stringent competition in the later months.

Economical Relations.

Remembering that the chief economical service of the robin is done before and after the midsummer wealth of fruits tempts it from the chase of insects, we find it not unreasonable that the cat-bird, coming later and departing earlier, scarcely anticipating the garden fruits in its arrival and disappearing when the vineyard and orchard are at their best, should be a much less useful bird than its companion. The credit I have given it must be still further reduced because of its serious depredations in the apple-orchard. I have often seen it busily scooping out the fairest side of the ripest early apples, unsurpassed in skill and industry at this employment by the red-headed woodpecker or the blue jay.

At the bottom of the table of food given on page 118 a set of percentages will be seen similar to those previously mentioned in the discussion of the food of the robin. The beneficial elements eaten by this bird, including fruits and the carnivorous insects, run as follows, from May to September:—13, 53, 75, 45 and 19, the average for the season being 41 per cent. The corresponding ratios of injurious elements are 29, 21, 7, 16, and 4, giving a general average of 15 per cent. for the year. Referring to the vertical column of figures at the right of the table we find the injurious insects of this bird's food as follows: saw-flies one per cent., Lepidoptera seven, leaf-chafers two, snout-beetles one, plant-beetles one, chinch-bugs one and Orthoptera three; while the beneficial insects in the same column are — predaceous beetles five, predaceous Hemiptera one, and Arachnida two. A careful comparison of these elements with each other will probably convince the intelligent reader that these insect averages balance each other fairly well, and that the injury done in the fruit-garden by these birds remains without compensation unless we shall find it in the food of the young. This statement is made upon the hypothesis that ants are to be regarded as neutral insects; and the entire question of the *immediate* value of this species, aside from the still unsettled question of the food of the young, may be reduced apparently to the following form: will the destruction of a given quantity of ants pay for three times that quantity of the smaller garden fruits?

TABLE OF THE FOOD OF MIMUS CAROLINUS, L. CAT-BIRD.

	Jan.	Feb.	March	April	May	June	July	August	Sept.	Oct.	Nov.	Dec.	TOTAL	Ratios of each element to whole of food.
Number of specimens	...	...	...	...	22	19	11	12	6	...	...	...	70	

KINDS OF FOOD. — Number of specimens in which each element of food was found.

	Jan.	Feb.	March	April	May	June	July	August	Sept.	Oct.	Nov.	Dec.	TOTAL	Ratios
I. INSECTS					22	19	10	12	6				69	
					.83	.49	.18	.46	.21					.43
1. Hymenoptera					20	13	7	12	6				58	
					.22	.12	.04	.21	.09					·13
Formicidæ					20	13	4	11	6				54	
					.18	.11	.02	.20	.09					.12
Ichneumonidæ								1					1	
								†						
Tenthredinidæ					2	1	1						4	
					.02	.01	.02							.01
2. Lepidoptera					11	8	1	3	1				24	
					.14	.10	.03	.04	.04					.07
Caterpillars					10	4		2					16	
					·.12	.05		.03						.04
Noctuidæ					1	2		1					4	
					.02	.02		.02						.01
3. Diptera					7	5							12	
					.20	.07								.05
Tipulidæ					6	3							9	
					·.19	.07								.05
Bibionidæ						1							1	
						†								
4. Coleoptera					13	18	8	6	4				49	
					.23	.17	.09	.06	.07					.12
Cicindela						1							1	
						†								
Carabidæ					7	3	6	2	1				1	
					.09	.05	.07	.01	.05					.05
Dytiscidæ					2								2	
					.01									
Hydrophilidæ					2								2	
					.03									.01
Staphylinidæ					1	1							2	
					†	†								
Phalacridæ					1								1	
					†									
Nitidulidæ						1	1						2	
						†	†							
Heteroceridæ					1								1	
					†									
Histeridæ					3	1							4	
					.01	†								
Scarabæidæ					1	9	3		1				14	
					.01	.10	.03		.01					.03

TABLE OF THE FOOD OF MIMUS CAROLINUS, L. CAT-BIRD.— Continued.

	Jan.	Feb.	March.	April.	May.	June.	July.	August.	Sept.	Oct.	Nov.	Dec.	TOTAL.	Ratios of each element to whole of food.
Number of specimens.....					22	19	11	12	6				70	
KINDS OF FOOD.					Number of specimens in which each element of food was found.									
Melolonthinæ........					1 .01	5 .07							6	.02
Euroymia					1 †									
Elateridæ............					1 †								1	
Lampyridæ..........					3 .01	1 †							4	
Tenebrionidæ........					1 †	1 †							2	
Meloidæ							1 .01						1	
Rhyncophora........					6 .05	2 .01							8	.01
Brevirostres					4 .04								4	.01
Longirostres........					2 .01								2	
Chrysomelidæ					3 .01	2 .01	2 .01						7	.01
Coccinellidæ........						1 †							1	
5. *Hemiptera*.........					2 †	2 .01	1 †	7 .07					12	.02
Blissus.								1 .05					1	.01
Cydnidæ					2 †	2 .01		2 .02					6	.01
6. *Orthoptera*........					6 .04	3 .02	1 .01	5 .07					15	.03
Diapheromera					1 †								1	
Gryllidæ...........					1 †								1	
Acrididæ...........					3 †	3 .12		4 .07					10	.02
7. *Neuroptera*					1 †				1 .01				2	
II. ARACHNIDA.......					7 .13	3 .03	2 .04	1 †					13	.02
III. MYRIAPODA......					6 .07	2 .03	4 .04	1 †	2 .03				15	.03
Chilopoda					1 †									

TABLE OF THE FOOD OF MIMUS CAROLINUS, L.　CAT-BIRD — Concluded

	Jan.	Feb.	March.	April.	May.	June.	July.	August.	Sept.	Oct.	Nov.	Dec.	TOTAL.	Ratios of each element to whole of food.
Number of specimens....	...	...	...	...	22	19	11	12	6	...	...	...	70	
KINDS OF FOOD.					Number of specimens in which each Element of food was found.								
Diplopoda					6	2	4	1	2				15	.03
					.07	.03	.04	†	.03					
IV. FRUITS					2	13	10	12	6				41	.52
					.07	.45	.74	.54	.76					
Strawberries					2								2	.01
					.03									
Raspberries					9								9	.04
					.21									
Blackberries							16	8					24	.20
							.64	.35						
Currants					1								1	.01
					.04									
Cherries					6			1	3				10	.12
					.16			.03	.40					
Grapes									1				1	.03
									.14					
Sumach					2								2	.01
					.07									
Ampelopsis								1	1				2	.01
								.02	.04					
Elderberries									3				3	.03
									.15					
Percentages for each Month. Beneficial elements					13	53	75	45	19					41
Injurious elements					29	21	07	16	04					15
Neutral elements					68	26	28	39	77					44

HARPORHYNCHUS RUFUS, L.　BROWN THRUSH.

The brown thrush, although not so common a bird as the two preceding species, is still abundant enough to make its habits a matter of economical interest, both to the gardener and the farmer. It is reported by Baird, Brewer and Ridgeway to reside and breed all over the United States east of the Rocky Mountains, but in this state it is, like the robin and cat-bird, practically a strict migrant. Mr. E. W. Nelson reports its occasional occurrence in southern Illinois in mid-summer. It reaches Bloomington a little earlier than the cat-bird, and, like that species, leaves us in September; it is a shyer bird than either of the preceding, shrubbery and thickets being its favorite haunts and nesting-places.

April.

The record opens with fourteen specimens taken from the 8th to the 28th of April. Five of them were from central Illinois and nine from the northern part of the state, in Lake and Jo Daviess counties. Fifty-one per cent. of the food of these birds consisted of insects, two per cent. of spiders and six per cent. of thousand-legs. Seven per cent. of the food was Hymenoptera, nearly all ants; five parts were caterpillars and five were grubs of Diptera,— apparently crane-flies. Beetles make about one-fourth of the food, and one-fifth of these were Carabidæ. Platynus, Agonoderus and Harpalus were the only genera recognized. A remarkable feature of the food was the occurrence of four per cent. of carrion beetles, chiefly *Silpha lapponica* and *S. americana.* Thirteen per cent. of the food of the month consisted of Scarabæidæ, about three-fourths of these belonging to the genus Euryomia, which eats the leaves of fruit trees later in the season. A few June beetles were also taken at this time. A trace of wire-worms, three per cent. of snout-beetles (about two-thirds of them Brevirostres), one per cent. Hemiptera and two per cent. Orthoptera were the remaining insect elements. We come next to the distinctive feature of the food of this bird among all the thrushes. Forty-one per cent. of the food consisted of seeds and fragments of grain, of which about one-seventh was acorns taken by woodland specimens, and nearly all the remainder corn. The appearance and odor of the contents of these stomachs left no doubt that the fragments mentioned were picked from the excrement of animals.

May.

The month of May is represented also by fourteen specimens, taken at various dates from the 1st to the 27th, chiefly early in the month. Eleven of these were shot in the northern part of the state, between Galena and Waukegan. The large percentage of insect food in May reminds us of the corresponding rise, in this month, of the insect averages of the food of the robin and the cat-bird. Seventy-nine per cent. of the food of these birds

consisted of insects proper, only one per cent. of spiders and three
per cent. of thousand-legs. Ants now amount to four per cent.,
caterpillars to twelve per cent. (one-third of them distinguishable
as cut-worms), and Coleoptera to precisely one-half the food, one-
tenth of it being Carabidæ.

Scarabæidæ rise to thirty-five per cent., chiefly June beetles of
the genus Lachnosterna, wire-worms to three per cent. and
Hemiptera and grasshoppers likewise to three per cent. The
Hemiptera were all soldier-bugs. Among the predaceous beetles
Pterostichus, Anisodactylus and *Harpalus* were recognized. A
single specimen of *Cytilus sericeus* was the only representative of
the family Byrrhidæ found in the food of any of these birds.
Corymbetes and *Monocrepidius auritus* were among the spring-
beetles taken. In this month, as in the preceding, the snout-
beetles were chiefly Brevirostres. The Scarabæidæ included
Onthophagus hecate, Aphodius fimetarius, inquinatus and *grana-
rius*, and *Euryomya inda*. Seventeen per cent. of the food of
the month consisted of the fragments of grain.

June.

The birds of June, fifteen in number, taken from the 1st to the
29th, all from the northern part of the state but two, had eaten
about equally of insects and vegetable substances. Ants rise in
this month to eleven per cent., caterpillars fall to three, about one-
third of these being cut-worms. Diptera fall to one, and Coleoptera to twenty-seven per cent., and Carabidæ drop likewise to four
per cent. Scarabæidæ return to seventeen, thirteen of these being leaf-chafers; wire-worms fall to one, snout-beetles rise to four,
and plant-beetles are represented by a single *Chrysomela suturalis*.
Among the snout-beetles occur *Sphenophorus parvulus*, and *S.
sculptilis*. Several specimens of *Graphorhinus vadosus* were
eaten by three birds. *Phanæus carnifex, Onthophagus hecate* and
Aphodius fimetarius appear among the Scarabæidæ. The commencement of the fruit season is here distinctly discernible. Twenty-two per cent. of the food of these birds consists of raspberries,
five per cent. of strawberries and one per cent. of cherries, making
a total of twenty-nine per cent. of fruits. Fragments of corn and
oats amount to nineteen per cent.

July.

But seven birds were examined in July; all from the vicinity of Normal. All of these had eaten insects which amounted to only about one-fourth of the food. Both ants and caterpillars were present in trifling quantity. Only about half as many Coleoptera had been taken as in the month preceding. Hemiptera and Orthoptera each make up four per cent. of the food, and Arachnida and Myriapoda are entirely wanting. Carabidæ stand at four per cent., as in June; spring-beetles continue at three and snout-beetles amount to two per cent. *Evarthrus colossus* was found among the Carabidæ. *Heteraspis pubescens, Colaspis brunnea* and *Diabrotica* 12-*guttata* represented the plant beetles. The fruits of July amounted to sixty-two per cent. of the food,—all blackberries. Twelve per cent. consisted of fragments of corn.

August.

Twelve birds were shot in August, all from McLean county, at various times in the month from the 7th to the 30th. The insect averages rally again in August, returning now to fifty-one per cent. Hymenoptera rise to fourteen per cent,—the highest average of the season, a fact due doubtless to the swarming of certain species of ants at this time of the year.

Caterpillars amount to eleven per cent. of the food; Coleoptera fall away to ten, and all but one of these are Carabidæ. *Cratacanthus dubius* seems to be especially abundant in the later summer and early autumnal months. Four per cent. of the food of these birds consists of this species, and it has likewise been found prominent in the food of the blue-bird and the cat-bird at the same season of the year. A small percentage of snout-beetles and plant-beetles call for no special remark. Hemiptera now make one-tenth of the food—an exceptional occurrence due to the fact that this was one of the chinch-bug years in central Illinois and that three of these birds had eaten freely of that insect. Orthoptera stand at six per cent., about equally distributed between the three families of the crickets, locusts and grasshoppers. A specimen of Tridactylus was noticed among the first and one of the common katydids among the second. The fruits of this month amount to thirty per cent., eaten by nine of the birds. Half or these were cherries, and the remainder were blackberries, grapes, elderberries, and the berries of the mountain ash. Fragments of corn amounted to eighteen per cent. of the food.

16

September.

But two birds were shot in September, too few to give any correct idea of the food of the month. It is only necessary to say that these had eaten more largely of grasshoppers than the birds of the preceding month, and to about the same extent of fruits, all of which were grapes.

Summary for the Year.

Taking the food of the year together, we find that almost precisely one-half of it consisted of insects. Spiders amounted to but one per cent. and thousand-legs to but three. The remainder of the food consisted equally of the smaller garden fruits and the fragments of seeds and grain. Thirteen per cent. of the food of these sixty-four birds consisted of blackberries, four per cent. of raspberries, one per cent. of strawberries and three per cent. of cherries. The ants of the year stand at seven per cent., caterpillars at six, and Diptera at only one. Coleoptera amounted to precisely one-fourth of the food, predaceous beetles to six per cent. and Scarabæidæ to thirteen per cent., nearly all of these being leaf-chafers. Spring-beetles and snout-beetles each average two per cent., and Hemiptera and Orthoptera each stand at four.

In the paper previously cited, published in the Transactions of the Illinois Horticultural Society for 1879, I gave a table of the food of this species based upon twenty-eight specimens shot in April, May, June and July. A test of the substantial correctness of the conclusions of the present paper may be made by comparing the averages of the table printed herewith with the table on page 150 of the Transactions cited. If the important ratios of the present table, covering the food of sixty-four specimens, shot during six months of the year, agree substantially with that table of the food of twenty-eight specimens, covering but four months of the year, this will be sufficient evidence of their general correctness. I will give these averages alternately, first for the former table and then for the present. The twenty-eight specimens of 1879 had eaten insects to the amount of fifty-nine per cent., and sixty-four specimens of the table of 1880 had eaten insects to the amount of fifty-seven per cent. Hymenoptera are seven in the first and eight in the second; ants are seven in the first and also in the second; Lepidoptera seven and seven, Diptera a trace and one, Coleoptera twenty-nine and twenty-five, Carabidæ six and six, Silphidæ two and one; leaf-chafers nine and ten, spring-

beetles one and two, snout-beetles three and two, Hemiptera two
and four, Orthoptera four and four, Arachnida one and one, Myriapoda four and three, and fruits twenty-two and twenty-four.
A larger percentage of Hemiptera is due to the much greater
abundance of chinch-bugs in 1880.

Recapitulation.

The brown-thrush, arriving in April, finds nearly one-half of its
food in fragments of corn and other grains and seeds picked from
the droppings of animals. This curious habit it maintains throughout the year, evidently taking this food from preference as well
as from necessity. In fact I have often found these vegetable
fragments associated with blackberries in the food.

After April this element averages about sixteen per cent.
throughout the season. Insects amount to about half the food
for each month, except in May when they rise to three-fourths and
in July when they drop to one-fourth. The excess in May occurs
at the time of the greatest number and activity of the beetles, and
the diminution in July coïncides with the period of the greatest
abundance of the small fruits. One-half the insects eaten are
beetles, which stand at one-fourth of the food in April and June,
rise to one-half in May and fall to about one-eighth in July and
August. Half the beetles of the year are Scarabæidæ, chiefly June
beetles and Euryomia, all taken previous to July. Nearly one-
fourth of the beetles are Carabidæ, which remain at about five per
cent. of the food, except in May when they rise to ten per cent.
Although the ratios of spring-beetles and snout-beetles are but
two per cent., the numbers eaten are of some significance. My
notes show that these birds were eating each at the daily rate of
about 1½ Curculios, and consequently had averaged a total of
about 250 to each thrush for the season. The brown-thrush takes
ants more freely than the robin, but eats comparatively few caterpillars; seven per cent. of each were found in the food of the year.
Diptera are taken in very trivial quantity and Hemiptera in moderate number only. This bird eats thousand-legs more freely than
the robin, especially in the early spring. In the garden it plays a
part very similar to that of the other thrushes, but is less mischievous, on the whole. Its average of the edible fruits for June, July
and August is thirty-eight per cent. as against sixty per cent. of the
robin and forty-nine per cent. of the cat-bird. It relishes the whole
list of garden fruits, and later in the season resorts, like the other

thrushes, to the wild fruits of the woods and thickets. Compared with the robin, this bird is seen to be especially peculiar in the coprophagous habit already mentioned as distinguishing it from all the other thrushes. It takes about one-half as many Lepidoptera, about half as many again Coleoptera, nearly twice as many Carabidæ and three times as many leaf-chafers; but eats comparatively few grapes and cherries. From the cat-bird it is further distinguished by taking half as many ants, a trivial number of Diptera, twice as many Coleoptera and twice as many Carabidæ, five times as many leaf-chafers and more spring-beetles, snout-beetles, Hemiptera and Orthoptera. It eats two-thirds as many berries and one-third as many cherries and grapes as the cat-bird.

ECONOMICAL VALUE.

Compared with the robin for corresponding months, this species seems to show very similar economical relations. In both the totals of beneficial elements eaten during this period are to the injurious about as four to three; but with the brown thrush as with the cat-bird, its later arrival and earlier departure are to its disadvantage. Balancing as carefully as I can its seven parts of Lepidoptera, ten of leaf-chafers, two of spring-beetles, two of snout-beetles, one of chinch-bugs and four of Orthoptera on the one hand, against its six parts of Carabidæ, two of predaceous Hemiptera, one of spiders, one of predaceous thousand-legs and twenty-one of small fruits on the other, I cannot see that, so far as the *immediate* consequences of its food habits are concerned, it does more good than harm. In short, its Orthoptera must pay for its garden fruits; that is to say, eliminating these two elements, I judge that the predaceous insects eaten would destroy during the year about as many injurious insects as the bird itself has taken. However, I must repeat the suggestion that they could hardly destroy the *same kinds* as the bird, and that, if allowed to live, they would probably decimate some species already sufficiently restricted by existing checks, and permit an unrestrained increase of others now kept down by the thrush. That the disturbances thus set up would soon lead us to regret this bird if its numbers were greatly lessened, is therefore very probable, and I believe the species should be preserved. We must not overlook the special services of the brown-thrush in devouring a much larger number of June beetles than any other of the species examined.

TABLE OF FOOD OF HARPORHYNCUS RUFUS, L.   BROWN THRUSH.

	Jan.	Feb.	March.	April.	May.	June.	July.	August.	Sept.	Oct.	Nov.	Dec.	TOTAL.	Ratio of each element to whole of food.
Number of specimens	...	...	...	14	14	15	7	12	2	...	...	...	64	

KINDS OF FOOD. — Number of specimens and ratios in which each element of food was found.

KINDS OF FOOD.	Jan.	Feb.	March.	April.	May.	June.	July.	August.	Sept.	Oct.	Nov.	Dec.	TOTAL.	Ratio
I. MOLLUSCA					1 / †			1 / †					2	†
II. INSECTA				14 / .51	14 / .79	15 / .49	7 / .26	12 / .51					62	51
1. *Hymenoptera*				9 / .07	10 / .04	11 / .12	4 / .02	11 / .14					45	08
Formicidæ				8 / .06	10 / .04	10 / .11	3 / .02	11 / .14					42	07
Ichneumonidæ				2 / †		1 / †							3	......
2. *Lepidoptera*				6 / .05	12 / .12	6 / .03	2 / .03	7 / .11					33	07
Caterpillars				6 / .05	12 / .12	4 / .02		7 / .11					29	06
Noctuidæ					4 / .04	2 / .01							6	01
3. *Diptera*				1 / .05		2 / .01							3	01
4. *Coleoptera*				12 / .26	14 / .50	14 / .27	7 / .13	9 / .10					56	25
Carabidæ				9 / .05	7 / .10	5 / .04	3 / .04	5 / .09					29	06
Silphidæ				3 / .04									3	01
Nitidulidæ								1 / .02					1	†
Staphylinidæ					1 / †								1	†
Histeridæ				2 / †	3 / .01	1 / †		1 / †					7	†
Byrrhidæ					1 / †								1	†
Scarabæidæ				9 / .13	11 / .35	8 / .17		1 / †					29	13
Melolonthinæ				2 / .01	6 / .26	4 / .13							12	08
Euryomya				2 / .09	2 / .03								4	02
Buprestidæ						1 / †							1	†
Elateridæ				1 / .01	8 / .03	3	3 / .03						15	02
Tenebrionidæ								1 / .01					1	†

TABLE OF FOOD OF HARPORHYNCUS RUFUS, L. BROWN THRUSH.
Continued.

	Jan.	Feb.	March.	April.	May.	June.	July.	August.	Sept.	Oct.	Nov.	Dec.	TOTAL.	Ratio of each element to whole of food.
Number of specimens....				14	14	15	7	12	2				64	

| KINDS OF FOOD. | Number of specimens and ratios in which each element of food was found. | | | | | | | | | | | | | |

Kinds of Food	Jan.	Feb.	March.	April.	May.	June.	July.	August.	Sept.	Oct.	Nov.	Dec.	TOTAL.	Ratio
Lampyridæ						1 †							1	†
Rhyncophora				7 .03	5 .01	7 .04	3 .02	3 .01					25	02
Brevirostres				3 .02		3 .02	1 .91	2 .01					9	01
Longirostres				3 .01	1 †	2 .01		1 †					7	†
Brenthidæ						1 †							1	†
Chrysomelidæ						1 †	2 .01	1 †					4	1
5. Hemiptera				3 .01	6 .03	1 .01	5 .04	6 .10					21	04
Blissus								3 .06					3	01
Cydnidæ					6 .03	1 .01	3 .02						10	01
6. Orthoptera				1 .02	4 .04	3 .05	2 .04	4 .06					14	04
Gryllidæ							1 .02	1 .02					2	01
Locustidæ								1 .01					1	†
Acrididæ				1 .01	3 .03	3 .04		1 .01					8	02
III. ARACHNIDA				3 .02	1 .01	1 .01							5	01
IV. MYRIAPODA				6 .06	7 .03	5 .03		1 .01					19	03
Geophilidæ				1 .01				1 .01					2	01
Iulidæ				5 .05	7 .03	5 .03							17	02
V. FRUITS					9 .29	6 .62		9 .30					24	24
Blackberries						6 .62		1 .05					7	13
Raspberries						6 .22							6	04
Strawberries						3 .05							3	01
Grapes								2 .02					2	†

TABLE OF FOOD OF HARPORHYNCUS RUFUS, L. BROWN THRUSH.
Concluded.

Number of specimens.....	Jan.	Feb.	March.	April.	May.	June.	July.	August.	Sept.	Oct.	Nov.	Dec.	TOTAL.	Ratio of each element to whole of food.
	...	...	...	14	14	15	7	12	2	...	...	...	64	
KINDS OF FOOD.	Number of specimens and ratios in which each element of food was found.													
Cherries.......					1 / .01		3 / .15						4	03
Elderberries.....							1 / .04						1	01
Mountain Ash.........							1 / .03						1	01
VI. SEEDS AND GRAIN...				12 / .41	7 / .17	9 / .18	1 / .12	5 / .18					34	21
Acorns......				2 / .16									2	01
Oats.........							2 / .01						2	†
Corn.				9 / .34		6 / .17	1 / .12	5 / .18					21	61
Wheat.........				1 / .01									1	†
Buckwheat........				1 / †									1	†
Percentages for each month — Beneficial elements				09	14	34	70	36						33
Injurious elements				21	49	25	13	24						26
Neutral elements...				70	37	41	17	40						41

TURDUS MUSTELINUS, Gm. WOOD THRUSH.

The remaining members of this family are much less important than the preceding species, and their food is of relatively little interest. I shall therefore treat them much more briefly, especially as I have comparatively few specimens of them. The wood-thrush is essentially a woodland bird, but occurs not unfrequently in groves and gardens and in other situations where trees and shrubbery are accessible. It reaches central Illinois in April, and retires usually in October, spending its winter in the southern states. I have studied the food of but twenty-two specimens of this species, ranging from April to September. Two of these birds were taken in April, five in May, six in June, six in July, two in August and one in September. I shall not attempt to follow the food of the species through these months, or to give its seasonal variations; but will content myself with a general statement of the food of the year as indicated by the contents of the stomachs of these twenty-two birds. Seventy-one per cent. of their food consisted of insects and twenty per cent. of fruit, a small ratio of spiders and mollusks and an unusually large percentage of Myriapoda making up the remainder. The four higher orders of insects occur in about equal quantities, the proportion of ants and crane-flies being extraordinary. Blackberries, strawberries, cherries and gooseberries appear among the fruits. Myriapoda amount to twelve per cent.—nearly all Polydesmus and Iulus. The two parts of Arachnida included a few harvest-men. Orthoptera and Hemiptera are respectively six and one per cent.; and snout-beetles and wire-worms thirteen per cent. A few June beetles had been taken, and one of the birds from northern Illinois had stuffed itself with rose-beetles (*Macrodactylus subspinosus*). Geotrupes and Onpthophagus were noticed among the other Scarabæidæ. The Carabidæ amounted to six per cent. of the food, including Evarthrus, Pterostichus, Harpalus, Anisodactylus and Bradycellus. Coleoptera make eighteen per cent. of the food and Diptera twelve per cent., chiefly crane-flies and the larvæ of *Bibio albipennis.* Lepidoptera were taken in about the same amount, one-third being recognized as cut-worms, while ants reached the unusual average of fifteen per cent. *Helix labyrinthica, Pupilla fallax* and a few other univalve mollusks made one per cent. of the food. Compared with other Turdidæ, we find the general insect average unusual, exceeding even that of the robin. It agrees with, and even surpasses, the cat-bird in its preference for ants; and with the robin in the ratios of Lepidoptera, Diptera, Coleoptera, Carabidæ and Scarabæidæ. It differs from the robin in its taste for ants and in the smaller ratio of fruits; and far surpasses all the other thrushes in the number of Myriapoda eaten in spring. In fact, the mid-summer fruits seem to replace these spring Myriapoda, instead of insects proper as in the species already discussed. This bird apparently contrasts more directly with the brown thrush in food than with any other member of the family. The large percentage of Orthoptera is misleading, being due to the fact that a single bird had taken nothing but grasshoppers and locusts. This species seems to do more good and less harm than the preceding thrushes, having the lowest fruit ratio and eating the highest number of insects, with only the average of predaceous species. Its advances, therefore, are to be cordially encouraged by the gardener and farmer—a fact which must be especially agreeable to every lover of bird music, who has learned to recognize the full, clear, rich and exquisite strains of this songster.

HYLOCICHLA PALLASI, Cab. HERMIT THRUSH.

The hermit thrush is strictly a migrant, passing us in May and October. It is reported by Mr. Ridgway as a rare winter resident in southern Illinois, but otherwise appears in the state only during its passage to and fro. Considering the fact, however, that all these birds travel slowly the whole length of the state, merely keeping pace with the advancing and retreating seasons, and also that the species is a very abundant one at the period of the migrations, it will be seen that its food has great economical significance. There is reason to suppose that these migrants, in passing north and south, follow, year after year, about the same route; do not vary, that is, far to the east or west. Consequently, occupying as we do a state that lies in five and one-half degrees of latitude, we can do much to protect this species in its wanderings, or can, if we choose, almost entirely eliminate that part of it passing over our territory. Twenty-one hermit thrushes were taken during the year, two in October and the remainder during the spring migrations. All but five of these birds were shot in extreme northern Illinois, at Waukegan, Evanston and Blue Island. Eighty-four per cent. of the food consisted of insects, four per cent. of spiders and twelve per cent. of thousand-legs. Ants amounted to fifteen per cent., Lepidoptera to nineteen per cent., including a few Phalænidæ, and Diptera only to three—chiefly the larvæ of Bibio. Coleoptera make thirty per cent. of the food, eleven per cent. being Carabidæ. *Dyschirius globulosus*, Platynus, Evarthrus, Pterostichus, Amara, *Anisodactylus discoideus*, Bradycellus and Stenolophus are mentioned in my notes. Four per cent. are water-beetles, five per cent. scavenger beetles, two per cent. Curculios and two per cent. plant beetles. Leaf-chafers and spring beetles amount to one per cent. each—the latter chiefly of the genus

17

Melanotus. *Lixus concavus* and *Listronotus inæqualipennis* occur among the Curculios, and *Chrysómela suturalis*, *Gastrophysa
dissimilis*, *Plagiodera viridis* among the plant-beetles. Eight
per cent. of the food was Hemiptera, nearly all of which were predaceous. *Podisus spinosus* was the only species determined.
Grasshoppers (Tettix and Tettigidea) make seven per cent. of the
food. Respecting the number of beetles eaten by this bird, we
have to remember that it passes us at the time of that great outpouring of insect life connected with the pairing of the spring
Coleoptera which we have already seen to have a very significant
relation to the food of birds. It rides northward, in fact, on the
crest of this Coleopterous wave, and we find the same excess of
predaceous Coleoptera in its food which occurs in the food of the
other thrushes at the same season. Concerning the two October
specimens taken in northern Illinois I need only say that they had
eaten ants, caterpillars, Carabidæ, Curculios, Cydnidæ and Orthoptera, spiders, Iulidæ and the larvæ of Bibio. The habits of this
bird suggest that the principal drain on the numbers of predaceous beetles may be due to the depredations of the migrants, at the
season of the greatest exposure of these insects; and that the complete destruction of resident birds would affect the number of these
carnivorous insects much less than would at first seem likely. The
reader curious to see the points in which this species contrasts
with the other thrushes, may consult the table of the food of the
family on page 136.

TURDUS ALICIÆ, Bd. ALICE THRUSH.

The Alice thrush is a bird of frequent occurrence during the migrations. It breeds far to the north, rare summer stragglers occurring in northern Illinois, according to Mr. E. W. Nelson, and
probably winters quite beyond our limits. By Dr. Cowes this is
regarded merely as a variety of the following species. I have ten
specimens of this bird shot in May, but none from the fall migration. This number is probably sufficient, however, to give a fairly
correct idea of its food in spring. Five per cent. of the food of
the month consisted of mollusks, chiefly Succinea and *Helix labyrinthica;* ninety-three per cent. was insects and nearly half of
these were ants, which reached the astonishing ratio of forty-three
per cent., eaten by every one of the birds. Fifteen per cent. of
the food was caterpillars; nine per cent. consisted of crane-flies

and their larvæ; Coleoptera amounted to eighteen per cent. (one-half Aphodiidæ) and the remainder were wire-worms, Curculios and plant-beetles. Carabidæ amounted only to one per cent., the lowest average of these beneficial insects found in the food of any thrush. Among the species of Coleoptera we find *Stelidota geminata, Onthaphagus janus, Conotrachelus anaglypticus, Chrysomela suturalis* and *C. similis.* Grasshoppers make three per cent. of the food and Myriapoda two per cent., all *Polydesmus serratus* and undetermined Iulides. Of spiders merely a trace was found in the stomachs of two birds. The striking feature of the food of this bird is evidently its enormous appetite for ants, its high insect average and the almost total absence of beneficial elements in its food, giving to this little thrush an enviable status in relation to the farm and garden.

TURDUS SWAINSONI, Cab.   SWAINSON'S THRUSH.

This is a migrant of which I have too few specimens for generalization. Six in April and May were taken at Warsaw, Waukegan and Normal, and five in September from the vicinity of Cairo, in extreme southern Illinois and northern Kentucky. The food in spring is very like that of the preceding species, its especial features being the large number of ants and caterpillars and Coleoptera. The September specimens, on the other hand, were feeding largely upon fruits, which constituted sixty per cent. of their food. Wild grapes, wild cherries, elderberries and blackberries were all eaten by them, grapes alone making more than half their food. Hymenoptera amounted to nineteen per cent. of the whole; ants to seven, caterpillars to twelve, crane-flies to four, and Coleoptera to eighteen per cent.; five per cent. were Carabidæ (including Anisodactylus), three per cent. were leaf-chafers and two per cent were Curculios. One of the birds, taken at Warsaw in April, had eaten little else than *Scolyts muticus.* Two per cent. of the food was Hemiptera, chiefly Cydnidæ and Reduviidæ; Rhyncophora and Hemiptera made two and one per cent. respectively. Of spring beetles and Aphodiidæ, only a trace had been eaten by two of the birds.

MIMUS POLYGLOTTUS, L.   MOCKING-BIRD.

This famous bird, not many years ago regarded as a rarity in the state, is evidently becoming more abundant, and is also extending its habitat northward. Collectors in the southern part of the state agree to its increasing numbers there. Three specimens were seen this year in the vicinity of Bloomington, two of which were secured. One of these, shot in August, was of this year's brood, and as the other two seemed thoroughly habituated, it is likely that they had nested in this vicinity this season. It may be worth while to note that sixty per cent. of the food of these two specimens consisted of Orthoptera, including the climbing cricket (Œcanthus). Besides these, they had eaten spiders and harvest-men, Coleoptera, Hemiptera and ants. Among the Coleoptera were specimens of *Onthophagus*, *Epicauta vittata* and long-snouted Curculios. The Hemiptera were undetermined Coreidæ and Cydnidæ. These birds had not eaten fruit, although the species is reported to be especially fond of grapes.

CONCLUSION.

As a very general statement of the peculiarities of the food of the resident species, we may say that the robin is characterized by its destruction of caterpillars (especially cut-worms) and the larvæ of Bibio, by its neglect of ants, spiders and Myriapods, and by its taste for blackberries, grapes, and especially cherries; that the cat-bird is distinguished by the large number of ants, blackberries and cherries eaten, and by the small number of insects generally, and of Lepidoptera, Coleoptera and Hemiptera in particular; that the brown-thrush is noted for its coprophagous habit, for the small number of caterpillars and Diptera taken, for the large percentage of phytophagous Scarabæidæ and the moderate ratio of small fruits; and that the wood-thrush differs from the others chiefly in the large percentage of insects (especially ants, caterpillars and crane-flies), its indifference to Hemiptera and preference for Orthoptera and Myriapoda, and its smaller ratios of fruits.

The migrants can be properly compared only with the residents during the migrating season. I have consequently made a table of the percentages of the food of the four resident species for April and May in comparison with the spring food of the three migrants. From this we learn that the hermit-thrush is distinguished at this season by the moderate ratio of ants and Coleoptera, the large number of Lepidoptera, Hemiptera, Orthoptera, spiders and Myriapoda, and the small percentage of Diptera taken. The Alice thrush eats mollusks, an enormous number of ants, a moderate number of Lepidoptera, Diptera and Scarabæidæ, and a small number of Carabidæ and Coleoptera generally, while Hemiptera are almost wanting in its food. Swainson's thrush takes large ratios of ants, Lepidoptera and Coleoptera, and small ratios of Hemiptera, Orthoptera, Arachnida and Myriapoda. It is not to be supposed that the number examined of the last two species is sufficient to give more than an approximate and doubtful outline of the food.

Indeed the reader may not unlikely receive with incredulity the precise statements made concerning the food characteristics of the resident species, and ask how it can be known that these peculiarities are specific and constant instead of local and accidental. To this very reasonable query I am able to make a definite answer. In the paper already frequently cited, I published a comparative table of food of the species of this family, based on the contents of the stomachs of one hundred and forty-nine birds,* upon which table certain differences of food are clearly shown. Now, if these differences were local and accidental, they would undoubtedly tend to disappear when larger numbers of specimens were examined; but if they are specific and constant, they should be made the more evident, on the whole, the larger the number of specimens taken. The table on page 136 presents data derived from three hundred and fifteen specimens, covering considerably more time and area than the table in the Transactions. If the difference between the food records of the various species are now greater than before, we may conclude that the differences noted are real and not artificial. If they are less, on the other hand, the whole question is still unsettled. The differences apparent in the later table may be specific, but there is no proof of it. In order to apply this crucial test as fully as possible, I have selected twelve food elements in which the differences were most apparent, and, taking the species in pairs, have ascertained the sum of the differences of the ratios of these elements for each pair separately, first from the old table and then from the new. In every case but one the sum of these differences has been much larger by the new table than by the old, thus proving conclusively that the species appear to diverge in food habits the more widely the greater the number of specimens studied. For example, the differences of the se-

* Trans. Ill. Hort. Society, 1879, N. S. Vol. 13, p. 163.

lected elements as shown in the original table of seventy-eight robins and cat-birds, amounted to sixty-four per cent.; and by the new table of one hundred and eighty-four birds, to eighty-two per cent. A similar comparison of the food of the catbird and hermit thrush gives one hundred and twenty-five as the sum of the differences of the old table of fifty-five birds, and one hundred and fifty-five as the sum of the differences of the new table of ninety-one birds. Taking the catbird and the brown thrush, we have sixty-four and ninety-nine parts for the old and new tables respectively, the first for sixty-five birds and the second for one hundred and thirty-four; while the brown thrush and wood thrush give seventy-eight and eighty-eight parts for thirty-nine and eighty-six birds respectively, and the catbird and wood thrush give seventy parts for eighty-five birds and eighty-three parts for ninety-two birds. It is not until we reach the last two migrants that we find any exception to these results; and of these, as already said, probably too few have been examined, even yet, to justify settled conclusions.

Finally, we must consider the family as a unit, must discuss the actual effect of the thrushes as a group upon the plants and animals of the state. A determination of this interesting question involves three elements: the average character of the food of each species as shown by the preceding calculations, the comparative abundance of the species, and the length of its stay in Illinois. I find the estimates of the second of these elements, as made by various collectors, to differ rather widely; and on this account only an approximate conclusion can be reached. Using the figures most satisfactory to myself, I present the following as a tolerably fair statement of the general food of the family: Sixty-one per cent. of the food consists of insects, one per cent. of spiders, two per cent. of Myriapods, and thirty-two per cent. of fruits, eleven per cent. being blackberries, eight per cent. cherries, one per cent. currants and five per cent. grapes. The fragments of grain eaten by the brown thrush will amount to four per cent. of the food of the family, and ants compose eight per cent. Lepidoptera, Diptera and Coleoptera are eaten in about equal ratios, the first forming thirteen, the second eleven and the third twelve per cent. of the entire food. Carabidæ amount to five per cent., June beetles to four per cent., wire worms to two per cent. and snout-beetles to two per cent. Hemiptera stand at three per cent., about two-thirds of them predaceous, and Orthoptera at four per cent. Five per cent. of the food was recognized as cut-worms. More briefly, thirty parts of the food consist of injurious insects, including the larvæ of Bibio, and eight parts of beneficial species, while twenty-six parts consist of edible fruits; or we may say that injurious insects compose about one-third, the edible fruits about one-fourth and the beneficial insects about one-twelfth of the food of the family, the remaining elements being of neutral value.

TABLE OF FOOD OF THRUSHES IN APRIL AND MAY.

	Robin.	Cat-bird.	Brown Thrush.	Wood Thrush.	Hermit Thrush.	Alice Thrush.	Swainson's Thrush
Number of specimens examined	31	22	28	8	18	10	6

KINDS OF FOOD.	Ratios in which each element of food was found.						
1. MOLLUSCA	01	......	....	†	......	05	......
2. INSECTS	93	83	65	84	87	95	98
Hymenoptera	03	22	05	20	16	47	31
Ants	03	18	05	20	13	43	28
Lepidoptera	24	14	08	21	19	15	22
Noctuidæ	09	02	04	08	......	02	......
Diptera	12	20	03	15	01	09	07
Tipulidæ	03	19	......	15		08	07
Bibionidæ	04	......	....				
Coleoptera	43	23	38	23	30	18	30
Carabidæ	11	09	07	09	11	01	05
Scarabæidæ	19	01	24	06	06	10	06
Coprophagous	......	......	......	......	05	10.	......
Phytophagous	12	01	19	02	01	......	06
Elateridæ	04	†	02	06	01	02	†
Rhyncophora	03	05	02	02	02	03	03
Chrysomelidæ	01	01	....		02	01	02
Hemiptera	04	†	02	01	08	......	......
Predaceous	03	†	02	......	06		
Herbivorous							
Orthoptera	05	04	03	03	....	03	......
3. ARACHNIDA	01	03	02	01	04	†	01
4. MYRIAPODA	01	07	04	13	09	02	01
5. FRUITS AND SEEDS	03	07	....		......	......	......
6. FRAGMENTS OF GRAIN	......		29	......	......	......	......

TABLE OF FOOD OF FAMILY TURDIDÆ.　　(THE THRUSHES.)

	Robin.	Cat-bird.	Brown Thrush.	Wood Thrush.	Hermit Thrush.	Alice Thrush.	Swainson's Thrush.	Mocking Bird.	Total.	Corrected average.
Number of specimens examined..	114	70	64	22	21	11	11	2	315	....

KINDS OF FOOD.	Ratios in which each element of food was found.									
1. MOLLUSCA	....	....	....	....	....	.05	....	....	....	....
2. INSECTS	.65	.43	.51	.72	.84	.93	.62	....	....	.61
Hymenoptera	.04	.13	.08	.15	.16	.47	.19	....	....	.09
Ants	.04	.12	.07	.15	.13	.43	·17	....	....	.08
Lepidoptera	.17	.07	.07	.13	.18	.15	.12	....	....	.13
Noctuidæ	.08	.01	.01	.04	....	.04	....	....	....	.05
Diptera	.17	.05	.01	.12	.04	.09	.04	....	....	.11
Tipulidæ	.01	.05	....	.12	....	.08	.04	....	....	.03
Bibionidæ	.15	....	....	....	....	....	....	....	....	.07
Coleoptera	.18	.12	.25	.18	.29	.18	....	....	....	.12
Carabidæ	.06	.04	.06	.06	.14	.01	.05	....	....	.05
Melolonthidæ	.03	.02	.10	.03	.01	....	.03	....	....	.04
Elateridæ	.02	....	.02	.03	.01	.02	†	....	....	02
Rhyncophora	.02	.01	.02	.03	.03	.10	....	....	....	.02
Chrysomelidæ	....	.01	†	....	.02	.01	....	....	....	....
Hemiptera	.03	.01	.04	.01	.10	....	·02	....	....	.03
Predaceous	.02	.01	.01	.01	.07	....	.01	....	....	.02
Herbivorous	....	.01	.01	....	....	....	.01	....	....	....
Orthoptera	.04	.03	.04	.06	.07	.03	.01	....	....	.04
3. ARACHNIDA	.01	.02	.01	.01	.04	....	.02	....	....	.01
4. MYRIAPODA	....	.03	.03	.07	·12	.02	.01	....	....	.02
5. FRUITS	.34	.51	.24	.19	....	....	.35	....	....	.32
Strawberries	....	.01	....	....	....	....	....	....	....	....
Blackberries	.09	.24	.17	....	....	....	†	....	....	.11
Cherries	.11	.12	.03	....	....	....	.06	....	....	.08
Currants	.02	.01	....	....	....	....	....	....	....	.01
Grapes	.07	·03	†	....	....	....	.27	....	....	.05
6. FRAGMENTS OF GRAIN	....	....	.21	....	....	....	....	....	....	.04

Family SAXICOLIDÆ. (The Stonechats.)

SIALIA SIALIS, L. THE BLUEBIRD.

This beautiful and beloved bird, endeared to the student of nature by every particular of its plumage, song and way of life, is also one of the most popular of all birds with farmers and gardeners. Living under the eyes of men from the first yielding days of the later winter until the year grows chill and dark with the retreat of autumn, it has been praised most warmly for its tireless service of man by those who knew it best. A cursory observation of its feeding habits will strongly support the general impression of its usefulness. Most frequently it takes a short, quick flight to the ground from a fence-post, or a low branch of a tree, and, after a moment's pause, returns to its perch with a caterpillar or a grasshopper or some other insect in its beak, which it devours at its leisure, repeating this operation so frequently that none can doubt its enormous destructiveness to insect life.

It is true that a little reflection will suggest that, as it evidently sees its prey before it leaves its perch, it must usually take only the most conspicuous and the most active insects, and that there is no security that these will be the most injurious—that they may not be, in fact, among the most beneficial; but this consideration does not seem to have made any impression, and the Bluebird remains to this day substantially without reproach.

I have now examined carefully, with the microscope, the contents of one hundred and eight stomachs of this species, of which ten were taken in February, twenty-one in March, thirteen in April, nine in May, ten in June, nine in July, twelve in August, ten in September, two in October and twelve in December (in southern Illinois). I propose to present the data for each of these months; to summarize them for the year; to estimate the benefit and injury indicated to farm and garden, and to make a comparison of the food of this bird with that of the robin, and of the thrushes generally.

February.

The ten birds of this month were all shot at Normal, Ill., from the 24th to the 29th of the month, in the present year. These stomachs, with those obtained from Galena, in early March, represent the first food of the season.

The record opens with a bird shot on the 24th. Thirty per

18

cent. of its food had been grass-eating cut-worms, forty per cent. crickets (*Gryllus abbreviatus*), five per cent. Ichneumonidæ (*Arenetra nigrita* Cress), and twenty-five per cent. the larvæ of the two-lined soldier-beetle (*Telephorus bilineatus*). Now, the Ichneumons are doubtless parasitic, although about the habits of the genus Arenetra, I have at present but little specific information; and the soldier-beetles are reported by Prof. Riley and others to be highly useful insects, noted especially for the destruction of the Apple-worm and the eggs of grasshoppers.*

Taking the month together, we find that the most important elements of the food were cut-worms and ichneumons—twenty-four per cent. of the former to twenty-two per cent. of the latter. The larvæ of the soldier-beetles amount to eight per cent., locusts (chiefly the young of *Tragocephala viridifasciata*) to nine per cent., Carabid beetles and their larvæ (including Amara and Aniso-dactylus) to five per cent., Cydnidæ or soldier-bugs (chiefly *Euschistus servus*) to seven per cent., spiders to four per cent., and Iulidæ (thousand-legs) to three per cent. Other items are, two per cent. caterpillars of Arctians (*Callimorpha lecontei*), four per cent. crickets, and nine per cent. dung beetles (*Aphodius fimetarius* and *A. inquinatus*). The ichneumons, Carabid beetles, soldier-bugs and spiders thus make up forty-six per cent. of beneficial insects, while the caterpillars and Orthoptera amount to but forty-one per cent. of injurious species. Or, if we drop the Cydnidæ from the former category, on account of the supposed trifling injuries to vegetation done by some of them (hence often called "plant bugs"), the figures will stand, beneficial insects thirty-nine to forty-one injurious.

March.

Twenty-one specimens were examined which had been shot in this month, in 1880, ranging from the 7th to the 31st. Seven of these were shot at Normal, nine at Heyworth (fifteen miles south) and five at Galena, in extreme northwestern Illinois. These latter differed from the central Illinois specimens chiefly in the presence of the dried and sometimes mouldy fruit of the sumach (*Rhus glabra*) in their stomachs, indicating a scarcity of desirable food at that early season. One of these, unfortunately for the record of the month, had stuffed itself with the larvæ of Harpalus, which made ninety-three per cent. of its food.

*See 4th Rept. State Entomol. of Mo., p. 29, and Rept. U. S. Ent. Com. 1877, p. 302.

Ichneumonidæ (Arenetra) appear again (four per cent.) for the last time during the season.

Harpalid beetles and their larvæ were unusually abundant, making up eleven per cent. of the food of the month. Among these, Platynus, Evarthrus, Pterostichus, Amara, *Chlænius tomentosus*, Agonoderus and Harpalus were recognized. The larvæ of soldier-beetles also occur, constituting four per cent. of the food, but do not appear again throughout the year. Four birds had eaten a predaceous bug (*Coriscus*, near *ferus*),* which is too minute to figure in the ratios; and four per cent. of the food was Cydnidæ, of which only *Peribalus modestus* was recognizable. Sixteen of the twenty-one birds had eaten spiders, making five per cent. of the food. The beneficial insects thus amount to twenty-eight per cent. On the other hand, thirty-eight per cent. was caterpillars, chiefly Noctuidæ,† including *Callimorpha lecontei* and the army worm (*Leucania unipuncta*); one per cent. was *Euryomia inda*, and twenty-one per cent. was Orthoptera (crickets and grasshoppers), the injurious species thus rising to sixty per cent. One bird had also eaten a minute curculio. Among neutral elements we enumerate Aphodii three per cent., Iulidæ three per cent., and sumach berries four per cent. Two birds had eaten ants, but in trivial quantity.

In order to determine the number of specimens which it is necessary to examine in each month, to reach reliable averages of benefit and injury, I divided my notes on twenty of the specimens for March, into two groups of ten each, so selected that all the localities and all parts of the month were equally represented in each group; and then averaged each ten separately and compared the averages. In the first group beneficial insects composed twenty-nine per cent. of the food, and injurious insects fifty-nine per cent.; in the second group beneficial insects composed twenty-seven per cent. of the food and injurious insects sixty-one per cent. The close correspondence of these averages shows that, on this question, ten specimens would have given as accurate information as twenty, and indicate that ten birds a month will usually afford a fair basis for an opinion.

---

*Kindly identified for me by Mr. Uhler.

†I have thus reported all smooth caterpillars in which the cervical and anal shields, common to most cut-worms, were distinguished. A few such caterpillars are not Noctuids, but are equally injurious.

140   *The Food of Birds.*

April.

The food of April, as shown by the thirteen specimens of that month (from Normal, Evanston, Waukegan, and Elizabeth, in 1876 and 1880), was remarkable for the number of Aphodii (dung beetles) it included; twenty-one per cent. of the food of the month was *Aphodius inquinatus*, nine per cent. *A. fimetarius*, and one per cent. undetermined Aphodii. This peculiarity is accounted for, in harmony with what has been said above respecting the feeding habits of the Bluebird, by the fact that this is the month when the Aphodii fly most actively in the latitude of Northern Illinois. Carabidæ now stand at eight per cent., including *Carabus palustris*, Pterostichus, Evarthrus, and other Pterostichi, Platynus, *Chlænius tomentosus*, *Anisodactylus rusticus*, *Amphasia interstitialis*, and Harpalus; four per cent. of Hemiptera include Coriscus and *Hymenarcys nervosa*, while spiders rise to nine per cent. Caterpillars are twenty-one per cent. (seventeen per cent. Noctuids), June beetles (Phyllophaga) two per cent., Curculionidæ one per cent., and grasshoppers (Tettigidea sp. and *Tettix ornata*) eight per cent.; a total of thirty-two per cent. of injurious insects against twenty-one per cent. of predaceous species. Among the neutral elements we find a sprinkling of ants (two per cent.), larvæ of a Tenebrionid (*Merucantha contracta*[*]) four per cent., and thousand-legs (Iulidæ) one per cent. Long strips of grass, in pieces much too large to have been eaten by any of the insects present, were found in the stomachs of two of these birds, and also occurred during each of the three following months. I am in doubt whether these were taken as food; but, since I have found them in no other bird, and since a species which feeds so largely on cut-worms and grasshoppers may have acquired the power of digesting the very considerable quantities of grass contained in the intestines of these insects, I have thought it best to include them in the percentages of food. It is probable, however, that they were swallowed accidentally with insects taken from the ground.

It will be noticed that the excess of Coleoptera in April is largely compensated by the diminished quantities of Orthoptera and caterpillars.

---

[*]For the determination of this species and most of the other larvæ which have been identified specifically, I am under obligations to Prof. Riley.

May.

In this month nine birds were taken, from six localities in central and northern Illinois, in 1876-'80. The Lepidoptera, Coleoptera and Orthoptera return to about their normal ratios, but spiders rise to the excessive figure of twenty-one per cent. This ratio is, however, partly misleading, as, although six of the nine birds had eaten spiders, yet eleven per cent. is due to a single bird, which had eaten nothing else. In such a case a larger number of specimens is required to restore the balance, so violently disturbed. Two birds of this month had eaten moths, and five had eaten cut-worms. The averages stand fifty-five per cent. of moths, caterpillars, June beetles, curculios and orthoptera, opposed to thirty-five per cent. of Carabidæ, soldier bugs and spiders. The Carabidæ include *Cratacanthus dubius, Agonoderus comma,* Anisodactylus, and Harpalus. Other details may be obtained from the table at the close of this paper.

June.

In June, ten birds---one from Mount Carroll, the others from Normal - -had taken a somewhat unusual diet. The ratio of spiders (eighteen per cent.) falls little short of that for May; but an examination of the notes show that here, too, a single bird had eaten nothing else. Ants rise suddenly from two per cent. in May, to twenty per cent. in June, taken by six of the birds. Most of these, however, were of the winged forms, and their number is evidently due to the same cause which rendered the Aphodii so abundant in April. Three of the birds of June proved, to my surprise, to have eaten raspberries, and one gooseberries - -these fruits amounting to eight per cent. of the food of the month. No cut-worms were recognized in June, but measuring worms (Phalænidæ) replaced them, composing six per cent. of the food. While all the cut-worms found in any month whose food was at all distinguishable had eaten nothing but grass---or endogenous foliage, more accurately speaking—several of these Phalænidæ had been feeding on netted-veined leaves. The Harpalinæ (six per cent.) include Evarthrus, sp., *Pterostichus lucublandus* and *Anisodactylus baltimorensis.* June beetles (Phyllophaga) had been eaten by one bird, a Melanotus, a curculio, and a long-horn beetle (*Tetraopes tetraophthalmus*), each by one. Cydnidæ reach

five per cent., chiefly *Hymenarcys nervosa*, and Orthoptera fall to
three per cent. The excess of ants is therefore taken, like the
excess of Aphodii, from the caterpillars and grasshoppers.

The averages of beneficial and injurious species stand thirty
per cent. to twenty-six per cent., respectively. Regarding ants, I
find such conflict of opinion among good authorities, that I am
not able to give them a definite place on either side the line. The
injury to fruits is probably too insignificant to be taken into account, except as evidence that the species is not strictly insectivorous, even in midsummer.

July.

The nine birds of this month were all shot in central Illinois,
during four successive years. Besides the return of the percentages of Hymenoptera, Coleoptera, Lepidoptera and Arachnida to
about their usual figure, we notice the large ratios of June-beetles
(twelve per cent.) and Orthoptera (twenty-seven per cent.). The
latter includes seven per cent. of *Udeopsylla nigra*, a large
cricket-like locust. We find also a trace of raspberries in the
food of two individuals. The caterpillars eaten by these birds
were unrecognizable, except those from a single stomach, which
Prof. Riley has identified as *Nephelodes violans*, Guen. The
record of benefit and injury is now more favorable to the species —
sixty-seven per cent. of injurious insects, and only fourteen per
cent. beneficial,— the latter Carabidæ and spiders.

August.

Twelve specimens were obtained in August, at Normal, three
early in the month and the others on the 29th and 30th. The
bluebirds were at this time most abundant in meadows and pastures; and the contents of their stomachs indicate that the chief
business of the month was the pursuit of locusts, crickets and
grasshoppers, moths and caterpillars. The Orthoptera eaten by
these birds amounted to fifty-eight per cent. of their food, and the
Lepidoptera to twenty-seven per cent. About half of the former
were Gryllidæ (Gryllus and Nemobius), and the remaining half
were equally Locustidæ and Acrididæ (*Xiphidium fasciatum* and
ensifer, Colopternus femur-rubrum and *bivittata*, and *Œdipoda
sordida*). Half of the Lepidoptera were unrecognizable moths, and
the remainder caterpillars—five per cent. being Noctuidæ. Ants

were about one per cent. of the food, Coleoptera only five per
cent. (including three per cent. Harpalidæ), Cydnidæ (*Cœnus
delia*) one per cent. and spiders six per cent. A few wild cherries
and elderberries were the only fruits taken. The beneficial elements thus amounted to nine or ten per cent. of the food and the
injurious elements to about eighty-five per cent.

September.

All but one of the ten specimens upon which the account of the
September food is based were shot at Normal, and all but two on
the 29th of the month. The chief peculiarity of the month is the
almost total disappearance of Coleoptera, which were represented
only by a few small Harpalids and a single minute Atænius. The
Lepidoptera rise to thirty-seven per cent., chiefly through the
abundance of the larvæ of *Prodenia lineatella*, Harvey. The Orthoptera make nearly half the food, the species differing from those
of the preceding month mainly in the greater number of red-
legged grasshoppers. Spiders were only two per cent. of the food;
and some unknown wild fruits formed seven per cent. It will be
seen that a striking change in the food of this species attends that
increase of the Orthoptera in numbers and activity, which occurs
in the late summer and early autumnal months, these insects being almost entirely substituted for Coleoptera, Hemiptera and
Arachnida. The Coleoptera of the six preceding months averaged
twenty-seven per cent. of the food, while this order amounts to
but three per cent. in August and September. The Orthoptera of
the foregoing months averaged but fourteen per cent., while those
of the two months in question rise to fifty-four per cent. It is
evident, from the foregoing, that Orthoptera and smooth caterpillars are the favorite autumnal food of this bird, and as the first
of these remain abundant until frost, it is not likely that the food
of October is much less favorable to the bird than that of September. The two specimens taken in the former month were well
filled with winged ants.

December.

To learn the food of the bluebird in midwinter, I went to extreme Southern Illinois in December, 1879, and shot a number of
specimens, some from the heavy forests in the bottoms of the Ohio
river, and others from the wooded and cultivated highlands in

Pulaski county. The weather at this time was sometimes above
and sometimes below freezing, and bluebirds were abundant and
very much at home. The principal food of the twelve specimens
examined consisted chiefly of various wild fruits (eighty-four per
cent.), of which the berries of the mistletoe (*Phoradendron flores-
cens*) were the most abundant (fifty-eight per cent.). Grapes, the
berries of sumach, scarlet thorn (Crataegus) and holly (*Ilex
decidua*) were also found. Sixteen per cent. of the food was
insects, of which the larger part (ten per cent.) was the larvæ of
Harpalinæ, eaten, however, by but two of the birds. Prominent
among these was the larva figured and described by Prof. Riley
in the Report of the United States Entomological Commission for
1877, p. 290, and there doubtfully referred to *Harpalus herbivagus*. The remaining kinds were *Geotrupes blackburnii*, *Podisus
spinosus*, a single spider, and one unknown caterpillar. Even in
the dead of winter, therefore, this bird does not cease its warfare
on our predaceous bugs and beetles.

Summary for the Year.

To these figures, giving the averages for all the months mentioned taken together (except October), I invite special attention.
Being derived from a much larger number of specimens than any
of the monthly averages, they are much less likely to be affected
by accident or error. They give, furthermore, the basis for an estimate of the total effect of the bird, year after year; and from
this we should be able to predict the probable effect of a destruction
or diminution of the species.

Taking up first the injurious insects destroyed, we find that
these include twenty-six per cent. of Lepidoptera, nearly two-
thirds of which were recognized as *Noctuidæ*, three per cent. of
leaf-chafers and twenty-one per cent. of Orthoptera, a total of fifty
per cent. on this side of the account. On the other hand, the
ichneumons amount to three per cent., the Carabidæ to seven per
cent., soldier-beetles to one per cent., soldier bugs to three per
cent. and spiders to eight per cent,- a total of twenty-two per
cent. of predaceous and parasitic forms. Other elements are ants
four per cent., Diptera only a trace, Aphodii six per cent., Iulidæ
one per cent. and vegetable food thirteen per cent. The edible
fruits amount only to about one per cent. of the food of these one
hundred and eight specimens. Comparing with the Turdidæ, we

find that the bluebird is essentially a thrush in food. From the robin it differs principally in the larger number of Hymenoptera (seven to four) and Lepidoptera (twenty-six to seventeen), the lack of Diptera (robin seventeen per cent.), the excess of *Aphodii* (six to two) of *Cydnidæ* (robin one per cent.) of *Orthoptera* (twenty-one to four) and of spiders (eight to a fraction); but especially in the matter of edible fruits (one to thirty-four). These differencse are but little greater, however, than those among the thrushes themselves. Compared with the thrush family as a whole, its salient peculiarties are its neglect of Diptera and garden fruits and its preference for Lepidoptera, Orthoptera and spiders.

ECONOMICAL RELATIONS.

Mr. B. D. Walsh, the first state entomologist of Illinois, reasoning from the comparative numbers of injurious and beneficial insects, concludes that a bird must be shown to eat at least thirty times as many injurious individuals as beneficial before it can be considered useful.[*]

According to this estimate, the bluebird does at least thirteen times as much harm as good; that is to say, the beneficial insects eaten would themselves have destroyed thirteen times as many injurious insects as the birds have eaten. This conclusion is so unexpected and astonishing that it certainly cannot pass without careful examination. In the first place we should bear in mind that nothing has yet been learned of the food of the young, and there is some reason for supposing that birds select the softer insects for their young. Whatever deficiency of credit may be due to this neglect of the food of the young, is compensated in part, at least, by the fact that the number of caterpillars eaten is doubtless over-estimated in comparison with hard insects, as their flexible skins remain in the stomachs of birds longer than the hard structures of insects. This is exactly contrary to the usual supposition, but the frequent occurrence of the empty and twisted skins of cutworms in the stomachs of these birds, still recognizable as Noctuidæ when not even a fragment of a single head remains, is sufficient evidence that the hard parts break up and disappear before these delicate but yielding skins. Secondly, while our knowledge of the food of Arctians, cut-worms and grasshoppers is sufficiently definite and full to enable us to predict with certainty exactly what would happen if those eaten by bluebirds were allowed to

[*] Birds vs. Insects. Practical Entomologist, Vol. II., pp. 44–47.

live and multiply, we have not the same complete and certain knowledge of the food-habits of the different genera of Ichneumonidæ, the ground-beetles, the soldier-bugs and soldier-beetles.

One hundred bluebirds, at thirty insects each a day, would eat in eight months about 670,000 insects. If this number of birds were destroyed, the result would be the preservation, on the area supervised by them, of about 70,000 moths and caterpillars, (80,000 of them cut-worms), 12,000 leaf-chafers, 10,000 curculios and 65,000 crickets, locusts and grasshoppers. How this frightful horde of marauders would busy itself if left undisturbed, no one can doubt. It would eat grass and clover and corn and cabbage, inflicting an immense injury itself, and leaving a progeny which would multiply that injury indefinitely. On the other hand, would the 160,000 predaceous beetles and bugs, spiders and ichneumons either prevent or compensate these injuries? I do not believe that we can say positively whether they would or not.

In a discussion of the natural checks upon the cut-worm, Prof. Riley, in his first report as State Entomologist of Missouri, mentions two species of Ichneumon that parasitize the larva, credits the Spined soldier-bug and the carabid larva, *Calosoma calidum*, with its destruction, and says that some kinds of spiders are known to prey upon it.

From the report of the United States Entomological Commission for 1877, we learn that the grasshopper is preyed upon at one or the other stage by Agonoderus, Harpalus, Amara and other Carabids; by soldier-beetles, soldier-bugs and spiders; and that certain Ichneumonidæ parasitize the egg. It seems *probable*, therefore, that the beneficial insects eaten by bluebirds include the special enemies of the cut-worms and grasshoppers it destroys; but he who knows best the small number of reliable observations upon which our general statements of the food of predaceous insects rest, will have the most hesitation in trusting them without reserve.

I would also call attention to the fact that we do not yet know that the normal rate of increase among these carnivorous and parasitic insects is not sufficient to keep their numbers full to the limit of their food supply, and to furnish also a *surplus* for destruction by birds. Just as a tree puts forth more leaves than it needs, and sets more fruit than it can possibly mature, as an offset to the constant, normal depredations of insects, so there is much reason to suppose that our insect friends have become adjusted to this steady drain on their numbers.

TABLE OF THE FOOD OF SIALIA SIALIS, L. THE BLUEBIRD.

Kinds of Food	Jan.	Feb.	March	April	May	June	July	August	Sept.	Oct.	Nov.	Dec.	Total	Ratio of each element to whole of food
Number of specimens		10	21	13	9	10	9	12	10	2		12	108	

Number of specimens and ratios in which each element of food was found.

Kinds of Food	Jan.	Feb.	March	April	May	June	July	August	Sept.	Oct.	Nov.	Dec.	Total	Ratio
I. INSECTS		10	21	13	9	9	10	12	10			5	101	.78
		.92	.88	.88	.76	.71	.89	.91	.91			.16		
1. Hymenoptera		7	5	5	3	7	6	4	4				41	.07
		.22	.04	.02	.02	.21	.04	.01	.04					
Formicidæ		2	5	3	6	6	4	4					30	.04
			.02	.02	.20	.04	.01	.04						
Ichneumonidæ		6	2										8	.03
		.22	.04											
2. Lepidoptera		8	20	9	7	5	7	7	7			1	69	.26
		.28	.38	.21	.39	.13	.27	.27	.37			.02		
Arctiinæ		1	4										5	.01
		.02	.04											
Noctuidæ		6	11	5	3		1	2	4				32	.12
		.24	.39	.17	.19		.10	.05	.14					
Phalænidæ						2							2	.01
						.06								
3. Diptera		1	1	1		1							4	......
		—	.01	—		—								
4. Coleoptera		9	16	13	8	7	10	7	4			4	78	.20
		.22	.19	.51	.10	.28	.26	.05	.02			.13		
Carabidæ		6	16	10	5	3	4	6	2			2	55	.07
		.05	.10	.08	.12	.06	.09	.03	.01			.10		
Dytiscidæ						1							1	......
						.01								
Staphylinidæ				1									1	......
				—										
Histeridæ		1	1										2	......
		—												
Byrrhidæ				1									1	......
				.01										
Scarabæidæ		3	4	10	2	5	3	3	1			1	32	.09
		.09	.04	.33	.05	.12	.14	.02	†			.03		
Aphodius		3	2	10	1	4			1				21	.06
		.09	.03	.31	.03	.04			†					
Geotrupes												1	1	......
												.03		
Phyllophaga				1	1	1	2						5	.03
				.02	.02	.08	.12							
Euryomia		1											1	......
		.01												
Elateridæ		1			1								2	......
		—			.01									
Tenebrionidæ			1										1	......
			.04											

TABLE OF THE FOOD OF SIALIA SIALIS, L. THE BLUEBIRD — Concluded.

	Jan.	Feb.	March.	April.	May.	June.	July.	August.	Sept.	Oct.	Nov.	Dec.	TOTAL.	Ratio of each element to whole of food.
Number of specimens.....	...	10	21	13	9	10	9	12	10	2	...	12	108	
KINDS OF FOOD.				Number of specimens and ratios in which each element of food was found.										
Telephorus	3 .08	2 .04											5	.01
Curculionidæ	1 —	1 —	3 .01	2 .01	1 .01	3 .01							11	
Cerambycidæ					1 .01	1 .02							2	
Tetraopes					1 .01	1 .02							2	
5. Hemiptera	4 .07	10 .04	7 .04	2 .02	3 .05	2 .05	1 .01	1 †				1 .11	31	.04
Coriscus		4 —	1 .01										5	
Alydus						1 .04							1	
Cydnidæ	3 .07	7 .04	4 .02	2 .02	3 .05	1 .01	1 .01					1 .01	22	.03
6. Orthoptera	7 .13	13 .21	5 .08	6 .13	2 .03	5 .27	12 .57	9 .48					59	.21
Gryllidæ	1 .04	2 .03		1 .02			1 .01	6 .28	2 .11				13	05
Locustidæ							1 .07	4 .15	1 .02				6	.03
Acrididæ	6 .09	11 .18	5 .08	5 .11	2 .03	5 .19	4 .14	6 .33					44	.13
II. ARACHNIDA	6 .04	16 .05	9 .09	6 .21	5 .18	2 .05	6 .06	3 .02			1	—	54	.08
III. IULIDÆ	6 .04	8 .03	2 .01	1 .01			2 .02						19	.01
IV. VEGETABLE FOOD		3 .04	4 .02	3 .02	5 .11	6 .04	3 .03	1 .07				12 .84	37	.13
Percentages for each Month. Beneficial elements	.46	.28	.21	.35	¹.38	.14	.10	.03				.11	.22	Totals for the Species
Injurious elements	.41	.60	.23	.55	.26	.67	.80	.85				.02	.49	
Neutral elements	.13	.12	.56	.10	.34	.19	.10	.12				.87	.29	

¹ Includes 8 per cent. fruit.

NOTES ON
INSECTIVOROUS COLEOPTERA

Stephen Alfred Forbes

NOTES ON INSECTIVOROUS COLEOPTERA

BY S. A. FORBES.

Mouth Structures of Carabidæ.

In studying the food of birds, I found it necessary to construct a key to the genera of the Carabidæ, based primarily upon the mouth structures, and prepared for this purpose a large number of slides of the mouth parts of Illinois species. In studying these, two characters were noted, which proved to be of considerable service for classification. The first of these is the frequent obliteration of the suture between the mentum and the gula (called the "gular suture", by Dr. LeConte, in his Classification, Pt. I., pp. X., XIII., 14, 15 and 16), the mentum being, in such cases, connate with the gula. This is true of Blechrus, although in Trechicus and Metabletus of the same group the suture is distinct. The mentum is again connate in many species, at least, of several genera of Dapti and Eurytrichi; viz., Geopinus, Anisodactylus, Xestonotus, Spongopus and Amphasia; but is not connate in Nothopus, Piosoma, Discoderes or Anisotarsus. This character was noticed nowhere else except in *Amara angustata*, which differs in this respect from all the other Amaræ in the Laboratory collection. This species is also peculiar in the very great development of the muscular ridges on the upper surface of the mentum. In the Lebiæ this mental suture is distinct in the middle but obsolete at the ends·

The second character referred to is found in the stipes of the maxilla. This body is covered with three plates—an outer, closely connected with the palpus, a lower, from which the two lobes of the maxilla spring, and an upper plate, which is applied to the under surface of the mandible. The last of these usually presents, in the Harpalidæ, a

more or less prominent angle at about the anterior third of
the outer margin, although this margin is sometimes regularly curved. In two genera, Agonoderus and Stenolophus, this plate is produced forward and outward beyond
the articulation with the palpus (which thus seems to
spring from beneath it), forming an oblique lamina with a
rounded outer angle and an acute tip. This character
seems to distinguish Stenolophus from Harpalus, as far as
I have been able to compare the species.

FOOD OF THE CARABIDÆ.

The large numbers of Carabidæ eaten by several of our
common birds make it important that the somewhat doubtful food habits of this family should be more thoroughly
studied; and I have undertaken the microscopic examination of the contents of stomachs and intestines as one
branch of this investigation. The facts thus obtainable
perhaps cannot give us a complete idea of the food of these
insects, but should probably be taken in connection with
field observations, as these beetles are said frequently only
to suck the juices of their prey, rejecting the solid parts;
and where this has been done the fact will be only obscurely indicated by the contents of the alimentary canal.
Where this contained an abundance of fatty chyme with
no solid tissues to fix its source, I have sometimes doubtfully inferred such an event; but usually liquid food will
escape detection.

The results of the examinations thus far made are so interesting that I am impelled to give the method I have
found most successful and convenient, with the hope that
others may turn their attention to the same subject. The
dissection should be made as soon as possible after the
beetle is taken,—within a few days at farthest,—as the
more unstable elements of the food are apparently soon
changed, even in strong alcohol. If the beetle is as large
as *Megilla maculata*, the elytra and wings may be cut off
and then, while the insect is held between the thumb and
finger of the left hand, the edges of the abdomen may be
carefully trimmed away with a pair of fine scissors (those
with curved blades are best) leaving the soft dorsal cover-

ing attached only at the base and tip. If one blade of the scissors be now carefully passed under this dorsal integument, it may be cut across and reflected (with the forceps and a mounted needle) forwards and backwards and cut entirely away. It will next be necessary to unroof the meso- and meta-thoracic segments, which usually contain at least a part of the crop. It will not be difficult to cut through the crusts of these segments at each side with the scissors-points. The terga may then be removed, as before, with forceps and needle. The specimen (if not too large) should now be transferred to a watch crystal, covered with glycerine and placed on the stage of the microscope; (a dissecting microscope is a convenience, but not indispensable). With mounted needles the reproductive organs, urinary tubes, etc., can be pushed out of the way, when the crop, stomach and intestine will be seen, variously arranged according to the family and genus. It is an easy matter to cut the alimentary canal loose at either end and to remove it from the body, placing it upon a slide in a shallow cell, with glycerine enough to mount the contents. Here the superfluous structures should be picked away, as far as possible, and then the stomach and intestines may be torn open with needles, and their contents spread out and picked in pieces upon the slide. After the removal of the remnants, the cell may be covered and the contents studied with any power necessary. The cover should, of course, be finally cemented down and the slide preserved for verification and repeated examination.

Galerita janus.—A specimen of this insect, taken at Bloomington, in September, contained but little food. All that was recognized consisted of insect fragments, one of which was a spinose tibia. It was impossible even to tell the order of the insect eaten.

Loxopeza atriventris.—Four specimens of this species were examined, three of which were taken in June and the other in September. The alimentary canal of the first was entirely empty. The second, sent me by Mr. A. S. McBride, from DeKalb county, had eaten immense numbers of minute, oval bi-nucleate cells, which, believing them to be spores of fungi, I referred to Prof. T. J. Burrill,

of the Illinois Industrial University. He reported them
to be "spores of Sphæronemei, probably Phoma"—a fungus which forms small, black specks on dead wood, stems
of weeds, etc. A third specimen from the same source
had eaten some undetermined, insect and about equal quantities of three elements; viz., the above spores of Phoma,
pollen and the anthers of grass (doubtless blue-grass upon
which the insect was taken). A few clavate bodies were
also noticed, consisting of a single row of nucleated cells,—
evidently the acrospores of some fungus. A September
specimen was taken at Normal. Its crop was distended
with an oily liquid, but contained no other visible food
except a few acrospores of a fungus. This specimen had
evidently been feeding upon animal food of some sort.

Calathus gregarius.—Three individuals of this species
were examined, all caught on blue-grass in blossom, by
Mr. Webster, of Waterman, and Mr. McBride, of Freeland.
The crop and œsophagus of the first were distended with a
brown mass which proved to be wholly made up of the
pollen and fragments of the anthers of grass. A second
specimen contained a smaller amount of pollen and anthers of blue-grass, with minute fragments of a black and
sparsely hairy insect. An antenna proved that it was a
larva—probably a young caterpillar. The third contained
traces of a similar larva and the fragments of the cornea
of a perfect insect—evidently a remnant of some former
repast.

Anisodactylus baltimorensis.—The single specimen of
this species had not recently taken food. The stomach
was empty; but in the intestine was a large amount of
chyme which possibly indicated liquid animal food. A
specimen of *A. rusticus* gave only similar negative results.

Anisodactylus sericeus.—A specimen taken in June
showed fragments of anthers and pollen of grass, with
other vegetable tissues, apparently derived from the seeds
of grass. A small insect had also been eaten, as shown by
particles much too few and minute for determination. A
second specimen had taken precisely similar food—the insect here being represented by a few facets of the cornea.

Amara angustata.—One of this species, likewise taken

in June, had also fed on vegetation, as indicated by a few particles of parenchyma too far digested for recognition; but fully nine-tenths of its food consisted of spherical eggs, in different stages of development, many of them easily recognizable as the eggs of mites. The most advanced embryos had six legs and a pair of large palpi; and, by the shape of the abdomen and the position of the legs, recalled the larvæ of the spinning mites (Tetranychi).

Harpalus pennsylvanicus De G.—A specimen of this species taken running in the road, at Normal, August 31st, had the alimentary canal well filled with vegetable tissues, some of which were evidently derived from the ovules and roots of grass. Among these were the tips of an ovule with the styles unbroken and the tip of a rootlet with the root-cap entire. A single mite was found, and a few acrospores of fungi. This beetle was infested by a large number of intestinal parasites of the genus Gregarina. A second specimen had eaten similar vegetable food. Here a piece of the epidermis of a rootlet, still covered with trichomes, was noted, as well as several root-tips and fragments from the growing tips of grass. Pieces of the epidermis of grass with their peculiar zigzag cell boundaries, confirmed these determinations. A detached stigma of a grass floret and a few stylospores completed the food. A third specimen, taken at Normal on the 5th of Septem. ber, contained some vegetable tissues with spiral cells, the mandible and maxilla of an ant and vast numbers of minute, spherical corpuscles, which Professor Burrill regarded as forms of bacteria such as occur on stagnant water. This beetle had apparently skimmed this minute vegetation from the surface of some pool. The fourth specimen of this species, received in September, from Mr. Webster, who collected it from the blossoms of ragweed, I found to have eaten large quantities of vegetable tissue, the fragments of which showed branched bundles of spiral ducts with parenchyma between. These were evidently the bracts or other floral organs of the ragweed.

Harpalus caliginosus.—A single individual, running free upon the ground, had gorged itself with plant and animal food,—apparently about three times as much of the former as of the latter. In the crop were a few hairs of a caterpillar and much half-digested muscle, with spores of fungi, a little epidermis of some graminaceous plant and a few pollen grains of Compositæ. In the stomach was a great deal of chyme, with fragments of the wings and tarsi of some minute dipter, more pollen of Compositæ and some vegetable parenchyma, apparently derived from unripe seeds of grass. In the ileum and colon these last mentioned tissues predominated, although the latter contained also a large quantity of pollen of Compositæ indistinguishable from that of ragweed (Ambrosia). Here were also found two feet of a larva,—possibly of the previously mentioned caterpillar. It is worthy of notice that these Harpali were full of eggs, of which there were about six in each abdomen. The crop of the second specimen, taken at Normal, in September, was distended with a brown, oily fluid, containing no recognizable material. In the intestine was a small mite and considerable vegetable parenchyma, apparently derived from some young seeds or ovules of plants. A little parallel-veined vegetable tissue was also seen, evidently derived from grass.

Harpalus herbivagus.—A specimen of this beetle, taken by Mr. McBride, in July, was filled with cryptogamic vegetation which had the form of a dense mat of slender branching tubes enclosing many spherical cells. This, Professor Burrill, to whom one of the slides of this material was referred, regarded as a fleshy or cartilaginous fungus with Palmella cells, although he thought that it might have been derived from a lichen. A second specimen, obtained by Mr. Webster, in March, had evidently been feeding on the young shoots of grass.

Cratacanthus dubius.—One of this species, taken at Normal, in August, contained no apparent food except a few spores of fungi. In the stomach were great numbers of Gregarina, apparently of the same species as those found in *Harpalus pennsylvanicus.* In the colon, especially, scores of these parasites in the "resting state" formed considerable masses which half filled the intestines.

Evarthrus colossus.—One of this species, taken in September, had eaten a brown beetle of medium size, the fragments of which filled the whole alimentary canal. From the general appearance of these, from the tips of one anterior and one middle tibia and from a maxillary palpus, it was inferred that this beetle was one of the Scarabæidæ· A fragment of a mandible showing a ridged masticatory surface, made it likely that it was a vegetable feeder. There was no trace of vegetable food in this Evarthrus. Another specimen, taken at Normal, in September, had eaten a large Coleopterous larva and two minute, indeterminable insects. Traces of confervoid Algæ were also discovered in the intestine.

Pterostichus sayi.—A specimen of this species, taken at Normal, in September, was full of the remains of an unrecognized hairy insect with two tarsal claws.

Pterostichus lucublandus.—This specimen, taken likewise at Normal, in September, contained a multitude of fragments of some Hymenopterous insect, including a maxillary palpus and a labrum nearly entire, with pieces of the legs and tarsi. This beetle had also eaten a small mite and a few acrospores of fungi.

Chlænius tomentosus.—One of this species, taken at Normal, in September, contained traces of insect food not otherwise determined, and a nematoid parasite.

Chlænius diffinis.—A specimen of this species, taken under a log, near Normal, in September, contained traces of some crustaceous insect, with pieces of vegetable tissue (apparently wood) penetrated by the mycelium of a fungus. Large vegetable fragments were also seen, which Professor Burrill determined as pieces of a large, fleshy fungus. The stomach likewise contained acrospores of Dematiei.

Bradycellus dichrous.—A specimen, taken at Bloomington, in September, had eaten insect food not otherwise determinable.

Twenty-eight specimens of Carabidæ, representing seventeen species, are here reported. It will be seen that twenty-one specimens, belonging to fifteen species, had eaten animal food, and that twenty specimens, belonging to eleven species, had eaten vegetation of some sort. I estimated as carefully as possible the relative amounts of these two kinds of food in the alimentary canal of each insect, and from these data concluded that about half the food of these twenty-eight specimens consisted of vegetation, and that one-third of it consisted certainly of insects,—the remainder being made up of doubtful animal matter. About one-third of the vegetable food had been derived from cryptogamic plants and another third from the different structures of grasses, Compositæ and other miscellaneous vegetation making up the remainder. Considering the fact, however, that the commonest species were found feeding upon vegetation far the most generally, it is likely that, taking the Carabidæ as a group, not more than one-third or one-fourth of their average food consists of animal matter.

Food of Podabrus.

The contents of three stomachs of *Podabrus tomentosus* were examined; and all these had eaten only the spores of Phoma mentioned under Loxopeza. The specimens were all sent me in July, by Mr. A. S. McBride, of Freeland, Ill.

Food of Coccinellidæ.

Coccinella novem-notata.—Two specimens which were taken at Normal, in August, were examined, agreeing very closely in their food, each having eaten various spores of fungi (about ninety per cent.) and plant-lice (ten per cent.). Among the fungus spores, Professor Burrill, to whom they were submitted, recognized spores of *Ustilago and Helminthosporium*; and a few lichen spores were also noticed.

Brachyacantha ursina.—The stomach of one individual of this species contained only a few fungus spores.*

Hippodamia convergens.—A specimen, captured in August, at Normal, had eaten great quantities of fungus spores, which composed about three-fourths of its food. Fragments of a mite and a plant-louse and a little pollen of Compositæ were also found. In a second specimen, taken in September, the remains of a myriapod belonging to the family Geophilidæ, acrospores of a fungus, the pol-

---

*I have assured myself that none of the fungi found in the alimentary canals of these beetles were entophytes.

len of Compositæ and the remains of a plant-louse were the only elements noticed.

Megilla maculata.—Three specimens of this species were dissected,—one received from Mr. Webster in May, one from Mr. McBride in July, and one taken at Normal in September. The specimen from Mr. Webster was captured on the flowers of dandelions. Its entire alimentary canal was closely packed with hexagonal, spinose pollen cells, doubtless taken from that plant. A second had eaten the anthers and pollen of grass with a few spores of Myxogastres.* The third specimen contained pollen and fungus spores in about equal quantities. While these Coccinellidæ had made good their usual reputation as enemies of plant-lice, it should be noticed that these constituted only about ten per cent. of their food.

If these specimens of the various families of predaceous beetles are fair examples of their class, the above facts imply that the individual carnivorous insect is much less valuable than has usually been supposed, while predaceous insects as a class are much more beneficial. If these species are predaceous, as a rule, not more than from one-fourth to one-third of the time, the injury done by the destruction of one of them is very much less than if they were, as is usually supposed, almost wholly carnivorous. But, on the other hand, if they can live on the soft parts of plants when animal food becomes scarce, their numbers will be maintained at a far higher figure than would be possible if they were dependent upon animal food alone. Preferring animal food to vegetable, as they doubtless do when equally obtainable, they operate as a much more effective check on the undue increase of other insects than if their number were at all times strictly limited by the numbers of their food species. We should remember, in this connection, that we cannot ordinarily expect of any predaceous animal that it will do more than to eliminate the excess of the species it preys upon, keeping their numbers down within certain constant limits. As a prudent sovereign finds it worth while to maintain a much larger

*Burrill.

fighting force than is necessary to the ordinary adminis-
tration of his government, in order that he may have always a reserve of power with which to meet aspiring rebellion, so it is to the general advantage that carnivorous
insects should abound in larger numbers than could find
sustenance in the ordinary surplus of insect reproduction.
They will then be prepared to concentrate an overwhelming attack upon any group of insects which becomes suddenly superabundant. It is evidently impossible, however, that this *reserve* of predaceous species should be
maintained unless they could be supported, at least in
part, upon food derived from other sources than the bodies
of living animals.

THE REGULATIVE ACTION OF BIRDS
UPON INSECT OSCILLATIONS

Stephen Alfred Forbes

THE REGULATIVE ACTION OF BIRDS UPON INSECT OSCILLATIONS.

By S. A. FORBES.

Attention has already been repeatedly called in these studies to the fact (fundamental to this investigation) that the principal injuries due to insects are done by a few species, existing, for a time, in numbers far above the average, and soon to retire again to a much lower limit. As the number of a species which reach maturity is determined by the checks on its multiplication, it follows that these oscillating species are held in check by variable forces, and to the variations in these checks we must look for an explanation of their oscillations. On the other hand, we must expect to find that those insects whose numbers remain relatively constant from year to year are under the control of restraining influences of a much more uniform character than the preceding class.

Concerning the effects of birds upon insect life, and through this upon the interests of agriculture, there are therefore three questions to answer :—

1. Do birds originate any oscillations among the species of insects upon which they feed ? That is, are their food habits ever so inconstant from year to year that species which are at one time principal elements of their food, are at other times neglected and allowed to multiply without restraint ?

2. Do birds prevent or restrain any oscillations of insects now noxious, or capable of becoming so if permitted to increase more freely ? That is, do they bring to bear upon any such species a constant pressure so great that those insects would increase unduly if this pressure were removed by the destruction of the birds ?

3. Do they do anything to reduce existing oscillations of injurious insects ? Do they sometimes vary their food habits so far as to neglect their more usual food and take extraordinary numbers

of those species which, for any reason, became superabundant for a time ?

For the purpose of answering these questions, two separate lines of investigation are necessary. For the first two we require a knowledge of the food habits of the various species of birds under ordinary circumstances, when the conditions of life are of average character, and especially when no species of insects are unusually and excessively abundant. On the other hand, for an answer to the third question we must look to the food habits of the birds under extraordinary circumstances, where the opposite condition of affairs prevails. We must learn to what extent birds depart from their usual practices when confronted by an uprising of some insect species. If they concentrate for its suppression, they must assist more or less effectively to reduce to order the disturbed balance of life; but if they remain indifferent to this condition of things, their influence is *nil*.

The present paper is a contribution to a discussion of the last of the above questions. As a striking and conclusive example of an extraordinary condition of insect life, and of the food of birds in the presence of a disturbed balance of nature, I selected an orchard which had been for some years badly infested by canker-worms, shot a considerable number of birds therein for two successive years, representing nearly all the kinds seen in the orchard, made full notes of the relative abundance of the species, examined carefully the contents of all the stomachs obtained, with reference not only to the presence of canker-worms but of all other insects as well, and tabulated the results as the basis of this paper. Besides preparing as full an account of the food of these birds as practicable, I have brought the summaries on these tables into comparison with those derived from birds of the same species shot in ordinary situations during the same month. These comparisons have been confined to a few of the kinds obtained in the orchard, for the reason that most were not found there in sufficient number to give a fair idea of the average food of the species. The collections were made in an orchard of forty-five acres of bearing apple-trees (belonging to Mr. J. W. Robison) in Tazewell County, Ill., which had been infested by canker-worms for about six years. As a result of their depredations, a considerable part of the orchard had the appearance, from a little distance, of having been ruined by fire. Closer examination of the trees most affected showed that the branches, stripped of every vestige of green, were festooned with the webbing left by the worms. To the webs the withered remnants of the leaves adhered as they fell, the very petioles having been gnawed off at the twigs. Not one per cent. of the trees were uninjured, and these were invariably on the outer part of the orchard. Those which had been attacked several years in succession were killed; and there was a large area in the midst of the orchard from which such trees had been removed. One did not need to enter the enclosure to learn that the birds were present in extraordinary numbers and variety. From every part of it arose a chorus of song more varied than I had ever heard in any similar area at that season of the year. Most of the common summer residents were found there; and upon a second visit in 1882 many of the migrant species likewise occurred. The first collection was made on the 24th of May, 1881, and the second on the 20th of the same month im the following year. The season was less advanced at the time of the second collection than at the first, so that the actual difference between the two was probably not less than two weeks. At the first visit fifty-four birds were taken, representing twenty-four species, and seven other species were noted in the orchard of which no specimens were obtained. On the second visit ninety-two birds were shot, representing thirty-one species, and four other species were seen. In 1881 the worms were nearly all fully grown, and many of them had already entered the ground for their transformation, so that the larvæ were less abundant than they had been earlier. In 1882 most of them were about half-grown, only a few having reached adult size. They were distinguishable with difficulty upon the leaves of the trees; but when a large branch was shaken or jarred, from a dozen to twenty would expose themselves by spinning down and hanging at the end of a thread. The owner of the orchard informed me that they were about twice as abundant the preceding season.

TURDIDÆ. Thrushes.

TURDUS MIGRATORIUS, L. ROBIN.

This species was abundant and nesting in the orchard. Nine specimens were obtained in all, three in 1881 and six in the fol-

lowing year. The food was wholly animal, neither fruit nor any
other kind of vegetation having been taken by any of the birds.
Only three of the above number had eaten canker-worms, which
composed, as nearly as could be estimated, about one-fifth of the
food of the entire group. Insects made ninety-three per
cent., the remainder consisting of a common species of myriapod (five per cent.), earth-worms, and gasteropod mollusks. Ants were eaten by these birds only in trivial
numbers. Diptera, Orthoptera and spiders were conspicuous
by their entire absence. Cut-worms were extraordinarily
prominent in the food, making twenty-eight per cent. of the
whole. Half of them consisted of a single large, injurious
species (*Nephelodes violans*). Among the Coleoptera, which
amounted to thirty-six per cent. of the whole, the Scarabæidæ
and Elateridæ were the principal elements, the former represented
by eighteen per cent., and the latter by eleven. Among the Scarabæidæ was a species known as a vine leaf-chafer (*Anomala
binotata*), which made fourteen per cent. of the food. This insect was scarcely less abundant than the canker-worm, and
appeared in extraordinary numbers in the food of nearly all the
species of birds examined, although it had not attracted the attention of the owner of the grounds. I searched a small vineyard adjacent, but saw no signs of unusual injury to the leaves. Carabidæ,
although common in the orchard, had scarcely been touched by
the robins, only a single specimen of the family occurring.
Hemiptera were found but in trivial numbers, representing about
equally the families Coreidæ and Cydnidæ. Hymenoptera were
still less abundant, composing only one per cent. of the food.

MIMUS CAROLINENSIS, L. CATBIRD.

This species was very common, and thoroughly at home among
the trees, where it was doubtless nesting. Fourteen specimens
were taken, three at the first visit and eleven at the second.
With the exception of two per cent. of myriapods, their food consisted entirely of insects. Canker-worms had been eaten by eight
of the birds, but not in any great number, as they composed but fifteen per cent. of the food of the species. A few cut-worms had
been taken, and a larger number of other caterpillars, bringing
the total for Lepidoptera up to about one-fourth of the food.

The catbird had shown its usual preference for ants, eating fourteen per cent. of these insects. These birds had taken an unusual number of Coleoptera, which made more than half the food, chiefly Scarabæidæ. About two-thirds of them belonged to the single species (*Anomala binotata*) mentioned above under the food of the robin. Three of these birds had likewise eaten large June bugs. Elateridæ and their larvæ occurred only in trivial quantities, while Carabidæ amounted to four per cent., chiefly Anisodactylus. As in the robin, Diptera, Orthoptera, and Arachnida, were not represented in the food.

HARPORHYNCHUS RUFUS, L. BROWN THRUSH.

This bird was not common in the orchard, and only four specimens were taken. The food of these was entirely animal, an unexpected circumstance, as the brown thrush usually feeds largely upon grain. Six per cent. of the food consisted of thousand-legs, and insects made the entire remainder. Lepidoptera were about one-fifth of the food, and half of these were canker-worms. Like the preceding species, this bird had eaten an enormous number of beetles, which amounted to two-thirds of its food. Twelve per cent. of the whole was Carabidæ, chiefly a species of Chlænius. Scarabæidæ stand at forty-four per cent., largely Diplotaxis, Melolontha, and Anomala. Six per cent. were Elateridæ, and three per cent. Rhynchophora. No specimens of the remaining orders had been eaten by these birds.

Summary of the Family.

Treating, now, of the twenty-seven thrushes mentioned as one group, we find that none of them had eaten any vegetation whatever; that ninety-six per cent. of their food consisted of insects (myriapods and earth-worms making up the remaining four per cent.); that sixteen per cent. was canker-worms; and only four per cent. predaceous beetles. The Anomala previously mentioned made just a fourth of their entire food, other Scarabæidæ bringing up the average of that family to thirty-eight per cent. Click beetles (Elateridæ) with their larvæ were five per cent. of the whole, and snout beetles (Rhynchophora) two per cent.

SAXICOLIDÆ.   Bluebirds.

SIALIA SIALIS, L.   BLUEBIRD.

This species was not at all abundant in the orchard in either year. Only one was taken in 1881, and four in 1882. All but two per cent. of the food of these five specimens consisted of insects, spiders making the remainder. Canker-worms were twelve per cent. of the food, and other Lepidoptera five per cent. additional. Two-thirds of the food consisted of Coleoptera. Carabidæ made more than one-third (twenty-three per cent.), belonging chiefly to a species (*Anisodactylus baltimorensis*) which depends largely upon vegetable food. Four of the birds had eaten *Anomala binotata*, which made thirty-six per cent. of the food of the whole. Five per cent. was Chrysomelidæ, and fifteen per cent. Hemiptera, all belonging to the family Cydnidæ.

PARIDÆ.   Chickadees.

PARUS ATRICAPILLUS, L.   BLACK-CAPPED CHICKADEE.

This little bird, unfortunately, was not at all common in the orchard; and only two specimens were taken, one in each year. Sixty-one per cent. of their food consisted of canker-worms, eaten by both the birds, and Coleoptera made the entire remainder. These were nearly all Cerambycidæ (*Psenocerus supernotatus*) and Rhynchophora of undetermined species, twenty-five per cent. of the former, and ten of the latter.

TROGLODYTIDÆ.   Wrens.

TROGLODYTES DOMESTICUS, Bartr.   HOUSE WREN.

Several specimens of this little species were observed, some of them evidently nesting. The food was chiefly insects,—all, in fact, but six per cent. of spiders and one of thousand-legs. Nearly half the food of these birds consisted of canker-worms, and other Lepidoptera and their larvæ brought the average of the order up to fifty-nine per cent. A few gnats and other Diptera (four per cent.) and five per cent. of ants were also noted. Coleoptera and Hemiptera were taken in nearly equal quantities, thirteen per cent. of the former and ten of the latter. Two of the birds had eaten *Psenocerus supernotatus*, amounting to four per cent. of the food, and the other Coleoptera were scattered through the families Carabidæ, Nitidulidæ, Scarabæidæ, Elateridæ and Calandridæ. The Hemiptera were represented by trivial numbers of four families, including a few chinch bugs.

MNIOTILTIDÆ.   Warblers.

HELMINTHOPHAGA PEREGRINA, Wils.   TENNESSEE WARBLER.

A single specimen of this little warbler was taken in 1882. Four-fifths of its food consisted of canker-worms, and all the remainder of a single species of beetle (*Telephorus bilineatus*).

DENDRŒCA ÆSTIVA, Gmel.   SUMMER YELLOW BIRD.

This bird, common every where at this season, was also abundant in the orchard. Five specimens were shot in all. The food was insects, excepting six per cent. of spiders. Two-thirds of the total amount eaten by all of the birds consisted of canker-worms. Coleoptera were twenty-three per cent. of the whole amount, six per cent. being Aphodius, and twelve per cent. *Psenocerus supernotatus*, already frequently mentioned. Carabidæ and Calandridæ were represented by insignificant ratios, and Lampyridæ by a single Telephorus eaten by one of the birds. One per cent. of Hemiptera, and two of Hymenoptera complete the record.

DENDRŒCA PENNSYLVANICA, L.   CHESTNUT-SIDED WARBLER.

Two specimens of this abundant migrant were shot in the orchard in 1882. Like the preceding warbler, two-thirds of their food consisted of canker-worms, and an additional ten per cent. of other caterpillars. A few ants were eaten by both of the birds. Eleven per cent. of Coleoptera, likewise eaten by the two, was about equally divided between some undetermined Scarabæidæ and *Psenocerus supernotatus*. One of the birds had eaten plant-lice, which amounted to five per cent. of the food; and both had taken ants to the amount of six per cent.

DENDRŒCA STRIATA, Forst.   BLACK-POLL WARBLER.

Four of these birds were shot in 1882. Some undetermined seeds found in the crop of one of them reduced the insect ratio to ninety-five.   Again two-thirds of the food consisted of canker-worms.   The same little borer (*Psenocerus*) eaten by so many of the smaller birds in this orchard, made fifteen per cent. of the food; and an Aphodius and an undetermined carabid bring up the ratio of the Coleoptera to nineteen per cent.   Four per cent. of ants, a few gnats (five per cent.), and traces of Hemiptera and mites were the only other elements detected.

DENDRŒCA VIRENS, Gm.   BLACK-THROATED GREEN WARBLER.

A single specimen of this migrant was shot in 1882.   Seventy per cent. of its food consisted of canker-worms, fifteen per cent. of Psenocerus, and five of undetermined Hemiptera.   The remaining ten per cent. was made up of trivial numbers of Hymenoptera, gnats, coleopterous larvæ and mites.

GEOTHLYPIS TRICHAS, L.   MARYLAND YELLOW-THROAT.

This resident warbler occurred but sparingly in the orchard. One specimen was seen in 1881, and two were obtained in 1882. Lepidoptera made four-fifths of their food, about equally canker-worms and undetermined caterpillars.   A few Staphylinidæ and some specimens of Psenocerus composed the eight per cent. of Coleoptera.   A small hemipter (*Piesma cinerea*) amounted to five per cent., and four per cent. was gnats.

Summary of the Family.

Of the warbler family as a whole, as represented by these fifteen specimens, I need only remark that fourteen of the birds had eaten canker-worms, which composed nearly or quite two-thirds of the food of the group; that ten per cent. consisted of *Psenocerus supernotatus;* and that the remaining averages, with the exception of six per cent. of undetermined caterpillars, were so much subdivided as to have little or no significance.

VIREONIDÆ.   Vireos.

VIREO GILVUS, V.   WARBLING VIREO.

Three specimens of this little bird were shot, of purely insectivrous habit.   They had eaten canker-worms to the amount of forty-four per cent.; and other caterpillars made thirty-five per

cent. additional.   A few Coleoptera (fifteen per cent.) of which
one-third were carabid larvæ, and three per cent. of Cydnidæ
(*Podisus*), were the only other important elements.   *Anomala
binotata* (eight per cent.), Telephorus, and an undetermined long-
horn, were the other Coleoptera.

AMPELIDÆ.   Wax-wings.

AMPELIS CEDRORUM, V.   CEDAR WAX-WING.

A flock of about thirty of these birds was repeatedly started
in the orchard during the first visit, but none were seen in 1882.
Seven of the flock were shot, and the contents of their stomachs
carefully studied.   With the exception of a few Aphodii eaten
by three of the birds in numbers too insignificant to figure in the
ratios, the entire food of all these birds consisted of canker-worms,
which therefore stand at an average of one hundred per cent.   The
number in each stomach, determined by actual count, ranged from
seventy to one hundred and one, and was usually nearly a hundred.   Assuming that these constituted a whole day's food, the
thirty birds were destroying three thousand worms a day, or
ninety thousand for the month during which the caterpillar is
exposed.

HIRUNDINIDÆ.   Swallows.

PETROCHELIDON LUNIFRONS, Say.   CLIFF SWALLOW.

This species was nesting in great numbers under the eaves of a
barn at the edge of the orchard, and many of the birds were
continually circling through the air.   A single specimen was shot,
and found to contain nothing but the very abundant scavenger
beetle (*Aphodius inquinatus*), with about two per cent. of 'undetermined Hemiptera.

FRINGILLIDÆ.   Finches.

ASTRAGALINUS TRISTIS, L.   AMERICAN GOLDFINCH.

A flock of these birds passed through the orchard, but only a
single one was shot.   No canker-worms had been eaten by it;
but about seventy per cent. of its food consisted of undetermined
seeds, and the remainder of a harpalid beetle.

2

COTURNICULUS PASSERINUS, Wils. YELLOW-WINGED SPARROW.

A single specimen of this bird, shot in 1881, contained spiders thirty per cent., seeds of pigeon grass (*Setaria*) fifteen per cent., an unrecognized beetle five per cent., and some undetermined caterpillars, certainly not canker-worms.

SPIZELLA DOMESTICA, Bart. CHIPPING SPARROW.

This species was not common in the orchard in 1881, and only a single specimen was obtained; but in the following year it was found much more abundant, and seven additional were taken. About one-third of the food consisted of caterpillars, half of which were recognizable as canker-worms. A large number of gnats (twenty-eight per cent.), nearly as many Coleoptera, (principally Scarabæidæ, including nine per cent. of Anomala), and six per cent. of Hemiptera, are all the other noteworthy items.

SPIZELLA AGRESTIS, Bart. FIELD SPARROW.

This species was less abundant than the preceding, and was represented by only three specimens. With the exception of five per cent. of gnats, and one of Hemiptera, the food of this bird was equally divided between Lepidoptera and Coleoptera. Nearly half the former consisted of canker-worms, while the Coleoptera were represented by Histeridæ, Scarabæidæ (chiefly the scavengers), Monocrepidius and Rhynchophora.

SPIZA AMERICANA, Gmel. BLACK-THROATED BUNTING.

This bird was the most abundant species in 1881, though but few were seen during the following May. Eleven were shot at the first visit and three at the second. With the exception of a little wheat eaten by two of the birds, and a trace of undetermined seeds, the food consisted almost entirely of insects and mollusks, eighty-eight per cent. of the former and six of the latter (Helix). Ten of these birds had eaten canker-worms, which made forty-three per cent. of the food of the entire group; Lepidoptera as a whole composing two-thirds of the food. Among the twenty-two per cent. of Coleoptera, we note Harpalus and Histeridæ, each four per cent., Aphodius and Anomala likewise each four per cent., and Sphenophorus and other Rhynchophora, two per cent.

ZAMELODIA LUDOVICIANA, L.   ROSE-BREASTED GROSBEAK.

Only two were seen, and both were killed. A very few canker-worms were found (five per cent.) with fifty-eight per cent. of other caterpillars. About half the fifteen per cent. of Coleoptera were Rhynchophora, the remainder being *Anomala binotata*, one of the Lampyridæ, and undetermined specimens. One-fifth of the food consisted of seeds not recognized.

PASSERINA CYANEA, L.   INDIGO BIRD.

This bird, noted as common in 1881, was by far the most abundant species in the orchard at the second visit. Eighteen specimens were shot, two in the first and the remainder in the second year. Although this bird is one of the typical finches, only three per cent. of its food consisted of seeds, chiefly Setaria and Compositæ. Canker-worms made fifty-nine per cent., eaten by all the birds but one, and other caterpillars an additional eight per cent. With the exception of a trace of Hymenoptera, the remainder of the food consisted entirely of beetles, about one-third of which were *Anomala binotata*.

S u m m a r y   o f   t h e   F a m i l y .

Only seven per cent. of the food of the forty-seven members ot this family (commonly called seed-eaters) consisted in fact of seeds; and insects made up all but two per cent. of. the remainder. The most interesting items on the general list are canker-worms forty per cent., predaceous beetles (Carabidæ) two per cent., and *Anomala binotata* six per cent.

ICTERIDÆ.   Blackbirds.

MOLOTHRUS ATER, Bodd.   COWBIRD.

A single wandering specimen of this bird contained only Scarabæidæ, including Aphodius, and a few other Coleoptera, with about sixty per cent. of corn and some seeds of Polygonum and other plants.

AGELÆUS PHŒNICEUS, L.   RED-WINGED BLACKBIRD.

Two specimens of this bird, which were also accidentally in the orchard, had fed about equally upon insects and upon wheat and other seeds. The Lepidoptera (twenty-seven per cent.) were nearly all the larvæ of *Nephelodes violans.* Of the Coleoptera (eleven per cent.), part were Anomala and Elateridæ, and the remainder consisted of specimens of *Tanymecus confertus,* eaten by one of the birds. A grasshopper had also been taken by one, making ten per cent. of the food; and traces of Hemiptera were recognized.

ICTERUS GALBULA, L.   BALTIMORE ORIOLE.

Not common. Three were shot. These had fed only on insects,—Lepidoptera forty per cent. and Coleoptera sixty per cent., the former all canker-worms, and the latter chiefly *Anomala binotata* (fifty per cent.). Six per cent. of Cerambycidæ and two of Rhynchophora should also be mentioned.

ICTERUS SPURIUS. L.   ORCHARD ORIOLE.

This bird was common in 1881, although but two were shot; but was not noticed the next year. More than three-fourths of the food of these consisted of canker-worms, and other caterpillars made an additional twenty per cent., leaving but three per cent. for ants.

QUISCALUS PURPUREUS ÆNEUS, Bartr.   BRONZED GRACKLE.

Wandering specimens of the grackle were seen, and a few were apparently roosting in the trees at night. But three were shot, all of which had fed chiefly upon corn, which amounted to sixty-two per cent. of their food. Fragments of a crawfish were found in the stomach of one. Half the thirty per cent. of Coleoptera were Carabidæ, including a specimen of *Calosoma calidum,* and the remainder were nearly all Lucanidæ (Dorcus, eight per cent.) and undetermined Elateridæ.

Summary of the Family.

The five species of this family mentioned were represented by but eleven specimens, which, taken together, were found to have made two-thirds of their food of insects, the remaining third of corn and wheat with a few seeds of weeds. Canker-worms, eaten by the orioles, only amounted to one-fourth of the food of the whole, and Coleoptera to a little more than another fourth. Of these, Carabidæ made four per cent., Cerambycidæ two, Rhynchophora one, and *Anomala binotata* fourteen.

TYRANNIDÆ.   Flycatchers.

TYRANNUS CAROLINENSIS, L.   KINGBIRD.

This species was not uncommon, but only three were shot. Two of these, to my surprise, were found to have eaten cankerworms, which made more than a fourth of the food of the whole. Five per cent. of the remainder consisted of undetermined Hemiptera, and all the balance was Coleoptera. Seven per cent. was Elateridæ, two Lampyridæ, and more than fifty-eight Scarabæidæ, all Anomala except thirteen per cent. of *Aphodius inquinatus*, eaten by one of the birds.

CONTOPUS VIRENS, L.   WOOD PEWEE.

Three of these were shot, none of which had taken cankerworms. Their food consisted chiefly of flies and gnats, which amounted to fifty-five per cent. Thirteen per cent. of Aphodius and ten per cent. of Ips, with a few ants and other Hymenoptera, are also worthy of mention.

EMPIDONAX TRAILLI, Aud.   TRAILL'S FLYCATCHER.

Two specimens, shot in 1882, had eaten only insects, one-fourth of which were canker-worms, and one-third Ichneumonidæ. Another fourth consisted of Coleoptera, nearly half of which were Anomala; and ten per cent. were ants and other Hymenoptera.

EMPIDONAX FLAVIVENTRIS, Bd.   YELLOW-BELLIED FLYCATCHER.

A single specimen had eaten a number of Lepidoptera and their larvæ, but no canker-worms. Half the food was Coleoptera, nearly all Aphodius and *Anomala binotata*,—fifteen per cent. and twenty-five per cent. respectively. The little Psenocerus was likewise taken by this bird, and a specimen of Hymenarcys (Hemiptera).

Summary of the Family.

The nine flycatchers taken had eaten only insects, of which nearly half were Coleoptera, and the remainder were about equally distributed between the Hemiptera, Lepidoptera, and Diptera. Canker-worms make fifteen per cent. of the whole, and *Anomala binotata* seventeen per cent. The Scarabæidæ include all but ten per cent. of the Coleoptera.

CUCULIDÆ. Cuckoos.

COCCYZUS ERYTHROPHTHALMUS, Wils. BLACK-BILLED CUCKOO.

Three-fourths of the food of a single specimen shot consisted of canker-worms, other caterpillars making an additional twenty per cent. *Anomala binotata* was the only remaining element.

PICIDÆ. Woodpeckers.

MELANERPES ERYTHROCEPHALUS, L. RED-HEADED WOODPECKER.

This bird was abundant in the orchard, evidently nesting in the trees, although but four specimens were shot. Two of these had eaten corn, which amounted to twenty per cent. of the food. Fifteen per cent. was canker-worms, and twenty-four per cent. Carabidæ (eaten by two of the birds), including Calosoma, Scarites, and several Harpalids. Twenty-nine per cent. of Scarabæidæ embraced a Canthon and some specimens of *Anomala binotata.* Melanotus and other spring-beetles were also eaten by two of the birds.

COLAPTES AURATUS, L. FLICKER.

A single specimen, killed in 1881, had fed only on ants, the usual aliment of the bird.

COLUMBIDÆ. Doves and Pigeons.

ZENAIDURA CAROLINENSIS, L. MOURNING DOVE.

Several mourning doves were seen, and a single specimen was taken. Three-fourths of the food of this was corn, and the remainder the seeds of some leguminous plant.

PERDICIDÆ.   Quails and Partridges.

ORTYX VIRGINIANA, L.   QUAIL.

Two quails were shot, among half a dozen seen.   All but four
per cent. of their food consisted of corn and other seeds, chiefly
those of Compositæ.   A single chrysomelid, a rhynchophorous
beetle, and a carabid, were the only insects found.

Besides the species of birds above mentioned, the following
were noted rarely in the orchard, but no specimens were secured:
and *Vireo olivaceus, Sturnella magna, Cyanurus cristatus,* and
Chætura pelasgica.   The blue jay was seen eating canker-worms in
the trees.   The total number of species observed in the orchard wa
therefore forty, and the number of specimens obtained and studied
was one hundred and forty-one, representing thirty-six of the
species.   Twenty-six of these species had been eating canker-
worms, which were found in the stomachs of eighty-five specimens.   That is to say, seventy-two per cent. of the species, and
sixty per cent. of the specimens, had eaten the worms.   Taking
the entire assemblage of one hundred and forty-one birds as one
group, we find that thirty-five per cent. of their food consisted of
canker-worms; and if we exclude the species evidently merely
accidental in the orchard, the average of canker-worms in the
food of those properly belonging there rises to about forty per
cent.

For a correct estimate of the probable effect of the birds
in limiting the increase of the canker-worm, it is necessary to
take into account some of the features of its natural history.
The larval life of the insect lasts about one month, after which it
enters the ground and pupates, where it remains until the following spring.   The imagos, the females of which are wingless,
emerge about the middle of April.   They lay their eggs upon
the bark of the trees, usually at night, remaining concealed upon
the ground by day under fallen leaves and other rubbish.   The
eggs remain upon the trees about a month before the worms
emerge, when the latter crawl up the trunk and commence their
attacks upon the leaves.   The pest is consequently exposed to
destruction from the time it emerges until it disappears again, the
adults falling an easy prey to birds which search the ground for

food, and the eggs to the small species which pry about the trunks of trees. The entire period during which the insect is doubtless fed upon by birds will usually amount to somewhat more than two months.

Besides the abundance of the canker-worms noted in the food of these birds, it is evident that two or three other species of insects occurred in this situation in extraordinary numbers, especially the vine leaf-chafer (*Anomala binotata*) and a small borer (*Psenocerus supernotatus*). The purple cut-worm (*Nephelodes violans*) was also somewhat commoner than usual. The Anomala was eaten by thirty-nine of the specimens, representing fifteen species, and amounted to eleven per cent. of the food of all the birds taken in the orchard. Many of these were too small to feed upon so large an insect, and a better illustration of the abundance of this beetle may be gathered from the food of the thrushes and blue-bird. Of thirty-two specimens of these families, nineteen had eaten the vine leaf-chafer, which amounted to twenty-seven per cent. of the food of all. Only fourteen of the same birds had eaten the canker-worm, which amounted to less than twenty per cent. of the food. It seems likely, therefore, that some of these birds were attracted to the orchard, not by the canker-worms, but by the superabundance of Anomala. The unusual frequency of *Psenocerus supernotatus*, a small long-horned beetle found upon the trees, is shown by the fact that of the twenty-five small arboreal birds (Paridæ, Troglodytidæ, and Mniotiltidæ), thirteen had eaten this beetle, which composed nearly one-tenth of their food.

We have next to make the comparison of the food taken in the orchard by the species most abundant there, with the food of the same species, taken elsewhere under ordinary circumstances. For the purpose of this compar-ison I have selected the robin, the catbird, the black-throated bunting (*Spiza americana*), and the indigo bird (*Passerina cyanea*). In the table of the ordinary food of the robin for May, published in Bulletin 3 of this series, as represented by fourteen specimens, caterpillars amounted to but twenty-three per cent., whereas in the orchard they rise to fifty-four per cent. This difference between the averages is almost exactly accounted for by the ratios of canker-worms and *Nephelodes violans* not appear-ing on the former table; these together amounting to thirty-five per cent. Notwithstanding the number of Anomala eaten in the orchard, the ratios of the Scarabæidæ are substantially the same, as the ordinary food of the robin in May consists largely of June beetles. The surplus of Lepidoptera seems to be balanced by a deficiency in all the other orders, no one of which rises to the average of its ordinary food in May. The loss is greatest, however, in the Diptera, which drop from eleven per cent. to nothing.

Comparing the record of the fourteen catbirds shot in the orchard with that of twenty-two obtained in miscellaneous situations, we note, first, that the caterpillars on the first table are more than twice those of the second,—twenty-six in the one, and twelve in the other; and that this difference is evidently due to the fifteen per cent. of canker-worms taken by the birds of the first group. This shows that the catbird, like the robin, had simply added the canker-worms eaten to its usual ratio of caterpillars. A more striking difference is shown in the totals of Coleoptera, which stand at fifty-six per cent. in the orchard birds, and twenty-three in the others. This, again, is evidently due to the abundance of *Anomala binotata;* for when the ratio of this insect is subtracted from the total of Coleoptera, the remainder is twenty per cent. as against twenty-three of the ordinary food. These excessive ratios of Lepidoptera and Coleoptera are compensated by deficiencies in the Diptera, Arachnida, Myriapoda and Orthoptera, especially in the three first named groups. The decided preference of this bird for ants is shown by the fact that the usual ratio of these insects is scarcely diminished, fourteen per cent. having been taken in the orchard and eighteen elsewhere.

Fourteen of the black-throated bunting (*Spiza americana*), killed in the orchard, are to be contrasted with twelve shot in May from various situations. A striking difference is seen at once in the insect ratios, which amount respectively to eighty-eight and forty-seven per cent. This surplus of insects eaten by the orchard birds is readily traced to the orders Lepidoptera and Coleoptera. Of the former these birds had eaten more than three times their ordinary average, and of the latter nearly four times the usual amount. The excess of Lepidoptera is clearly due, as usual, to the presence of the canker-worms, since the balance left

3

after subtracting the canker-worm ratio from the average of that order taken by the first group, differs by only three per cent. from the average taken by the second group. The discrepancy in the ratios of Coleoptera is not so easily explained, but is distributed among several genera of Scarabæidæ and the small scavenger beetles. The excess of these two orders is compensated principally by diminished ratios of vegetation, which amount to only six per cent. in the birds shot in the orchard, and fifty-two per cent. among those taken through the country at large. Diptera and all the lower orders of insects as well as Arachnida and Myriapoda, are also omitted from the food of the orchard birds.

Insects composed ninety-seven per cent. of the food of eighteen indigo birds (*Passerina cyanea*) shot in the orchard, and but fifty-seven per cent. of the food of fifteen individuals taken elsewhere, the balance in both cases being seeds, chiefly Setaria, Polygonum and wheat. The excess of insects in the orchard specimens appears under Lepidoptera and Coleoptera, the former sixty-seven per cent., the latter twenty-nine, as compared with twenty-eight and nineteen per cent. respectively, in the other group. The Lepidoptera of the orchard birds are nearly all canker-worms, as are likewise ten per cent. of those taken by the specimens from various situations. The difference in the ratio of Coleoptera taken by the two groups was exactly compensated by the ten per cent. of *Anomala binotata* eaten in the orchard. The excess of caterpillars and beetles taken by the former group, is partly compensated also by the almost total disappearance of all other insects from the food.

What, now, may we conclude, from the above data, respecting the influence of birds upon such entomological insurrections as are illustrated by the uprising of the canker-worms in Mr. Robison's orchard?

Three facts stand out very clearly as results of these investigations: 1. Birds of the most varied character and habits, migrant and resident, of all sizes, from the tiny wren to the bluejay, birds of the forest, garden and meadow, those of arboreal and those of terrestrial habit, were certainly either attracted or detained here by the bountiful supply of insect food, and were feeding freely upon the species most abundant. That thirty-five per cent. of the food of all the birds congregated in this orchard should have consisted of a single species of insect, is a fact so extraordinary that its meaning can not be mistaken. Whatever power the birds of this vicinity possessed as checks upon destructive irruptions of insect life, was being largely exerted here to restore the broken balance of organic nature. And while looking for their influence over one insect outbreak we stumbled upon at least two others, less marked, perhaps incipient, but evident enough to express themselves clearly in the changed food ratios of the birds.

2. The comparisons made show plainly that the reflex effect of this concentration on two or three unusually numerous insects was so widely distributed over the ordinary elements of their food that no especial chance was given for the rise of new fluctuations among the species commonly eaten. That is to say, the abnormal pressure put upon the canker-worm and vine chafer was compensated by a general diminution of the ratios of all the other elements, and not by a neglect of one or two alone. If the latter had been the case, the criticism might easily have been made that the birds, in helping to reduce one oscillation, were setting others on foot.

3. The fact that, with the exception of the indigo bird, the species whose records in the orchard were compared with those made elsewhere, had eaten in the former situation as many caterpillars other than canker-worms as usual, simply adding their canker-worm ratios to those of other caterpillars, goes to show that these insects are favorites with a majority of birds.

TABLES OF THE FOOD.

	Turdidæ				Sialidæ	Paridæ	Troglodytidæ	Mniotiltidæ						
	Robin	Catbird	Brown Thrush	Total	Bluebird	Black-capped Chickadee	House Wren	Tennessee Warbler	Summer Yellow Bird	Chestnut-sided Warbler	Black-poll Warbler	Black-throated Green Warbler	Maryland Yellowthroat	Total
Number of Birds	9	14	4	27	5	2	5	1	5	2	4	1	2	15

KINDS OF FOOD. — NUMBER OF SPECIMENS AND RATIOS IN WHICH EACH ELEMENT OF FOOD WAS FOUND.

Kinds of Food	Robin	Catbird	Brown Thrush	Total	Bluebird	Black-capped Chickadee	House Wren	Tennessee Warbler	Summer Yellow Bird	Chestnut-sided Warbler	Black-poll Warbler	Black-throated Green Warbler	Maryland Yellowthroat	Total
Animal Food	9	14	4	27	5	2	5	1	5	2	4	1	2	15
	1.00	1.00	1.00	1.00	1.00	1.00	1 00	1.00	1 00	1.00	.95	1 00	1.00	.99
I. Mollusca	3			3										
	.01			+										
II. Insecta	9	14	4	27	5	2	5	1	5	2	4	1	2	15
	.93	.98	.94	.96	.98	1 00	.91	1 00	.94	1.00	.95	1 00	1.00	.97
1. Hymenoptera	4	11	2	17					2	2	3	1		8
	.01	.14	.03	.08					.05	.02	07	.04	.02	.03
Formicidæ	4	11	2	17					2		2	3		5
	.01	.14	.03	.08					.05		.06	.04		.02
2. Lepidoptera	9	12	4	25	3	2	5	1	5	2	4	1	2	15
	.54	.26	.22	.34	.17	.61	.39	.80	.67	.75	.66	.70	.82	.71
Noctuidæ (larvæ)	6	5	1	12										
	.28	.04	.05	.12										
Nephelodes violans (larvæ)	3	1		4										
	.14	.01		.05										
Anisopteryx vernata	3	8	2	13	1	2	3	1	5	2	4	1	1	14
	.21	.15	.12	.16	.12	.61	.46	.80	.67	.65	.66	.70	.37	.64
3. Diptera							3	1	2	4	1		2	10
							.04	.01	.02	.05	.05		.04	.08
Gnats							2		1	4	1		2	8
							.03		.02	.05	.05		.04	.08
4. Coleoptera	9	14	4	27	5	2	5	1	5	2	4	1	1	14
	.36	.56	.67	.51	.66	.39	.13	.20	.23	.11	.19	.08	.06	.18
Carabidæ	4	5	3	12	3		1		1		1			2
	.01	.04	.12	.04	.23		.01		.01		.01			.01
Staphylinidæ	1			1									1	1
	.04			.01									.05	.01
Phalacridæ							1							
							+							
Histeridæ	5	2	1	8										
	.01	+	.02	.01										
Scarabæidæ	6	12	3	21	5		1		1	1	1			3
	.18	.49	.44	.38	.36		.01		.06	.05	.02			.03
Anomala binotata	3	10	2	15	4									
	.14	.36	.14	.25	.36									
Elateridæ	6	1	3	10			1							
	.11	.01	.06	.05			.01							
Lampyridæ								1	1					2
								.20	.04					.08
Cerambycidæ					1	2			3	2	3	1	1	10
					.25	.04			.12	.06	.15	.15	.03	.10
Psenocerus supernotatus									3	2	3	1	1	10
									.12	.06	.15	.15	.03	.10
Chrysomelidæ					1									
					.05									
Rhynchophora	3	3	1	7		1	1	1						+
	.01	.01	.03	.02		.10	.01	+						+
5. Hemiptera	3	6		9	3		4	2	1	1	1		2	7
	.02	.02		.02	.15		.10	.01	.05	.01	.05		.06	.02

TABLES OF THE FOOD—Continued.

	Robin	Catbird	Brown Thrush	Total	Bluebird	Black-capped Chickadee	House Wren	Tennessee Warbler	Summer Yellow Bird	Chestnut-sided Warbler	Black-poll Warbler	Black-throated Green Warbler	Maryland Yellow-throat	Total
Number of Birds	9	14	4	27	5	2	5	1	5	2	4	1	2	15
KINDS OF FOOD.	\multicolumn{14}{NUMBER OF SPECIMENS AND RATIOS IN WHICH EACH ELEMENT OF FOOD WAS FOUND.}													
Homoptera						1 .02			1 .05					....
Aphides									1 .05					....
Tettigonidæ						1 .02								
Heteroptera	3 .02	4 .02		6 .02	3 .15	1 .06							1 .05	1 †
Aradidæ						1 .02								
Lygæidæ						1 .02							1 .05	1 †
Chinchbugs						1 .02								
Coreidæ	1 .01			1 .01										
Cydnidæ	2 .01	3 .02		5 .01	3 .15									
III. ARACHNIDA					2 .02	3 .06		3 .06		1 †	1 †	5 .02		....
IV. MYRIAPODA	2 .05	2 .02	1 .06	5 .03		1 .01								
VI. VERMES (Lumbricus)	1 .01			1 .01										....
Vegetable Food (seeds)												1 .05	1 .01	....

TABLES OF THE FOOD — Continued.

Each cell shows the number of specimens (top) and the ratio (bottom) in which each element of food was found.

KINDS OF FOOD	Warbling Vireo (Vireonidæ)	Cedar Wax-wing (Ampelidæ)	Cliff Swallow (Hirundinidæ)	American Goldfinch	Yellow-winged Sparrow	Chipping Sparrow	Field Sparrow	Black-throated Bunting	Rose-breasted Grosbeak	Indigo Bird	Total	Cowbird	Red-winged Blackbird	Baltimore Oriole
				Fringillidæ								*Icteridæ*		
Number of Birds	3	7	1	1	1	8	3	14	2	18	47	1	2	3
Animal Food	3 1.00	7 1.00	1 1.00	1 .30	1 .65	8 .96	3 1.00	14 .94	2 .80	18 .97	47 .93	1 .30	2 .50	3 1.00
I. Mollusca								1 .06			1 .01			
II. Insecta	3 1.00	7 1.00	1 1.00	1 .30	1 .35	8 .95	3 1.00	14 .88	2 .80	18 .97	47 .91	1 .30	2 .50	3 1.00
1. Hymenoptera		+	.02			1 .03		1 .01	.02	.01	2 .01			
Formicidæ						1 .01		+			1 +			
Tenthredinidæ						1 .02					1 .01			
2. Lepidoptera	3 .79	7 1.00			.30	7 .32	3 .47	13 .65	2 .63	17 .67	43 .57		2 .29	3 .40
Noctuidæ								2 .14			2 .04		1 .25	
Nephelodes violans (larvæ)													1 .25	
Phalænidæ (larvæ)	3 .44	7 1.00				2 .16	1 .20	10 .46	1 .05	17 .60	31 .41		1 .01	3 .40
Anisopteryx vernata	3 .44	7 1.00				2 .16	1 .20	10 .43	1 .05	17 .59	31 .40		1 .01	3 .40
3. Diptera	1 .03					7 .28	1 .05				8 .05			
Gnats						7 .28	1 .05				8 .05			
Muscidæ	1 .03													
4. Coleoptera	2 .15	3 +	1 .98	1 .30	1 .05	7 .25	3 .47	11 .22	2 .15	18 .29	43 .26	1 .30	2 .11	2 .60
Carabidæ	2 .05			1 .30				3 .04			4 .02			
Nitidulidæ								1 .01			1 +			
Histeridæ						1 .03	3 .04			1 .01	5 .01			
Trogositidæ														1 .01
Scarabæidæ	1 .06	3 +	1 .98			3 .14	3 .08	7 .11	1 .02	9 .15	23 .12	1 .25	1 .03	1 .50
Anomala binotata	1 .08					2 .09		2 .04	1 .02	4 .10	9 .06		1 .03	3 .50
Elateridæ							1 .01		2 .01	3 .01	.01		1 .04	1 .01
Lampyridæ							1 +				1 +			
Cerambycidæ	1 .02													3 .06
Psenocerus supernotatus														1 .02
Chrysomelidæ						2 .01					2 +	1 .02		

TABLES OF THE FOOD — Continued.

Number of Birds	Warbling Vireo	Cedar Wax-wing	Cliff Swallow	American Goldfinch	Yellow-winged Sparrow	Chipping Sparrow	Field Sparrow	Black-throated Bunting	Rose-breasted Grosbeak	Indigo Bird	Total	Cowbird	Red-winged Blackbird	Baltimore Oriole
	3	7	1	1	1	8	3	14	2	18	47	1	2	3

KINDS OF FOOD.	NUMBER OF SPECIMENS AND RATIOS IN WHICH EACH ELEMENT OF FOOD WAS FOUND.													
Rhynchophora						1 / .01	3 / .08	6 / .02	2 / .08	6 / .03	18 / .03		1 / .04	1 / .02
5. *Hemiptera*	1 / .03					5 / .06	1 / .01			1 / +	7 / .01		1 / +	
Homoptera						1 / .03				1 / +	1 / +		1 / +	
Heteroptera						1 / .03					1 / +			
Lygæidæ	1 / .03					1 / .03					1 / .01			
Cydnidæ	1 / .03													
6. *Orthoptera* (Acrididæ)													1 / .10	
III. ARACHNIDA					1 / .30	6 / .01	2 / +				9 / .01			
Vegetable Food (seeds)				1 / .70	1 / .35	4 / .04		3 / .06	2 / .20	6 / .03	17 / .07	1 / .70	2 / .50	
Compositæ										2 / .01	2 / .01			
Polygonum											1 / .05	1 / .05	1 / .05	
Wheat							1 / .05				1 / .02		1 / .45	
Setaria					1 / .15	2 / .01		2 / +		2 / .01	7 / .01			
Corn												1 / .60		
Panicum						1 / .01					1 / +			

TABLES OF THE FOOD — Continued.

KINDS OF FOOD	Icteridæ			Tyrannidæ					Cuculidæ	Picidæ			Columbidæ	Perdicidæ
	Orchard Oriole	Bronzed Grackle	Total	King Bird	Wood Pewee	Traill's Flycatcher	Yellow-bellied Flycatcher	Total	Black-billed Cuckoo	Red-headed Woodpecker	Flicker	Total	Mourning Dove	Quail
Number of Birds	2	3	11	3	3	2	1	9	1	4	1	5	1	2

NUMBER OF SPECIMENS AND RATIOS IN WHICH EACH ELEMENT OF FOOD WAS FOUND. *(each cell shows specimens count and ratio)*

KINDS OF FOOD	Orchard Oriole	Bronzed Grackle	Total	King Bird	Wood Pewee	Traill's Flycatcher	Yellow-bellied Flycatcher	Total	Black-billed Cuckoo	Red-headed Woodpecker	Flicker	Total	Mourning Dove	Quail
Animal Food	2 / 1.00	3 / .38	11 / .68	3 / 1.00	3 / 1.00	2 / 1.00	1 / 1.00	9 / 1.00	1 / 1.00	4 / .80	1 / 1.00	5 / .84		2 / .04
II. INSECTA	2 / 1.00	8 / .30	11 / .66	3 / 1.00	3 / 1.00	2 / 1.00	1 / 1.00	9 / 1.00	1 / 1.00	4 / .80	1 / 1.00	5 / .84		2 / .04
1. *Hymenoptera*	1 / .03		1 / .01	1 / .05	2 / .12	2 / .48	1 / .02	6 / .16		1 / .01	1 / 1.00	2 / .21		
Formicidæ	1 / .03		1 / .01	1 / .04	1 / .05			2 / .02		1 / .01	1 / 1.00	2 / .21		
Ichneumonidæ							1 / .33	1 / .07						
2. *Lepidoptera*	2 / .97		7 / .34	2 / .28	1 / .05	1 / .25	1 / .30	5 / .20	1 / .95	1 / .15		1 / .12		
Noctuidæ (larvæ)			1 / .05											
Nephelodes violans (larvæ)			1 / .05											
Phalænidæ (larvæ)	2 / .77		6 / .25	2 / .28		1 / .25		3 / .15	1 / .75	1 / .15		1 / .12		
Anisopteryx vernata	2 / .77		6 / .25	2 / .28		1 / .25		3 / .15	1 / .75	1 / .15		1 / .12		
3. *Diptera*				3 / .55			1 / .08	4 / .19						
Tipulidæ							1 / .08	1 / .01						
Gnats				1 / .23				1 / .08						
Muscidæ				2 / .32				2 / .10						
4. *Coleoptera*		3 / .30	8 / .29	3 / .67	2 / .28	2 / .25	1 / .50	8 / .43	1 / .05	4 / .64		4 / .51		2 / .04
Carabidæ		3 / .16	3 / .04								2 / .24	2 / .19		1 / .01
Nitidulidæ						1 / .10		1 / .03						
Trogositidæ			1 / +											
Lucanidæ		1 / .08	1 / .02											
Scarabæidæ	1 / .01		6 / .17	3 / .58	2 / .18	2 / .10	1 / .40	8 / .33	1 / .05	3 / .29		3 / .23		
Anomala binotata			4 / .14				1 / .25	3 / .17			2 / .04	2 / .03		
Elateridæ	1 / .05		3 / .03	1 / .07				1 / .02			2 / .09	2 / .07		
Lampyridæ				1 / .02				1 / .01						
Cerambycidæ			3 / .02						1 / .10			1 / .01		
Psenocerus supernotatus			1 / .01						1 / .10			1 / .01		
Chrysomelidæ			1 / +											1 / .02

TABLES OF THE FOOD—Concluded.

	Orchard Oriole	Bronzed Grackle	Total	King Bird	Wood Pewee	Traill's Flycatcher	Yellow-bellied Flycatcher	Total	Black-billed Cuckoo	Red-headed Woodpecker	Flicker	Total	Mourning Dove	Quail
Number of Birds	2	3	11	3	3	2	1	9	1	4	1	5	1	2
KINDS OF FOOD.	NUMBER OF SPECIMENS AND RATIOS IN WHICH EACH ELEMENT OF FOOD WAS FOUND.													
Rhynchophora		1/†	3/.01											1/.01
5. Hemiptera		1/†							1/.10	1/.01				
Homoptera		1/†												
Cydnidæ									1/.10	1/.01				
6. Orthoptera (Acrididæ)			1/.02											
V. CRUSTACEA (Crawfish)		1/.08	1/.02											
Vegetable Food (Seeds)		3/.62	6/.32							2/.20	2/.16		1/1.00	2/.96
Leguminosæ													1/.25	1/.02
Compositæ														1/.32
Polygonum			2/.01											2/.03
Wheat			1/.08											
Setaria														1/.02
Corn		3/.62	4/.22							2/.20	2/.16		1/.75	2/.57

GENERA AND SPECIES RECOGNIZED IN THE FOOD.

The following lists are intended to supplement the preceding tables and, taken together with them, to present all the details concerning the food of the birds observed in the orchard, upon which the foregoing discussion is based. In the first list the genera and species recognized in the food of each kind of bird are given separately; in the second the food elements are systematically arranged, and against the name of each element the names of all the species of birds are placed in whose food that element was recognized. The figures preceding the names of the birds in the second list indicate the number of individuals in which the given element was found:

TURDIDÆ.

Turdus migratorius · Helix, Hyalina, Limnea humilis, Formica, Nephelodes violans, Anisopteryx vernata, Elaphrus ruscarius, Staphylinus badipes, Aphodius, A. inquinatus, Phyllo-

4

phaga, Anomala lucicola, A. binotata, Melanotus, Monocrepidius, Graphorhinus vadosus, Alydus eurinus, Cœnus delius,
Hymenarcys, Polydesmus serratus, Lumbricus.

Mimus carolinensis: Formica, F. fusca, Lasius, L. niger,
Nephelodes violans, Anisopteryx vernata, Clivina striatopunctata, Anisodactylus, Hister americanus, H. perplexus,
Onthophagus, Aphodius, A. inquinatus, Phyllophaga, Anomala binotata, Melanotus, Graphorhinus vadosus, Tanymecus
confertus, Baris, Sphenophorus, Cœnus delius, Podisus
spinosus, Iulus.

Harporhynchus rufus: Anisopteryx vernata, Chlænius, Stenolophus conjunctus, Hister americanus, H. perplexus, Aphodius,
Diplotaxis georgiæ, Anomala binotata, Melanotus, Monocrepidius, Baris confinis, Iulus.

SAXICOLIDÆ.

Sialia sialis: Anisopteryx vernata, Anisodactylus baltimorensis,
Aphodius, Anomala binotata, Chrysomela suturalis, Diabrotica vittata, Cœnus delius, Hymenarcys æqualis, Euschistus.

PARIDÆ.

Parus atricapillus: Anisopteryx vernata, Psenocerus supernotatus.

TROGLODYTIDÆ.

Troglodytes domesticus: Anisopteryx vernata, Olibrus, Aphodius, Monocrepidius auritus, Psenocerus supernotatus, Blissus
leucopterus, Iulus.

MNIOTILTIDÆ.

Helminthophaga peregrina: Anisopteryx vernata, Telephorus
bilineatus.

Dendrœca œstiva: Anisopteryx vernata, Aphodius, Telephorus
bilineatus, Psenocerus supernotatus.

Dendrœca pennsylvanica: Anisopteryx vernata, Psenocerus
supernotatus.

Dendrœca striata: Anisopteryx vernata, Aphodius, Psenocerus
supernotatus.

Dendrœca virens: Anisopteryx vernata, Psenocerus supernotatus.

Geothlypis trichas: Anisopteryx vernata, Psenocerus supernotatus, Piesma cinerea.

VIREONIDÆ.

Vireo gilvus: Anisopteryx vernata, Anomala binotata, Telephorus bilineatus, Euschistus.

AMPELIDÆ.

Ampelis cedrorum: Anisopteryx vernata, Aphodius inquinatus,
A. femoralis.

HIRUNDINIDÆ.

Petrochelidon lunifrons: Aphodius inquinatus.

FRINGILLIDÆ.

Coturniculus passerinus: Setaria.

Spizella domestica: Anisopteryx vernata, Anomala binotata, Baris, Setaria, Panicum.

Spizella agrestis: Anisopteryx vernata, Onthophagus, Aphodius A. inquinatus, Monocrepidius, Baris, Sphenophorus.

Spiza americana: Helix, Agapestemon, Anisopteryx vernata, Anisodactylus, Ips fasciatus, Aphodius, A. inquinatus, Anomala binotata, Sphenophorus, Wheat, Setaria.

Zamelodia ludoviciana: Anisopteryx vernata, Anomala binotata.

Passerina cyanea: Aphidius, Anisopteryx vernata, Onthophagus, Aphodius, Anomala binotata, Monocrepidius, Baris, Setaria.

ICTERIDÆ.

Molothrus ater: Aphodius, Dibolia aërea, Polygonum, Corn.

Agelæus phœniceus: Nephelodes violans, Anisopteryx vernata, Anoma'a binotata, Tanymecus confertus, Polygonum, Wheat.

Icterus galbula: Anisopteryx vernata, Anomala binotata, Phymatodes variabilis, Psenocerus supernotatus.

Icterus spurius: Camponotus, Anisopteryx vernata.

Quiscalus purpureus æneus: Calosoma calidum, Dorcus parallelus, Crawfish, Corn.

TYRANNIDÆ.

Tyrannus carolinensis: Anisopteryx vernata, Aphodius inquinatus, Anomala, A. binotata, Melanotus.

Contopus virens: Ips fasciatus, Aphodius, A. inquinatus.

Empidonax trailli: Anisopteryx vernata, Anomala.

Empidonax flaviventris: Aphodius, Anomala binotata, Psenocerus supernotatus, Hymenarcys.

CUCULIDÆ.

Coccyzus erythrophthalmus: Anisopteryx vernata, Anomala.

PICIDÆ.

Melanerpes erythrocephalus: Camponotus, Anisopteryx vernata, Calosoma calidum, Scarites substriatus, Canthon hudsonias, Anomala binotata, Melanotus, Corn.

COLUMBIDÆ.

Zenaidura carolinensis: Corn.

PERDICIDÆ.

Ortyx virginiana: Chrysomela suturalis, Polygonum, Setaria, Corn.

Helix :   1 Turdus migratorius, 1 Spiza americana.

Hyalina :   1 Turdus migratorius.

Limnœa humilis :   1 Turdus migratorius.

Agapestemon :   1 Spiza americana.

Formica sp.:   1 Turdus migratorius, 1 Mimus carolinensis.

F. fusca :   1 Mimus carolinensis.

Lasius sp.:   1 Mimus carolinensis.

L. niger :   3 Mimus carolinensis.

Camponotus :   1 Icterus spurius, 1 Melanerpes erythrocephalus.

Aphidius :   1 Passerina cyanea.

Nephelodes violans :   3 Turdus migratorius, 1 Mimus carolinensis, 1 Agelæus phœniceus.

Anisopteryx vernata :   3 Turdus migratorius, 8 Mimus carolinensis, 2 Harporhynchus rufus, 1 Sialia sialis, 2 Parus atricapillus, 3 Troglodytes domesticus, 1 Helminthophaga peregrina, 5 Dendrœca æstiva, 2 Dendrœca pennsylvanica, 4 Dendrœca striata, 1 Dendrœca virens, 1 Geothlypis trichas, 3 Vireo gilvus, 7 Ampelis cedrorum, 2 Spizella domestica, 1 Spizella agrestis, 10 Spiza americana, 1 Zamelodia ludoviciana, 17 Passerina cyanea, 1 Agelæus phœniceus, 3 Icterus galbula, 2 Icterus spurius, 2 Tyrannus carolinensis, 1 Empidonax trailli, 1 Coccyzus erythrophthalmus, 1 Melanerpes erythrocephalus.

Elaphrus ruscarius :   1 Turdus migratorius.

Clivina striatopunctata :   1 Mimus carolinensis.

Calosoma calidum :   1 Quiscalus purpureus æneus, 1 Melanerpes erythrocephalus.

Scarites substriatus :   1 Melanerpes erythrocephalus.

Chlænius :   1 Harporhynchus rufus.

Anisodactylus sp.:   1 Mimus carolinensis, 1 Spiza americana.

A. baltimorensis :   2 Sialia sialis.

Stenolophus conjunctus :   1 Harporhynchus rufus.

Staphylinus badipes :   1 Turdus migratorius.

Ips fasciatus :   1 Spiza americana, 1 Contopus virens.

Olibrus :   1 Troglodytes domesticus.

Hister americanus :   1 Mimus carolinensis, 1 Harporhynchus rufus.

H. perplexus :   1 Mimus carolinensis, 1 Harporhynchus rufus.

Dorcus parallelus :   1 Quiscalus purpureus æneus.

Canthon hudsonias :   1 Melanerpes erythrocephalus.

Onthophagus :   1 Mimus carolinensis, 1 Spizella agrestis, 1 Passerina cyanea.

Aphodius sp.: 1 Turdus migratorius, 1 Mimus carolinensis, 1 Harporhynchus rufus, 1 Sialia sialis, 1 Troglodytes domesticus, 1 Dendrœca æstiva, 1 Dendrœca striata, 1 Spizella agrestis, 3 Spiza americana, 1 Passerina cyanea, 1 Molothrus ater, 2 Contopus virens, 1 Empidonax flaviventris.

A. inquinatus: 1 Turdus migratorius, 1 Mimus carolinensis, 2 Ampelis cedrorum, 1 Petrochelidon lunifrons, 1 Spizella agrestis, 1 Spiza americana, 1 Tyrannus carolinensis, 1 Contopus virens.

A. femoralis: 1 Ampelis cedrorum.

Diplotaxis georgiæ: 1 Harporhynchus rufus.

Phyllophaga: 1 Turdus migratorius, 3 Mimus carolinensis.

Anomala sp.: 1 Tyrannus carolinensis, 2 Empidonax trailli, 1 Coccyzus erythrophthalmus.

A. lucicola: 1 Turdus migratorius.

A. binotata: 3 Turdus migratorius, 10 Mimus carolinensis, 2 Harporhynchus rufus, 4 Sialia sialis, 1 Vireo gilvus, 2 Spizella domestica, 2 Spiza americana, 1 Zamelodia ludoviciana, 4 Passerina cyanea, 1 Agelæus phœniceus, 3 Icterus galbula, 2 Tyrannus carolinensis, 1 Empidonax flaviventris, 2 Melanerpes erythrocephalus.

Melanotus: 1 Turdus migratorius, 1 Mimus carolinensis, 1 Harporhynchus rufus, 1 Tyrannus carolinensis, 1 Melanerpes erythrocephalus.

Monocrepidius: 1 Turdus migratorius, 1 Harporhynchus rufus, 1 Spizella agrestis, 1 Passerina cyanea.

M. auritus: 1 Troglodytes domesticus.

Telephorus bilineatus: 1 Helminthophaga peregrina, 1 Dendrœca æstiva, 1 Vireo gilvus.

Phymatodes variabilis: 1 Icterus galbula.

Psenocerus supernotatus: 1 Parus atricapillus, 2 Troglodytes domesticus, 3 Dendrœca æstiva, 2 Dendrœca pennsylvanica, 3 Dendrœca striata, 1 Dendrœca virens, 1 Geothlypis trichas, 1 Icterus galbula, 1 Empidonax flaviventris.

Chrysomela suturalis: 1 Sialia sialis, 1 Ortyx virginiana.

Diabrotica vittata: 1 Sialia sialis.

Dibolia aërea: 1 Molothrus ater.

Graphorhinus vadosus: 1 Turdus migratorius, 1 Mimus carolinensis.

Tanymecus confertus: 1 Mimus carolinensis, 1 Agelæus phœniceus.

Baris: 1 Mimus carolinensis, 1 Spizella domestica, 1 Spizella agrestis, 1 Passerina cyanea.

B. confinis: 1 Harporhynchus rufus.

Sphenophorus: 1 Mimus carolinensis, 1 Spizella agrestis, 1 Spiza americana.

Piesma cinerea: 1 Geothlypis trichas.

Blissus leucopterus: 1 Troglodytes domesticus.

Alydus eurinus: 1 Turdus migratorius.

Cœnus delius: 1 Turdus migratorius, 1 Mimus carolinensis, 1 Sialia sialis.

Hymenarcys: 1 Turdus migratorius, 1 Empidonax flaviventris.

H. æqualis: 2 Sialia sialis.

Euschistus: 1 Sialia sialis, 1 Vireo gilvus.

Podisus spinosus: 1 Mimus carolinensis.

Polydesmus serratus: 1 Turdus migratorius.

Iulus: 1 Mimus carolinensis, 1 Harporhynchus rufus, 1 Troglodytes domesticus.

Crawfish: 1 Quiscalus purpureus æneus.

Lumbricus: 1 Turdus migratorius.

Polygonum: 1 Molothrus ater, 1 Agelæus phœniceus, 2 Ortyx virginiana.

Wheat: 1 Spiza americana, 1 Agelæus phœniceus.

Setaria: 1 Coturniculus passerinus, 2 Spizella domestica, 1 Spiza americana, 1 Passerina cyanea, 1 Ortyx virginiana.

Corn: 1 Molothrus ater, 3 Quiscalus purpureus æneus, 2 Melanerpes erythrocephalus, 1 Zenaidura carolinensis, 2 Ortyx virginiana.

Panicum: 1 Spizella domestica.

THE FOOD RELATIONS OF THE
CARABIDAE AND COCINELLIDAE

Stephen Alfred Forbes

THE FOOD RELATIONS OF THE CARABIDÆ AND COCCINELLIDÆ.

By S. A. FORBES.

A group or association of animals or plants is like a single organism in the fact that it brings to bear upon the outer world only the surplus of forces remaining after all conflicts interior to itself have been adjusted. Whatever expenditure of energy is necessary to maintain the existing internal balance amounts to so much power locked up, and rendered unavailable for external use. In many groups this latent energy is so considerable and is liable to such fluctuations, that a knowledge of its amount and kinds, and of the laws governing its distribution, is extremely important to one interested in measuring or foreseeing the sum and character of the outward-tending activities of the class.

This seems especially true of the insect world. If the checks upon the multiplication of insects and upon their average length of life which are due to insects themselves were to be suddenly removed, there is much reason to suppose that the total external effect of the class would be very greatly intensified, at least for a time.

Whether our purpose be merely to understand the internal economy of insect life as a part of the general system of nature, or to apply such knowledge to a regulation of the depredations of insects upon plants and animals, it is equally necessary that we should know the character and extent of the conflicts which prevail within the class, and should understand how the various subordinate groups limit each other's numbers and activity, either indirectly by competition, or directly by destruction.

The following notes are a contribution to a more exact knowledge of this subject than has hitherto prevailed. The view of the functions of the two principal predaceous families of Coleoptera (Carabidæ and Coccinellidæ) which is common among entomologists, is largely due to a hasty generalization, based upon insufficient data. Observations of the food of these beetles have hitherto been left almost wholly to chance, and have nowhere been systematically pursued—from which it has resulted that we know their habits only in the most conspicuous situations, and have not a fair idea of the general average of their food. Neither have observations of any kind been numerous enough to enable us to detect clearly differences of food habit in different species or genera of these families; but, with slight occasional exceptions, all Carabidæ and Coccinellidæ have been classed together as essentially carnivorous.

Besides insufficient observation, a tendency to reason too confidently from structure to function is responsible for many mistaken notions—a tendency particularly liable to mislead when applied to the habits of animals. It is frequently assumed that the most prominent and peculiar adaptive structures are necessarily indicative of the most important and customary habits, and that structures especially fitted for one function are thereby incapacitated for every other.

The first of these assumptions ignores the fact that many adaptive structures are acquired for the sake of the advantage derived, not in ordinary, but in extraordinary circumstances. The struggle for existence is one of greatly varying intensity, and the really decisive moments of the conflict are often only brief and occasional. The time spent in actual combat by very belligerent and very powerful animals, is doubtless but a small fraction of their whole lives; and yet by far the most prominent and important of the structural peculiarities which serve to distinguish them from their more peaceful allies, may be those which enable them to triumph in these occasional but critical instants. Likewise the pinch of starvation must commonly be felt only at rare intervals, but no structures will be more thoroughly elaborated or carefully preserved than those serving to give the animal the advantage during these brief periods, since the continued existence of the species depends on these no less than on those of constant use. From the prominent adaptive structures we may safely infer, as a general rule, what the animal will do in the stress of a life and death struggle, but not necessarily what are its ordinary practices.

The second of the above assumptions is also negatived, occasionally, at least, by the principles of natural selection, especially as applied to the machinery of food prehension. Whatever departure from the primitive vegetarian habit of animals any group has acquired, was of course initiated to enable it to draw on other food resources than those previously open to it. But as animal food is usually less abundant and less generally distributed than vegetable, it would not, at first, be to the advantage of any that they should become exclusively dependent upon the former; their interests would be best served by such modifications of structure and habit as would enable them to draw upon one or the other store, according to circumstances. Acquiring some power to capture and masticate animal food, they would not wholly lose that of appropriating vegetable food also; and however well fitted their prehensile and digestive organs might become for the former function, we should expect that they would not altogether lose their fitness for the latter. It would be only as competition on this higher plane increased to the pressure point, that a few members of the differentiating group would be forced to the highest plane of complete dependence on animal food alone.

The first results of an attempt at a more exact and exhaustive investigation of this subject, were given by the writer in a brief paper published in Bulletin 3 of this series, in November, 1880.* In another paper by Mr. F. M. Webster in the same Bulletin,† a summary of previously recorded observations was given, together with many additional and original field notes. A few other items have since been published by others, but confined, as far as known to me, to chance observations on single insects.

The method here followed, as in the paper above mentioned, has been that of dissection. The alimentary canals of beetles taken in a great variety of situations, at various seasons and at different times of day, have been removed, placed in glycerine on microscope slides, and opened with small knives and mounted needles, so as to display the contents completely. These have then been studied with whatever power of the microscope was necessary, and mounted as microscope slides for permanent preservation and repeated examination. The amount of information

---

Notes on Insectivorous Coleoptera. By S. A. Forbes. Illinois State Laboratory of Natural History, Bulletin No. 3, pp. 153–160.
†Pp. 149–152.

which could thus be acquired by patient study, was often quite surprising. While it was of course rarely possible to distinguish species, or even genera, all the fragments could usually be classified with some fair degree of definiteness; and there was commonly no difficulty in making satisfactory estimates of the ratios of the different food elements present.

In some of the most important cases, the facts elicited were of the highest degree of exactness. Several collections of predaceous beetles were made in situations where some particular species of noxious insect was especially abundant, with a view to determining to what extent the latter was preyed upon by its supposed enemies. In such cases it was not difficult to tell with certainty, even from very minute fragments, whether the given insect had been eaten or not. Even where no solid structures were present, and the contents of the alimentary canal were entirely fluid, it was still usually possible to say whether these fluids had an animal or a vegetable origin. After many observations and some experiments, it was found that partially digested animal food in the stomach of a beetle was commonly bathed in a black juice, which, when examined under a high power of the microscope, was seen to contain nothing but a minutely divided flocculent matter, probably composed of irregular aggregations of fat droplets and other organic particles. This fluid was never found in connection with purely vegetable contents, but sometimes filled the stomach alone, and contained nothing to indicate its origin. In all the latter class of cases I have regarded it as proof that the food had been derived from animal sources, probably usually consisting of the juices of insects recently captured.

For the determinations of the fungi mentioned herein, I am indebted chiefly to Prof. T. J. Burrill, of the Industrial University at Champaign.

The insects dissected for this paper were partly obtained in the course of miscellaneous collecting, and partly secured for me especially for the purpose, by one of my entomological assistants, Mr. F. M. Webster, who kept careful notes of the situations in which the specimens were taken, the hour of the day when they were captured, and the objects upon which it seemed probable that they had lately fed. Examples of the latter were also frequently bottled with the specimens, for comparison. The special collections from the orchard infested by canker-worms, and the corn-fields at Jacksonville and Normal overrun by chinch-bugs, were made by myself.

In the following discussion, each genus is taken up separately, and the details of its food are given both under general circumstances, as shown by specimens from miscellaneous situations, and also under the various peculiar conditions illustrated by the special collections, made for the purpose of exhibiting the food of these insects as related to particularly injurious species, and these are followed by a summary and discussion of the food of each family, taken as a unit. The tables exhibit, first, the food of the family under ordinary circumstances; second, under peculiar conditions; and, third, under all the circumstances, taken together.

FAMILY CARABIDÆ.

My notes upon the food of this family are derived from the dissection and study of one hundred and seventy-five specimens, representing thirty-eight species and twenty genera. Eighty-two specimens were collected in miscellaneous situations, twelve were taken in a field infested by cabbage-worms, ten in a corn-field overrun by chinch-bugs, and seventy-one in an orchard which was being destroyed by canker-worms. The first collection of eighty-two specimens from various situations represented thirty-two species, belonging to eighteen genera. They were obtained in different parts of the State, from DeKalb County in the north to Union in the south, and at all seasons of the year, from April to October; and doubtless represent fairly well the food of the family in Illinois during the entire year. The collections illustrating the food of the Carabidæ as related to the cabbage-worm were made in a field of young plants at Normal, Ill., in April, 1882, where the larvæ of *Agrotis annexa* were abundant and destructive. The collection showing the food of this family in the presence of the chinch-bug, consisted of ten specimens of a single species found in July, 1882, very abundant about the roots of corn in a field where the bases of the stalks were largely covered by young chinch-bugs. The third special collection consisted of seventy-one insects, representing nineteen species, obtained in May of two successive years (1881 and 1882) in an orchard which had been infested for several years with the canker-worm to such an extent as to cause the total destruction of a large part of the trees.

Genus Calosoma.

This genus is represented by three specimens of *C. scrutator*, collected in the orchard with the canker-worms, and by nine of *C. calidum*, which were variously distributed.   The *C. scrutator* was found to have eaten only animal food, about two-thirds of which was recognizable as of insect origin.   The remaining third was due to the occurrence of liquid animal food, or the fluid to which I have given this interpretation.   In the stomach of one of the beetles the insect food consisted only of minute particles of a reddish brown crust which it was impossible to classify further. A single *C. calidum*, taken in May in Central Illinois, contained only liquid animal food.   Seven specimens, taken in the orchard above-mentioned, had likewise fed upon animal food alone, forty per cent. recognizable as insects, and the remainder not otherwise determinable.   As far as 'can be judged from the contents of the alimentary canal in these thirteen specimens, the species of this genus are strictly carnivorous, and have the habit either of sucking the juices of their prey, or of selecting only those parts most easily masticated, reducing these to indistinguishable fragments. Certainly there was not the slightest trace of vegetable food in any of these beetles.*

Genus Scarites.

Two specimens of *S. subterraneus*, taken in 1882, one at Normal and the other at Anna, in Southern Illinois, had eaten only animal food, one-half of which was unrecognizable, and the remainder insects.   Four specimens of the same species, taken in the cabbage-field, have a precisely similar record.

These nineteen specimens, belonging to three species, were the only examples of *Carabidæ proper* whose food was studied, and all agreed in a strictly carnivorous character.

---

*Mr. F. M. Webster has seen a *C. calidum* eating a small grasshopper.

Genus Brachynus.

A single specimen of *Brachynus fumans*, caught in Central Illinois, in May, had taken only liquid animal food.

Genus Galerita.

Seventeen specimens of *Galerita janus*, four collected in various situations, and thirteen in the orchard in Tazewell County, had made a much more varied record. All of the group first mentioned had eaten insects, which amounted to eighty-eight per cent. of their food, nearly all caterpillars of undetermined species. The remaining twelve per cent. consisted of vegetable food eaten by two of the specimens, and was apparently derived chiefly from the seeds of grass. A larger ratio of animal food is noticed in the thirteen taken where canker-worms abounded. Here vegetation amounted to only six per cent., all of exogenous origin, as shown by the branching bundles of spiral cells in the vegetable fragments noticed, while the animal food amounted to ninety-four per cent. Insects stand at eighty-five per cent., seven per cent. being Diptera, one per cent. unrecognizable insect larvæ, and the whole of the remainder caterpillars. The last were nearly all easily determined as canker-worms, which amounted to a little over half the food. Seven individuals of the thirteen had eaten these worms. Five per cent. of the food (taken by three of the specimens) consisted of spiders, and four per cent. (taken by a single specimen) was animal food, not otherwise determinable. The remains of a caterpillar in the stomach of a single beetle were clearly distinguished as those of a noctuid larva (cutworm).

If from the ratios of animal food taken by the examples from the orchard we subtract the ratio of canker-worms (fifty-two per cent.) the remainder is just seven times the ratio of vegetation eaten. Recalling the percentages of animal and vegetable food taken by the four specimens first mentioned, we find that here also the former is almost exactly seven times the latter. This shows beyond question that the canker-worms eaten were *in addition* to the ordinary ratio of animal food taken by this species under the usual conditions.

GENUS LOXOPEZA.

Three specimens of this genus were studied, all belonging to the species *L. atriventris,* collected in July and September in Northern and Central Illinois. One of these had eaten immense numbers of minute, oval, binucleate cells, determined by Prof. Burrill as spores of Sphæronemei, probably Phoma, a fungus which forms small black specks upon dead wood, stems of weeds, etc. A second specimen had eaten some undetermined insect, and about equal quantities of three elements, namely: the above spores of Phoma and pollen and anthers of grass,—doubtless blue grass, upon which the insect was taken. A few spores of Helminthosporium were likewise noticed. The crop of a third specimen, taken at Normal, was distended with an oily liquid, but contained nothing else except a few spores of Helminthosporium. This specimen had probably been sucking the juices of some insect. The ratios of animal and vegetable food, as nearly as I could estimate them, were as forty-four to fifty-six. A specimen of this species, captured in the orchard, had not recently taken food.

GENUS CALATHUS.

Six examples of *C. gregarius,* three from DeKalb County and three from the orchard, are the only representatives of this genus. One-third of the food of those first mentioned consisted of caterpillars, a second third of other insect larvæ, and the remainder of the pollen of grass. The food of the second group was extremely similar, a third consisting, as before, of vegetation, another third of canker-worms, and the remainder of insect fragments not further determinable.

GENUS PLATYNUS.

The stomach of a single *P. decorus,* taken in the orchard, contained only liquid animal food. Two examples of *P. limbatus,* both from Southern Illinois, in April, had derived about four-fifths of their food from the vegetable kingdom, partly seeds of grass and partly the parenchyma of exogenous plants. The remainder consisted entirely of Aphides (plant-lice). These specimens were doubtless too few to give a correct idea of the average food of the genus as a whole.

GENUS EVARTHRUS.

Five specimens of *E. colossus*, taken at various dates and places, had derived about one-tenth of their food from endogens, and the remainder wholly from insects. Twenty per cent., eaten by one of the beetles, was recognized as caterpillars. Scarabæidæ are credited with another twenty per cent., and undetermined larvæ of Coleoptera with about an equal ratio. Minute quantities of fungi were noticed in the stomachs of two of these beetles, and traces of undetermined Algæ in one.

Two examples of *E. sodalis*, taken in the Tazewell County orchard, had consumed only insects, all canker-worms, except traces of an ant and a single gnat.

The insect ratio of the genus as represented by these seven specimens, stands at ninety-three per cent.

GENUS PTEROSTICHUS.

Thirteen specimens were dissected, representing *P. permundus*, *P. sayi*, and *P. lucublandus*.

The number of each species is not sufficient to give distinctive food characters, and the genus may therefore best be treated as a whole. Seven of the specimens, taken in miscellaneous situations in Central Illinois in April, May, and September, had found about one-fourth of their food in the vegetable kingdom, about one-third of which consisted of undetermined fungi, and the remainder chiefly of exogenous plants. A few spores of Helminthosporium, probably accidental, were noticed in the stomach of a single beetle. Forty-three per cent. of the food consisted of insects, among which Hymenoptera only were recognized. A single mite occurred in one of the beetles. Three specimens taken in the orchard infested by canker-worms, had eaten endogenous vegetation, to the amount of about one-fifth of their food. Caterpillars made eleven per cent., and undetermined insects two per cent., the remaining ratio being accounted for by the presence of liquid animal food. Two-thirds of the contents of three specimens taken among the cabbages, consisted of animal matter, half of which was clearly recognized as the larvæ of *Agrotis annexa* infesting the field. The remaining third, composing the entire food of one of the beetles, consisted wholly of fragments of grass.*

GENUS AMARA.

Six specimens of this genus were dissected, three of *A. cari-nata*, one of *A. angustata*, and two of *A. impuncticollis*. Three specimens of *A. carinata*, taken in Southern Illinois in April, 1882, had eaten only vegetation, partly derived from gramina-ceous plants, and partly consisting of seeds and exogenous tissues. About one-fourth of the food was recognizable as fungi, chiefly of the genus Peronospora. Ninety per cent. of that of a single *A. an-gustata*, taken in June, consisted of mites, the remainder being fragments of grass. An *A. impuncticollis*, taken in the orchard with the canker-worms, had eaten only vegetable food, chiefly undetermined, but with traces of fungi. Another of the same species from the cabbage field, had derived its food about equally from plant and animal sources, that from the former consisting chiefly of grass.

GENUS DICÆLUS.

Three examples of *Dicælus elongatus* had taken only animal food, as indicated by the fluid contents of the stomachs. One of these was found in the orchard, and the other in Central Illinois.

GENUS CHLÆNIUS.

This genus is represented by twenty-three individuals, the next to the largest number studied of any genus of Carabidæ. Six examples from Southern Illinois, collected from April to Septem-ber, belong to the species *C. diffinis*, *C. nemoralis*, and *C. tomen-tosus*. The animal food of these was about three times the veg-etable. Two-thirds consisted of insects, of which caterpillars alone were determinable; and earth-worms eaten by one of the beetles made about eight per cent. More than half the vegeta-ble food consisted of fungi, which included fourteen per cent. of some fleshy fungus, apparently Coprinus, together with spores of Dematiei. Fragments of exogenous plants were recognized in one of the beetles. A single *C. diffinis*, taken among the cab-

---

*A specimen of *P. lucublandus* was seen by Mr. F. M. Webster making a meal from a dead *P. sayi*.

bage-worms, had eaten only insects, chiefly a caterpillar, and a larva of a beetle. A mere trace of endogenous vegetation was also detected. Of sixteen specimens collected among the canker-worms, three were *C. erythropus*, and thirteen *C. diffinis*. Cut-worms made about one-third of the food of the first, and earth-worms the remaining two-thirds. The latter were easily distin-guishable by the peculiar spines mixed with dirt in the stomachs of the beetles. About ninety per cent. of the food of the other species was of animal origin, and about half the vegetable food was fungi. Insects made seventy-two per cent., nearly half cater-pillars, of which the greater part (thirty-one per cent.) was canker-worms. Fragments of a fly were observed in one of the beetles, and another had eaten one of the *Telephoridœ*. Mites and myria-pods (Geophilus) had also been devoured by one.

Genus Agonoderus.

Fifteen specimens of Agonoderus were studied, ten of which were those already referred to as representing the food relations of these beetles to the chinch-bug. Fragments of that insect amounted to about one-fifth the food of all, and were found in four of the beetles; and plant-lice, taken by half that number, amounted to about eight per cent. A single ant, *Lasius flavus*, eaten by one, was rated at five per cent.; and other insects brought the general average of the class up to thirty-five per cent. Vegetation made just half the food, all fragments of the higher plants except one per cent. each of Helminthosporium and Peronospora. A single Agonoderus, taken among the cab-bages, had eaten only undeterminable animal food. Four speci-mens from various situations had made a similar record, differing only by the presence of a few mites in the stomach of one of the beetles. Eleven per cent. of fungi, taken by the group last mentioned, was derived from Ramularia and Coleosporium. The circumstances of capture, together with the contents of the stomach of one of these beetles, indicated that it had made its meal chiefly from the seeds of June grass, but the remainder of the vegetable food could not be more definitely classified.

6

GENUS ANISODACTYLUS.

This large and abundant genus is represented by thirty-one specimens, belonging to six species. Five specimens of *A. rusticus* were examined, captured in McLean and DeKalb Counties in May, June, and July. Two of these had taken only liquid animal food, but the remaining three had eaten no animal matter at all. Among the fungi found, Cladosporium and Peronospora were recognized, and fragments of Hepaticæ were noted in two of the beetles. Two specimens of *A. harrisi*, taken in Union County in April, 1882, had eaten only vegetation, all seeds of grass and of other plants. A single *A. discoideus* from McLean County in June, contained nothing but liquid food. Seven examples of *A. baltimorensis*, widely distributed in time and place, had derived only about fourteen per cent. of their food from the animal kingdom, all taken by one of the beetles, whose stomach contained only chyme. About half of the eighty-six per cent. of vegetation, composing the entire food of the remaining six specimens, was demonstrably obtained from the seeds of June grass, upon which several of the insects were taken. Two examples of *A. sericeus* from Northern Illinois had made about three-fourths of their food of grass, and the remainder of unrecognizable insects. In the stomachs of two specimens of *A. opaculus*, fragments of seeds and other vegetation were the only objects found.

Taking together the nineteen specimens of this genus above mentioned, collected in various places, we find that animal food made about one-fourth of the total, and that the vegetation as far as recognized was chiefly derived from June grass and other graminaceous plants.

The record of ten specimens taken from the canker-worm orchard, is not especially different from that of the foregoing group. Only one of these had eaten animal matter at all, ninety per cent. of the food of this consisting of undetermined Diptera. Here, again, the recognizable vegetation was chiefly graminaceous, only ten per cent. being clearly derived from exogenous plants. Two specimens from the cabbage field afford no occasion for special remark. The stomach of one was distended with liquid animal food; that of the other contained vegetation only.

Genus Amphasia.

Four examples of *A. interstitialis* indicated that this species is almost strictly vegetarian, only three per cent. of the food consisting of insects. Of the remaining ninety-seven per cent., little can be said except that it was certainly of vegetable origin.

Genus Bradycellus.

A single specimen of *B. dichrous* had eaten only insects, which could not be further classified.*

Genus Harpalus.

Nineteen specimens of Harpalus were studied, belonging to the three species *caliginosus, pennsylvanicus,* and *herbivagus.* Two individuals belonging to the first of these species, from Normal and Towanda in August and September, had taken about one-tenth of their food from insects (caterpillars and Diptera). Twenty per cent. of unrecognizable animal food and five per cent. of mites bring the general average up to thirty-five per cent. The sixty-five per cent. of vegetation eaten consisted chiefly of tissues of grass. A little pollen of Compositæ, and other exogenous structures were likewise recognized. Three per cent. was fungi, all spores of Helminthosporium. Seven specimens of *H. pennsylvanicus,* caught in Northern, Central, and Southern Illinois, in April, August, and September, had taken about one per cent. of their food from the animal kingdom. This included an ant eaten by one of the beetles, and a few mites taken by another. About half the vegetable food was not further recognizable. Twenty-nine per cent. was the pollen of rag-weed, taken by two beetles captured upon that plant, and fourteen per cent. was derived from June grass. Fungi made eight per cent. of the food of these beetles, a little of it Helminthosporium, but chiefly Peronospora. Three examples of *H. herbivagus,* taken in Northern Illinois, had eaten only vegetation, about one-third of it graminaceous, and another third fungi. Only seven per cent. of the food of the above twelve specimens of this genus, taken from

---

*Mr. Webster reports a specimen of *B. rupestris* taken in 1881 in the act of devouring an earth-worm.

ordinary situations, consisted of animal food, of which a little less
than half was insects. Fungi made thirteen per cent., and the
remaining vegetable food was about equally divided between
grasses and exogenous plants. Three specimens of *H. caliginosus* and *H. pennsylvanicus*, taken among the canker-worms, had
derived one-third of their food from those caterpillars, while the
other two-thirds consisted of vegetation, sixteen per cent. being
Peronospora, and the remainder chiefly seeds and exogenous
tissues. Four specimens of *H. herbivagus*, collected in the cabbage field, in April, had eaten none of the cabbage-worms, and
only ten per cent. of insects (Diptera). The remainder of the
food consisted apparently of fragments of seeds, as indicated by
the contents of the cells of the fragments and by other microscopic characters. A piece of the epidermis of grass was noticed
in one of the beetles. Taking the genus Harpalus as a whole, as
far as these nineteen specimens can be supposed to indicate its
food, we find that only about one-eighth of it consisted of animal
substances. Insects stand at nine per cent., two-thirds of them
caterpillars,—ants and Diptera making up the balance. Among
the items on the vegetable side of the account, we find fungi and
pollen of Compositæ each eleven per cent. and seeds and other
tissues of grasses, fourteen per cent.

GENUS PATROBUS.

Two specimens of *P. longicornis*, one from Central and the
other from Southern Illinois, had eaten nearly twice as much
vegetation as animal food. The latter consisted chiefly of caterpillars, and included in fact nothing else but traces of plant-lice,
eaten by one of the two. A little of the vegetation was derived
from grass, but the source of the remainder could not be satisfactorily traced.

THE FAMILY AS A UNIT.

We have now to treat the various collections of Carabidæ upon
which this paper is based, as distinct and unbroken groups, without reference to the genera of which they are composed. The
eighty-three specimens of all the species obtained in miscellaneous situations, are found to have derived forty-two per cent. of
their food from the animal kingdom, while the seventy specimens

captured in the orchard so often mentioned took seventy-seven per cent. of their food from the same sources. The individuals from the cabbage field, however, show no such excess of animal food as those just mentioned, the ratios standing for them at forty-one per cent. If we seek to account for this striking surplus shown by the second group, we shall find, in the first place, a difference of more than sixteen per cent. between the ratios of insects eaten by the first and second groups respectively—a fact clearly due to the presence of canker-worms where the second group was collected. This species was eaten by sixteen of the seventy beetles, and composed about one-fifth of the contents of all the alimentary canals. This accounts, however, for only about half the difference noted, the remainder appearing in the larger ratios of the other insects, of mollusks, of earth-worms, and of undetermined animal food.

This indicates either that other forms of animal life than the canker-worms were superabundant in the orchard, or else that the miscellaneous collections do not correctly represent the ordinary food of the Carabidæ. The truth probably lies between the two. The extraordinary wetness of the season, together with the amount of rubbish on the ground in the orchard, gave these beetles an unusual opportunity to capture slugs and earth-worms, and afforded excellent harborage for all sorts of insects. On the other hand, many of the beetles from other situations were preserved especially for dissection because the circumstances of their capture made it seem probable that they were feeding upon vegetation.

These tables indicate one interesting and important fact with regard to the preferences of this family, namely, that where an extraordinary abundance of any kind of animal food appeared, with a consequent increase in the percentage of that kind appropriated by the beetles, this increase was compensated, not by a decrease in the other animal elements, but in the ratios of vegetation only—a fact which clearly shows that the preferences of the Carabidæ are for animal food. It should be noticed, however, that this argument does not apply to all the genera, as is seen, for example, by recalling the record of Anisodactylus. The ten specimens of this genus taken in the orchard had eaten much more vegetation than the nineteen from various other places.

The combination of these various tables into the final one given will tend to correct the deficiencies of the separate exhibits, and the averages of that table will consequently be found to represent more closely the general food of the family than either of the others.

Continuing the comparison of the three separate tables, we find that the beetles represented by the first had taken insects to the amount of twenty-six per cent.; that those from the orchard had about doubled this ratio; while those from the cabbage field fell a little short of it. This last fact is probably related to the time of the year when these beetles were taken—the middle of April in a very late spring, when insect life in general was but just beginning to stir abroad. The ratios of Diptera, Coleoptera, and Hemiptera, were but trivial in all these groups, and not worth separate mention. The extraordinary difficulty of determining the elements of the vegetable food from the minute fragments found in the stomachs of these beetles, makes it impossible to enter into much detail with respect to this. The miscellaneous collections and those from the cabbage field had found a little over half their food in the structures of plants, while those from the orchard had obtained from this source somewhat less than a quarter. Pollen of exogenous plants, which will be found to form so large a ratio of the food of the family next to be considered, appeared here only in three of the specimens, and amounted to but three per cent. of the entire food of the first group. These beetles fed much more largely on graminaceous plants, the recognizable tissues of which amounted to about seventeen per cent. in the first group, and eight in each of the special collections. Fungi were reckoned at about one-tenth of the food of the beetles included in the first collection, and only two per cent. of those from the orchard. The spores of the omnipresent Helminthosporium make the most important contribution to this element of the food, but a number of other genera were recognized.

A few words will suffice for the final table, summarizing the data relating to all the collections, from whatever source derived. This table presents the ratios from one hundred and seventy-five specimens, and as already remarked, a little over half the food of all consisted of animal matter, about one-third being insects, while mollusks, earth-worms, myriapods and Arachnida make up the remainder.

All orders of insects are represented on the list, with the exception of Orthoptera and Neuroptera. The ratios of none of these are of any special importance, except that of the Lepidoptera, which stands at fifteen per cent. Hymenoptera and Diptera are each one per cent., and Coleoptera and Hemiptera each two. Among the Coleoptera, only Scarabæidæ and Telephoridæ were recognized; among the Hymenoptera only a single ant; and among the Hemiptera, plant-lice and chinch-bugs only. About half the vegetable food could be distinguished as exogenous or endogenous, the remainder being of too indefinite a character to be assigned to either class. As far as known, the endogenous food was more than twice as abundant as the exogenous, and consisted almost wholly of grass or grass-like plants. The fungi, which make somewhat more than a fourth of the food, require no further special mention.

If, discarding the ratios given above, we look only to the number of specimens in which the various food elements are detected, we reach similar results. One hundred and seventeen individuals of the one hundred and seventy-five represented by this final table had eaten animal food, and ninety-seven had taken vegetation. Insects were recognized in eighty-two, Lepidoptera in thirty-one (about one-half of which had eaten canker-worms), Diptera and Coleoptera in nine and four respectively, and Hemiptera in seven. Earth-worms were found in five, myriapods (Geophilus) in but one, and Arachnida (mites and spiders) in nine. Grass-like plants were taken by thirty-six, and fungi by twenty-nine.

Scanning the totals for each genus on this final table, a few results are noted which are worthy of special remark. First, we observe that at least two very abundant genera, represented by specimens enough to give us a fair probability that the average food is correctly exhibited, can hardly be classed as carnivorous insects at all, namely, Harpalus, with its nineteen specimens and twelve per cent. of animal food, and Anisodactylus, with its thirty-one specimens and twenty-one per cent. of the same. Amara and Amphasia should probably be placed in the same category, six specimens of the first and five of the second having taken but twenty-three per cent. and seven per cent., respectively, of food of animal origin. The excessively abundant Agonoderus ranks but little higher as a carnivorous insect, fifteen examples having derived only about one-third of their food from animal sources. On the other hand, twenty-three specimens of Chlænius, and seventeen of Galerita had taken about nine-tenths of their food from insects, mites, myriapods and earth-worms. Thirteen specimens of Pterostichus had obtained three-fourths of theirs from similar sources, while Evarthrus and Calathus, represented by seven and six specimens respectively, had averaged ninety-three per cent. and sixty-seven per cent.

The fact has already been alluded to that the Carabidæ proper had eaten only animal food, and that nearly all this was of a fluid character.

Second, we find the Carabidæ dividing into at least three tolerably distinct groups as respects their food: first, those which seem usually to seize their prey and suck its juices, and take vegetation rarely, if at all; second, those which take a much larger ratio of animal food than of vegetable, but masticate and swallow it, as a rule, including indigestible fragments; and third, those whose habit is essentially vegetarian, but which still take solid animal food in diminished ratios. A fourth group, consisting of Lebia and its allies, is perhaps obscurely indicated by the facts relating to the three specimens of *Loxopeza atriventris* studied. This will probably be found to feed largely upon pollen and fungus spores, after the manner of the Coccinellidæ; and the fossorial Carabidæ will, perhaps, constitute a fifth.

If we look now to the structures of these beetles for some explanation of their differences of habit, we shall find corresponding variations in the form and structure of the mandibles. Where the mandibles are long and curved, and are destitute of basal molar processes, but are provided at or near the middle of the cutting edge with processes relatively long and sharp, the beetle seems to feed substantially upon soft or liquid animal food. If they are of medium length, somewhat slender, broad at base and tapering distally, with the tip acute, and provided with basal processes which are not especially prominent or sharp, the food is chiefly animal, but solid structures are masticated and swallowed, and some vegetation appears in the alimentary canal; while, finally, if they are short and quadrate, blunt at the tips, and provided either with strong basal processes or broad opposed surfaces, vegetable food is found to predominate. Calosoma is an example of the first of these classes, Chlænius of the second, and Anisodactylus of the third. The seeming exceptions to this generalization shown by the tables at the close of the paper, are found among those genera of which too few specimens have been studied to warrant general conclusions respecting their food.

FAMILY COCCINELLIDÆ.

This family shares with the preceding the credit of limiting the multiplication of other insects, but was shown in the Bulletin of the Laboratory previously mentioned, apparently to depend largely while in the adult stage upon fungi and other vegetable food. The notes in the paper mentioned referred, however, to so small a number of specimens as to make this conclusion of doubtful value. Numerous dissections of Coccinellidæ made since that time have afforded the material for a much more comprehensive and thorough treatment of the subject, and the results of a careful study of thirty-nine slides are herewith given. The Aphis-eating habit of the Coccinellidæ is a fact of such easy observation, and is so thoroughly well known, that I have not thought it worth while to investigate especially the food of beetles of this family taken among plant-lice.

The collections from which the present notes are derived, are from a variety of miscellaneous situations, and also from a corn-field mentioned in the notes on the food of the preceding family, in which chinch-bugs were superabundant, the purpose of the latter collection being to determine the food relations of the Coccinellidæ to those insects. It so happened that the same field was infested by the corn Aphis in great numbers, and the specimens obtained therein consequently illustrate to some extent the food of the lady-bugs in the presence of plant-lice. It was in this last situation only that larvæ were collected, and the facts here given consequently relate almost wholly to the adult beetles.

GENUS HIPPODAMIA.

Eleven specimens of *H. maculata*, taken in Northern, Central, and Southern Illinois at various seasons of the year, from April to

7

September, give an average of forty-six per cent. of animal food, all insects excepting a few mites eaten by three of the beetles, and amounting to only one per cent. of the food. The insect ratio, as far as recognized, with the exception of a single Podura, consisted wholly of plant-lice, which amounted to thirty-five per cent., while the fifty-four per cent. of vegetable food contained only pollen of plants and spores of lichens and fungi, the pollen and spores occurring in about equal quantities. The former was chiefly from flowers of grass and composite plants, about seven per cent. of the first and fifteen per cent. of the second. One per cent. of the pollen of Polygonum, and a trace of the pollen of pine, both eaten by a single beetle, are the only other items under this head. Lichen spores, including Physcia, were reckoned at two per cent., and those of fungi at twenty-five per cent. At least two-thirds of the latter, eaten by nearly half the beetles, consisted of spores of Helminthosporium.

Three specimens of this species, taken in the corn-field at Jacksonville, had eaten much smaller ratios of animal food, which amounted to only thirteen per cent., all insects. Traces of plant-lice were recognized, but no structures of chinch-bugs occurred. All but five per cent. of the vegetable food was derived from spores of fungi, very largely Cladosporium. Helminthosporium amounted to nine per cent. Macrosporium and Septoria were also found. Three per cent. of the spores of Physcia and other lichens, and two per cent. of the pollen of rag-weed and other Compositæ, complete the record.

Four examples of *H. convergens*, all taken at Normal in August and September, had eaten about the same amount of animal food as the preceding species (forty per cent.), but differed in the distribution of it by the fact that one of the specimens had eaten a myriapod (Geophilus), and that a caterpillar had been taken by another. Insects proper amounted to but twenty-five per cent., over half plant-lice. The vegetable food of this species stands at fifty-six per cent., as compared with fifty-four of the preceding, and the ratios under this head are very similar to those just given for the other species. Pollen of Compositæ (dandelion) makes thirteen per cent. that of grass makes five per cent., spores of lichens two, and those of fungi thirty-three per cent. As in *H. maculata*, Helminthosporium was by far the most important fungus element. The other genera recognized were Septoria, Ustilago, Macrosporium, Coleosporium, Peronospora, Menispora, and some spores of Sphæronemei and Myxogastres.

Five adults, taken at Jacksonville, were found to have made about one-third of their food of insects, equally divided between plant-lice and chinch-bugs, each eaten by one of the beetles. The vegetation consisted, as usual, of pollen of Compositæ (eleven per cent.), spores of lichens (two per cent.), and of fungi (seventy-one per cent.). The list of the last includes Septoria, Ustilago, Helminthosporium, Macrosporium, Cladosporium, and Peronospora.

Two larvæ of this species, taken at the same place and time, differed but little in food, to my surprise, from the adults just mentioned. Chinch-bugs, plant-lice, and caterpillars, in about equal ratios, with traces of unrecognizable insects, amount to twenty-three per cent. Pollen of Compositæ stands at five per cent., lichen spores at seven, and spores of fungi at sixty-five, including the same genera as those just mentioned, except Peronospora and Septoria.

H. glacialis was represented by four specimens, taken in the corn-field. The differences between their food and that of *H. convergens* were purely trivial. Insects amount to thirty per cent., all chinch-bugs and plant-lice, twelve per cent. of the former and eighteen of the latter. The seventy per cent. of vegetable food is divided about as before, between pollen of Compositæ, seven per cent., and spores of fungi fifty-one per cent. Lichen spores were taken more freely, however, and were estimated at twelve per cent., eaten by all the beetles. The fungi were mostly Cladosporium (forty-three per cent.), but Septoria, Uredo, Helminthosporium and Peronospora likewise occur.

Genus Coccinella.

Six specimens of this genus were studied, three of *C. 9-notata*, and three of *C. 5-notata*. All were from Central Illinois except one, which was from Jacksonville. Excluding the last, the ratio of animal food eaten by these specimens was not far from two-thirds of the total, all plant-lice. Only a trace of pollen of Compositæ was noticed in one of the insects. Fungus spores amounted to thirty-two per cent. (about half Helminthosporium and Ustilago),

and lichen spores to four per cent. The Jacksonville specimen
had eaten only fungi.

GENUS CYCLONEDA.

In the corn-field with the chinch-bugs, three specimens of *C.
sanguinea* were collected, which had eaten plant-lice, pollen of
Compositæ, lichen spores and spores of fungi. The first made
about one-third of their food, the pollen grains were estimated at
nearly half, and lichen spores at three per cent. The eighteen
per cent. of fungi were of the usual character.

THE FAMILY AS A UNIT.

A summary and comparison of the food of these two groups,
taken singly without reference to their genera, develops some interesting and unexpected facts. Although the corn-field in which
the second collection was made was teeming with insects of the
kinds especially tempting to the Coccinellidæ, and although these
beetles themselves were there in truly surprising numbers, it is
not easy to believe, considering the tables upon which this discussion is based, that the Coccinellidæ were attracted to the field
by the abundance of insects available for their food. The beetles
of the first group are seen to have eaten nearly twice as many
insects as those from the field of corn, while the fungi eaten were
as thirty-six to fifty-six respectively. Only eighteen specimens
were dissected, out of the large number collected in the corn-field,
but the contents of their stomachs were of so uniform a character
that there was every reason to suppose that they illustrated correctly the food of the family at that time and place. It would
therefore seem possible that these beetles were attracted rather
by the stores of fungi in the field, than by the chinch-bugs and
Aphides. The condition of the leaves and stalks of the corn,
drained and deadened by insect depredations, was such as to
afford an excellent nidus for the development of those fungi
which spring up every where spontaneously upon dead and decaying vegetation, and these were in fact extremely abundant. An
alternative explanation is perhaps more probable. The condition
of the field gave abundant evidence that the plant-lice had been
very much more numerous some time before; and it is possible
that, as a consequence of this decrease of food, and the increase of

the Coccinellidæ themselves, the latter had reached an excessive number, for which the supply of plant-lice was really insufficient, and that for this reason they had resorted to fungi.

The chinch-bugs taken by the specimens of the second group amounted to only eight per cent. of their entire food, and plant-lice to fourteen per cent., — less than half those taken by the other specimens, which stand at thirty-six per cent. The pollen eaten by each group was thirteen per cent.,—the same in both. If we combine the two collections, and treat the thirty-nine specimens of both as a whole, we find that insect food is about a third of the entire amount, and that the other animal elements are only trivial. The function of the beetles of this family of limiting the multiplication of plant-lice is expressed by the fact that these insects compose a fourth of the food of this entire collection. The pollen of grasses and Compositæ make fourteen per cent., the spores of lichens four per cent., and those of fungi nearly half the whole (forty-five per cent.). The list of genera, as far as recognized, and the relative importance of these, may be found by reference to the tables at the end of this paper.

SUFFICIENCY OF DATA.

The food of the Coccinellidæ seems to be, on the whole, remarkably simple and uniform, consisting almost wholly of spores of the lower cryptogams, pollen grains, and plant-lice, and varying but little from one genus to another. This similarity is likewise reflected in the mouth parts, which agree as closely in form and structure as do the ratios of the food. I have consequently little doubt that the data derived from the thirty-nine specimens here discussed, will be found sufficient for a correct general idea of the food of the family under ordinary circumstances.

With respect to the Carabidæ, we have other proof. In the preliminary paper in Bulletin 3 already referred to, based on an examination of only twenty-eight specimens belonging to seventeen species, the conclusion was announced that about one-half of the food of this family consisted of vegetation, and one-third of insects; and the vegetation was thought to be about equally divided between cryptogams, grasses and exogens. If these figures or those of the present paper were far wrong, the probabilities would be very slight indeed that the two estimates would agree,

especially as no comparison whatever was made of the two
sets of data until the tables were completed in their present form.
When, therefore, we find that the one hundred and seventy-five
specimens of the present paper, belonging to thirty-eight species,
were estimated to have taken fifty-seven per cent. of animal food
and thirty-six of insects, and that the ratios of cryptogams, graminaceous plants and exogens are respectively five, eleven, and five,
we must conclude that these figures are a fair average of the
ordinary food of the family.

RELATIONS TO BIRDS.

The foregoing pages have set forth the relations of the Carabidæ and the Coccinellidæ to the species upon which they feed,
and a few general statements will now be proper concerning the animals which prey upon them in turn. Predaceous ground-beetles
are peculiarly exposed to birds which commonly seek their food upon
the ground, and we need not be surprised to find that they enter
largely into the food of such species as the thrushes and the bluebird. Carabidæ were found to furnish about five or six per cent.
of the food of four hundred and twenty-three specimens of these
birds, as stated in a paper on that subject in the third Bulletin of
this series, but Coccinellidæ did not occur at all. Indeed, in the
food of more than four hundred other birds, of various families,
Coccinellidæ were found only in Regulus, where a single species
was reckoned at one per cent. of the food.

The great differences in the food of the Carabidæ, disclosed by
this paper, give considerable importance to the question of the
kinds of these beetles most freely eaten by birds, and the following list of species and genera recognized in the food of the collection of thrushes and bluebirds above mentioned is given as an
answer.

It will be seen that there is a very wide difference between the
number of Carabidæ proper taken by these birds, and the number
of Harpalidæ, representatives of the former group occurring in
only six specimens, and of the latter in one hundred and sixteen.
On the other hand, fifty-nine of the birds had taken Harpalids
which may be fairly classed with the second group established in
this paper, and fifty-seven had taken those belonging to the third
group, or phytophagous Carabidæ. The genera most preyed

upon are Harpalus, taken by twenty-eight of the birds, Anisodactylus by eighteen, Agonoderus by fourteen, Cratacanthus by thirteen, Pterostichus by twelve, and Evarthrus by eleven; numbers which represent fairly well the relative abundance of individuals, taking the entire season through. We note, however, a remarkable deficiency of the highly colored genera, — such as Galerita, Brachynus, Lebia, Platynus, Chlænius, etc., which are either absent, or found but rarely in these birds' food. Evidently these more showy beetles are protected by some more effective means than obscurity of color. In the following list, the figure preceding the name of each species of bird denotes the number of specimens in which the insect mentioned was found:

LIST OF GENERA AND SPECIES OF CICINDELIDÆ AND CARABIDÆ EATEN BY 423 OF THE THRUSHES AMD THE BLUEBIRD.

1. Cicindela lecontei: 1 Mimus carolinensis.
2. Carabus palustris: 1 Sialia sialis.
3. Scarites, sp.: 1 Harporhyncus rufus.
4. Dischyrius globulosus: 1 Turdus pallasi.
5. Aspidoglossa subangulata: 2 Mimus carolinensis.
6. Clivina bipustulata: 1 Harporhyncus rufus.
7. Platynus, sp.: 1 Mimus carolinensis, 1 Harporhyncus rufus, 1 Sialia sialis.
8. Evarthrus, sp.: 1 Turdus mustelinus, 1 T. migratorius, 2 Mimus carolinensis, 5 Sialia sialis.
9. E. colossus: 1 Turdus pallasi, 1 Harporhyncus rufus.
10. Pterostichus sp.: 1 Turdus mustelinus, 1 T. pallasi, 1 T. migratorius, 1 Harporhyncus rufus, 5 Sialia sialis.
11. P. lucublandus: 1 Turdus mustelinus, 1 Sialia sialis.
12. P. sayi: 1 Mimus carolinensis.
13. Amara, sp.: 1 Turdus pallasi, 1 T. swainsoni, 1 T. migratorius, 2 Sialia sialis.
14. Brachylobus lithophilus: 3 Turdus migratorius.
15. Chlænius, sp.: 1 Turdus migratorius, 1 Sialia sialis.
16. C. tomentosus: 1 Sialia sialis.
17. Lachnocrepis parallelus: 1 Turdus migratorius.
18. Geopinus incrassatus: 1 Turdus migratorius.
19. Cratacanthus dubius: 1 Turdus mustelinus, 3 T. migratorius, 2 Mimus carolinensis, 2 Harporhyncus rufus, 5 Sialia sialis.

20. Agonoderus, sp.: 2 Turdus migratorius, 1 Mimus carolinensis, 1 Harporhyncus rufus, 1 Sialia sialis.

21. A. pallipes: 2 Turdus migratorius, 1 Mimus carolinensis, 1 Harporhyncus rufus, 2 Sialia sialis.

22. A. partiarius: 1 Turdus migratorius.

23. Anisodactylus, sp.: 1 Turdus mustelinus, 1 T. swainsoni, 2 T. migratorius, 2 Harporhyncus rufus, 6 Sialia sialis.

24. A. rusticus: 1 Sialia sialis.

25. A. discoideus: 1 Turdus pallasi.

26. A. baltimorensis: 2 Turdus migratorius, 1 Mimus carolinensis, 1 Sialia sialis.

27. Xestonotus lugubris: 1 Turdus mustelinus, 1 Sialia sialis.

28. Harpalus, sp.: 1 Turdus mustelinus, 7 T. migratorius, 4 Mimus carolinensis, 8 Harporhyncus rufus, 6 Sialia sialis.

29. H. herbivagus: 1 Turdus migratorius.

30. H. pennsylvanicus: 1 Mimus carolinensis.

31. Stenolophus, sp.: 2 Turdus pallasi.

In the following tables, the elements of the food, arranged in systematic order, are placed at the left of each page, while a vertical column of the table is assigned to each genus of beetle. The upper figure of each couple indicates the number of specimens in which the given element was found, while the lower figures (decimal) show the ratio of the element to the entire food of of all the examples of the genus. The dagger has been used to indicate a trace too small to figure in the percentages, usually less than one-half per cent., and an asterisk denotes that the element against which it is placed was present in the food, but that the ratio was not estimated.

CARABIDÆ. — MISCELLANEOUS COLLECTIONS.

	Calosoma.	Scarites.	Brachynus.	Galerita.	Lebia.	Calathus.	Platynus.	Evarthrus.	Pterostichus.	Amara.	Dicælus.	Chlænius.	Cratacanthus.	Agonoderus.	Anisodactylus.	Amphasia.	Bradycellus.	Harpalus.	Patrobus.	Summary.
No. of Specimens......	1	2	1	4	3	3	2	5	7	4	2	6	1	4	19	4	1	12	2	83

KINDS OF FOOD.	NUMBER OF SPECIMENS AND RATIOS IN WHICH EACH ELEMENT OF FOOD WAS FOUND.																			
ANIMAL FOOD....	1 / 1.00	2 / 1.00	1 / 1.00	4 / .88	2 / .44	2 / .65	1 / .17	5 / .90	6 / .73	1 / .22	2 / 1.00	5 / .75		1 / .01	3 / .24	1 / .03	1 / 1.00	4 / .07	1 / .33	43 / .42
			1	4 / .88	1 / .11	2 / .65	1 / .17	5 / .90	4 / .43			5 / .67			2 / .03	1 / .03	1 / 1.00	2 / .03	1 / .33	30 / .26
I. INSECTA............		.50		.88	.11	.65	.17	.90	.43			.67			.03	.03	1.00	.03	.33	.26
						1		1				1								3 / .02
Larvæ................						.32		.10				.08								2
1. Hymenoptera								.13		1								1 / .01		1 / .01
Formicidæ............																		1 / .01		+
				3	1	1						1						1	1	8
2. Lepidoptera (larvæ).				.63	.33	.20						.17						.02	.30	.08
3. Diptera...........						2												1 / +		1 / +
								39												2
4. Coleoptera								1												.02
Larvæ......								19												1 / .01
								1												1
Scarabæidæ...........								20										1		1 / .01
5. Hemiptera(Aphides)					1 / .17															2 / .03
II. ARACHNIDA (Mites)...									1 / .01	1 / .22				1 / .01				2 / .01		5 / .01
III. VERMES (Lumbricus)..													1 / .08							1 / .01
VEGETABLE FOOD....				2 / .12	3 / .56	2 / .35	2 / .83	1 / .10	4 / 27	4 / .78			2 / 25	1 / 1.00	4 / .99	15 / .76	4 / 97		12 / .93	2 / .67	58 / .58
															2 / 48	3 / .16	1 / .22		1 / .08	1 / .50	8 / 10
Seeds...............						1 / .50	2 / .14	1 / .25			1 / .08							3 / .17		8 / .07
1. Exogens.........							1 / +													1
Seeds....																		3 / .17		3 / +
Compositæ (Pollen)...																		2 / .17		2 / .03
Ambrosia............															1 / .7			4 / 23	1 / .10	20 / .03
2. Endogens.........				1 / 10	1 / .11	2 / .35	1 / 30	1 / .10		1 / 28				1 / 25	7 / .34			4 / 23	1 / .10	18 / .17
Gramineæ............				1 / 10	1 / .11	2 / .35	1 / 30			1 / 28				1 / 25	7 / .34					4 / .02
Seeds...						1 / 30				1 / *				1 / .05				1 / +		4 / .04
Pollen..............				1 / .11	2 / 35										•					2 / .02
Phleum (seeds)															1 / 25	4 / .21				5 / .06
3. Hepaticæ.........															2 / .01					2 / +
4. Algæ.....·.......						1 / +			+						+					+
Protococcus........									+						+					+
5. Fungi............				1 / +	3 / .45		1 / .03	1 / +	3 / 07	1 / .21		1 / 14	1 / 1.00	2 / .11	3 / .04			5 / .13	1 / .07	23 / .09
												1 / .14								1 / .01
Coprinus...........					2 / .45															1 / .02
Phoma					2 / .45															1 / +
Coleosporium															1 / +					1 / +
Dematiei............															1 / +					1 / +
Helminthosporium...					2 / +		1 / +					1 / 1.00						2 / +		6 / .01
Cladosporium															1 / .01	1 / .01		1 / .08		3 / +
Peronospora									1 / .19						1 / .01					1 / .01
Ramularia															1 / .10					1 / .01

CARABIDÆ AND CANKER-WORMS.

KINDS OF FOOD.	Calosoma.	Pasimachus.	Scarites.	Galerita.	Calathus.	Platynus.	Evarthrus.	Pterostichus.	Amara.	Dicælus.	Chlænius.	Anisodactylus.	Amphasia.	Harpalus.	Summary.
No. of Specimens.	10	2	4	13	3	1	2	3	1	1	16	10	1	3	70
NUMBER OF SPECIMENS AND RATIOS IN WHICH EACH ELEMENT OF FOOD WAS FOUND.															
ANIMAL FOOD	10 / 1.00	2 / 1.00	4 / 1.00	13 / .94	3 / .68	1 / 1.00	2 / 1.00	3 / .80		1 / 1.00	16 / .92	1 / .09	1 / .25	2 / .35	59 / .77
I. MOLLUSCA										1 / 1.00					1 / .02
II. INSECTA	5 / .50		2 / .50	13 / .85	3 / .68		2 / 1.00	1 / .13			12 / .64	1 / .09	1 / .25	2 / .35	42 / .50
Larvæ				2 / .11							1 / .06				3 / .04
1. *Hymenoptera* (ants)							1 / .05								1 / +
2. *Lepidoptera* (larvæ)				8 / .56	1 / .33		2 / .40	1 / .11			8 / .41			1 / .33	21 / .26
Anisopteryx vernata				7 / .52	1 / .33		2 / .90				5 / .26			1 / .33	16 / .21
Noctuidæ				1 / .01							2 / .12				3 / .03
3. *Diptera*				3 / .07	1 / .02		1 / .05				1 / .01	1 / .09			7 / .03
Culicidæ							1 / .05								1 / +
4. *Coleoptera*				.							1 / .06				1 / .01
Telephoridæ											1 / .06				1 / .01
III. ARACHNIDA				3 / .05							1 / .02				4 / .01
Spiders				1 / .05											1 / .01
Mites				2 / +							1 / .02				3 / +
IV. MYRIAPODA (Geophilus)											1 / .01				1 / +
V. VERMES (Lumbricus)											4 / .25				4 / .06
VEGETABLE FOOD				2 / .06	1 / .32			1 / .20	1 / 1.00		5 / .08	10 / .91	1 / .75	2 / .65	23 / .23
Seeds											3 / .20			1 / .32	5 / .04
1. *Exogens*			2 / .06								1 / .10			1 / .17	4 / .03
2. *Endogens*							1 / .20				5 / .01	5 / .50		1 / +	10 / .08
Gramineæ												4 / .40			4 / .06
Seeds												4 / .30			4 / .04
3. *Fungi*								1 / +			2 / .04			1 / .16	4 / .02
Peronospora														1 / .16	1 / .01
Ascomycetes											1 / .02				1 / +

CARABIDÆ, AND CABBAGE WORMS AND CHINCH-BUGS.

	Pterostichus.	Amara.	Chlænius.	Agonoderus.	Anisodactylus.	Harpalus.	Summary.	Agonoderus.
Number of Specimens Examined..	3	1	1	1	2	4	12	10
KINDS OF FOOD.	NUMBER OF SPECIMENS AND RATIOS IN WHICH EACH ELEMENT OF FOOD WAS FOUND.							
ANIMAL FOOD................	3 / .67	1 / .50	1 / 1.00	.....	1 / .50	1 / .10	7 / .41	8 / .51
I. INSECTA	2 / .33	.....	1 / 1.00	.....	.....	1 / .10	4 / .20	8 / .36
Larvæ.....................								1 / .02
1. Hymenoptera								1 / .05
Lasius flavus..............								1 / .05
2. Lepidoptera..............	1 / .33	.....	1 / .60	.....	.....	.....	2 / .13	.....
Caterpillars	.....	.....	1 / .60	.....	.....	.....	1 / .05	.....
Agrotis annexa.............	1 / .33	.....	.....	.....	.....	.....	1 / .08	
3. Diptera.................	.....	.....	.....	.....	.....	1 / .10	1 / .03	
4. Coleoptera (larvæ)	.....	.....	1 / .30	.....	.....	.....	1 / .03	.....
5. Hemiptera	.....							5 / .29
Aphides								2 / .06
Chinch-bugs..................								4 / .21
VEGETABLE FOOD..............	1 / .33	1 / .50	1 / +	1 / 1.00	1 / .50	4 / .90	9 / .59	7 / .49
Seeds.......................	.....	1 / *	.....	.....	1 / *	4 / .90	6 / .30	.....
1. Exogens.................								2 / .13
2. Endogens....................	1 / .33	1 / *	1 / +	.....	.....	1 / +	4 / .08	2 / .11
Grass......................	1 / .33	1 / *	.....	.....	.....	1 / +	3 / .08	.....
3. Fungi....................	.....							2 / .02
Helminthosporium								1 / .01
Peronospora..................								1 / .01

CARABIDÆ. GENERAL TABLE.

KINDS OF FOOD.	Calosoma.	Pasimachus.	Scarites.	Brachynus.	Galerita.	Loxopeza.	Calathus.	Platynus.	Evarthrus.	Pterostichus.	Amara.	Dicœlus.	Chlenius.	Cratacanthus.	Agonoderus.	Anisodactylus.	Amphasia.	Bradycellus.	Harpalus.	Patrobus.	Summary.
No. of Specimens	11	2	6	1	17	3	6	3	7	13	6	3	23	1	15	31	5	1	19	2	175

NUMBER OF SPECIMENS AND RATIOS IN WHICH EACH ELEMENT OF FOOD WAS FOUND.

KINDS OF FOOD.	Calosoma.	Pasimachus.	Scarites.	Brachynus.	Galerita.	Loxopeza.	Calathus.	Platynus.	Evarthrus.	Pterostichus.	Amara.	Dicœlus.	Chlenius.	Cratacanthus.	Agonoderus.	Anisodactylus.	Amphasia.	Bradycellus.	Harpalus.	Patrobus.	Summary.
ANIMAL FOOD	11	2	6	1	17	2	5	2	7	12	2	3	22		9	5	2	1	7	1	117
	1.00	1.00	1.00	1.00	.93	.44	.67	.45	.93	.73	.23	1.00	.88		.34	.21	.07	1.00	.12	.33	.57
I. MOLLUSCA (slugs)													1								1
													.33								.01
II. INSECTA	5		3		17	1	5	1	7	7			18		6	8	2	1	5	1	82
	.45		.50		.86	.11	.67	.12	.93	.34			.67		.24	.05	.07	1.00	.09	.33	.36
Larvæ					2		1		1				2		1						7
					.09		.16		.07				.06		.02						.03
1. Hymenoptera									1	1									1		4
									.02	.07			.03						.01		.01
Formicidæ													1						1		3
													.03						.01		†
Lasius flavus													1								1
													.03								†
2. Lepidoptera	11				2		3	2					10						2	1	31
	.58		.33		.40	.10		.36											.06	.30	.15
Larvæ	5				1		1	1					3						1	1	13
	.18		.17		.14	.03		.05											.01	.30	.04
Anisopteryx vernata	7				1		2						4						1		16
	.39		.16		.26			.18											.05		.08
Noctuidæ (larvæ)	1												2								4
	.01												.09								.02
Agrotis annexa										1											1
										.07											.01
3. Diptera	3				1		1						1		1				1		9
	.05		.01		.01								†		.03				.02		.01
Culicidæ					1																1
					.01																†
4. Coleoptera					2								2								4
					.28								.06								.02
Larvæ					1								1								2
					.14								.01								.01
Scarabæidæ					1																1
					.14																†
Telephoridæ													1								1
													.05								.01
5. Hemiptera					1								5						1		7
					.12								.19						.03		.02
Aphides					1								2						1		4
					.12								.05						.03		.01
Chinch-bugs													4								4
													.14								.01
III. ARACHNIDA	3						1	1		1			1						2		9
	.04						.01	.15		.01			†						.01		.01
Spiders	1																				1
	.04																				†
Mites	2						1	1		1			1						2		8
	†						.01	.15		.01			†						.01		.01
IV. MYRIAPODA (Geophilus)													1								1
													.01								†
V. VERMES (Lumbricus)													5								5
													.19								.02
VEGETABLE FOOD					4	3	3	2	1	6	5		8	1	12	26	5		18	2	97
					.07	.56	.33	.55	.07	.27	.77		.12	1.00	.56	.79	.93		.88	.67	.43
Seeds															2	7	1		6	1	19
															.13	.16	.18		.29	.50	.08
1. Exogens	2				1		2	1					1		2	1			4		14
	.04		.33		.08	.16		.02					.09	.03					.14		.05
Seeds					1																1
					*																*

CARABIDÆ. GENERAL TABLE—Concluded.

	Calosoma	Pasimachus	Scarites	Brachynus	Galerita	Loxopeza	Calathus	Platynus	Evarthrus	Pterostichus	Amara	Dicælus	Chlænius	Cratacanthus	Agonoderus	Anisodactylus	Amphasia	Bradycellus	Harpalus	Patrobus	Summary
No. of Specimens	11	2	6	1	17	3	6	3	7	13	6	3	23	1	15	31	5	1	19	2	175

KINDS OF FOOD. — NUMBER OF SPECIMENS AND RATIOS IN WHICH EACH ELEMENT OF FOOD WAS FOUND.

KINDS OF FOOD	Calosoma	Pasimachus	Scarites	Brachynus	Galerita	Loxopeza	Calathus	Platynus	Evarthrus	Pterostichus	Amara	Dicælus	Chlænius	Cratacanthus	Agonoderus	Anisodactylus	Amphasia	Bradycellus	Harpalus	Patrobus	Summary
Composite (pollen)																			3		3 / .01
Ambrosia																			2 / .11		2 / .01
2. Endogens					1 / .02	1 / .11	2 / .17	1 / .20	1 / .07	2 / .12	1 / *		4 / .01		3 / .14	12 / .37			6 / .14	1 / .10	36 / .13
Gramineæ					1 / .02	1 / .11	2 / .17	1 / .20	1 / .08		1 / *				2 / .10	/ .34			5 / .14	1 / .10	25 / .11
Seeds								1 / .20											1 / +		8 / .03
Pollen							1 / .11	2 / .17													5 / .01
Phleum (seeds)															1 / .07	4 / .13					2 / .03
3. Hepaticæ								1					1				+				2 / †
4. Algæ								+					+								1 / †
Protococcus													+								1 / †
5. Fungi					1 / +	3 / .45	1 / .02	1 / +	3 / .04	1 / .14			3 / .06	1 / 1.00	4 / .04	3 / .03			6 / .11	1 / .07	29 / .05
Coprinus												1 / .04									1 / †
Phoma						2 / .45															1 / .01
Coleosporium																	1 / +				1 / †
Dematiei												1 / +									1 / †
Helminthosporium						2 / +								1 / 1.00	1 / .01				2 / +		6 / .01
Cladosporium															1 / +	1 / .01					5 / †
Peronospora									1 / .12						+	+			2 / .05		2 / .01
Ascomycetes						1 / +						1 / .01			1 / +						1 / †
Ramularia															1 / .03						1 / †

COCCINELLIDÆ.

KINDS OF FOOD.	Miscellaneous. Hippodamia.	Coccinella.	Brachyacantha.	Summary.	Chinch-bugs. Hippodamia.	Coccinella.	Cycloneda.	Summary.	General. Hippodamia.	Coccinella.	Cycloneda.	Brachyacantha.	Summary.
Number of Specimens	15	5	1	21	14	1	3	18	29	6	3	1	39

NUMBER OF SPECIMENS AND RATIOS IN WHICH EACH ELEMENT OF FOOD WAS FOUND.

KINDS OF FOOD.	Hipp.	Cocc.	Brach.	Summ.	Hipp.	Cocc.	Cycl.	Summ.	Hipp.	Cocc.	Cycl.	Brach.	Summ.
ANIMAL FOOD	12/.44	5/.64		17/.47	12/.25		2/.32	14/.25	24/.35	5/.53	2/.32		31/.37
I. INSECTA	11/.39	5/.64		16/.43	12/.25		2/.32	14/.25	23/.32	5/.53	2/.32		30/.35
1. Lepidoptera	1/†			1/†	1/.01		1/.01	1/.01	2/.01				2/.01
Larvæ					1/.01		1/.01	1/.01	1/.01				1/.01
2. Hemiptera	8/.29	5/.64		13/.36	7/.21		2/.32	9/.22	15/.25	5/.53	2/.32		22/.29
Aphides	8/.29	5/.64		13/.36	4/.11		2/.32	6/.14	12/.20	5/.53	2/.32		19/.26
Siphonophora granariæ					2/.05		1/.24	3/.08	2/.02		1/.24		3/.08
Chinch-bugs	1/*			1/*	4/.10		4/*	4/.08	4/.05				4/.03
3. Neuroptera (Podura)	1/.01			1/.01					1/†				1/†
II. ARACHNIDA (Mites)	4/.01			4/.01	2/†			2/†	6/.01				6/†
III. MYRIAPODA (Geophilus)	1/.04			1/.03					1/.02				1/.02
VEGETABLE FOOD	13/.56	2/.86	1/1.00	16/.53	14/.75	1/1.00	3/.68	18/.75	27/.65	3/.47	3/.68	1/1.00	34/.63
Pollen	1/.03	1/†		2/.02					1/.01	1/†			2/.01
1. Exogens (Pollen)	7/.13			7/.09	13/.07		3/.47	16/.13	20/.10		3/.47		23/.11
Compositæ	6/.13			6/.09	13/.07		3/.47	16/.13	19/.10		3/.47		22/.11
Taraxacum	1/.07			1/.05					1/.04				1/.03
Ambrosia					2/*		2/†	2/†	2/*				2/*
Polygonum	1/†			1/†	1/*			1/†	1/†				1/*
Coniferæ	1/†			1/†					1/†				1/†
2. Endogens	1/.05			1/.04					1/.03				1/.02
Gramineæ	1/.05			1/.04					1/.03				1/.02
Pollen	1/.05			1/.04					1/.03				1/.02
3. Lichenes	4/.02	1/.04		5/.02	11/.06		2/.03	13/.05	15/.04	1/.03	2/.03		18/.04
Physcia	1/.01			1/†	6/*		1/†	7/†	7/.01		1/*		8/.01
4. Fungi	11/.33	2/.32	1/1.00	14/.36	14/.61	1/1.00	2/.18	17/.56	25/.46	3/.44	2/.18	1/1.00	31/.45
Myxogastres	1/*			1/†					1/†				1/*
Sphæronemei	1/*			1/.01					1/†				1/*
Septoria	1/*			1/.01	7/*		1/*	8/*	8/*		1/*		9/*
Ustilago	1/*	2/.08		3/.02	2/*		2/*	3/*	3/*	2/.07			5/*
Uredo					1/*		1/*	1/*	1/†				1/*
Helminthosporium	6/*	1/.10	1/1.00	8/.17	11/*		1/*	12/*	17/*	1/.08	1/*	1/1.00	20/*
Macrosporium	3/*			3/.04	7/*		1/*	8/*	10/*				11/*
Cladosporium	3/*			3/.02	13/*		2/*	15/*	16/*				18/*
Peronospora	3/*			3/.01	7/*			7/*	10/*				10/*
Menispora	1/*			1/†					1/*				1/*

THE FOOD OF THE
SMALLER FRESH-WATER FISHES

Stephen Alfred Forbes

THE FOOD OF THE SMALLER FRESH-WATER FISHES.

BY S. A. FORBES.

In a paper on the food of fishes, published in 1880,* I characterized the food of all the Illinois Acanthopteri, with the exception of the Aphredoderidæ; and in the present article, which is to be regarded as a continuation of that just mentioned, I propose to summarize my observations on all the smaller fishes occurring in the waters of the State, with the exception of the darters (Etheostomatinæ), which were treated in the preceding paper.

The purposes and methods of the investigation upon which the following discussion is based, are so similar to those already described, that they will not need any especial present explanation.

The data for it have been obtained by a minute and careful study of the contents of the alimentary canals of 319 specimens, belonging to twenty-five species, representing twenty-two genera and seven families, namely: Aphredoderidæ, Cottidæ, Gasterosteidæ, Atherinidæ, Cyprinodontidæ, Umbridæ, and Cyprinidæ.

An additional feature is the description of the structures subsidiary to alimentation, given, in this paper, for each genus, in order to furnish a basis for a more exact discussion of the relations of structure to food-habits than I attempted formerly. Under this head I have included the length and complication of the alimentary canal, the character of the pharyngeal structures, the number and development of the gill-rakers, and the presence of any peculiar prehensile apparatus about the mouth.

First giving for each species a brief account of its numbers and distribution throughout the State, I shall add for each genus a description of these alimentary structures, following this by a detailed statement of the observations made upon its food, and closing with a summary of such observations, and a discussion of the correlations of structure to food characters, given sometimes

---

* Bulletin No. 3, Ill. State Lab. Nat. Hist., pp. 18–65.

under the genus and sometimes under the special group to which the genus is assigned.

FAMILY APHREDODERIDÆ.

This family is represented by a single peculiar species (*Aphredoderus sayanus*), resembling the sun-fishes in most of its characters, but remarkably distinguished by the fact that the vent, although occupying the normal position in the young, opens in the adult far forward under the head, moving gradually to the front with increasing size. This fish is not over three inches in length. It occurs in rivers and smaller streams, as well as in lakes and ponds throughout the State. We have collected it from the Illinois River and various tributaries, as well as from the lakes connected with that stream, and from ponds and creeks throughout Southern Illinois. It has also been taken in the Calumet River near Chicago, and from lakes in that vicinity, but is not known to occur in Lake Michigan. It is said to be nocturnal in its habits, by Dr. C. C. Abbott, who kept specimens in an aquarium for some time.* The same author reports that in confinement it feeds voraciously upon small fishes, especially immature Cyprinidæ; and for this reason he bestowed upon it the name of *pirate perch,* by which it has become generally known among ichthyologists. The observations presently to be detailed will show, however, that his specimens were doubtless forced to feed so largely upon fishes for want of food more natural to them, since in their native haunts fishes make but a small percentage of their ordinary food.

The intestine of this species is short and simple, less than the length of the head and body without the tail, and distinguished only by the character previously mentioned. The gill apparatus is ineffective, the rakers being very short, thick, blunt, and few, and covered with short spinules. The pharyngeal jaws consist of small plates, covered with short, sharp spinulose teeth, similar to those of the sun-fishes. The mouth is large, but not remarkably protractile.

The specimens dissected number nineteen, representing seven different dates and localities, throughout Central and Southern Illinois. Some were taken from small temporary ponds left by

---

*Proc. Phil. Acad. Nat. Sci., 1861, p. 95.

the retreating overflow of streams, others from permanent lakes, and still others from creeks and rivers. The food from the different localities varies but little, on the whole, and it is scarcely worth while to discuss the separate collections. That of these nineteen specimens was almost purely animal, traces of a minute flowering plant (Wolffia), and small quantities of filamentous Algæ only being taken by two of the specimens. Fishes were eaten by but two, and were reckoned at two per cent. of the food of the whole. One of these found was recognizable as a Cyprinoid, but the other could not be determined. Insects amounted to more than ninety per cent., all of them aquatic, with the exception of a few gnats (Culicidæ) taken by eight of the fishes. Nearly half of the food consisted of larvæ of Chironomus and Corethra. Aquatic coleopterous larvæ were reckoned at eleven per cent., and specimens of Corixa, taken by three of the fishes, at two. A single fish had also eaten Galgulus. A fourth of the food consisted of neuropterous larvæ (Ephemeridæ and Libellulidæ). Crustaceans, though captured by more than half the fishes, made but four per cent. of the food. As far as recognized, this element consisted chiefly of the amphipod, *Allorchestes dentata*, and the common isopod, Asellus. A few specimens of Cyprididæ were noticed in two of the fishes, and Cyclops and other Copepoda were taken by five. One fish had eaten a Lumbriculus, a species closely allied to the common earthworm.

A careful comparison was made of the food of specimens of various ages — those, consequently, in which the situation of the vent was widely different — but no differences of food whatever were distinguishable. It is highly probable, consequently, that the explanation of this peculiar character must be sought elsewhere than in the food. With respect to the other relations of food to structure, we have at present only to note the coincidence of fishes and aquatic insects as the principal elements of the food with the large mouth and inferior development of the gill and pharyngeal apparatus, and short and simple intestine.

FAMILY COTTIDÆ.

This curious family, chiefly marine, is represented in the State by several species from Lake Michigan, mostly from its deeper waters, and by a single one recently discovered in our streams.

POTAMOCOTTUS MERIDIONALIS, Gill. GOBLIN, BLOB.

Although this fish has not hitherto been recorded from the
State, we have found it abundant in small streams in Southern
Illinois, and a single specimen has been sent us from McHenry
County, near our northern limits. The first of these situations is
in a limestone region, where small caves are not infrequent; but
the second is in an area deeply covered by drift, with rock nowhere exposed.

The general appearance of this fish is not unlike that of a catfish, the head being broad and flat, the mouth very large, and the
skin smooth. The gill-rakers are few, short and thick, and of insignificant character; the pharyngeals are similar to those of Aphredoderus, but form thicker and larger plates; the intestine is short
and simple, its entire length being less than that of the head and
body.

Six specimens of this species, taken in Southern Illinois, had
eaten only animal food, about one-fourth of which consisted of
fishes, one of which was furnished with ctenoid scales. Undetermined aquatic larvæ (thirty-six per cent.) and other insects, were
estimated at forty-four per cent. of the food. Crustacea, all belonging to the genus Asellus, eaten by two of the fishes, composed the
remaining twenty-nine per cent. The general resemblance of the
food of this species to that of Aphredoderus seemingly corresponds to the similar character of their alimentary structures.

FAMILY GASTEROSTEIDÆ.

Of the interesting little stickle-backs, two species were studied,
only one of which is common in the State.

EUCALIA INCONSTANS, Kirt. BLACK STICKLEBACK.

This fish is abundant in streams and lakes in the northern part
of Illinois, but has not been taken by us south of Rock River.

Its mouth is small; the gill-rakers are long and slender (about
half as long as the corresponding filaments), but are not unusually
numerous; the pharyngeal apparatus is insignificant or wanting;
and the intestine is short and simple, not longer than the head and
body together.

Four specimens from Rock River, and one from Cedar Lake, in Lake County, had divided their food about equally between plant and animal substances: the former, consisting wholly of filamentous Algæ, taken by four of the specimens in quantities to make it certain that they were ingested purposely. The animal food was about equally insects and crustaceans, the former nearly all aquatic larvæ of Diptera (Chironomus being the commonest form), and the latter chiefly Entomostraca, of which Cladocera were the most abundant. One of the specimens had eaten Cypris—some of them *Cypris vidua.* Cyclops was also noticed in three of the fishes, and amounted to three per cent. of the food.

The herbivorous character of this fish seems not to be related to any structural facts; but the occurrence of the large ratio of Entomostraca is at once accounted for by the well-developed gill-rakers, these serving as a straining apparatus by means of which the fishes possessed of it are able to appropriate minuter organisms than would otherwise be available for their food.

PYGOSTEUS PUNGITIUS, Lac. MANY-SPINED STICKLEBACK.

This species has hitherto been found by us only in Lake Michigan, and in Calumet River near its mouth.

But two specimens were dissected; and these had fed wholly on larvæ of Chironomus and Simulium (sixty per cent.), and on Chydorus and other Cladocera (forty per cent.).

With so small an amount of material to illustrate the food of the family, we can only say that it evidently consists chiefly of aquatic larvæ and Entomostraca, together with a considerable percentage of vegetable substances. In the absence of any apparatus for mastication, the latter will doubtless be found to consist of Algæ, as in the cases examined.

FAMILY ATHERINIDÆ.

LABIDESTHES SICCULUS, Cope. SILVERSIDES.

This elegant little fish, the only fresh-water representative of its family, is generally abundant throughout the State, and has been collected by us in a great variety of situations, from the northern lakes to the Wabash River.

It is long and slender, the mouth small and well furnished with teeth, while the throat is destitute of special pharyngeal apparatus. The gill-rakers are unusually well developed, being numerous, slender, finely toothed, and longer than the corresponding filaments of the gills. Taking into account the small size of the fish, and the consequently small diameter of the apertures of the mouth and gills, it will be seen that it is provided with an especially effective straining apparatus. The intestine is unusually short, the entire alimentary canal measuring considerably less than the length of the body without the head.

The following account of its food is derived from the dissection of twenty-five specimens, obtained from Crystal Lake, Fox River, and Calumet River in Northern Illinois, from Peoria and Mackinaw Creek in the central part of the State, and from Little Fox River in the Wabash Valley. The food of these specimens was purely animal, a little over half consisting of insects, and a little less than half of crustaceans. The larvæ of Chironomus were among the most important elements of the food, standing at thirty per cent. of the whole. The crustaceans were all Entomostraca, and represented a great variety of both Copepoda and Cladocera, although none of the specimens examined happened to have eaten Ostracoda. Among the Cladocera recognized were *Daphnia pulex, retrocurva* and *hyalina, Simocephalus americanus,* Bosmina, Chydorus, Pleuroxus, Alona, and Eurycercus; and among the Copepoda were *Cyclops thomasi,* Canthocamptus, Diaptomus, Limnocalanus, and *Epischura lacustris.* Spiders and terrestrial insects, accidentally washed or fallen into the water (the latter including Chalcididæ, various Diptera, plant-lice, Tettigonidæ, Thrips, and Podura), amounted to twelve per cent. of the food. The only peculiarities of food corresponding to differences of locality were found among the group from the northern lakes, in which the Chironomus larvæ were present in diminished ratios, while the Cladocera were more abundant.

FAMILY CYPRINODONTIDÆ.

This family consists, in Illinois, of four species, one of Fundulus and three of Zygonectes.* The family is divided into two sections, *carnivorous* and *herbivorous,* by Dr. Günther in his "In-

*I do not consider *Fundulus menona,* Jor. and Cope., as distinct.

troduction to the Study of Fishes." Although our genera both belong to the carnivorous section, it will be seen that they are not by any means strictly confined to animal food, vegetation making about one-fifth of their usual nutriment.

FUNDULUS DIAPHANUS, LeS. BARRED KILLIFISH.

This species is very abundant in the northern part of the State, especially in lakes or in clear and sandy streams, but we have not taken it anywhere in Central or Southern Illinois. Most of our collections were made in the lakes of Lake and McHenry Counties.

The intestine is shorter than the body, the gill-rakers are short, obtuse, and few in number, the pharyngeal jaws are of the pavement type, set with fine, sharp teeth, and the mouth is small, but extraordinarily protractile.

Eight specimens were studied, from Crystal and Cedar Lakes. About four-fifths of the food consisted of animal substances, the remaining fifth of vegetation. Except a few filamentous Algæ taken by one of the specimens, the latter consisted wholly of seeds of various plants fallen into the water. Eighty per cent. of the food of two of the specimens, and twenty per cent. of that of a third consisted of such seeds; ratios evidently too large to have been taken accidentally. Two of the specimens had eaten Planorbis, and all had eaten insects, which made about forty per cent. of the food; terrestrial species, including spiders, making twelve per cent. Among the aquatic forms were Chironomus larvæ, Hydrophilidæ, and larvæ of Ephemeridæ, the latter eleven per cent. Crustacea were a fifth of the food, chiefly the abundant amphipod, *Allorchestes dentata.* Cypris and Candona were likewise noticed in considerable quantity (seven per cent.), and a few specimens of various Cladocera occurred.

ZYGONECTES NOTATUS, Raf. TOP MINNOW.

This species ranges in ponds and sluggish streams throughout the State, but is most abundant southwards. Here it may commonly be seen swimming slowly about in stagnant pools, with the head at the surface of the water, as if interested in the phenomena of the weather, or possibly watching for the appearance of terrestrial insects. The alimentary structures are in all respects similar to those of Fundulus, except that the intestine is possibly a little longer, being about equal to the head and body. The only striking peculiarity is the depressed head, with the mouth placed at the upper angle and opening obliquely upward. This, with the surface-swimming habit of the fish, has given rise to the supposition that it feeds largely upon surface insects; but I did not find this to be the case, as the seventeen specimens studied contain no example of an insect of this character.

These specimens were taken from a considerable variety of situations throughout Central and Southern Illinois, and at various times of the year. The animal food amounted to about ninety per cent. of the whole. Vegetation, almost wholly filamentous Algæ, was taken by ten of the specimens, but in such quantities by various individuals as to make it certain that its presence was not accidental. In one, for example, the intestine was packed with these Algæ to the exclusion of all other food, and in three others this made more than half the whole. One specimen had also eaten Wolffia. Mollusks (Physa) had been eaten by three, and insects amounted to seventy-three per cent. Spiders and various terrestrial insects made fully a fourth of the food. Philhydrus, taken by three of the specimens, was reckoned at eight per cent. Corixa and other aquatic Hemiptera amounted to eleven per cent., and larvæ of Agrion to three. Crustacea were estimated at only six per cent. They included *Crangonyx gracilis*, and various Cladocera, Ostracoda and Copepoda. Among the Entomostraca recognized were Daphnia, Chydorus, Pleuroxus, Acroperus, Cypris, and Cyclops. Chironomus larvæ were about one per cent., taken by only two of the specimens.

ZYGONECTES INURUS, Jor. and Gilb.   BLACK-EYED TOP MINNOW.

ZYGONECTES DISPAR, Ag.   STRIPED TOP MINNOW.

The first of these species is peculiar in this State, as far as known, to Southern Illinois, not having been taken by us north of White County. The second ranges throughout.

Six specimens of the first and two of the second were studied. The food characters presented do not differ sufficiently from those of *Zygonectes notatus* to make it worth while to treat them separately, and a summary for the genus will be given instead.

Four-fifths of the food of the genus consisted of animal matter, nearly one-quarter being Mollusca, including Physa, Planorbis, and *Valvata sincera*. Insects make less than half, and nearly half of these were of terrestrial origin. Chironomus larvæ, usually so abundant in the food of insectivorous minnows, occurred here in only trivial quantity. Specimens of Philhydrus were eaten by three of the fishes. *Corixa alternata* amounted to five per cent. of their food, Agrion larvæ and case worms (Leptoceridæ) to two per cent. . Crustaceans were only four per cent. of the whole, partly Amphipoda, but chiefly Entomostraca. The vegetable food (sixteen per cent.) was chiefly Wolffia, taken by five of the specimens from southern lakes. Ten individuals had, however, eaten filamentous Algæ.

Summary.

The only essential difference between these two genera exhibited by the specimens studied, is the much larger ratios of terrestrial insects captured by Zygonectes, this genus eating nearly twice as many as the other. This fact is possibly related to the surface-swimming habit already mentioned, but is more likely due to the smaller bodies of water in which the top minnows occur. Concerning the food of the family as a whole, the salient characters are the presence of a considerable quantity of vegetable food, (about twenty per cent.) the occurrence of fifteen per cent. of Mollusca, the insignificant quantity of Crustacea eaten (four per cent.), and the importance of terrestrial insects as a source of support.

FAMILY UMBRIDÆ.

UMBRA LIMI, Kirt. MUD MINNOW.

This species, the only one of its family in Illinois, is very abundant in muddy ponds and ditches, and has been collected by us from Lake to Union Counties.

The intestine is short, less than the body in length; the gill-rakers are thick and rather long, about one-half the length of the filaments, and the pharyngeal apparatus is wholly insignificant.

Ten specimens were studied, from six localities, all from Southern Illinois but one, which was taken in Calumet River. Vegetable food amounted to forty per cent., chiefly Wolffia, eaten by seven of the specimens from Southern Illinois lakes. A considerable quantity of unicellular Algæ was also taken by one. Mollusks, eaten by two, were reckoned at five per cent., all Physa. Insects drop to fourteen per cent., chiefly undetermined larvæ. No terrestrial forms were recognized. Corresponding to the greater development of the gill-rakers, we find the Entomostraca assuming greater importance in the food. These were reckoned at ten per cent.; three per cent. additional consisting of *Crangonyx gracilis.*

FAMILY CYPRINIDÆ

This family includes all the fishes properly known as "minnows," embracing, in fact, by far the larger part of the smaller fishes of the State. Both in number and in variety of species it is much the most important family of fresh-water fishes. It includes, in Illinois, about forty species, nearly or quite one-fourth of the whole number known to occur in our territory. They occur in all waters from the Mississippi River and Lake Michigan to the smallest streams and ponds; but are much the most abundant in creeks and rivulets. The species differ greatly with respect to their favorite haunts, some affecting the principal lakes and larger rivers, others occurring most commonly in clear and rapid brooks, while still others are most frequent in the sluggish and muddy streams of prairie regions. The principal economic interest of the fishes of this family is due to the well-known fact that they furnish an important part of the food supply of larger species.

But little has hitherto been done upon their food in the United States. In fact, I have seen nothing more accurate or comprehensive than the following general statement made by Prof. Cope, in his paper on the Cyprinidæ of Pennsylvania:*

" These differences of habit are associated with peculiarities of food and of the structure of the digestive system. Few families of vertebrates embrace as great a variety in these respects as the present one. There are carnivorous, insectivorous, and graminivorous genera, which are distinguished as among mammalia, the former by the abbreviation, the last by the elongation of the ali-

---

*Trans. Amer. Philosophical Society, Vol. 13, New Series, page 353.

mentary canal, in the former the teeth are usually sharp-edged or hooked, in the latter truncate, hammer, or spoon-shaped."

"In the American genera, as far as included in the scope of this essay, the peculiarities of the intestines correspond with the food. In the *Alburnellus rubrifrons*,[1] they are but four-fifths the length of head and body (excluding caudal fin). In *Hypsilepis kentukiensis*,[2] *Photogenis leucops*, *Argyreus atronasus*[3] and *nasutus*,[3] *Ericymba buccata*, and *Exoglossum maxillingua*, about seven-ninths; the food of the last five species is insects and crustaceans, the last depending largely on mollusca. In the species of Ceratichthys, Semotilus, and Hybopsis, with *Hypsilepis cornutus*, fifteen-sixteenths to equal the length; the habits insectivorous. The genera with longer intestines are, first, Stilbe[4] one and two-fifths to one and three-fourths the length; Chrosomus, Hyborhynchus, and Pimephales two and two-fifths to two and two-thirds, and Hybognathus four times. The intestines in these are generally filled with a soft, dark-colored slime, without remains of insects, but of vegetable origin. In the remarkable genus Campostoma the canal extends to between eight and nine times the length, and, like that of other vegetable feeders, is usually found occupied by the ingesta for a considerable part of its length."

This statement is in the main correct as far as it goes, but it will be seen from the following data, and from the discussion of the food of the family, that it is far from the truth with respect to the genus Campostoma and its allies.

If we examine the alimentary structures of the Cyprinidæ, to which reference has been made in describing the food of the preceding families, we shall find these fishes easily divided into at least four tolerably distinct groups, defined by characters drawn from the gill-rakers, the pharyngeal teeth and the intestines. In all but two of the genera of this paper* the gill-rakers are short and insignificant. The pharyngeal teeth may be either hooked or plain, and with or without grinding surface, while the intestine varies in length from less than that of the body without the head

---

[1]Minnilus or Notropis. [2]Photogenis analostanus. [3]Rhinichthys. [4]Luxilus cornutus. [5]Notemigonus.

*I have used here, for convenience' sake, the nomenclature of the Catalogue of the Fishes of Illinois, published in our third bulletin.

to seven or eight times the length of the head and body together. For convenience' sake I have grouped the genera as follows:

Group I.—Intestine long. Pharyngeal teeth not or slightly hooked, with grinding surface.

> *Campostoma, Pimephales, Hyborhynchus, Hybognathus.*

Group II.—Intestine rather long. Teeth hooked, with grinding surface.

> *Notemigonus, Chrosomus.*

Group III.—Intestine short. Teeth hooked, with grinding surface.

> *Hybopsis, Luxilus, Lythrurus, Hemitremia, Platygobio.*

Group IV.—Intestine short. Teeth hooked, without grinding surface.

> *Minnilus, Photogenis, Ericymba, Phenacobius, Semotilus, Ceratichthys, Rhinichthys.*

The second group, consisting of Notemigonus and Chrosomus, may be again divided according to the development of the gillrakers, which are numerous, long, and slender in Notemigonus; few, short, and insignificant in Chrosomus.

FOOD OF THE YOUNG.

The genera and species of Cyprinidæ are not easily recognized, even in the adult, the characters upon which they are based being often either trivial or extremely variable; and when one has to do with individuals small enough to show the earliest food of the family, it is commonly quite impossible to identify even the genus. In the few specimens which I have studied, I have not attempted such determination, although I have reason to believe that most of those examined belong to some species of Minnilus.

Their food was so far peculiar, as compared with the young of other families, that I will describe in detail that found by dissecting six specimens under an inch in length. The first of these, three-eighths of an inch long, taken in Fox River on the 8th of July, had eaten only a small Chironomus larva, and a single example of Bosmina. Two specimens, six-tenths of an inch long, captured in August in a creek in Central Illinois, had derived their food from quite different sources. Filaments of Spirogyra and other filamentous Algæ, cells of Cosmarium and Closterium, and Cymatopleura and other diatoms, and spores of Ustilago, were the vegetable elements, while the head of a Chironomus larva and great numbers of the ciliate infusorian *Euglena viridis*, and a few specimens of *Euglena acus*, represented the animal kingdom. Full half the contents of these intestines consisted of the Protozoa mentioned. A third specimen of the same length, taken from the Illinois River in June, had derived about eight-tenths of its food from Bosmina, the remainder consisting of a small Chironomus larva and a minute larval hydrachnid. In a specimen seven-tenths of an inch long, taken in Mackinaw Creek in August, *Euglena viridis* was the most abundant object, making about six-tenths of the food; and *Euglena acus* and a species of Phacus also occurred. Various filamentous Algæ, specimens of Closterium and Cosmarium, and numbers of diatoms were the remaining elements. In another specimen, taken at the same time and place, about three-fourths of an inch in length, fungi and fungus spores amounted to more than half the food, although the same forms of Algæ occurred as before, together with a few examples of *Euglena viridis* and Difflugia. A Chironomus larva, a plant-louse, and some other insect not determined, had also been eaten.

From the above we may conclude that the young Cyprinidæ draw almost indiscriminately, for their food supply, upon Protozoa, Algæ, and Entomostraca, differing in this respect from the young of all the other families which I have studied, with the exception of the Catostomidæ. It is worthy of note, as a suggestive coincidence at least, that the other families just mentioned which were found to take Entomostraca and Chironomus larvæ as their earliest food, were all possessed of raptatorial teeth on the jaws when very young; whereas in young suckers and Cyprinidæ, the mouth is unarmed at all ages.

GROUP I.

Intestine long. Pharyngeal teeth not hooked, with grinding surface.

CAMPOSTOMA ANOMALUM, Raf. STONE LUGGER.

This very peculiar fish is exceedingly abundant everywhere except in the great lakes. I have taken it in streams of all magnitudes, from the Illinois River to the smallest creeks, but have not yet encountered it in Lake Michigan or in stagnant pools. It is commonest, however, in swift creeks of medium size.

It is distinguished from all other species by the great length of the intestine, which is from six to nine times the length of the body, and is spirally coiled about the air bladder. The gill-rakers are numerous, about twenty in number to each gill, but are very short, scarcely projecting beyond the anterior margin of the arch. They are evidently almost totally inefficient as a straining apparatus.

Of the great number of specimens available for dissection, only nine were studied, since the contents of the intestines were found so uniform in character that it was not deemed worth while to multiply instances. These were from both extremes and also from the center of the State, but were all taken in July, August and September. The intestine was invariably filled from end to end with a black and slimy matter, which, when examined under the microscope, was found to consist almost wholly of fine mud. When the intestine was emptied and the contents stirred up in alcohol and repeatedly decanted so as to separate the coarser fragments, the organic matter was easily distinguished. It made on an average, only about one-fourth of the contents of the intestine, the remainder consisting of the finest particles of sand and clay. Not far from one-fifth of the whole amount was of vegetable origin, consisting chiefly of filamentous Algæ, mingled with a few diatoms, but comprising occasionally minute fragments of other kinds of vegetation also. The only animal objects noted were occasional Chironomus larvæ and Difflugia. Sometimes the intestine was wholly filled with almost pure mud, in which no organic structures whatever could be detected. Date and locality seemed to make no material difference in the food of this fish, which should evidently be classed as limophagous. The ratios of animal to vegetable food were scarcely different from what one would expect to find in the intestine of a fish which had the habit of swallowing mud rich in organic matter, the greater ratios of vegetation being apparently due to the fact that plants are more abundant in the water than animals.

PIMEPHALES PROMELAS, Raf.   BLACK HEAD.

This species is generally distributed throughout Central and Northern Illinois, but is not very abundant. We have taken it only in rivers and larger creeks, but have not found it south of Jersey County.

The alimentary canal is two or three times the length of the body, and the gill-rakers are fifteen in number and somewhat more prominent than usual, those on the posterior part of the first arch being about one-third the length of the corresponding filaments.

Only four specimens were studied, one from the Pecatonica River at Freeport, and three from Otter Creek in Jersey County. With this fish as with the preceding, about three-fourths of the contents of the intestine consisted of mud, the remainder being almost wholly insects. These were partly terrestrial species, occurring accidentally in the water, and partly aquatic larvæ of Diptera. The vegetable food of these specimens amounted only to about one per cent., chiefly various unicellular Algæ.

HYBORHYNCHUS NOTATUS, Raf. BLUNT-NOSED MINNOW.

This extremely abundant minnow occurs in streams and rivers throughout the State, but has not been found by us in ponds. Specimens were taken, however, in the small lakes of Northern Illinois.

The intestine is about two and one-half times the length of the head and body. The gill-rakers are few, short and thick, being about one-fifth of the length of the corresponding filaments.

Nine specimens were studied from all parts of the State, when their food proved to be so uniform in character that further observations were deemed unnecessary. Mud made about eighty per cent. of the contents of the alimentary canal, the remainder consisting of unrecognizable vegetable debris, with a few filaments of Algæ. Undeterminable insects occurred in one, and a single specimen of Cypris in another.

HYBOGNATHUS NUCHALIS, Ag. BLUNT-JAWED MINNOW.

This species is likewise generally distributed in rivers, creeks and ponds, occurring in our collections from Galena to Cairo, and at a great number of points intermediate.

The alimentary canal in this genus is elongate, being about four times the length of the body. The gill-rakers are few and rather short, triangular in form, and about one-fourth to one-fifth the length of their corresponding filaments.

Eight specimens of this species were dissected, with results in all respects similar to those given for the other members of this group. Filamentous Algæ, diatoms, and a few accidental fungus spores, were the only objects found imbedded in the quantities of mud which filled each intestine.

SUMMARY OF THE GROUP.

If we average the results of the four species studied, belonging to this first group, we shall find that about three-fourths of the contents of the stomach and intestine consist of soft, black mud, the remaining fourth being derived from both animal and vegetable substances, about three times as much from the latter as from the former. The animal food is chiefly insects, both terrestrial and aquatic, and the vegetation is divided about equally between Algæ and miscellaneous fragments of higher plants. This group, with long intestine and grinding pharyngeals, is consequently to be considered as essentially limophagous. We find this peculiar form of pharyngeal teeth associated only with intestines of this type.

GROUP II.

Intestines moderately long; pharyngeal teeth hooked, with grinding surface.

CHROSOMUS ERYTHROGASTER, Raf.   RED-BELLIED DACE.

This species is locally abundant, although not generally common. It occurs in clear streams in the northern part of the State, but has not been taken by us in Central or Southern Illinois.

The length of the fish is contained one and two-thirds times in the length of the intestine; the gill-rakers are few and rather short, triangular, acute, and about one-fifth the length of the corresponding filaments.

I examined carefully but three specimens of this species, derived from two localities. These were alike in the presence of great quantities of mud, which amounted to about eighty-seven per cent. of the contents of the intestine. The animal food was confined to a trace of Cladocera. The vegetation amounted to thirteen per cent., partly tissues of aquatic plants, with traces of fungi, but chiefly Algæ of various forms, including a little Oscillatoria.

NOTEMIGONUS CHRYSOLEUCUS, Mitch. SHINER.

This extremely abundant minnow, commonly called the shiner, occurs in all waters throughout the State, from the largest rivers to the smallest creeks, and from Lake Michigan to small stagnant ponds.

The intestine is shorter than in any of the preceding species, although still rather long, the head and body being contained one and one-third times in its length. The gill-rakers are long, fine, and numerous, about twenty in number on the anterior arch, and fully one-third the length of the corresponding filaments, making, therefore, an effective apparatus for the separation of the Entomostraca from the water. As this fish presents a peculiar combination of alimentary structures, and as its food was found unusually various, a larger number of specimens were studied than of any of the species already discussed.

Twenty-five fishes were dissected, from a great variety of situations in all parts of the State, and representing various dates from May to September inclusive. As the food differed widely according to situation, that of specimens from certain localities being more widely different than the food of different species has usually been found, it will be best to mention the most conspicuous differences depending upon situation.

Specimens taken from the Pecatonica River at Freeport, an extraordinarily muddy stream, noted for the abundance of its mollusks, had eaten no other food than univalve Mollusca, chiefly *Valvata tricarinata* and *Planorbis deflectus*. Another, from the Illinois River at Pekin, had also eaten largely of mollusks, while three taken in Otter Creek in Jersey County, in almost stagnant reaches of the stream, extremely muddy, and green with Algæ, had filled their intestines with mud, like Campostoma; and still others from ponds near Normal had eaten only Entomostraca, about equally Cladocera and Copepoda. Another specimen from the Illinois River had taken similar food, all Daphnias. One specimen from Nippersink Lake, in the northern part of the State, was full of wild rice (Zizania). Taking all these groups together, and considering the species as a whole, besides the mud already mentioned, about fourteen per cent. of the food consisted of mollusks, and only six per cent. of insects, nearly all of which were of terrestrial species. Crustaceans amounted to fifteen per cent., all Entomostraca. Vegetation stands at fifty per cent., more than half of it accidental vegetable debris, partly from aquatic and partly from terrestrial plants. About one-fifth of the food consisted of Algæ, half of which was filamentous in character, and the remainder desmids, including Closterium, and various diatoms.

The peculiar character of the alimentary structures of this species are very clearly reflected in this summary of its food, the elongate intestine corresponding to the presence of mud, and the well developed gill-rakers to the occurrence of Entomostraca. I have not yet noticed any structural peculiarity of the Cyprinidæ related to the habit of feeding upon mollusks.

Summary for the Group.

The two species foregoing agree only in their mud-eating propensity, — probably habitual in one and occasional in the other,— the first having the longer intestine, and the second the longer gill-rakers. To this last difference we doubtless must trace the different relations of these fishes to Entomostraca.

I find nothing whatever, by comparison of the food of these specimens with those of the preceding group, to show the meaning of the hooked form of the pharyngeal teeth.

Group III.

Intestine short, teeth hooked, with grinding surface.

This group includes Hybopsis, Luxilus, Lythrurus, Hemitremia, and Platygobio. My studies were limited to three genera: Hybopsis, Luxilus and Hemitremia.

Hybopsis hudsonius. Clint. Spawn-eater.

This fine minnow is common everywhere to the northward, especially in Lake Michigan and the other lakes of Northern Illinois, but not abundant south of the central part of the State, although it has been taken to its extreme southern limit. It has never occurred in our collections in the smaller streams, but is confined to the lakes, rivers, and creeks of some magnitude.

The gill-rakers of this minnow are short and few.

Seventeen specimens were studied, from Lake Michigan,

Nippersink Lake and the Illinois River. Mud was found in notice-able quantities only in a single specimen, and there in small amount. About seventy per cent. of the food consisted of animal substances, three per cent. being fishes, taken by two of the minnows. One had also eaten a small bivalve mollusk. Insects made half the food, about one-third of them of terrestrial species (Rhynchophora), the remainder being chiefly larvæ of ephemerids. A few Chironomus larvæ and an aquatic hemipter, were the only other kinds determined. Crustacea amounted to thirteen per cent., nearly all Ostracoda (*Cypris vidua*) taken by two of the speci-mens from Chicago. Vegetable food stands at thirty-one per cent., eaten by ten of the specimens. One-third of this consisted of Algæ, chiefly of the filamentous forms, the remainder being miscellaneous fragments of exogenous plants, chiefly evidently aquatic.

Local and individual peculiarities.—The general summaries of the food of so many individuals from so great a variety of situations often disguise interesting and important facts relating to the food resources of the species, since an element taken in large quantity by one or two specimens may figure in the general average in such an insignificant ratio as to lead to the inference that its occurrence is merely accidental. In other words, general averages for a variety of situations will not necessarily indicate all the food resources open to the species. These can only be demonstrated by exhibiting the *peculiarities* of the record as well as its general average characters. For example, the fact that only eleven per cent. of the food of this species consisted of Algæ has a somewhat different aspect when we learn that one of the speci-mens had eaten nothing else, and that they made three-fourths of the food of another. Three specimens had eaten only insects, and these made ninety per cent. or more of the food of three others. Two had eaten nothing but Entomostraca, all the *Cypris vidua* previously mentioned. Vegetable structures made the entire food of four, and ninety per cent. or more of the food of three other specimens. Three out of four individuals taken at Nippersink Lake in May, had derived from ninety to one hundred per cent. of their food from terrestrial beetles of a single family (Rhynchoph-ora), while ephemerid larvæ occurred in the food of three others in ratios exceeding seventy-five per cent.

HYBOPSIS STRAMINEUS, Cope. STRAW-COLORED MINNOW.

This insignificant species has been found by us in rivers and small streams throughout the State.

The gill-rakers were few and short.

Only five specimens were studied, all from rivers in Central Illinois. About three-fourths of their food consisted of animal matter, nearly all neuropterous larvæ (fifty-eight per cent.), Ephemeridæ standing at forty-eight per cent., and case-worms at ten. Crustacea were ten per cent., all Cyclops except a trace of Canthocamptus. About one-fourth of the food was vegetation, chiefly seeds of grasses, occurring, of course, only accidentally in the water. Two had derived from ninety to one hundred per cent. of their food from ephemerid larvæ, and four of the five had eaten vegetation amounting to as much as eighty per cent.

LUXILUS CORNUTUS, Raf. SHINER.

This large and fine minnow is probably the commonest fish in Illinois, occurring in lakes and streams of all sizes everywhere throughout our limits.

The gill-rakers are short and few, and of insignificant development, and the intestine is shorter than the head and body.

Twenty-one specimens were studied, from all parts of the State and at various seasons of the year. Animal food amounted to two-thirds of the whole, fourteen per cent. being fishes, eaten, however, by only one of the specimens. Insects, eaten by nineteen, were reckoned at forty-five per cent., only one-fourth of them terrestrial. Gyrinid larvæ, Corixa, and larvæ of *Palingenia bilineata* were among the forms recognized. The crustacean ratio was insignificant, standing at only three per cent., all the abundant amphipod, *Allorchestes dentata*, with the exception of traces of a considerable variety of Entomostraca, including Chydorus, *Acroperus leucocephalus*, and Cypris. One of the water-worms (Lumbriculus) was noticed in a single specimen. Vegetable food was reckoned at thirty-eight per cent., only about one-third of it consisting of Algæ, and the rest of accidental fragments, including the seeds, anthers, and pollen of plants, with a little Potamogeton and various forms of fungus spores. One of the commonest of the Algæ was *Cladophora glomerata*,* taken

---

*Kindly determined for me by Rev. Francis Wolle, Bethlehem, Pa.

by those from Effingham. The fact has already been noted that one of the specimens had eaten only fishes. Five had confined themselves to an insect diet, while twelve had derived more than half their food from the vegetable kingdom, one of them eating ninety-five per cent. and another one hundred.

HEMITREMIA HETERODON, Cope.   NORTHERN HEMITREMIA.

This species, extremely abundant in Northern Illinois, has not been taken by us south of the central part of the State. North of Rock River it has been generally found in streams and lakes of all descriptions, from Lake Michigan down.

The gill-rakers are few in number, but thick, triangular, and rather long, those on the posterior part of the arch being from a fourth to a third the length of the filaments. The intestine is contained one and one-fourth times in the length of the head and body.

Eighteen specimens were studied, suitably distributed as to time and place. A little mud was found in the stomach of one. Only about one-tenth of the food consisted of vegetation, chiefly flowers and seeds. Traces of filamentous Algæ occurred in two of the specimens. Univalve Mollusca were noticed in one, and insects in twelve, amounting to more than a fourth of the entire food. These were chiefly larvæ of Chironomus (twenty per cent.), ephemerid larvæ occurring in but one. Crustacea were reckoned at fifty-eight per cent., all Entomostraca, with the exception of a single *Allorchestes dentata*. About two-thirds of these were Cladocera, the remainder being Ostracoda and Copepoda. Rotifers and Protozoa also rarely occurred, the latter including Centropyxis and Difflugia. Five of the specimens had eaten Entomostraca only, and two others ninety per cent. or more. Only two had derived more than half their food from vegetable sources.

It will be seen that the peculiar fact with respect to this speciès was the large per cent. of Entomostraca appropriated. I find nothing in the structure of the fish to explain this circumstance, other than the somewhat unusual development of the gill-rakers and the small size of the species. The latter probably had more to do with it than anything else. It should be noted, however, that nearly half the specimens were derived from places where Entomostraca were excessively abundaut at the time of their capture.

SUMMARY OF THE GROUP.

Taking now this group as a whole, we remark, first, the absence of mud mingled with their food, as related to the greatly diminished length of the alimentary canal. We have now also a decided predominance of animal food, which is about three-fourths of the entire amount, and note likewise the first occurrence of fishes. Although Mollusca occur in this group, it is in quantity too small to appear in the ratios. Insects make about half the food of all, nine per cent. being terrestrial forms. The larvæ of Neuroptera are by far the most important insect species, and stand at twenty-five per cent. Entomostraca make a fifth of the whole food, distributed among all the orders. The vegetation eaten was largely of a purely miscellaneous and incidental character, only about a third of it being derived from aquatic plants.

GROUP IV.

Intestine short; teeth hooked, without grinding surface.

This group, organized more strictly for predatory purposes than any of the preceding, contains also the largest number of genera, embracing nine of those occurring in Illinois. It was not thought necessary to study all of these, and my dissections were confined to five of them, namely: to Minnilus, Photogenis, Phenacobius, Semotilus and Ceratichthys.

MINNILUS ATHERINOIDES, Raf. EMERALD MINNOW.

This species is everywhere abundant in streams and lakes, but does not occur in ponds. It is most common northward, swarming in summer along the shores of Lake Michigan.

The gill-rakers are short, triangular, and about one-fourth the length of the filaments; and the intestine is less than the length of the head and body.

Eighteen specimens were studied, all from the northern half of the State. The food was almost strictly animal, but five per cent. consisting of vegetation, and this chiefly of accidental character, occurring in trivial ratios. Only a single specimen had taken about forty per cent. of its food from filamentous Algæ. A minute fish had been eaten by one of these minnows. Insects made two-thirds of the food, nearly two-thirds of them being terrestrial. Neuropterous larvæ were the principal aquatic forms, chiefly case-worms and larvæ of ephemerids. The Crustacea (twenty-two per cent.) were all Entomostraca, about two-thirds of them Cladocera, the remainder Copepoda. Among the former Bosmina and Chydorus were recognized, and Diaptomus among the latter.

Six of this species had eaten only insects, and these made ninety per cent. of the food of two others. One had filled itself with the larvæ of *Bibio albipennis*, a terrestrial grub abundant in early spring. Three from Peoria Lake, captured in October, had eaten Cladocera only, nearly all a single species, *Bosmina longirostris*.

PHOTOGENIS ANALOSTANUS, Grd. SILVER FIN.

Excessively abundant in streams of all sizes.

The gill-rakers are short, triangular, about one-fourth of the length of the filaments. The intestine is shorter than the head and body.

Thirty-three specimens of this species were examined. Two-thirds of the food was insects, seven per cent. fishes, taken by three individuals, and one per cent. spiders, bringing the ratio of animal food up to seventy-one per cent. Besides these, a Limnæa was eaten by one, and traces of Cladocera and Copepoda occur in three. Nearly half the insects were terrestrial, Corixa and neuropterous larvæ being the most important aquatic forms. The vegetable food (twenty-nine per cent.) was nearly all of terrestrial origin, about one-third consisting of Algæ, both filamentous and unicellular, including Spirogyra and Glœocystis. Seeds, anthers and pollen of plants, and fragments of grass-like vegetation were noticed.

Eight of the specimens had taken only insects, and in two others these amounted to ninety-five per cent. Two had fed upon terrestrial species only. Corixa made ninety-five per cent. of the food of one. One had fed solely upon filamentous Algæ, and ninety per cent. or more of the food of three others consisted of vegetable structures in general.

PHENACOBIUS SCOPIFERUS, Cope.

This species occurs not very abundantly throughout the State, from Galena to extreme Southern Illinois. It has been taken by us almost invariably in swift and shallow streams.

The mouth is small and inferior, provided with fleshy lips somewhat resembling a sucker's in form. The gill-rakers and pharyngeal teeth are as usual in this group and the intestine is contained once and a half in the length of the head and body.

The nine specimens studied were from five localities, distributed from Galena to Union county. The food was almost purely insects, only two per cent. being unrecognized vegetation. Seventy-six per cent. consisted solely of Chironomus larvæ, and six per cent. of case-worms. Adult chironomids, taken by two of the specimens, amounted to two per cent. A few Cyclops found in a single specimen were the only Crustacea eaten by these fishes.

The peculiar character of this food, almost precisely that of a darter, is evidently related to the habitat of the fish.*

SEMOTILUS CORPORALIS, Mitch. CHUB.

This is a widely distributed and very abundant fish, perhaps the commonest species in the small creeks; but is less abundant in lakes and ponds.

The head and mouth are unusually large for a minnow; the intestine is six-sevenths the length of the head and body; and the gill-rakers are of the usual form.

Twenty-two specimens, from widely separated localities, give a ratio of seventy-six per cent. of animal food, four per cent. being fishes (partly Cyprinidæ), thirteen per cent. vegetation, and three per cent. worms. Insects make a little over half the whole, about one-half of them terrestrial. No Chironomus larvæ were found in the food of these fishes. Of neuropterous larvæ only a trace occurred, aquatic Coleoptera were noted in two, and Corixa in one. Grasshoppers (Acrididæ) made ten per cent. of the whole and were eaten by three of the specimens. Five had taken crawfishes, which made twelve per cent. of the entire food. No Entomostraca were noted, with the exception of one per . cent. of Cyclops

---

*For a discussion of this matter, see Bulletin 3 of this series, p. 25.

occurring in two of the specimens. Numerous examples of Gordius were found in two, and were reckoned at three per cent. of the food.* The vegetable food (twenty-four per cent.) was half Algæ, the remainder miscellaneous vegetable debris.

Eight had eaten only insects, two having filled themselves with grasshoppers. Three from a prairie stream near Normal had taken only crawfishes, while of four specimens captured in McLean County in July, filamentous Algæ composed ninety-four per cent. of the food.

CERATICHTHYS BIGUTTATUS, Kirt. HORNED CHUB.

This species is everywhere abundant northward, chiefly, like Semotilus, in the smaller streams, but preferring swifter waters. We have not taken it, however, south of the center of the State.

It differs from the preceding members of the group by the greater length of its alimentary canal, which considerably exceeds the head and body, the latter being contained in the intestine about one and one-fourth times. The gill-rakers are not peculiar.

Thirteen specimens from Northern and Central Illinois had derived less than half their food from the animal kingdom. Only about one-fourth of it consisted of insects, largely case-worms and other neuropterous larvæ, another fourth being Crustaceans (crawfishes), eaten, however, by only two of the specimens. The vegetable food (fifty-four per cent.) was about equally divided between filamentous Algæ and seeds of Setaria and other grass-like plants.

Notwithstanding the small ratio of insects figured out, it is worthy of note that two specimens out of four captured in a creek in September had eaten only insects, chiefly case-worms, while these composed ninety-five per cent. of the food of another. As the intestines of these fishes contained a considerable quantity of gravel, it is evident that they had fed upon the bottom in rather swift water. On the other hand, two specimens had derived all their food from vegetable sources, and three others had eaten eighty per cent. or more of vegetation. The extraordinary amount of vegetation in the food of this fish is possibly related to the increased length of the alimentary canal.

---

*These were not from the same specimens as those containing the grasshoppers.

SUMMARY FOR THE GROUP.

Ninety-five specimens of Group IV examined, representing five genera, had derived about three-fourths of their food from the animal kingdom, three per cent. of it being fishes, sixty-one per cent. insects, one per cent. Arachnida, and eleven per cent. Crustacea. One-third of the insects and spiders belong to terrestrial species. Chironomus larvæ are among the most important aquatic elements, amounting to sixteen per cent.; neuropterous larvæ coming next (eleven per cent.). About two-thirds of the crustaceans were crawfishes, the remainder being Cladocera and Copepoda. The vegetation (nearly one-fourth of the entire food) was chiefly of miscellaneous origin, nine per cent. only being recognizable as of aquatic forms. This was almost entirely filamentous Algæ.

Concerning this fourth group it may consequently be said, roughly, that the food consists of insects, crustaceans, and vegetable debris, about two-thirds of it the first, one-fourth of it the last, and one tenth, the other.

Summary for the Family.

If we regard the two hundred and fourteen specimens of fourteen genera which I have studied, as fairly representative of the family Cyprinidæ, and strike a separate balance of their food, we shall find that about thirty per cent. of the contents of the alimentary canal consists of mud; that one-half of it, or a little less, is animal matter, and that vegetation amounts to about one-fourth. Insects make one-third of the entire food, about ten per cent. being terrestrial species, eight per cent. Chironomus larvæ, and an equal number larvæ of Neuroptera. Of aquatic Coleoptera we have only a trace, and of aquatic Hemiptera (Corixa) but one per cent. Crustacea stand at ten per cent., nearly half of them Cladocera, Entomostraca as a whole amounting to about three-fourths of the crustacean ratio. Fishes are only two per cent., and mollusks less than one. Nearly half the vegetable food consists of Algæ (chiefly filamentous forms), the remainder being miscellaneous structures, derived from a great variety of plants, mostly terrestrial.

Summing up, in a word, the characteristics of the food of the family as thus indicated, we may say that about one-half of it consists of animal substances, one-third being insects, and one-third of these of terrestrial species, and ten per cent. being crustaceans; that one-fourth consisted of vegetation, about equally aquatic and terrestrial, and that the remainder is mud, probably containing more or less fluid organic matter.

COMPARISON OF THE GROUPS.

It will be remembered that the groups were based upon differences in the structures relating to the appropriation and mastication of food. It is consequently from a comparison of the ratios of these groups that we shall derive the most interesting facts relating to the correspondence of food and structure. The most conspicuous result is the great preponderance of mud in the intestines of the fishes of the first group, characterized by an extraordinarily elongate intestine, and by pharyngeal teeth destitute of hooks and provided with a broad grinding surface. Here, as already noted, mud, sand, and gravel amounted to about three-fourths of the matter ingested, while in the third and fourth groups only trivial and accidental quantities occurred. In the second group, on the other hand, with intestines intermediate in length, mud was still abundant, but much less so than in the first, averaging less than half the whole. If we exclude this indigestible matter, however, we shall find the first group still further distinguished by the predominance of vegetation as compared with animal matter, the latter being only about one-third the former, while in Groups III and IV, on the other hand, vegetation amounts to about one-third the animal food. The groups last mentioned, distinguished from each other as they are, only by the presence of a masticatory surface on the pharyngeal teeth in the first, and its absence in the second, differ scarcely at all in their general food characters, and this structural feature seems therefore to be of little significance. In both the animal ratio amounts to seventy-five per cent., and vegetation stands in each at twenty-five; while insects are respectively fifty and sixty-one. It is true that we find neuropterous larvæ greatly predominant in the first group, making one-fourth of their food, and Chironomus larvæ in the second amounting to sixteen per cent. The second of

these facts we find upon analysis to be evidently due to Phenacobius, by which genus nearly all the Chironomus larvæ were taken;
and this, as already shown, is explained not by any structural
feature, but by its peculiar habitat; and when we note that
aquatic larvæ together amount in Group III to twenty-five per
cent., and in Group IV to twenty-seven, we see that the significance of the difference mentioned disappears. A similar explanation is found of the difference in the ratios of Entomostraca,—
that of the first group amounting to twenty per cent. and that of
the second only to four. An examination of the tables shows that
this predominance in the group first mentioned is nearly all tracea-
ble to Hemitremia, a very small fish with rather elongate gill-rakers.

The importance of these gill structures is still more clearly
indicated, as already noticed, by the difference between Notemigonus and Chrosomus of the second group, and clearly far out-
weighs the structure of the teeth as an indication of the food
habits of the fish.

The general conclusions reached may be thus briefly stated:
An extraordinarily elongate intestine indicates the limophagous
habit, rather than an especial preference for vegetable food. The
length or number of the gill-rakers has much to do with the
abundance of Entomostraca and other minute animal forms in the
food of the fish, while the presence or absence of the terminal
hook or the masticatory surface to the pharyngeal teeth is not thus
far shown to have any sensible influence upon the general average
of the food. Finally, a species may depart widely in food characters from those more nearly allied to it in structure, if its
favorite haunts are peculiar.

TABLE OF FOOD: APHREDODERIDÆ TO UMBRIDÆ.

	Aphredoderus sayanus.	Potamocottus meridionalis.	Gasterosteidae.	Labidesthes sicculus.	Fundulus diaphanus.	Zygonectes notatus.	Zygonectes inurus.	Zygonectes dispar.	Summary of Cyprinodontidae.	Umbra limi.
No. of Specimens Examined	19	6	7	25	8	17	6	2	33	9
KINDS OF FOOD.	NUMBER OF SPECIMENS AND RATIOS IN WHICH EACH ELEMENT OF FOOD WAS FOUND.									
ANIMAL FOOD	.99	1.00	.75	1.00	.81	.90	.63	1.00	.70	.59
I. FISHES	.02	.27		.01	....	*		*		
II. MOLLUSCA				.01	.02	.03	.42	*	.15	.05
Univalves					.02	.03	.42	*	.15	.05
Bivalves				.01						
III. INSECTS	.91	.44	.44	.54	.37	.73	.18	*	.43	.14
Aquatic larvæ	.08	.36								
1. Hymenoptera				.01	.04	.08	.02		.05	
2. Diptera	.45	.01	.43	.46	.14	.21	.07	*	.14	*
Terrestrial				.30	.01	.02			.01	
Brachycera				.04		.01			.01	
Chironomidæ				.25	.01	.01			†	
Aquatic (larvæ)	.43		.37	.05	.02	*			.01	
Culicidæ	†					*			*	
Corethra	.07									
Chironomus	.36		.17	.05	.02				.01	
Simulium			.20							
3. Coleoptera	.11				.06	.17			.08	
Terrestrial	.11									
Aquatic larvæ					.03	.08			.04	
Hydrophilidæ						.08			.03	
Philhydrus					.03	.04			.02	
4. Hemiptera	.02			.01	.02	.17	.06	*	.09	.08
Terrestrial				.01	.02	.05	.05		.04	
Aquatic						.11	.01	*	.04	
Corixa	.02					.10		*	.03	
5. Neuroptera (larvæ)	.25			.06	.11	.03	.01		.05	
Ephemeridæ	.14				.11				.04	
Palingenia	.07									
Odonata					.06		.03		.01	
Agrion					.01		.03		.01	
Libellulidæ	.11									
Leptoceridæ							.01		†	
6 Thysanura (Podura)					†					
IV. ARACHNIDA				.01	.03	.02	.03		.02	*
Terrestrial				.01	.03	.02	.03		.02	
Hydrachnidæ					†	†				*
V. CRUSTACEA	.04	.29	.30	.40	.21	.06	.03	*	.10	.13
1. Amphipoda	.03				.13	*	.02		.05	.03
2. Isopoda (Asellus)	.01	.29								
3. Cladocera				.28	.24	.01	.02	†	*	.01	.06
Daphnidæ				.14	.23		.01			
Lynceidæ				.14	.01	.01	†			
4. Ostracoda	†			.01		.07	.02		.03	.04
5. Copepoda	†			.01	.16		.02	.01	.01	
VI. VERMES (Chætopoda)	.02									
VII. Protozoa						†		†		
VEGETABLE FOOD	.01			.25		.19	.10	.37	.17	.41
Seeds						.18	.01		.05	
1. Endogens	†								.09	.33
2. Algæ	.01			.25		.01	.09	.03	.34	.03	.06

TABLE OF FOOD OF CYPRINIDÆ.

KINDS OF FOOD.	Campostoma anomalum.	Pimephales promelas.	Hyborhynchus notatus.	Hybognathus nuchalis.	Summary of Group I.	Notemigonus chrysoleucus.	Chrosomus erythrogaster.	Summary of Group II.	Hybopsis hudsonius.	Hybopsis stramineus.	Luxilus cornutus.	Hemitremia heterolon.	Summary of Group III.	Minnilus atherinoides.	Photogenis analostanus.	Phenacobius scopiferus.	Semotilus corporalis.	Ceratichthys biguttatus.	Summary of Group IV.	Summary of Cyprinidæ.
No. of Specimens....	9	4	9	8	30	25	3	28	17	5	21	18	61	18	33	9	22	13	95	
	NUMBER OF SPECIMENS AND RATIOS IN WHICH EACH ELEMENT OF FOOD WAS FOUND.																			
ANIMAL FOOD....		.27		.07	.85	†		.18	.69	.76	.62	.87	.73	.95	.71	.98	.76	.46	.77	.48
I. FISHES									.03			.14		.04	.03		.04			.02
II. MOLLUSCA........						.14	.07	.01				†	†		†					
Univalves........						.14	.07					†	†							
Bivalves..........								.01						†						
III. INSECTS..........	†	.25	†		.06	.06		.03	.51	.66	.45	.29	.48	.67	.63	.98	.56	.24	.61	.34
Terrestrial........									.03			.01		.02	*			.01		.10
Aquatic..........									.02	†	†			.01		.12	.08	.03		
1. *Hymenoptera*									.02			.01	.05	.04		10		.04		
2. *Lepidoptera*.......	*	.15			.04	.03		.01			†		†	.01	.05		.01	.03	.02	
3. *Diptera*..........	†	.10			.02	†		†	.03	.05	.04	.24	.09	.14	.26	.78	.08	.01	.25	
Terrestrial........												.01	†	.11	.18	.02	†		.06	
Chironomidæ												.01	†			.02				
Aquatic (larvæ) ...	†				†	†		†				.20	.06	.02	.06	.76		.01	.17	
Chironomus:	†				†	†		†	.03			.20	.06		.03	.76		.01	.16	.08
Simulium														.02	.02					
4. *Coleoptera*......					.02			.01	.18		.06		.06	.13	.02		.10	.01	.05	
Terrestrial........					.02			.01	.18		.05		.06	.13	.01		.05	.01	.04	
Carabidæ........														.04			.01		.01	
Aquatic												.01	†				.03		.01	†
Haliplus ..																	.02		.01	
Gyrinidæ (larvæ)...												.01	†							
Hydrophilidæ......														.01			.03		†	
5. *Hemiptera*......					.01			.01	.01	♪	.02		†	.11	.10		.03		.05	
Terrestrial........											†		†	.05	.05				.01	
Aquatic					.01			01	.01		.01		.01	.05	.08		.03		.03	.01
Corixa									.01		.01		.01	.05	.08		.03		.02	
6. *Orthoptera*																	.12		.02	
7. *Neuroptera*									.28	.58	.10	.03	.25	.17	.10	.14		.16	.11	.08
Larvæ............											.04		.01	.01	.08	-..	.03	.02		
Pupæ......														.03					.01	
Ephemeridæ									.28	.48	.06	.03	.21	.10	.07	.06		.01	.02	
Phryganeidæ......										.10			.03	.07	.05	.06		.12	.06	
IV. ARACHNIDA ...												†	†	.02	.01				.01	
V. CRUSTACEA			†	†	.15	†		.08	.13	.10	.03	.58	.21	.22	†	†	.13	.22	.11	.10
1. Decapoda (Crawfish)																	.12	.22	.07	
2. *Amphipoda*									.03		.01	.01								
3. *Cladocera*						.10	†	.05	.01		†	.36	.09	.15	†			.03	.04	
Sididæ (Daphnella)...												.01	†					.02		
Daphniidæ..........						.10		.06			†			.12	†			†		
Lynceidæ						*		*	.01		†	.35	.09	†	†			†		
4. *Ostracoda*			†	†					.12			.04	.04	.04					.01	
5. *Copepoda*......						.05		.03	.10			.17	.07	.07	†	†	.01	.01	.02	
VI. VERMES		.02			.01							†	†				.03	.01		
Lumbriculus...;												†	†							
Naidæ............		.02			.01															
Gordius																	.03	.01		
Rotifera.	†											†	†							
VII. PROTOZOA	†				†							†	†							
VEGETABLE FOOD	.20	.01	.20	.25	.18	.50	.13	.31	.31	.24	.38	.11	.26	.05	.29	.02	.24	.54	.23	.23
Miscellaneous........	.02	†	.15		.09	.21	.02	.11	.10	.01	.05	.03	.05	.01	.06	.02	.10	.02	.04	
1. *Fungi*..........			†									†	†	†						
2. *Algæ*..........	.18	.01	.05	.25	.09	.19	.09	.14	.11		.12	.01	.05	.01	.09		.12	.23	.09	.10
MUD AND GRAVEL	.75	.72	.80	.75	.75	.15	.87	.51	†			.02	.01			†		†		.29

THE FIRST FOOD OF THE
COMMON WHITE-FISH

Stephen Alfred Forbes

THE FIRST FOOD OF THE COMMON WHITE-FISH.

(Coregonus clupeiformis, Mitch.)

By S. A. FORBES.

In a very large lake the conditions of life are remarkably uniform. The volume of water remains, of course, nearly constant from season to season and from year to year, and the extremes of summer heat and winter cold have but a moderate effect upon the temperature of the lake as a whole. Consequently both plant and animal life exhibit there a regularity and stability which are in remarkable contrast to their fluctuations in smaller bodies of water and on the surrounding land. Not only do the relative numbers of individuals in the various species remain about the same, but the absolute number of each must necessarily change but little, as a rule.

Such a state of affairs is eminently favorable to an exact and *economical* balance of supply and demand, of income and expenditure, of multiplication and destruction, among the inhabitants of the lake. Here, every species of animal, whether predaceous or vegetarian, must find, in the surplus products of growth and reproduction among the species upon which it depends for food, a far more constant and unvarying supply for its needs than elsewhere; and the species fed upon must be subject to a far more regular drain upon their surplus numbers or unessential structures. Where there is little fluctuation there is little waste.

A system of life like this, running on with relatively even tenor for centuries, must of course be much less *flexible* than one where wide and violent fluctuation and continual readjustment are the rule; and a species in any way deeply affected will here have within itself far less recuperative power than one which has been forced again and again—each year, perhaps—to rally against the most destructive attacks as the price of its continued existence. Disturbances of the natural balance of life, of the primitive and spontaneous system of reäctions by which the different groups of organisms are related, will therefore be unusually serious and lasting; and where such disturbances result from human interference, as by the yearly capture of large numbers of any important fish, it is especially desirable that artificial means of compensation be taken to restore the disturbed balance as nearly as possible. Excessive loss will be made good by natural reäctions far more slowly than if it occurred to a pond or river species, accustomed, as most of the latter are, to fill up rapidly enormous gaps in their numbers.

On the other hand, to multiply *unduly* by artificial measures any species naturally abundant in such a lake, will have scarcely a less disturbing influence than to *diminish* its numbers in the same ratio. The relatively nice balance between the demand for food and food supply which here naturally obtains, is such that an extraordinary increase in a species must soon reäct to diminish greatly its food resources — a fact which will then take effect on the species itself, reducing it below its natural, original level; and if both excessive capture and excessive multiplication go on side by side we shall have this result finally aggravated to an extreme degree.

As fishes are caught before the end of their natural lives, but planted by the fish culturist when young, it is evidently the food of the young which will be first and most seriously affected by over-production. Only a part of the adults, perhaps a small fraction, will live a life of ordinary natural length, many being captured before they have attained even the average size; but a far greater number, perhaps nearly every one, must survive the earliest period and must consequently draw most heavily upon the earliest food resources of the species when these differ from those of the adult.

The above considerations are brought forward here to show the especial importance, to us, of a study of the system of natural interactions by which the animals of our great lakes affect each other, if we would avoid the necessarily injurious consequences of our own interference with the natural order there obtaining, and above all to show the extraordinary value of a knowledge of the food habits and food capital of the *young.* They apply perhaps more forcibly to the white-fish than to any other species in the lakes; because this is for several reasons the most important purely fresh-water fish of the great lake region, and proves to have a distinctly different food when young from that upon which it is dependent later.

According to the recent census report,* more than twenty-one million pounds of white-fish were taken in the Great Lakes in 1879, valued at over three-quarters of a million of dollars, and representing nearly half the total sum derived from the lake fisheries of all kinds. These fisheries employ over five thousand men, and a fixed capital of one million three hundred and fory-six thousand dollars. When we reflect that this enormous drain upon the number of the species is necessarily, to a considerable extent, an addition to the natural tax levied upon it by its enemies other than man, we see that there must be an artificial supply provided, or the fisheries will gradually fail.

The importance of the knowledge of the food of so valuable a species needs no demonstration, especially when we consider that, consistently with what has been said above, it may not be difficult to overdo the work of propagation.

If the white-fish were to be multiplied indefinitely, without any attention to the character or abundance of its food supply, it would soon reach such a number that it must infringe upon its own food capital, diminish the average number of the animals upon which it depends for subsistence, and so finally indirectly cripple itself. Then the money and labor expended in its culture would be principally lost, and the last state of the species would be worse than the first. An acquaintance with the food of. the young is especially necessary, because they are planted by the fish-culturist when, having already absorbed the egg-sac (the supply of food by which they are under natural conditions supported until they have time to scatter themselves widely through the water), they are in a peculiarly helpless condition, unable to wander far in search of subsistence, and compelled to find food speedily or perish. One would say, therefore, that their alimentary resources and habits should be well and thoroughly known, that_the range, period and abundance of the organisms upon

---

*Census Bulletin No. 261, Sept. 1, 1881.

which they feed should be carefully determined, and that each
locality where the young are deposited should be closely searched
for the purpose of ascertaining whether their food species occur
there at the time in sufficient quantity to prevent immediate starvation.

Previous studies of the food of young fishes of a variety of families, reported in the third Bulletin of this series, had showed that,
with exceptions presently to be mentioned, the earliest food of all
the families studied consisted almost wholly of various species of
Entomostraca and some equally minute and delicate dipterous
larvæ.   When that paper was prepared, I had, however, no opportunity to study the food of the young of any members of the family
Salmonidæ, to which the white-fish belongs, neither could I learn
that any such studies had been made by others; and I could only
infer the same fact with regard to this family from the general
character of the results obtained by the study of the other groups.
Even this inference, however, was rendered doubtful by the discovery that the youngest individuals of two of the toothless families (Catostomidæ and Cyprinidæ) were not strictly dependent
upon the food elements above mentioned, but were likewise able
to draw upon much smaller organisms, namely: the minutest
Protozoa and unicellular Algæ; and as the adult white-fish is likewise destitute of teeth, it was not by any means certain that their
young would not fall under the latter category.   Upon looking up
the literature of the subject, I found that although the food of
the adult had been very well made out in a general way,* only
two items had been published respecting the food of the young.
In the report of the United States Fish Commission for 1872-3,
an assistant commissioner, Mr. J. W. Milner, made some experiments on young white-fish hatched artificially, supplying them
with a number of articles of food, in the hope of finding something suitable for their nourishment.

"A few crawfish," he says, "were procured and pounded to a
paste, and small portions put into jar No. 1; the young fish ate
it readily.   They were fed at night, and the next morning every
one of them was found to be dead.   Jar No. 2 was supplied with
bread crumbs, and the fish were seen to take small particles in

---

*Report of the U. S. Fish Commission for 1872-3, pp. 44-46.

their mouths; they did not die so suddenly. Jar No. 3 was supplied with sweet cream, but no evidence was afforded that the occupants fed upon it. A quantity of rain-water was exposed to the rays of the sun for the purpose of generating minute forms of life, and a teaspoonful was poured into jar No. 4, morning and evening, in the hopes that their proper food was of this character. In jar No. 5 a variety of food was provided, dry, fresh beef, milk, boiled potato, and bread. The crumbs of bread and the scrapings from the beef were all that the fish were seen to take into their mouths. They died, one after another, very rapidly, and in a few days all were dead." He further remarks: "This difficulty of procuring a suitable food for the young white-fish has been the experience of the few fish-culturists who have hatched them."

With the hope of ascertaining the natural food of these fishes, a few specimens, representing young captured in the Detroit River, and others from the hatchery, were submitted by Mr. Milner to Mr. S. A. Briggs, a microscopist, of Chicago. Four examples were examined by Mr. Briggs, two from each of the above situations. Those from the hatchery contained nothing whatever, while those from Detroit River contained numerous specimens of two species of Diatomaceæ, viz., *Fragilaria capucina* and *Stephanodiscus niagaræ*. The only fact at that time known would consequently indicate that the earliest food of the species consisted of Diatomaceæ.

The white-fish, as is well known, lays its eggs in the open lake in autumn, the young not appearing until early in the following spring. At this cold and stormy season in the exposed situations where they are to be sought, it is practically impossible to find the young fish; a fact which rendered the study of their earliest food a subject of unusual difficulty. There seemed, in fact, no practicable way to reach satisfactory conclusions upon it except by experiment upon individuals artificially hatched.

In December, 1880, I made an arrangement, through the kindness of Prof. Baird, of the Smithsonian Institution, with Mr. F. N. Clark, superintendent of the U. S. fish hatchery at Northville, Mich., for a supply of young white-fish to be sent me at intervals from the hatchery under his control. The specimens furnished were taken from two lots. The fishes of one lot, hatched January 18, were kept in a tank in the hatchery, where they were supplied with water from a spring, which had been cooled by exposure to the air in artificial ponds before entering the hatchery, in order to retard the development of the fry. The ordinary range of temperature in the tank, was from thirty-five to thirty-nine degrees. These fishes were fed daily with a paste made by grinding small amphipod crustaceans (Gammarus) in a mortar.

The second lot, hatched January 20, was kept, unfed, in a perforated tin box, in a rivulet flowing from a spring, about sixty feet from its source. The water had a uniform temperature of forty-seven degrees.

Those in the spring being in warmer water than the others, developed much more rapidly, and it was believed that the character and source of this water was such as to furnish them at least a small supply of such food as young fishes are accustomed to appropriate.

Ninety specimens were received from the hatchery February 9, at which time they were three weeks old. They were thirteen mm. (half an inch) in length by one in depth. The egg-sac was but partially absorbed in most of the lot, but in those most advanced was represented by an oil globule back of the head. The pectoral fins were well developed, but no trace of the ventrals had as yet appeared. The single median fin extended well in front of the vent, and forwards on the back nearly to the head. The opercles did not fully cover the gills. The most highly developed specimens — those whose gill-sacs had nearly disappeared — had, at a short distance on either side of the symphysis of the lower jaw, a sharp, strong, raptatorial tooth, curved backwards and slightly inwards. The base of this tooth was very broad, and the point acute and slender. At a point behind each of these teeth about half their distance from each other, was a second much smaller tooth, directed almost exactly inwards. The upper jaw was, however, wholly toothless.

These fishes were all passed under the microscope, after having been rendered transparent, but only four of them contained anything whatever; three a little dirt, and the fourth a minute fragment of the crust of the Gammarus, with which they had been fed.

Of one hundred and eleven specimens received February 17, seventeen had taken food. I dissected nine of these and found fragments of Gammarus and nothing else.   Ninety specimens from the same lot were examined February 25, and food was found in fourteen.   Four of these had eaten Gammarus fragments; two, larvæ of gnats; one, a small Cypris, and eight contained small fragments of the leaves and stems of vascular plants, including a bit of a netted-veined leaf and a little piece of pine wood.   Thirty-nine specimens, the last of the lot, were received March 15, and food was found in fourteen.   I dissected nine of these, finding fragments of Gammarus in four, a larva of a gnat, a Chironomus larva, a larva of some undetermined fly, a minute vegetable fragment, a Cyclops, a Cypris, and an undetermined Entomostracan each in one.   Three hundred and forty fry from the hatching house were examined in all, in forty-seven of which (fourteen per cent.) more or less food was discernible.   Of the thirty-five dissected, eighteen had eaten Gammarus fragments; five, minute insect larvæ; four, Entomostraca, and eight, small particles of vegetation.

Only four lots were received from the spring, on the 9th, 14th, 17th, and 25th of February, after which all died of starvation.   In the first hundred only one was found which had taken food, and this had eaten a trace of filamentous Algæ and a minute fragment of the parenchyma of some higher plant, with a few diatoms. But one of the second hundred contained even a trace of food, a minute quantity of some thread-like Alga, the cells of which still contained a little chlorophyll.   In the third hundred likewise, food was found in but one.   This consisted of a few particles of vegetable parenchyma, doubtless derived from the decaying plant structure in or around the water.   In the third lot of only forty-two specimens, six showed traces of food, consisting almost entirely of a few filamentous Algæ (including a fragment of Oscillatoria) and a little vegetable parenchyma.   Desmids and diatoms were observed in trivial numbers.

The total number received from the spring was two hundred and forty-two, of which but eight were found to have eaten anything (a little over three per cent. of the whole), and these had taken only Algæ and vegetable fragments.

An example of the water of the spring sent me contained many Algæ but no animals larger than rotifers.   The water of the hatchery, being exposed in ponds of considerable size, afforded a better opportunity for the development of animal life, to which fact was doubtless due the occurrence of insect larvæ and Entomostraca in the intestines of the fishes reared in it. The situation of the spring, on the other hand, was particularly unfavorable, as it was under the hatchery, and consequently in the dark.

The observations above described on the specimens kept in spring water, have but little value for the reason that evidently very little food was contained in the water flowing through their cage. The vegetation in the streams being chiefly filamentous Algæ and the number of Entomostraca apparently trivial, very little of either vegetable or animal food could reach the little prisoners. It is not surprising, therefore, that notwithstanding their greater age and the higher temperature of the water in which they were kept, a much smaller ratio of the specimens had taken food than of those captured in the hatchery. From the contents of their intestines we can only infer that these fishes, reduced to a desperate strait by starvation, will snatch at almost anything contained in the water. The result obtained by a study of those from the hatching house was more significant, but still unsatisfactory. It seemed to indicate that in confinement white-fish fry will feed upon both animal and vegetable structures to some extent, and that they can be induced to take minute fragments of the higher crustaceans, but not in sufficient quantity to keep them alive. The fact that animal food was more abundant than vegetable in this last lot, indicates nothing of their natural preference, since it was doubtless also more abundant in the water containing them.

More light was thrown upon the earliest food habits of these fishes by the discovery of raptatorial teeth upon the lower jaw, than by these dissections of their alimentary canals. All the families of fishes which I had previously studied whose young were provided with teeth were found strictly dependent at first upon Entomostraca and the minuter insect larvæ; while only those whose young were toothless fed to any considerable extent upon other forms. The discovery of teeth in the young white-fish, therefore, placed this species definately in the group of those carnivorous when young. The fact that the adult was itself toothless interfered in no way with this inference, because other toothless fishes (Dorsoma) whose young were furnished with teeth, had been found carnivorous at an early age.

The inconclusive character of the results thus far obtained, made it necessary to attempt to imitate more closely the natural conditions of the young when hatched in the lake. In February, 1881, I obtained, through the kindness of Mr. Clarke, twenty-five specimens of living young white-fish, saved from a lot which he was planting in the waters of Lake Michigan, off Racine, Wisconsin. I succeeded in conveying these to the laboratory without loss, and there kept them for several days in a glass aquarium and supplied them with an abundance of the living objects to be obtained by drawing a fine muslin net through the stagnant pools of the vicinity. These consisted of many diatoms and filamentous fresh-water Algæ, of two or three species of Cyclops, of *Canthocamptus illinoisensis*, and *Diaptomus sanguineus* among the Copepoda, and of two rather large Cladocera, *Simocephalus vetulus* and *S. americanus*. These little fishes were kept under careful observation for several days, the water in the aquarium being frequently aërated by pouring. Many of them had, however, been injured by handling, and eleven of the specimens died without taking food. It was soon evident that the larger Entomostraca (the Simocephalus, and even the Diaptomus), were quite beyond the size and strength of these little fishes, and that only the smaller Copepoda among the animals available, could afford them any food at first. These they followed about from the beginning with signs of peculiar interest, occasionally making irresolute attempts to capture them. Two days after their arrival, one of the young white-fish had evidently taken food, which proved, on dissection, to be a small Cyclops. During the next two days nine others began to eat, dividing their attentions between the Cyclops above mentioned and the Canthocamptus, and on the 22d two others took a Cyclops each and a third a Canthocamptus. One of these fishes contained still a large remnant of the egg-sac, showing that the propensity to capture prey must antedate the sensation of hunger. On the 25th the fourteenth and last remaining fish captured its Cyclops and was itself sacrificed in turn. As an indication of the efficiency of the raptatorial teeth, it may be worth while to note that I saw one of the smallest fishes make a spring at a Cyclops, catch it, give three or four violent wriggles, and drop it dead to the bottom of tank.

As a general statement of the result of the observations made on these fourteen fishes, we may say that eight of them ate a single Cyclops each, that one took two, and another three of the same, that one took a single Canthocamptus, that two specimens captured two each of this genus, and that finally, a single fish ate Cyclops and Canthocamptus both.   The final conclusion was a highly probable inference that the smallest Entomostraca occurring in the lake would prove to be the natural first food of the species.

In order to test this conclusion with precision, I arranged a similar experiment on a larger scale and under more natural conditions.   Through the generosity of the Exposition company, of Chicago, I was allowed the use of one of the large aquarium tanks in the exposition building on the lake shore, and by the repeated kindness of Mr. Clarke, of Northville, Michigan, I was furnished with a much larger number of living white-fish.   Five thousand fry were shipped to me in a can of water, but through unfortunate delays in changing cars at intermediate points, about two-thirds of these were dead when they reached my hands.   Those living were immediately transferred to the tank, through which the water, taken from the city pipes, had already been allowed to run for several hours.   As this water is derived from Lake Michigan at a distance of two miles from the shore, and had at this time the exact temperature of the open lake, the conditions for experiment were as favorable as artificial arrangements could well be made.

Sending a man with a towing net out upon the lake with a boat, or upon the remotest breakwaters, immense numbers of all organic objects in the water were easily obtained. After enclosing the exit of the tank with a fine wire screen, to prevent the escape of objects placed in it, we poured these collections of all descriptions indiscriminately into the water from day to day, thus keeping the fishes profusely supplied with all the various kinds of food which could possibly be accessible to them in their native haunts. From this tank one hundred fishes were taken daily and placed in alcohol for dissection and microscopic study, to determine precisely the objects preferred by them for food.   These were examined at a later date, and all contents of the intestines were mounted entire as microscopic slides, and permanently preserved. A careful study, was of course made of the organisms of the lake, as shown by the product of the towing net, and when the experiment was finally ended, an equally careful examination followed of the living contents of the water of the tank at that time.

These ·fishes, like those previously described, had already reached the age and condition at which it is customary to "plant" them in the lake.   The ventrals were still undeveloped, the egg-sac had nearly disappeared, the four mandibular teeth were present, and the median fin extended from the tips of the pectorals on the belly to a point opposite the middle of the same fins on the back. In most the egg-sac did not protrude externally, being reduced in some to a droplet of oil, but remaining in a few of a size at least as great as that of the head.   The alimentary canal was of course a simple straight tube, without any distinction of stomach and intestine.

The sufferings of these fry in transit had doubtless weakened the vitality of the survivors, and although every care was taken to keep the water of the tank fresh and pure, about one-third of those remaining died during the progress of the experiment. The aquarium in which they were confined was built of glass, and had a capacity of about one hundred cubic feet.   The temperature, tried repeatedly, stood at forty-two degrees Fah.   A steady current of the water of the lake was maintained through this tank, entering through a rose, from which it fell in a spray, thus insuring perfect aëration.

By far the greater part of the organic contents of the water of the lake, as shown by the product of the towing net, consisted of diatoms in immense variety, which formed always a greenish mucilaginous coating upon the interior of the muslin net.   In this were entangled, a variety of rotifers, occasional filamentous Algæ, and many Entomostraca, the latter belonging chiefly to the genera Cyclops, Diaptomus, and Limnocalanus among the Copepoda, and to Daphnia among the Cladocera.

As the Entomostraca proved to be far the most important elements of this food supply, the particulars respecting them may be properly more fully given.   The smallest of all was a Cyclops, then new, but since described by me under the name of *Cyclops thomasi.**   This little Entomostracan is only .04 inch long, by .011

---

*On some Entomostraca of Lake Michigan and Adjacent Waters. American Naturalist, Vol. XVI., No. VIII, August, 1882, pp. 640 and 649.

wide. The next in size, and by far the most abundant member of
this group was a Diaptomus, likewise new, described in the paper
just cited under the name of *Diaptomus sicilis.*   This appears in
two forms, one evidently young in the stage just preceding the
adult.   Full grown individuals were .065 inch long, by one-fourth
that depth. The Limnocalanus was a much larger form, evidently
preying, to a considerable extent, upon the two just mentioned.
All the Cladocera noticed were *Daphnia hyalina*, an elegant and
extremely transparent species, occurring likewise in the lakes of
Europe.   A single insect larval form (Chironomus)·should likewise
be mentioned in this connection, since it had about the same size
and consistence of the Entomostraca, and was consequently equally
available for food.

The specimens of each of the above species from a certain
quantity of these collections were counted, in order to give a definite idea of their relative abundance in the lake.   The Diaptomus
numbered 225, the Cyclops 75, Limnocalanus 7, Daphnia 3, and
Chironomus larvæ 1.   It was a curious fact, however, that when
the water was drawn off at the end of the experiment, more than
half the Entomostraca were Limnocalanus; a fact partly to be explained by the predaceous habit of the latter, and partly by the
facts relating to the food of the fishes themselves, which are presently to be detailed.

The fry were placed in the tank and supplied with their first
food on the evening of the 12th of March.   On the 14th, one
hundred specimens were removed, and twenty-seven of these were
dissected.   Twenty were empty, but the remaining seven had
already taken food, all Cyclops or Diaptomus.   Three had eaten
Cyclops only, and six Diaptomus, while two had eaten both.
Fourteen of these Entomostraca, seven of each genus, were taken
by these seven fishes.   From those captured the next day, twenty-
five specimens were examined, of which nineteen were without
food.   Of the remaining six, three had eaten Diaptomus and three
Cyclops; five of the former being taken in all, and ten of the latter.   Three specimens were next examined from those caught on
the 19th of March, two of which had devoured Diaptomus, and a
third a single *Cyclops thomasi* and a shelled rotifer, *Anuræa striata.*
The character of the food at these earliest stages was so well settled by these observations that I deemed it unnecessary to exam-

ine the subsequent lots in detail, but passed at once to the specimens taken on the 23d. Twenty-six of these were examined, and found to have eaten thirty-three individuals of *Cyclops thomasi*, fourteen of *Diaptomus sicilis*, and fourteen of the minute rotifer already mentioned (*Anurœa striata*). Two had taken a few diatoms (*Bacillaria*) and one had eaten a filament of an Alga. Cyclops was found in sixteen of the specimens, Diaptomus in nine, and Anurœa in eight, only two of them being empty. The amount of food now taken by individual fishes was much greater than before, one specimen dissected having eaten two Cyclops and six *Diaptomus sicilis*, male and female. Another had taken five Cyclops, one Diaptomus and five examples of *Anurœa striata*. Still another had eaten four of the Cyclops, four Diaptomus, and one Anurœa.

Twenty-five specimens were examined from those removed on the 24th of the month, at which time the water of the tank was drawn off and all the remaining fishes bottled. Four of these had not eaten, but the twenty-one others had devoured fifty specimens of *Diaptomus sicilis*, forty-seven of *Cyclops thomasi*, fourteen of *Anurœa striata*, and a single *Daphnia hyalina*, the latter being the largest object eaten by any of the fishes. A few examples of their capacity may well be given. The ninth example had eaten six Diaptomus, two *Cyclops thomasi* and one Anurœa; the tenth had taken eight Diaptomus, two Cyclops and an Anurœa; and the twentieth, seven Diaptomus and three *Cyclops thomasi*. In two of these examples were small clusters of orange globules, probably representing unicellular Algæ.

Summarizing these data briefly, we find that of the 106 specimens dissected, sixty-three had taken food, and that the ratio of those which were eating increased rapidly, the longer the fishes were kept in the aquarium. Only one-fourth of those examined on the fourteenth of the month had taken food, while more than five-sixths of those bottled ten days later had already eaten. The entire number of objects appropriated by these sixty-three fishes was as follows: *Cyclops thomasi*, ninety-seven; *Diaptomus sicilis*, seventy-eight; *Anurœa striata*, twenty-nine; *Daphnia hyalina*, one. Seven of the fishes had eaten unicellular Algæ, two had eaten diatoms, and one, filamentous Algæ.

From the above data we are compelled to conclude that the earliest food of the white-fish consists almost wholly of the smallest species of Entomostraca occurring in the lake, since the other elements in their alimentary canals were evidently either taken accidentally, or else appeared in such trivial quantity as to contribute nothing of importance to their support.   In fact, two species of Copepoda, *Cyclops thomasi* and *Diaptomus sicilis*, are certainly very much more important to the maintenance of the white-fish in this earliest stage of independent life than all the other organisms in the lake combined.   As the fishes increase in size, vigor, and activity, they doubtless enlarge their regimen by capturing larger species of Entomostraca, especially Daphnia and Limnocalanus.

A few words respecting the relative abundance of these species at different seasons of the year and their distribution in the lake, will have some practical value.   We may observe here an excellent illustration of the remarkable uniformity of the life of the lake as contrasted with that of smaller bodies of water already referred to, in the introduction to this paper.   While in ponds minute animal life is largely destroyed or suspended during the winter, the opening spring being attended by an enormous increase in numbers and rate of multiplication, in Lake Michigan there is but little difference in the products of the collecting apparatus at different seasons of the year.*   There is a slight increase in the number of individuals during spring and early summer, but scarcely enough appreciably to affect the food supply of fishes dependent upon them.   They are not by any means equally distributed, however, throughout the lake, my own observations tending to show that there are relatively very few of these minute crustaceans to be found at a distance of a few miles from shore, and that in fact by far the greater part of them usually occur within a distance of two or three miles out.   Indeed, the mouths of the rivers flowing into the lake are ordinarily much

---

*For definite assurance of this fact, I am indebted less to my own observations (which are, however, consistent with it as far as they go) than to the statements of B. W. Thomas, Esq., of Chicago, who, while making a specialty of the Diatomaceæ of the lake, has collected and studied all its organic forms for several years, obtaining them from the city water by attaching a strainer to a hydrant many times during every month throughout the year.

more densely populated by these animals than the lake itself, as has been particularly evident at Racine and South Chicago. Neither are they commonly equally distributed throughout the waters in which they are most abundant, but like most other aquatic animals, occur in shoals. In the deeper portions of the lake, many species shift their level according to the time of day, coming to the surface by night, and sinking again when the sun is bright.

These facts make it important to the fish-culturist that the particular situation where it is proposed to plant the fry should be searched at the time when these are to be liberated, to determine whether they will find at once sufficient food for their support. A little experience will easily enable one to estimate the relative abundance of the Entomostraca at any given time and place, and they require nothing for their capture more complicated or difficult of management than a simple ring net of cheese-cloth or similar material, towed behind a boat. This may be weighted and sunk to any desired depth, so that the contents of the water either at the surface or at the bottom, may be ascertained by a few minutes' rowing.

In conclusion, I wish again to express my great obligation to the United States Fish Commissioner, Prof. S. F. Baird, and to Frank N. Clark, Superintendent of the United States Hatchery at Northville, Mich., through whom, as already stated, the specimens were derived upon which these studies were made. My best thanks are also due to the Exposition company of Chicago, and especially to their secretary, the Hon. John P. Reynolds, for the use of a tank in the Exposition building, and for many courtesies received while the experiment there was in progress.

STUDIES ON THE
CONTAGIOUS DISEASES OF INSECTS

Stephen Alfred Forbes

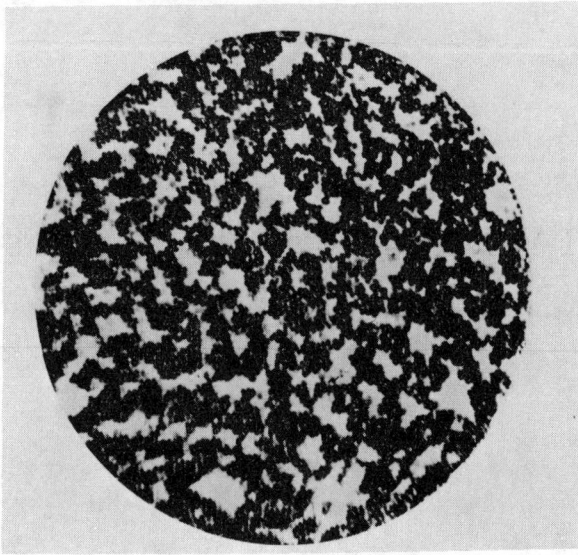

FIG. 1.

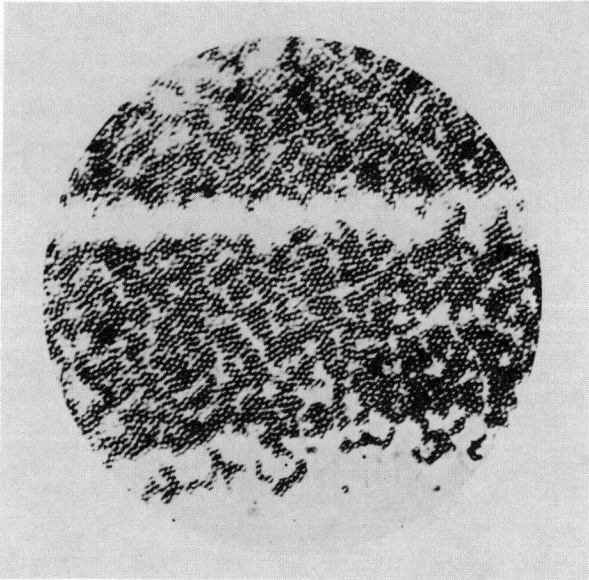

FIG. 2.

Culture of MICROCOCCUS from diseased Cabbage Worm (*Pieris rapæ* L.), in test tubes of sterilized beef broth, commenced October 20, 1883 (see p, 275). Preserved over winter in plugged test tube, and slides mounted April 10, 1884, in carbolized water, after staining with brown aniline.

Photographed with lamplight by Dr. H. J. Detmers, with Spencer $\frac{1}{15}$ homogeneous immersion, $\times$ 1000.

Both figures are from different parts of the same slide. differing only in focal adjustment,—figure 1 being a "positive," and figure 2 a "negative."

Article IV.— Studies on the Contagious Diseases of Insects.— I.

By S. A. FORBES.

Since August, 1883, the writer has used such opportunities as came in his way for observation of the diseases of insects, and for more or less careful and systematic work upon them, directed especially to the point of artificially propagating them for the destruction of injurious insect species. While these researches are not by any means completed, lacking especially critical study of the bacterial forms dealt with, as botanical species, and imperfect also on the side of field experiments on a large scale, I have thought them worthy of present report as a contribution to progress on a difficult but interesting subject, especially as opportunity for further continuance of some of these studies may not soon return.

My main object has been experimental and economical, and I have adopted such methods of study as seemed to me to offer the simplest means of surely ascertaining whether some of the common diseases of our insects were of bacterial origin, whether their germs were readily and conveniently cultivable, and whether such cultures could be used to convey the original affections to healthy insects.

This will serve to explain what may seem to some an excessive reliance on fluid cultures,—much more convenient for my purpose in these preliminary studies than "solid cultures" with gelatine films or tubes, and quite conclusive as to the identity of the forms dealt with, if the cultures are often enough repeated and the results are closely scanned.

Without attempting at this time to summarize the literature of the subject,—scattered and chiefly fragmentary, except as relates to the silkworm and the honey-bee,—I give here only a brief account of my own earlier notes and observations.

The appearance of what seemed to be an epidemic of contagious disease among the chinch bugs of Central Illinois in the latter part of the summer of the above year, gave rise to an article on this subject, published in 1883 in my first report as

State Entomologist of Illinois (pp. 45-57). This article contained an account of a considerable series of microscopic observations on the fluids of chinch bugs apparently affected with disease, and described some successful attempts at the culture of the Micrococcus found invariably characterizing this insect affection. Time failed for further experiments, and the chinch bug has since been so scarce in my vicinity that no further opportunity has offered to complete the study of the subject. The observations made amounted to a practical demonstration of the occurrence of a "germ disease" in this insect species, identified the germ as a Micrococcus, since described as *Micrococcus insectorum*, Burrill, and proved that this was easily and freely cultivable in beef broth. The Micrococcus was shown to have its seat in the alimentary canal of the insects, occurring most abundantly in the posterior part of the same, to infest pupæ and adults more seriously than the younger stages, and to have the apparent effect to retard the development of the brood as well as to destroy a large percentage of them before they reached maturity. This disease was apparently the representative of *flacherie* or *schlaffsucht* in caterpillars, as described by previous authors and in the following pages.

Next there appeared early in August, 1883, in our breeding cages of *Datana ministra* (the yellow-necked apple caterpillar), an outbreak of disease characterized by the occurrence, at first in the alimentary canal and later in the blood, of immense numbers of micrococci of a form very different from the above, and evidently quite readily conveyed from one insect to another. Elaborate studies of this disease were made during the remainder of the season and the following spring, the bacteria associated with it were repeatedly cultivated with success in animal infusions, and several experiments were made to convey the disease by their means to still healthy larvæ. Tubes of the culture fluids were sealed up for preservation over winter, their contents were cultivated again in June, 1884, and the resulting cultures were used to infect the food of larvæ of *Mamestra picta*, with the hope of thus reproducing the original disease of the Datana larvæ of the preceding year.

Parallel with these experiments was a similar series made on a frightfully contagious and destructive disease of the European cabbage worm (*Pieris rapæ*), first observed by us at Normal, September 11, 1883. The bacterial character of this disease was ascertained, many attempts at cultures were made, some of them successful, and the possibility of conveying the disease to a distance by means of affected cabbage worms was tested by us in Western Illinois and Iowa. Many of the observations and experiments relating to *flacherie* in this insect were repeated by me in 1884, and in the early summer of 1885 admirable photographs of several of the slides were made for me by Dr. H. J. Detmers, of Champaign.

"Jaundice" of the silkworm appearing in an experimental nursery of this species, under the charge of Professor Burrill at the State Industrial University, at Champaign, in June, 1884, an opportunity was afforded me to study this affection. Many successful cultures were made of the bacteria involved, and several experiments were undertaken for the infection of healthy cabbage worms with the contagion from these artificial cultures. Succeeding in the laboratory, these experiments were carried into the field, and attempted on the large scale of actual practice.

An epidemic of muscardine appearing in certain breeding cages of the forest tent-caterpillar (*Clisiocampa sylvatica*) in June, 1884, this disease was studied by us as there illustrated, and connected more or less certainly with a destructive epidemic of the preceding year, which had swept away vast numbers of that species under my observation in Southern Illinois.

With the exception of the *flacherie* of the chinch bug, these observations have not hitherto been anywhere fully reported, although brief notices and general accounts of a more or less popular character have been printed in the scientific journals and in some economic publications.

The chinch bug observations were published, as already mentioned, in the Twelfth Report of the State Entomologist of Illinois, the species of Micrococcus concerned having been previously described by Prof. Burrill in the Report of the Trustees of the Illinois Industrial University for 1882, and in the "American Naturalist" for March, 1883.

A brief preliminary paper on *flacherie* of Datana was read to the entomological club of the American Association for the

Advancement of Science at its Minneapolis meeting in August, 1883, and of this a synopsis appeared in the "Canadian Entomologist" for September, 1883. In the "Prairie Farmer" (Chicago) for October 6. 1883, and in "Science," also, for October 5, 1883, brief notes occur with reference to this disease in the cabbage worm.

In the Transactions of the Illinois State Horticultural Society for 1883 (printed February, 1884) is a somewhat elaborate paper on the Contagious Diseases of Caterpillars, read before this Society December 18, 1883, giving a general and rather popular account of the character of the work done by me on this subject, up to that time ; and a still more elaborate paper (never published) was read before the State Natural History Society of Illinois, at its meeting in Peoria, July 8, 1884, in which a classification of insect diseases was presented, and a full *résumé* of methods and results, up to that date, was given. At a meeting of the State Horticultural Society, held at Champaign, December, 1884, I added some further items relating to cultures and experiments, especially those affecting the cabbage worm, and these notes were published in April, 1885, in the Transactions for the year preceding.

It is my purpose, in this paper, to present the principal results of the above studies,— both the successful and the unsucessful issues,— the latter so far as they have any significance or value.

Disregarding the chronological order of my observations, I shall first discuss *flacherie* of the cabbage worm, and jaundice of the silkworm, with experiments upon the former insect with the artificial cultures derived from the latter. I will then take up the longer and more complicated record of *flacherie* in our Datana larvæ and the experiments drawn from it, and will conclude with a brief account of the *muscardine* of the forest tent-caterpillar.

EUROPEAN CABBAGE WORM *(Pieris rapœ, L.)*

In studying experimentally an insect disease, it is necessary, in the present state of our knowledge, (1) to determine precisely the symptoms and character of the disease itself, in order that it may be subsequently recognized with certainty; (2) to learn whether it is characterized by bacteria; and (3) whether it is practically contagious. Determining these questions affirmatively, (4) cultures of the bacteria must be made artificially, and (5) these cultures must be used to produce, in healthy insects of the same or other species, a disease characterized by the symptoms and results of the original affection. It is further desirable that (6) second cultures should be prepared from these cases of disease artificially produced, in order that a strict comparison may be made of the bacteria concerned, as they occur both in the bodies of the insects and in artificial culture fluids.

I propose to take up these points *seriatim*, (first with respect to *flacherie* in the cabbage worm), presenting separately the facts bearing upon each, only premising that the proof of one proposition is sometimes partly contained in the data relating chiefly to another, so that some repetition will be necessitated by this mode of discussion; but this disadvantage will doubtless be found insignificant, compared with the gain in clearness and cogency.

DESCRIPTION OF FLACHERIE IN THE CABBAGE WORM.

In this insect *flacherie* is distinguishable with great ease and certainty by conspicuous external symptoms, the color alone of affected larvæ being, in fact, entirely characteristic and unmistakable. The natural color of a healthy cabbage worm is a light lively green, sometimes slightly tinged with yellowish, but without any approach to an ashy or milky hue. As the first symptom of *flacherie*, however, the larva commences to turn pale, this paleness increasing more or less rapidly until the color is almost milky white, only slightly tinged with greenish. This discoloration is uniform and simple, no other tint usually appearing until after death. Then, however, the color deepens to a sooty gray, commonly uniform, but sometimes appearing first about the center of the length of the larva. Occasionally this deeper color appears a little before death, but it is not then of equal depth over the whole surface.

In the actions of the insect there is little to indicate any change of state, except a gradually increasing sluggishness, slowness of movement, and loss of appetite. These are later to appear than the pale discoloration above mentioned, and even shortly before death a larva may show considerable impatience if roughly handled. When the disease is well developed, the caterpillar is very feeble, and will remain motionless for a long time ; or if it attempt to crawl where some strength is needed, as horizontally on a vertical surface, it may lose its hold with its jointed limbs and cling only by its central prolegs, the fore and hinder parts hanging limp and helpless at right angles to the remainder of the body.

Most commonly an escape of fluid from the vent is among the earlier symptoms of the affection, at first greenish or whitish, and later a dirty gray, or even a chocolate brown. Rarely this fluid exudes also from the mouth. The amount of it is usually sufficient to stain considerably the surfaces over which the larva crawls: but sometimes this symptom is wholly absent. Occasionally the stomach is found empty after death, but almost invariably it is well filled with food, much of which has not yet lost its native color, digestion being, in fact, evidently suspended during the course of the disease. I have found in only a single instance an appearance of bubbles of gas in the alimentary canal, such as Pasteur describes in the *flacherie* of the silkworm. Usually the mass of the alimentary contents seems to lie inert in the stomach, undergoing neither digestion nor decay.

The color of the fluids of the healthy larva is a very pale transparent green, the blood containing only lymphoid corpuscles ·in greater or lesser number; but if a proleg of a diseased specimen be snipped off, and a cover glass be pressed against the cut surface, the droplet exuding will be of almost milky whiteness, or, in the latest stages of the disease, a dirty gray. Rarely, where there has been much escape of fluid from the vent, the juices of the larva will be thick and scanty, so that it requires some pressure to force out a very small quantity. If a minute droplet of the milky fluid obtained by snipping off a proleg be examined under a high power of the microscope, it will found to contain innumerable myriads of very minute spherules, varying in diameter, according to the individual, from .5 μ to 1 μ. Usually their average size does not surpass .7 μ. It is the infinite multitude of these which gives to the fluids of the diseased caterpillar their milky look, and likewise, unquestionably it is they which cause the ashy appearance of the surface, the skin being thin and delicate, so that the color of the fluid contents shows through. The diseased blood is so thick with these minute corpuscles that little else can be ordinarily seen in it. Sometimes, however, degenerated lymphoid corpuscles of the blood will be noticed, recognizable by their size and spherical contour, but differing from the normal corpuscles in their darker tint and coarsely and irregularly granular structure. These darker, granular corpuscles are always dead, no longer exhibiting amœboid movement, and have usually a spherical form. Not infrequently *débris* of the fatty bodies is apparent in the form of large irregular cells, floating freely in the fluid, but these cells themselves will be found to contain immense numbers of the minute spheres already mentioned. In fact, if a little portion of the soft remnant of the fatty bodies be removed, spread upon a cover, and examined with a power of a thousand diameters, it will be seen that the cells of these organs are the seat of an extreme degeneration, the entire contents of many of them being wholly replaced by the spherical granules mentioned above. Occasionally a cell containing a nucleus will be found, but more commonly all distinction of contents has disappeared.*

---

* As an example of the condition of the fatty bodies, I will describe those of a larva examined October 9, whitish in color and nearly dead, making little effort to escape. A droplet of the blood exuding from a small cut made in the back was alive with the minute spherules already mentioned, and contained also noticeable numbers of dead blood corpuscles in a dark, spherical, granular condition, together with a few unaltered examples still capable of amœboid movement.

A fragment of the fatty bodies examined. consisted chiefly of pale spherical cells, 1.5μ to 7.5μ in diameter, resembling oil globules, except that they had not the high refractive index of fat. A few of these globular cells were very pale and indistinct, the contents very indefinitely granular and often with a large spherical nucleus likewise very pale ; but most of them were more or less completely filled with dancing spherules, slightly different in size in different cells, these differences having, however, no relation to the proper size of the cells. Sometimes there were not more than twenty-five or thirty such granules in an optical section of a large cell, the con-

If the body of a diseased larva be cut across and a cover glass be pressed against the cut end of the intestine, or, still better, if the larva be opened lengthwise, the stomach removed and laid open separately, so that a droplet of the pure contents of the alimentary canal may be obtained, the fluid portion of these contents will be seen to swarm with infinitesimal granules identical in appearance with those found in the blood, except that they are, on an average, often appreciably larger and are occasionally more or less oval in outline. These same forms may also be found in the fluid excreta escaping from the vent of the still living larva. If the specimen has been dead some time, so that the sooty discoloration of the surface has occurred, the fluids both of the alimentary canal and of the body at large will often be found to contain, besides myriads of the above spherules, various other forms clearly recognizable as septic bacteria,—among these, members of the genus Bacterium, easily distinguishable by their oval form and by the manner in which they actively propel themselves across the field of the microscope. Rod-like bacilli may also appear in the fluids at this time, equally active, and evidently moving by means of flagella, especially in the vicinity of the bubbles of air which may be included in the fluid under the cover glass. Occasionally these latter bacterial forms may be found in smaller numbers even before death, very rarely in the perivisceral fluids, but not very uncommonly in the contents of the alimentary canal. Still they are infinitely less abundant than the Micrococcus-like spheres already mentioned, even long after the death of the larva.

The most characteristic *post mortem* phenomenon is the rapid softening, decay, and deliquescence of the body, the whole of which may be converted, in an hour or two after death, into a dirty fluid mass which the rotten skin is barely sufficient to hold together. This breaks at a touch, allowing the fluid contents to escape.

---

tents being otherwise fluid. Many of the cells were not full, areas occuring which the dancing particles did not invade. Occasionally an unaltered nucleus would be seen in the midst of the corpuscular contents of the cell. The fat globules intermingled were easily distinguished from the above cells by their very different refracting power, and were always free from the spherical granules. They were less than half as numerous as the pale cells. The average size of the granules was not far from .66 μ.

THE CHARACTERISTIC BACTERIA.

As implied in the foregoing, I have no doubt that a large percentage, at least, of the minute spherical granules abundant in the fluids of the body and alimentary canal of the diseased larvæ are genuine bacteria, belonging to the genus Micrococcus I cannot hope to convey verbally the same conclusive conviction of this fact which I have myself derived from long study of these little forms under a great variety of conditions, and the preparation and examination of a multitude of slides, both recent, and permanently mounted. These latter were sometimes unstained, and again stained with a considerable variety of aniline dyes,—brown, blue, violet, and magenta. Several successful cultures have also assisted to confirm this view, the products of the cultures being unmistakably the form originally taken from the caterpillars under experiment.

In form the micrococci of the cabbage worm are usually strictly spherical, although in the alimentary canal a patch will occasionally occur in which they are of a slightly oval outline. The micrococci of the fluids of the diseased larvæ seen in the field of the microscope are mostly separate spheres, but a considerable percentage of them are attached in pairs, as if in process of division. Rarely a short chain of four, six, or eight may be seen. In the stomach they occur not infrequently in compact patches or zoöglœa-like masses. In size the individuals vary from .5 μ to 1.25 μ in diameter, the small forms being those in the blood and the larger those in the stomach. Individual larvæ differ, in fact, with respect to the size of their micrococci, in some the average of those found in the blood being not far from .75 μ to 1 μ, while in others they barely reach .5 μ. Commonly, those of the stomach average 1 μ.

In addition to the direct evidence above adduced, the close resemblance of these corpuscles to those occuring in other larvæ affected similarly to the cabbage worm, in which the bacterial character was even less obscure. gave indirect and cumulative evidence with respect to the nature of those forms in the cabbage worm. Their reaction to the usual staining fluids was such as the hypothesis of their bacterial character would require. Although staining with some difficulty, many slides were prepared in which the individual bacteria were beautifully stained and distinctly differentiated from an uncolored film, by brown aniline, methyl violet, and magenta.

Although several of the attempts at artificial culture were abortive, and although the cultures resulting were sometimes impure and occasionally doubtful, enough cases of unquestionable success occurred to give full effect to this mode of proof. The details supporting this statement will be given under another head. It is worthy of special remark that in no case did the beef broth in which these cultures were made, although it became densely milky with bacteria, give off the slightest smell of decomposition. Only a faint, indescribable odor was perceptible, but little different from that of the fresh liquid.

To the above we may add the association of these spherules with diseased conditions and with *post mortem* phenomena which could scarcely be accounted for at all, except on the supposition of the bacterial character of these excessively abundant forms.

The proof of the contagious character of the disease in question, next to be adduced, must also be taken as indirect, or, at least, *prima facie* evidence, in the present state of our knowledge, of the living organic character of these multitudinous particles,—the only forms present which could in any way be connected with the disease as agents of the contagion.

CONTAGIOUS CHARACTER OF THE DISEASE.

Most of the considerations brought forward in the preceding section apply with some force to the subject of this, for if the fluids of the diseased and dead larvæ swarm with micrococci so minute (these appearing in the blood long before death), and if these are shown to escape from the body by way of the excrement and the fluids exuding from the vent, the presumption is strong that the disease which they characterize would be conveyed to healthy individuals by their instrumentality. But we must look for proof of contagion chiefly to the conditions of the occurrence of the disease, to the phenomena of its spread, and especially to the results of experiments for conveying it artificially to localities or regions where it had not before appeared.

That this affection, or one very similar to it, attacks the cabbage worms of the old world, is made likely by a chance remark in Curtis's "Farm Insects" (p. 96), where he says of several larvæ of an allied species, *Pieris brassicæ:* "On the 20th they appeared healthy, but inclining rather to a yellow color; it rained during the night, and on looking at them in the afternoon of the following day, I saw they had removed to a leaf, to which they stuck by four of their hinder legs, and, to my surprise, they were of a dirty color, and rotten, the skins being lax, and lying just as the wind blew them about. I found they only contained some cream-colored fluid, a portion of which was scattered upon the leaves."

In this country the disease seems to have been first noticed in the vicinity of Washington, in 1879, although little attention was paid to it, and its bacterial character was not then ascertained. In Bulletin 3, of the United States Entomological Commission, (pp. 69, 70), Dr. Riley remarks, while discussing some experiments made with yeast on the cabbage worm:

"An incident connected with these experiments which I made is, however, well worthy of being mentioned, because it shows how very easily single experiments may lead to false hopes and conclusions. A certain proportion of the last-named larvæ — the proportions differing in the different lots treated — perished before or while transforming to the chrysalis state. They became flaccid and discolored, and after death were little more more than a bag of black putrescent liquid. I should have at once concluded that the yeast remedy was a success, had I not experienced the very same kind of mortality in previous rearing of this larvæ, and had I not, upon returning to the field from which the larvæ in question were obtained, found a large proportion similarly dying there."

No other notices of it have occurred in my reading, previous to those of its appearance in Illinois, already mentioned (October 5, 1883). That it did not occur at Normal in 1882 is made certain by the fact that the cabbage fields there were frequently visited in autumn by myself and my assistants during the progress of a series of experiments with insecticides upon the cabbage worm, and that nothing of the sort was seen by us.

When first noticed there, its distribution was peculiarly irregular. In certain small fields, for example, not one half mile distant from those in which the disease was raging violently, affecting one fourth to one half of the worms in sight, not a single dead larva could be found by very careful search. A few weeks later (October 4), larvæ in these fields were suffering as severely as the others, 20 per cent. of the worms, on an average, showing signs of illness.

September 27, at Rosehill, near Chicago, I visited fields in which, although the worms were fairly abundant, I could not find a single diseased larva during a careful examination of more than a hundred individuals; while across a road and a half mile away, the disease was fully at work in four adjacent fields, and fully one fourth of the worms had been attacked. These were in all stages of the disease, many of them being dead and rotten. The identity of the affection with that observed at Normal was established by careful microscopic examination.

From correspondents to whom I had described the cabbage worm mortality at Normal, I received various reports. Prof. A. J. Cook, of the State Agricultural College of Michigan, wrote me, October 2, that about 10 per cent. of the cabbage worms near Lansing were affected by it. On the other hand, Prof. Lintner, State Entomologist of New York, informed me, November 3, that it had not been noticed with him. Dr. E. R. Boardman, in Stark county, sixty miles northwest of Normal, reported, September 29, that the cabbage worm was there very destructive, but that no appearance of the disease in question was discoverable. October 5 he repeated this observation, but on the 13th of that month he finally found a very few affected larvæ.

D. S. Harris, of Cuba, in Fulton county, nearly south of Dr. Boardman, first wrote me on the 13th of October that no disease had appeared among the cabbage worms about his place, nor at adjacent towns, as he had learned by careful search and inquiry, but on the 25th of the month he wrote: "That disease

attacking the cabbage worm has made its appearance in Cuba
at last. On the 21st I found one full-grown worm sick (head
downward), and in about five hours it was dead and decom-
posed, and several others were affected. To-day it is a difficult
matter to find a sound worm on the plants, while the remains
of dead worms are numerous."

From Prof. G. H. French, at Carbondale, and Mr. Frank
Earle, at Cobden, I learned that the disease had not appeared in
Southern Illinois as late as October 29, nor did it occur there
during the season. From Champaign, east of me, Prof. Burrill
wrote me, October 25, that he had not yet seen any of it in his
small garden patch of cabbages, although watching carefully
for it; but that an intelligent student had described it as occurring in fields near the town.

In Iowa, to the westward, it seems not to have occurred
spontaneously that year, the only appearance of it noted by
Prof. Osborn, of the Agricultural College of that State, being
the result of an experiment, the material for which I furnished
him from Normal. Wherever it once occurred it continued to
prevail throughout the season, as far as our observations went.

The facts clearly and positively negatived the supposition
that there was anything in the weather or local conditions to
explain either the presence or the absence of the disease, and
all bore out the hypothesis of a gradual progress from the east
westward. The same phenomena of irregular local distribution
were manifest the next year (1884). In certain large fields
almost daily observed, it was impossible to find a single diseased
larva at a time when, half a mile away, the cabbage worms of
small patches had been almost wholly destroyed, their blackened bodies, or the shriveled remnants of the same, being scattered everywhere on the leaves.

I may say, incidentally, that the effect of the epidemic in limiting the ravages of the worms, was very evident last year. For
the first time in several seasons large fields of late cabbage were
brought to full maturity without the loss or serious damage of
a head.

From the foregoing the conclusion is unavoidable that all
the circumstances of the natural occurrence and spread of the
disease are consistent with the hypothesis of its contagious
character, and wholly inconsistent with any other.

Two attempts were made to convey the contagion by means of diseased larvæ to localities not reached by it,—one lot being sent October 3, to Dr. Boardmam, at Elmira, and one to Prof. Osborn, at Ames, Iowa. The experiment of Dr. Boardman was not wholly satisfactory, for the reason that through an unfortunate delay of the package the worms which I sent him did not arrive until October 22, at which time the disease had appeared spontaneously, in a small way, in his vicinity. Nevertheless he selected, October 23, two lots of twenty-five worms each, all perfectly healthy to appearance, fed them regularly, but exposed all of them to the contagion by enclosing them in two boxes with the dead and sick caterpillars which I had sent him. At the same time he secured ten healthy larvæ in a box by themselves and kept them free from infection. The latter lot all pupated without accident, but were not followed further. The first two lots commenced to show symptoms of disease on the fifth day, and by the eighth day all of both lots were dead, except three, only one of which finally reached pupation. Even this pupa, in fact, afterwards died and decayed. By this time, however, the disease was so violently raging in the open fields that no great value can be attached to this experiment, especially as the fluids were not microscopically examined.

The material sent Professor Osborn, of Iowa, including dead and dying worms and a mounted slide of the micrococci, arrived October 5, and two cabbage heads were at once infected. On the 7th one of the worms " had evidently succumbed to the disease." The gathering of the cabbages under observation during the temporary absence of Prof. Osborn necessarily interfered with the further progress of the experiment, but he collected such worms as he could from the stumps and fed them in confinement. A number of these larvæ died, and December 28 he wrote me that he had "found micrococci in a number of sick and dead cabbage worms, which must certainly have taken the disease from the ones sent."

Although these experiments, taken alone, could scarcely be regarded as conclusive as to the contagious character of *flacherie*, taken in connection with the other facts mentioned, we must at least allow them some weight as cumulative evidence.

ARTIFICIAL CULTURES OF BACTERIA.

Methods of Culture.—The modes of culture used in all the
experiments reported in this paper were based, unless otherwise
specified, on those of Klein, as described in his paper " On the
Relation of Pathogenic to Septic Bacteria," in the Journal of
the Royal Microscopical Society for January, 1883, differing
only by modifications which will appear in the description.
The cultures were usually made in beef broth (rarely in infusion of cabbage) in test tubes plugged with sterilized cotton,
or in ordinary flasks similarly closed.   The broth was prepared
by boiling lean beef from a half hour to an hour in a
porcelain-lined vessel, and then filtering and carefully neu-
tralizing with caustic soda.   The tubes, flasks, and cotton
were sterilized by heating in a tin oven, over a gasoline stove,
for several hours, at a temperature of not less than 275 degrees
or more than 300, as determined by a thermometer inserted in
the oven.   The heat was sufficient to considerably scorch the
cotton without actually charring it.   While still within the hot
oven the tubes and flasks were securely plugged by means of
steel forceps freshly heated in the flame of an alcohol lamp, the
cotton plugs, from two to three inches in depth, being pushed
firmly in.   Most frequently the mouths of these plugged
vessels were covered with a cap of sterilized cotton, held in
place by an inverted beaker also carefully sterilized by dry heat.
In charging the vessels with the fluids the plugs were rapidly
withdrawn and as promptly returned, the infusions being introduced boiling hot and afterwards boiled for several minutes to
destroy any germs which might have entered during the instant
before the plug was replaced.   It was not found necessary to
test the sterilization of the tubes by protracted incubation, as
all the check tubes and the stock flasks in which the store of
prepared infusion was preserved, remained unchanged throughout the entire season.   Neither was any incubator required, the
ordinary temperature of the air during the weeks when these
experiments were in progress being never below 60° by day, and
ranging commonly above 70° both day and night.
    When a culture tube or flask was infected with the fluids
of a larva, the following process was invariably used.   From

small glass tubes, about a quarter of an inch in diameter, pipettes were made in the flame of an alcohol lamp by drawing out each end to a capillary filament, the tips being closed by melting at the time of making. To charge these pipettes, the end of a proleg of a caterpillar was usually cut off with sterilized scissors, the point of the capillary tube broken off with forceps just flamed in an alcohol lamp, one of these points pushed into the cavity of the proleg, and the pipette partially filled by exhaustion of the air from the other end. To introduce the droplet of fluid so obtained into the test tube we invariably, removing first the beaker and the cap of cotton contained within it, carefully forced down through the cotton plug the capillary tube of the pipette containing the infection material, without loosening at all the plug itself. Sometimes the tip of the tube containing the fluid was broken off inside the test tube before the withdrawal of the pipette, at other times the contents were carefully forced out with the breath, pains being taken not wholly to expel the fluid contents of the pipette. After withdrawal of the tube, the cotton plug was grasped with sterilized forceps and slightly twisted within the mouth of the test-tube to close effectually any small opening through the plug which might have been made by the introduction of the pipette. After this the cap of cotton and the beaker glass were restored, and the tube was set aside with a companion precisely like it in all respects, except that it had not been infected. During the latter part of our investigations these check tubes were themselves operated upon with capillary pipettes, distilled water only being introduced, at the time that the experimental tube was infected.

In withdrawing portions of the products of the culture for inspection a similar process was used, the freshly made pipettes being introduced as for infection and partially filled by exhausting the air from the upper end. After withdrawal, the cotton plug was again twisted as already described and the cap returned. The cover glasses used in the examination and preparation of the material, whether this was derived directly from the larvæ or from artificial cultures, were flamed with an alcohol lamp immediately before using, after being thoroughly cleaned by rubbing with a linen cloth. Slides were similarly treated for study with the microscope. A droplet of the fluid was allowed to flow from the tip of the pipette upon the cover glass, spread in a thin film by means of the capillary glass tube, and either placed at once upon the slide for immediate examination, or laid aside under a glass shade to dry. After drying, if it was desired to stain and permanently mount the specimen, the cover with the film attached was passed repeatedly through the flame of an alcohol lamp, covered for some minutes with a drop of the staining fluid (the glycerine aniline colors recommended by Prof. T. J. Burrill[*]), and then thoroughly washed with distilled water. The covers thus prepared were often mounted in balsam, but most frequently, at first, in carbolized water in very shallow cells made with white zinc cement, this cement being also used to fasten the covers to the slides. For microscopic study of the material, my principal reliance was a superb $\frac{1}{16}$-inch homogeneous immersion objective, made to order for the purpose by Herbert R. Spencer & Company, of Geneva, New York. This objective was used with Bulloch oculars, giving powers ranging from 500 to 1,450 diameters. Some of the more interesting or difficult slides were also studied under a $\frac{1}{18}$-inch homogenous immersion of Zeiss.

The measurements here reported were originally made by means of an eye-piece micrometer so graduated that with the highest powers used, each space equalled 2 μ, the micrococci being commonly measured in doubles and chaplets. Many of these measurements were verified by repetition with a more finely divided micrometer, the spaces of which, with a power of 1,000 diameters, had a value of .3 μ; but practice with the more coarsely spaced scale enabled me to measure as accurately with this as with the other, and with much greater convenience.

The products and results of the fluid cultures were commonly so satisfactory that I rarely resorted to solid cultures upon gelatine films. A few of these were made, however, but not with the micrococci of the cabbage worm; and they will be described under the head of the Datana larvæ. As I was primarily interested only in the disease and secondarily in the bacteria, cultures on films were less essential to my purpose

[*] Proceedings Amer. Society of Microscopists, 1883, p. 79.

than if I had wished to discriminate and describe the various forms appearing. I depended upon frequent repetition of the experiments and uniformity of results, rather than upon the more critically exact cultures and continuous observation of current methods with gelatine films and masses.

Culture Experiments.—Concerning our first cultures, the fact should be remembered that one could rarely expect to find a perfectly pure culture in the body of a diseased insect, exposed as it is by way of the food ingested to invasion by bacteria in great variety. I consequently did not find it remarkable that several of our unquestionably sucessful *infections* were not really pure *cultures*, other bacteria developing than those most abundant in the original fluids. For example, in the very first culture made,—one in beef broth begun September 16, when the infection process was very carefully managed without the slightest accident, and when the check tube remained clear indefinitely,— the culture became turbid the following day, and by October 3, was nearly as yellow as cream, with a thick yellowish felt on top and an abundant precipitate. The greater part of the product of this culture consisted of micrococci like the larger of those of the cabbage worm, the spherules, in singles and doubles, averaging 1 μ in diameter; but the surface film consisted largely of a Saccharomyces embedded in the Micrococcus. Individuals of Bacterium also occurred in the slides. The check tube, as already mentioned, was quite clear to the end.

The second culture was still less conclusive and satisfactory. A cabbage worm which, on the afternoon of the 17th September, was noticeably paler than its companions, was isolated and watched. At 9 p. m. it seemed a little stupid, but otherwise unchanged. At 9 a. m. of the 18th, however, it was dead, blackened, and very soft; the contents evidently little better than fluid. These fluids contained two micrococci,—one a larger spherical or slightly quadrate form, 1 μ in diameter, and the other a minute spherule .5 μ to .75 μ.

A small flask of rather weak beef infusion was infected from these fluids in the usual manner, and the next day, the 19th, it was already decidedly milky. Examined, it was found to contain Bacterium and the larger Micrococcus above men-

tioned, a slender Bacillus, and another Bacillus-like form, of
which there will be further question hereafter,—a short, broad
form with rounded ends and a paler center,—but nothing resembling the smaller Micrococcus. These organisms were quite
possibly all septic bacteria, derived from the decaying body of
the caterpillar.

The first unmistakable culture of the Micrococcus of the
cabbage worm was made October 20, in a test tube of beef
broth infected from the blood of a larva about half grown, decidedly pale, but far from dead. The slide representing the
blood of this larva is not stained, but is in good condition.
There were two bacterial forms visible in it,—a spherical Micrococcus .7 μ in diameter, and, very rarely, a slender Bacillus.
The flask in which the culture was made was poured from a
stock flask into a sterilized tube from which the plug of cotton
had just been removed, plugged again, boiled thoroughly about
three minutes and left to cool completely. The blood was
obtained by snipping the skin of the back with sterilized
scissors, and drawing up with a fresh pipette a little of the
thick fluid exuding. The tube was infected in the usual manner, and not examined until two days thereafter, when it was
found decidedly turbid, although not extremely so. The Micrococci were strictly spherical, 1 μ in diameter, very uniform
and abundant, usually in doubles, but often single. The slides
made were excellent and well stained, some violet and some
brown. The bacteria differed from their originals only in being
somewhat larger. Still they were not larger than the Micrococcus of the cabbage worm is often found, especially in the intestines. The check flask remained wholly clear.

Four other successful cultures of this Micrococcus were
made, so similar in all respects to the preceding that it is not
worth while to repeat details. It must be admitted, however,
that the minute blood form did not certainly reappear in its
original size in any of my cultures, if we except one case where
its numbers were relatively so few (about 100 to the field with
a power of 1,000) that it is barely possible that all were introduced in the original infection. This fact is capable of either
one of three interpretations : (1) We may suppose that the
proof is incomplete that these smallest spherules from the blood

were micrococci at all, notwithstanding their uniform shape, size, and character, and the fact that they were repeatedly distinctly stained ; or (2), taking for granted their bacterial nature, we may suppose them insusceptible of culture under the conditions supplied ; or, finally (3), we may assume that the conditions of tube culture in beef broth were so different from those occurring within the blood of the insect as to increase the size or even modify the form of the Micrococcus in question. In favor of the latter hypothesis we have the fact of the generally larger size and often slightly oval form of the micrococci found in the intestinal fluids, as compared with those in the blood of the same specimen.

These considerations apply, however, only to the minute blood form, and not at all to the intestinal Micrococcus. This I have cultivated repeatedly with indubitable success in this insect, and, still more frequently, forms indistinguishable from it occurring in other species. I venture to add that the frequency with which certain bacteria, different from the infection material, appeared in the test tubes when these were infected from the cabbage worm, suggested repeatedly the hypothesis of an alternation of certain forms which were in this way frequently connected,—a point on which I shall have more to say when describing the Datana bacteria. Especially was this true of the larger Micrococcus and of the short, broad Bacillus (?) with pale center and rounded ends (here called, for convenience, *Bacillus intrapallens*). The latter, I shall presently show, behaved precisely like a pathogenic form,—giving no odor of putrefaction in fluids swarming with it, killing insect larvæ whose food was treated with it, and certainly multiplying for some days within their living bodies.

The late period at which successful cultures of the cabbage worm Micrococcus were made precluded attempts at artificial infection by their means, and with respect to this particular insect this part of the proof is consequently wholly wanting.

When the evidence is given respecting the reproduction of what was clearly the same disease in other insects, I think that no reasonable doubt will remain that *flacherie* of the cabbage worm may be conveyed through artificial cultures of its Micrococcus.

THE SILKWORM *(Bombyx mori, L.)*

Late in July, 1884, I heard from Professor Burrill, of the State Industrial University, that a lot of silkworms which were being reared under his direction for experimental purposes were dying rapidly from an apparently contagious disease resembling the *flacherie* of the old world, and wishing to improve the opportunity thus afforded to determine the possibility of conveying this affection to our native Lepidoptera, I had, July 30, some of the dead and dying larvæ sent me by mail from Champaign. From our correspondence at the time and from an account of the experiment by Prof. Burrill, published in the Twelfth Report of the Board of Trustees of the Illinois Industrial University, 1884, we learn that the lot of worms (about 80,000 in number) in which this disease broke forth were raised from eggs derived from a perfectly healthy brood of the preceding year ; that they commenced to hatch June 21 ; that they were kept in a clean and thoroughly ventilated building set aside for their use on the University grounds ; that they began to spin July 25 ; that between this date and the 29th 183 cocoons were produced, but that in consequence of the outbreak of this disease among them only a single additional cocoon was made during the season. The entire remainder of the 80,000 worms perished,—commencing to die July 23, and continuing until the latter part of August.

DESCRIPTION OF THE DISEASE.

In a note of July 23, Prof. Burrill says of the affected larvæ that they "become yellow, shorten up; the skins become very tender so that they can hardly be picked up without bursting; body flaccid; the blood loses its clearness and becomes thick with a dirty yellowish color." Again, July 26, he writes: "They first refuse food and uneasily creep around, then become yellowish and flabby."

In the article above cited, Prof. Burrill distinguishes two forms of disease among the larvæ, as follows:

"In one case the affected larvæ became restless, ceased eating, the skin assumed a decidedly yellowish tint and ultimately became very tender and easily ruptured, while the blood, unusually copious, was thin and yellow instead of its normal limpid or grayish color. Other larvæ became sluggish, continued to eat, but consumed only a small quantity of food, the body gradually became flaccid, the skin wrinkled and tough, and the color a grayish or leaden tint, and finally nearly black. These, hours or even one or two days before their death, adhered by their prolegs, or some of them, to a support, and remained quiet, at length only showing signs of vitality when touched, and at last dying while still firmly anchored to the limb or other object upon which they rested. After, and for some time before, death, the flaccid body hung directly downward from the point of attachment. If this latter happened to be near the middle of the body, the two ends hung down, the parts nearly parallel with each other. From these dead and blackened worms a decided and characteristic odor of putrescence was perceptible, tainting, when numerous, the air of the well-ventilated room."

The first of these diseases was also characterized in the Statistical Record of the State Board of Agriculture for August, 1885, by Mr. Woodworth, who conducted the experiment for Prof. Burrill. "This disease," he says, "does not make its appearance until the worms are about ready to spin, that is, near the end of the last age. The body of the affected worms assumes a somewhat granular, yellow color, instead of the natural, bright semi-transparent hue. This change of color also differs from the normal change, in that the yellow is first on the middle of the segment instead of at the ends. The skin becomes soft and tender, breaking at the least fall, and allowing the yellow body fluid to escape more readily than wounds of equal size would in healthy worms. The affected worms become very restless, crawling about and shrinking in size from loss of blood until they finally die. A few spins cocoons, which are generally soft, often bright orange, and sometimes so thin that the pupa or dead worm may be seen within. Some of the worms even pupate without spinning, and from these pupæ moths may emerge, which will sometimes deposit their eggs. When a brood of worms is attacked by this disease generally very few survive."

Several lots of the larvæ were sent me in July and August, representing both the above-described affections, the difference between which was easily discernible. The former disease was apparently that known to the French as *jaunes* (sometimes called jaundice by the English writers and by some considered the same as *grasserie*), and the latter was unquestionably *flacherie* or *morts-flats* of the French — the *schlaffsucht* of the Germans.

The yellow color of the "jaundiced" worms was evidently due to the tint of the blood, and this, again, was as clearly derived from the great numbers of peculiar cellular bodies with which the blood was always loaded, these originating chiefly, if not wholly, in the fatty bodies, as a result of that form of degeneration of those organs in the larva which attends pupation. These bodies, when entire, consisted usually of masses of spheres, each 4 μ or 5 μ in diameter, the aggregate attaining a diameter of 30 μ — 40 μ. The individual spheres often presented a slightly angular outline, as if modified by mutual pressure, and they took no aniline color with which I tried to stain them. These bodies are evidently the mulberry cells and granules of Viallanes, as described in his admirable memoir on the histolysis of insects.* That they originated chiefly in the fatty bodies, I demonstrated by finding masses of them in portions of the fatty bodies themselves and by determining the substantially unaltered condition of all the other tissues of the affected silkworms.

In the blood of these larvæ no bacteria were found, as a general rule, although Professor Burrill occasionally recognized a Bacillus in it ; but in the alimentary canal I never failed to discover great numbers of micrococci and often also numerous examples of Bacterium and Bacillus.† These bacterial forms

---

* Ann. Sci. Nat., Zool., xiv. 1,—Art. 1. August, 1882.

† A transverse section of a jaundiced larva mounted in balsam without staining, shows great numbers of spherical micrococci, somewhat unevenly distributed throughout the entire thickness of the wall of the intestine, and fully as abundant in the outer portion of this wall as within. The same micrococci occur in the perivisceral spaces, being accumulated especially upon the free surface of the organs contained therein. A very few are apparent also in the sections of the fatty bodies, and occasionally in the muscles, but none occur in the skin

were not different from those observed in cases of undoubted *flacherie*, but they were usually far less numerous,—a fact which has suggested to me the following theoretical explanation of the supposed jaundice of the silkworms at the University. Assuming that the mortality was originally caused by the intestinal bacteria, we may suppose that this infection was not sufficiently overwhelming to destroy life by direct action, as seems to be the case in *flacherie*, but that it nevertheless had the effect to so disturb the balance of physiological functions as to retard the development and preparation for pupation of some of the organs, while the fatty bodies, being special stores of material accumulated for use in pupation, and so less promptly and easily affected by causes attacking the general health of the larva, went on to pupation and experienced the histolysis characteristic of that phenomenon. In other words, we may suppose, quite consistently with all the facts, that a relatively slight bacterial attack took *uneven* effect on the various parts of the animal and not immediately destructive effect on any ; that it retarded the preparations for pupation of the great vital organs, but that the fatty bodies, as if unaware of this fact, continued their course of maturation and histolysis, reaching a condition of pupal disorganization before pupation had actually occurred.

The condition of the fatty bodies of the larva affected by the supposed jaundice is well illustrated by slide 4732 of our collections, containing portions of the fatty bodies of larvæ received from Prof. Burrill on the 30th July. The cells of these organs, when examined under a power of 500 diameters, were found, nearly all of them, to have undergone a remarkable change. The contents of a few still remained minutely granular, a large nucleus being also occasionally visible, but the con-

---

or in the silk tubes. These micrococci are very distinctly visible, shining with a reddish light when slightly out of focus, not being rendered transparent by the mounting medium as are the tissues of the larva. They are arranged in patches and strings, the former of irregular shape, the latter sometimes containing as many as eight or ten spherules. The fatty bodies of this larva are almost solid masses of mulberry granules. The Malpighian tubules of another specimen show also, besides their normal crystalline contents, great numbers of these mulberry granules, formed within the cells or derived from outside sources.

tents of the greater number had been converted into very distinct pale granules, varying in size in the different cells from 2 μ in diameter to 4 μ or 5 μ. About 20 or 25 of the larger size were usually contained in a single cell, and a multitude—too numerous to count—of the smaller ones. Here and there in the area of the object were large irregular lacunæ evidently filled with liquid fat, as shown by the slightly crystalline character of their contents.

Whatever we may assume with respect to the bacteria infesting these worms as a cause of the premature pupal degeneration, I do not know that we have any reason to suppose that they are the only possible cause of such a catastrophe to the insect. Other influences tending to disturb seriously the balance of functions at the critical period when larval life is about to terminate in pupation might not impossibly have the same effect.

Additional details respecting this peculiar catastrophe to maturing larvæ will be given further on, under the head of *Mamestra picta.*

The Characteristic Bacteria.

As an illustration of some of the conditions characteristic of this disease, I give descriptions of well-mounted slides prepared from the fluids of one of the larvæ received from Prof. Burrill on the 30th July. The larva was dead when examined, but perfectly fresh. In the blood I found only the mulberry granules, some free and others still enclosed in their mother cells, as already described, together with blood corpuscles in various stages of degeneration. My notes at the time and a recent examination of carefully prepared slides show that no bacteria occurred in the blood.

In slide 4603, material for which was obtained by touching a cover glass to the cut end of a divided worm, I find great numbers of the mulberry granules, varying in size from 2.5 μ or 3 μ to 6 μ, the more usual diameter being, however, 4.5 μ to 5 μ. With these occurred, everywhere, myriads of micrococci, probably one fifth of the area of the field of the microscope being occupied by them where the film is of moderate thickness. These micrococci vary in form from exact spheres, usually in doubles, to broad ovals, with the transverse diameter about three fourths the longitudinal, these likewise usually in doubles. Occasionally pairs of doubles are joined end to end in four's, but longer chains than these were not observed. The micrococci frequently occurred upon the slide in patches of fifty to one hundred, in which most of the individuals were seemingly single. The ovals above mentioned have the same transverse diameter as the spheres, differing only in length. This diameter varies but little from .75 μ, although slightly smaller singles are not infrequently found. Many of these small, as well as larger, singles are scattered separately through the field. Besides the ovals above described, occasional ovals larger than these are seen, closely resembling, in fact, *Bacterium termo*, and probably to be considered as belonging to that genus. These are about 1.5 μ in length (doubles 3 μ) by 1 μ in transverse diameter.

In the thicker part of the film very considerable numbers of excessively minute spherules were discernible, deeply stained, .5 μ in diameter, apparently identical with those described under *Pieris rapæ*. on a preceding page*, and clearly the same as those appearing in the culture described on page 286.

The slide from which the above description is taken was deeply stained with methyl violet July 30, and mounted in dammar.

Another slide, 4612, derived from the same lot of worms and similarly treated, differs only in the fact that the micrococci average somewhat smaller; that nearly every one is almost strictly spherical; and that an occasional small Bacillus occurs, 2 μ to 3 μ in length by about .66 μ in width. The ends are broadly rounded, the sides parallel, except in the shorter specimens where they are slightly convex. These bacilli are sometimes single, more commonly attached endwise in pairs. The smaller oval forms, possibly distinct, frequently show a pale center with ends heavily stained.† In this slide are a considerable

---

*Is it perhaps possible that the silkworm affection had its exciting cause in the disease of the cabbage worm, which made its first appearance in this region the year before?

† To this form a peculiar interest attaches in some of my other studies, reported on a later page.

number of large, regularly elliptical bodies, about 5 μ in length
by 3 μ in transverse diameter. As they do not stain, they are
probably crystalline, especially as it is well known that larvæ
about to moult or pupate often have the blood loaded with crys-
tals of uric acid of which the form is often not different from
that here noted.

As characteristic of the second form of disease, *flacherie*,
that distinguishable in the living larvæ by the pale color of the
surface as compared with the lemon-yellow of *jaunes*, I have
selected slide 4727, derived from the fluids of a freshly dead
larva. In the blood of this specimen no bacteria were discernible, but in this slide, prepared from the mingled blood and
alimentary fluids, they occur in innumerable myriads. The
slides are, however, instantly distinguishable from those derived
from the yellow-skinned larvæ, by the complete absence of the
mulberry granules. The bacteria from the selected slide are
not by any means so uniform as those in the one previously
described, but vary from perfectly spherical micrococci to ovals,
double ovals, and elongate bacillar rods. The spherical and
oval forms of micrococci are, however, the predominant bacteria. The spheres in this slide are commonly wider than the
ovals, measuring about .75 μ, while the smaller ovals are not
more than .5 μ in their shorter diameter. The spheres vary in
arrangement from singles to chains of considerable length, but
the latter aggregates may be due to an accidental running
together in the drying film. The bacilli are not distinguishably
different from those described for the other form of disease.
Besides the above, occasional larger broad ovals appear, similar
to those doubtfully determined above as *Bacillus intrapallens*.
Judging, in short, from this representative slide, one would
say that the bacteria of *flacherie* of the silkworm consist of a
varied mixture of round and oval micrococci of different sizes,
of species of bacteria, and of small bacilli. Some of them, however, may have been of *post mortem* origin. The slide in question is beautifully stained with methyl violet, and mounted in
dammar.

CONTAGIOUS CHARACTER OF THE DISEASES.

I had no opportunity to observe the progress of these diseases in the silkworm, but Professor Burrill was entirely confident of their contagious character as exhibited under his observation. On this point he says * : "That the worms came from good eggs, and were, for a considerable time, perfectly healthy and wholly free from the malady which finally overtook them, we have the best of evidence. The disease which carried them off was not hereditary. It was not lurking unobserved during the more favorable weather in the living or dying worms. Its introduction occurred about, and probably at, the time of the first heavy rains spoken of, but we confidently know that it could have been artificially introduced without the rains or the wet weather at all. Moreover, the worms continued to die after the weather cleared up, and after every precaution had been taken to put them under the best possible conditions. We constructed new racks in a room not previously used, picked out the healthiest worms and moved them to the new and clean quarters, where, afterward, the temperature and other conditions were as favorable as could be desired ; but the ravages of the disease continued with no perceptible abatement. To further test the matter, other apparently healthy worms, voracious feeders, growing rapidly, were put out upon the open hedge, where they were watched from daylight until dark to keep off the birds, and where, for a time, they seemed to thrive under the favorable skies and wide isolation ; but here, too, they gradually fell victims to the destroyer. In each of these places about five hundred worms were placed, from which, as was before said, one cocoon only was secured, and this from the out-of-door lot. The latter did live longer than any of the others, but at length as surely succumbed. Another experiment proved equally futile ; viz. that of spraying the food with an aqueous solution of carbolic acid. No apparent improvement followed this treatment.

It may be said that our disaster followed in consequence of retarding too long the hatching of the eggs by keeping them in

---

* Twelfth Rep. of the Board of Trustees of the Illinois Industrial University, pp. 90, 91.

an ice-house, thus pushing the feeding season out of the
natural time and subjecting the worms to unfavorable summer
heat, or providing them with leaves too far advanced towards
maturity.   This might, indeed, seem plausible had not several
other lots, fed in the vicinity, but not so retarded, died in the
same way.   It is interesting to note that in some of these small
and isolated experiments in silkworm feeding, certain lots from
the same kind of eggs as our own, produced from the same lot
of moths, fed on the same kind of food, remained perfectly
healthy and produced good cocoons, while others totally failed.
It seemed that in every case where what appeared to be the
disease called in this paper *flacherie* became once introduced,
few or none of the worms lived to spin passably good cocoons.
Most of them died after the third or fourth moults, and after,
therefore, no little care had been bestowed upon them."

My own observations on this phase of the subject were of
an experimental character, and will be found in detail under the
head of Experiments for Artificial Infection.   Here I need
only say that they demonstrated the possibility of affecting with
disease healthy larvæ of the common cabbage butterfly *(Pieris
rapæ)* by means of artificial cultures of the bacteria occurring
in the sick silkworms,—these cultures being made in beef broth
and applied to the cabbage worms in confinement by sprinkling
or spraying their food.

ARTIFICIAL CULTURES.

Our first cultures of the bacteria of the silkworm were
made July 30, in test tubes of beef broth, by the methods
described above, in my account of the cabbage worm disease,
the material for infection having been obtained from a yellow-
skinned larva (affected by jaundice) received on the same date
from Professor Burrill, of Champaign.   The larva used was
recently dead, but still perfectly fresh.   Two cultures were
made, one from the blood and one from the alimentary fluids.
No bacteria were discernible in the blood, either in fresh
preparations or in mounted films, the latter presenting only
numerous and excellent examples of the mulberry cells and
granules characteristic of the disease.   The slides prepared from
the fluids of the alimentary canal, however, exhibit numerous

specimens of a strictly spherical Micrococcus, occurring usually in doubles, measuring 1 µ in diameter, with an occasional oval example apparently elongating for division, and then about 1.5 µ in length. These micrococci stained readily with methyl violet.

In the test tube infected from the blood, curiously enough this Micrococcus reappeared in a perfectly pure culture. The fluid, infected July 30, was seen to be milky on the 1st of August, and many micrococci were visible in doubles and chains, the latter being unusually abundant. On the prepared slides, less heavily stained than the originals from the silkworm, these micrococci measured a little less than those of the alimentary canal, the diameter usually falling between .75 µ and 1 µ, rarely attaining the latter dimension. Chains of six or eight were not uncommon.

The culture derived from the alimentary canal of this larva was unexpectedly impure and not altogether comprehensible. The fluid was observed to be milky August 1, and many micrococci appeared in fresh slides, both in doubles and chains. A perfect film, distinctly stained, but rather pale, shows, however, a variety of forms. Most conspicuous, but not the most abundant, are doubles and short chains of three to six of a strictly spherical Micrococcus, deeply stained, entirely similar to those above described, but averaging smaller, their mean diameter being a scant .75 µ. Besides these are short, broad ovals, a little less deeply colored than the above, of the same transverse diameter, but a fair 1 µ in length, some, indeed, falling scarcely short of 1.25 µ. In addition to these and of the same transverse diameter, we see, rarely, rod-like forms, apparently bacilli, measuring from 3 µ to 4 µ in length ; and, finally, thickly scattered, everywhere more abundant than any oval form, are very minute spherules, always in singles (except in now and then an instance seemingly accidental), measuring a scant .5 µ in diameter. These are well stained and conspicuous, and unquestionably do not belong to the film. They are extremely like the smaller form of cabbage-worm micrococci which I have already described. Their appearance under the circumstances suggests the possibility of their being bacillar spores, but the bacilli in the film are far too few to permit this explanation ; nor did any of those noticed seem to be spore-bearing. The impurity of this culture makes the supposition plausible that some of the bacteria of the orignal infection were introduced by accident and not derived from the silk-worm. The check tube, however, remained unaltered, as usual; and it seems to me more likely that the originals of all these forms were really derived from the alimentary canal. It is not to be supposed that the alimentary contents of a larva long diseased, and, indeed, actually dead, should remain wholly free from invasion by bacteria other than those strictly characteristic of its disease.

The cultivation of bacteria from the blood, although none were microscopically demonstrable in the latter itself, seems to me not a remarkable phenomenon (especially as the fluid was derived from a dead larva), since it could scarcely be credible that the circulatory fluids should, under such circumstances, be entirely free from the peculiar germs of the disease to which the larva had succumbed. It must be remembered that a single individual Micrococcus would be sufficient to start the culture in the tube, and that the quantity introduced into the beef broth was much greater than that represented by the films microscopically examined. Furthermore, an occasional Micrococcus in a stained film may readily be overlooked or passed as doubtful, since the difficulty of distinguishing single individuals from accidental granulations of the film itself forbids positive identification of the micrococci unless they occur in numbers sufficient to make their character unmistakable.

Another culture, commenced July 30, from the silkworm 4603, the bacteria from which were described under this number on page 281, was examined August 1, at which time the fluid was observed to be milky and found swarming with micrococci and a few examples of Bacterium (?). (The latter, it will be remembered, were also observed in the original material.) The resultant culture was possibly impure, the two forms appearing on the slides being distinguished, however, only by the positive strong stain of one and the very delicate stain of the other, shapes and sizes not being appreciably different.* That distinctly

---

*Those lightly stained were probably the empty walls of dead examples.

stained unquestionably agreed in every particular with the common spherical Micrococcus of the original silkworm material, except that it measured a trifle smaller, scarcely averaging 1 μ, although many individuals and doubles were fully that size. This culture was preserved for experiment and used as an infection fluid on the 9th August. The results of this attempted infection will appear under another head.

Still another culture, commenced and examined upon the same dates, yielded an abundance of the spherical Micrococcus most frequently mentioned above, together with occasional examples of a Bacillus 3 μ or 4 μ in length and about 1 μ in transverse diameter. These last were, however, too rare to have any special significance, except as a slight adulteration of the culture.

The next culture attempted, commenced July 31 and examined August 4, is of especial interest, as it resulted in the complete displacement of the normal Micrococcus of the silkworm by another organism present in its fluids (the questionable *Bacillus intrapallens* already mentioned*), but in small numbers.

This culture was made from a silkworm of the original lot received from Professor Burrill, July 30, the beef infusion being infected from a dead worm. The fluids of this larva contained vast numbers of the ordinary silkworm Micrococcus, somewhat under the usual size, averaging, indeed, only about .75 μ. An occasional large Bacillus, 4.5 μ long and 1 μ wide, also occurs on the slides made from this individual. Besides the above is the organism already mentioned, varying in form from a broad oval to a Bacillus-like rod, characterized by a pale center staining little or none, and heavily stained extremities. The culture examined August 4 contained vast numbers of this organism and apparently nothing else. Most of those appearing in the films from this culture were much smaller than the original, all the stages, in fact, appearing, from a simple sphere scarcely, if at all, distinguishable from a Micrococcus, to the

---

* This organism displaced similar cultures made from the larva of *Datana angusi* presently to be reported on, was preserved through the winter, cultivated the following season, and then applied effectively to the destruction of larvæ of other species.

rod-like form or double elongate oval, the paler centers commencing to appear in the oval and becoming more conspicuous
as this elongates.

A single somewhat later culture, commenced August 4, did
not differ materially in results from those preceding. No
bacteria were discoverable in the blood of the larva, used by
prolonged and careful search, but the alimentary fluid contained
the usual Micrococcus. Five days later the infected infusion
was decidedly turbid, but without either film or sediment.
Besides an occasional short Bacillus in active movement, it contained only the spherical Micrococcus of the usual size.

The slides of these various cultures clearly demonstrate the
presence of a spherical Micrococcus, varying in diameter from
.75 μ to 1 μ, as the characteristic Bacterium of the disease from
which these silkworms were perishing, and likewise the prac-
ticability of artificially cultivating this Micrococcus in neutralized beef broth by infections from the alimentary canal and
from the blood. Although the Micrococcus itself was not
demonstrable in the blood by the microscope, it was obtained
therefrom by cultures in which it appeared without admixture
of other forms. Intestinal cultures were, however, liable to
contamination by other bacteria but doubtfully connected with
the disease, among which was the form last described.

INFECTION EXPERIMENTS.

I found it by no means easy to provide means for testing
satisfactorily the possibility of conveying the disease of the silkworm above described to other demonstrably healthy insects.
The late period of the occurrence of the disease under my observation made it impossible to use other lots of the silkworm
itself in the experiment, and no other lepidopterous larva
was sufficiently abundant at the time, except the cabbage
worm. This, however, had alread been found, the previous
year, to suffer extensively from an extremely destructive disease of its own, and although at the time the experimental
stage of my studies of the sick silkworms had been reached, no
evidence of disease among the cabbage worms had yet appeared
in the fields, I had every reason to anticipate its outbreak among
them,—a fact which made me very doubtful of really bringing

the matter to a decisive test on that species. The occurrence of a spontaneous outbreak of the common cabbage worm *flacherie* among the lot under experiment, would of course arrest the progress of the experiment, and might even so mask the result as to mislead.

This accident, in fact, occurred to my first two experiments, begun August 9 and 10. Not only did the cabbage worm affection appear in both the experimental breeding cages and the checks, but the latter lot as well as the former gave evidence of infection from our silkworm material. The latter fact convinced me that my arrangements were inadequate for the protection of my check lots against accidental infection with the experimental material. These lots were placed at a distance from those purposely exposed to disease, but in another part of the same large hall, and were attended by the same assistant. In previous experiments, not yet detailed, with other larvæ, I had already had evidence of slight unintentional infection of the check lots by this too close association with those under treatment, and now arranged another experiment on a wholly different plan.

Careful examination was made of all the cabbage fields near Normal, and one was selected which showed no trace of the proper disease of the cabbage worm. From this field two lots of caterpillars were selected, twenty-five in each, those for experiment by the assistant whose duty it was to make the infections, and the check lot, by an intelligent student of the Normal school, who did not visit my zoölogical laboratory at all. The first lot was brought to the office and placed in a clean and disinfected cage in the usual place, but the second or check lot was taken by the student mentioned directly to his own home and confined in a new breeding cage. Care was taken that both lots should be fed and treated alike, except for the infection, but no opportunity was given for any communication between them. The results in this case were more satisfactory, and confirmed my suspicion that our check lots had not before been sufficiently isolated.

History of the Infected Lot.—The food of the twenty-five cabbage worms selected especially for experiment, was sprayed on the 6th September with beef broth infected nearly a month previously from the fluids of a silkworm recently dead from jaundice. Unfortunately, from some oversight, neither slides nor detailed notes were made of this culture until the experiment upon the cabbage worms was instituted. The beef broth, nearly a gallon in quantity, contained in a large receiver, the tube of which was closed with a sterilized cotton plug nearly six inches in length, had promptly become turbid, as usual, and was soon opaque with bacteria. By the 6th September the development of the bacteria had apparently nearly ceased, a thick deposit covering the bottom of the jar. The fluids at this time contained vast numbers of spherical micrococci .7 μ to .8 μ in diameter, mostly in doubles, apparently identical with those occurring in the silkworm. The culture, however, which had been several times opened for examination, was not at this time wholly pure, but contained likewise bacteria and large and small bacilli. These occurred, however, in relatively insignificant numbers, and the fluids when poured out presented no odor of putrefaction, but had, on the contrary, only the faint indescribable smell characteristic of the cultures of all our insect bacteria.

After infection on September 6. the cabbage worms were fed with fresh food collected for them daily. Their cage was kept in a large room, before an open south window, was thoroughly cleaned each day, the paper covering the floor of the cage being removed and burned, all the litter and *débris* destroyed, and the larvæ carefully transferred to fresh food upon clean paper.

A single individual died September 8, evidently from accidental injury. Three of the larvæ pupated on the 10th. On the 11th two died, apparently of disease. The fluids of these were carefully examined and found to swarm with micrococci. Of these covers were prepared in the usual form. The first slide, made from the blood, contains large spherical micrococci, nearly all in doubles, 1 μ in diameter, excellently stained with violet. The bacteria of the second slide, representing the contents of the alimentary canal, were more various in form. In addition to the above large Micrococcus, 1 μ in diameter, many slightly double ovals of about the same transverse diameter occurred, together with several .7 μ wide, most commonly arranged in small groups; occasionally, also, an unsegmented rod, possibly Bacillus. Nothing representing the minute spherical micrococci characteristic of the native disease of the cabbage worm occurred in this specimen. The next day, September 12, one larva pupated and four perished. The first of these examined was already blackened and deliquescent. It contained nothing but large and small micrococci strictly spherical in form, the large one 1 μ in diameter, the other about .6 or .7 μ. Both occurred usually in doubles, but not unfrequently in singles or short chains. Both stained well in methyl violet, and good slides were prepared. The smaller form of the above micrococci was found only in the blood, and the larger only in the intestine, as indicated by the stained slides from these two sources.

The second larva studied was soft and grayish green, but the skin was tougher than usual, and showed little tendency to the characteristic deliquescence of the cabbage worm disease. The fluids were yellowish white, and contained great numbers of large and small spherical micrococci, the larger 1 μ in diameter, the smaller .6 or .7 μ.

The third specimen, smaller than the preceding one, was a little darker in color, the fluids yellowish green and containing identical micrococci. Both forms were spherical and of the same dimensions as those just described. A single Bacillus was also noted, 2.5 μ in length, and an occasional double oval occurs upon the slides (probably Bacterium) each oval element about .8 μ long.

The fourth specimen was flaccid, but bright green, its fluids thick and milky white. It contained a moderate number of large spherical micrococci, identical in appearance with those described above, varying in character from .8 μ to 1 μ. Besides these, the blood was literally loaded with large spheres, evidently mulberry granules, occurring singly and in masses, the diameter varying from 2 to 4 μ. A close correspondence in the condition of this larva to that of the silkworm affected with jaundice will at once be noted.

Four other larvæ, two of which died September 13 and two on the day following, were briefly examined, but not carefully studied. Their fluids presented no considerable differences from those already treated. On the 15th another larva pupated, and a second died during the night which had been reported sluggish the previous day. The body was shrunken, not very soft, a little brown, but the general color was still the usual green. The fluids of the specimen were very white and thick, and contained vast numbers of mulberry granules, both singly and in clusters, together with great quantities of oval micrococci (some in chaplets of four) and occasional individuals of Bacterium, some of the latter in actual motion. The mulberry granules were strictly spherical, and varied in size from 1.5 μ to 3 μ in diameter.

Another larva which died was originally paler than natural, but not white. Before examination it had blackened and turned very soft, but was not deliquescent. Slides prepared from it contained *débris* of tissues, muscular and other, and vast numbers of minute spherical micrococci from .5 μ to .7 μ in diameter. No flagellar motion was detected in the fresh slides, and no other forms are apparent in the stained mounts.

Another example, small and shrunken, a little discolored, dried up in a few hours, and became hard and brittle. It was not especially studied. On the 17th of the month the last remaining larva died. It was not discolored, and I could find no bacteria in the blood or other fluids. The cause of its death, in fact, was not apparent. At this date a blackened pupa from the cage, evidently not long dead, was found full of a blackish fluid, which contained vast numbers of a small spherical Micrococcus (.6 μ in average diameter, commonly in doubles) and nothing else, except occasional mulberry granules 2 μ in average diameter. Of the individuals which pupated, six emerged successfully, three were deformed, and two failed to complete their transformations.

History of the Check Lot.—This lot, placed in a new breeding cage September 10 with fresh cabbage, was kept under continued observation until the 28th. One of the specimens died the first day from an accidental injury; one pupated on the 12th; and two others were necessarily crushed in opening the cage, having commenced to pupate on its sliding glass front. On the 14th four examples pupated, and two more upon the 15th. at which time fifteen healthy larvæ remained. The more rapid pupation of these specimens will be noticed, as compared with those treated with the infection material, — a fact consistent with what I have uniformly observed with regard to the effect of these diseases.

On the 17th four worms were drowned in a dish of water containing the food plant in the breeding cage. The fluids of these worms were carefully examined with a microscope, and careful studies were made of stained covers of their blood and alimentary contents, but no possible bacteria of any sort were detected in them. On the 21st three more larva pupated, and on the 23d three died. Unfortunately, the latter fact was not reported by the assistant in charge in time to permit an examination of these dead worms. All the remaining larvæ pupated, the imagos commencing to emerge on the 26th.

Although the results of the foregoing experiments were somewhat less definite than might be desired, yet they clearly indicate the transference of the disease affecting the silkworm to healthy larvæ of *Pieris rapæ*. It would perhaps have been difficult to establish by a study of the bacteria alone any marked difference between the disease resulting from this experiment and that native to the cabbage worm, but the symptoms of the two diseases were so unlike as to make it impossible to confound them. The general absence of the peculiar discoloration of the common *flacherie* of the cabbage worm, and of that rapid *post mortem* deliquescence even more characteristic of it, leave no doubt as to the actual difference between this induced disease and the spontaneous affection. That the artificial disease was identical with that of the silkworm, differing only in such a degree as was to be expected when attacking such widely different larvæ, is rendered probable, not only by all the attending circumstances, but also by the occurrence in the cabbage worm of the myriads of mulberry granules characteristic of the affection in the silkworm. This fact is especially significant, since in all our numerous examinations of the native *flacherie* of the cabbage worm this condition of the fluids was not once observed.

I followed this experiment with a similar one in the field, applying the same fluid to a number of cabbages infected by the worms and selecting others as a check on those treated, but the appearance in this field, at about this time, of the common *flacherie* of the cabbage worm, and the death, from this cause, of several of both lots of larvæ interrupted the experiment. The general outbreak, also, of the same spontaneous affection of the Pieris larvæ elsewhere in the vicinity, precluded all attempts at a repetition of these field experiments.

THE YELLOW-NECKED APPLE CATERPILLAR.
(*Datana ministra*, Drury.)

On this species my first studies of the bacterial diseases of caterpillars were made in the autumn of 1883. The affection which attracted my attention broke out in our breeding room shortly after the larvæ were collected, but was not seen among the species anywhere in the field. It probably was not different from the disease well known to entomologists who rear caterpillars to the imago, especially liable to appear in close and sultry weather, and when the breeding cages are insufficiently ventilated.

A lot of the larvæ, two or three hundred in number, obtained July 23, was reported to me, August 1, to have been mysteriously dying for several days at the rate of two or three a day. The small room in which they were kept was open to the south by a large window, and breeding cages of ample size were used, so placed as to be well ventilated. The larvæ were fed and the cages cleaned daily.

DESCRIPTION OF THE DISEASE.

Except that no change of color was usually perceptible, the symptoms of this disease were not especially different from those which have been already given for the silkworm and cabbage worm. Sluggishness and evident weakness and loss of appetite were the first noticeable phenomena. A larva while resting upon a vertical surface would often partly lose its hold, and hang only by a few of the legs,— this occurring long before the power of active locomotion was lost. As a very common thing a discharge of a brownish fluid from the vent occurred early in the disease, but occasionally this symptom was not observed. As a consequence of this purging, the body would become soft and flaccid and somewhat shrunken,—an appearance not presented by those in which the purging did not occur. Occasionally some portion of the body, usually the central or posterior part, became darker before death, but much more commonly the larva retained its natural hue. The approach of death was gradual, the affected insect becoming more and more sluggish and insensible to irritation. *Post mortem* changes were neither so rapid nor so extreme as in the cabbage worm, owing probably, in part, to the thicker and tougher skin.

The fluids escaping from the vent were microscopically examined, and found always swarming with bacteria,—many of them not infrequently having the flagellate motion of Bacterium proper, but the greater number of them being clearly Micrococcus. If a droplet of the blood were obtained before death, it rarely gave any evidence of bacterial affection, the only cases in which this was seen being those in which an *ante-mortem* blackening of the body was observed. After death, however, the blood invariably swarmed with the same bacterial forms which were found earlier in the intestine, the ordinary septic species soon developing rapidly. The alimentary canal usually contained, both before and after death, vast numbers of Micrococcus, and also, not infrequently, true Bacterium, but bacilli or other bacterial forms were rarely found. The micrococci occurring were not by any means as uniform as in the cabbage worm and silkworm, both spherical and oval species of various sizes often appearing on the same slides. The intestine was commonly filled with food little, if at all, digested. In only one instance was the alimentary canal empty and partly filled with gas.

THE CHARACTERISTIC BACTERIA.

The bacteria which, from their abundance and uniform presence, must be regarded as characteristic of this affection, occurring as they did in the still living larvæ almost to the exclusion of other forms, were oval and spherical micrococci,— sometimes one, sometimes the other, and sometimes both commingled in variable proportions. The oval micrococci were usually in singles and doubles, the spherical ones commonly in doubles and short chains of four to six ; in the latter case, often taking on a quadrate form. The ovals varied in length from 1 μ to 1.4 μ, and in transverse diameter from .8 μ to 1 μ. The spherical and quadrate forms were nearly always under 1 μ in diameter, usually averaging about .8 μ. Both forms stained readily with both methyl violet and brown, and occurred frequently in patches or colonies in the intestinal canal.

I mention here a point of especial interest in relation to subsequent attempts at culture and infection. I studied on the morning of the 5th August the fluids of a larva which had died during the night. The blood obtained by snipping a proleg was thick and gray with bacteria, as were also the intestinal fluids, many in both blood and alimentary canal having the form and flagellate movement of Bacterium. Occasionally a string of four, attached end to end, would be seen in serpentine movement across the field. Well-stained and permanently mounted slides of their fluids show three bacterial forms: one large oval, undoubtedly *Bacterium termo;* one a smaller oval (the Micrococcus already described); and the third a somewhat peculiar oval form which might be understood as a single oval 1.5 μ long, with a pale center, or as a short double oval whose division was indicated, not by indentations of its margins, but by a thinning of its central part. The study of slides subsequently made under other circumstances enables me to say that this form last mentioned is really a developing Bacillus of a peculiar character which, matured, is short, broad, and quadrate, its central portion pale when stained, and the ends contrasting by a positively darker tint. Unable to identify this form with anything described, or to obtain through my botanical friends any specific determination of it, I shall refer to it in this paper, merely for convenience sake, under the provisional name of *Bacillus intrapallens.**

---

*I do not know that this is a distinct species, or intend so to imply. *Bacillus subtilis* sometimes presents the peculiar segregation of its contents here described, under what peculiarities of circumstance I do not know, but never, as far as I have observed or can learn, until the full size of the cell has been reached. In the above Bacillus, on the other hand, it was usually evident as soon as the young cell was large enough to show it.

CONTAGIOUS CHARACTER OF THE DISEASE.

I made no effort to determine experimentally the question of the contagious character of this disease in *Datana ministra*, and can only report that it gradually invaded all the breeding cages of this and an allied species, *Datana angusi*, which we found during the season. Many of these were kept at a distance from those suffering from the disease, either as reserve or check lots, with the hope of protecting them from its operations ; but as they were, at farthest, in adjacent rooms, and as we passed freely from one to the other, none of them can be said to have been *isolated*. The bacteria appearing in the walnut Datana *(D. angusi)* were not different from those infesting the other species, except that in our observations the spherical form was usually the characteristic one for this species. Still, both spherical and oval micrococci were noted in a multitude of instances.

ARTIFICIAL CULTURES.

Our first culture illustrating this disease was commenced September 6, 1883, with material obtained from an example of *Datana angusi* seriously affected, but not yet dead. The slide made from the fluids of this larva is not by any means pure. It shows in nearly equal quantities the spherical and oval micrococci described in the preceding section, the oval form mostly in doubles, each pair varying from 2.5 μ to 3 μ in length, and being .75 μ in transverse diameter. The spherules were mostly in doubles (the pairs somewhat under 2 μ in length) and in chains of four or more, the elements of which were sometimes quadrate. Many of both ovals and spheres were aggregated in large, dense patches. Very rarely, also, occurred a larger form, not measured, apparently a Bacterium.

Sterilized and neutralized infusion of beef was infected with fluids from this larva, by the methods and with all the precautions already described. This infusion speedily became milky, and slides made a few days after the culture was begun show clearly a reproduction of the spherical Micrococcus of the original fluid, but of no other form. In size, general appearance, and reaction to staining fluids, this differed in no particular from the original. Singles occurred occasionally, but most of the specimens were in doubles, no chains being noticed.

Additional slides, mounted October 2, show likewise the same spherical Micrococcus without admixture, or change in size or mode of aggregation ; and still another series mounted from the same tubes, April 9, 1884, represent a still pure culture of what was probably this same Micrococcus. The specimens differ only by the somewhat smaller size, rarely surpassing .8 μ, — a difference probably to be accounted for by an exhaustion of the nutritive fluids, certain to have occurred during the seven months which had elapsed since the culture was begun. It should be said, also, that the slides of this last stage are less distinctly stained than the preceding, the micrococci very probably being dead.

After a careful re-examination of these materials I do not doubt that this was a successful culture of the spherical Micrococcus, preserved through the winter, practically unaltered, in a test tube plugged with cotton. It should be added that the check tube remained throughout unchanged.

An interesting culture was begun September 8, the material being obtained from a larva of *Datana ministra* dead several hours. The slides representing this larva are impure, the fluids from the alimentary canal containing not only spherical micrococci, but also a few ovals, and great numbers of bacilli. The spherical micrococci range in diameter from 1 μ to 1.25 μ, and are occasionally indistinctly quadrate, especially when occurring in chaplets (as they frequently do). A few doubles measure 3 μ. The bacilli are all slender, varying greatly in length (from 3 μ to 5 μ), but all .7 μ in transverse diameter.

The beef broth infected with this material on the 8th September was observed on the 15th to have become slightly milky, and, examined, was found to contain micrococci in couples and chaplets, chiefly arranged in the latter form. The slides made from this culture contain no bacilli, but only spherical or subquadrate micrococci in doubles and strings. These average a scant micro in diameter, some, however, reaching 1.25 μ. October 2 these fluids were found to contain only the same Micrococcus, not distinguishable in any way from

L. of C.

those on the slide already described ; and even on April 24 of the following year, the test tube, which had been preserved over winter, yielded only the same Micrococcus, as shown by well-stained and mounted slides prepared at that time. Magnified 1400 diameters and carefully measured, the single spherules vary from 1 μ to 1.25 μ in diameter.

From the foregoing I infer a verification of the experiment just reported, by a second successful culture of the spherical Micrococcus of the Datana larva and its preservation, uncontaminated, until the following year.

The only gelatine film cultures made with this material were begun September 8. Six films of solid beef gelatine, touched with a needle point dipped into the fluids of a larva of *Datana ministra* and inverted over a deep cell containing a droplet of distilled water to prevent drying out, exhibited September 10 a rapid growth of the infection,—each, originally a mere point, being now about the diameter of a pin head, and some having penetrated upwards the thickness of the film. The growth of this mass was in the form of thick finger-like processes, extending upwards through the gelatine film,—the marginal increase however being uniform and continuous. When warmed, these gelatine-film bacteria took on the flagellate motion of Bacillus, and the stained slides made from them strongly indicate that they are young individuals of *Bacillus intrapallens.*

INFECTION EXPERIMENTS.

A few experiments with cultivated material were made upon other Datana larvæ obtained from time to time out of doors, these being divided into experimental and check lots, and the food of the former treated with infusions containing the cultivated bacteria. These were among our first experiments, and the control cages were evidently imperfectly isolated. As a consequence, the experiments were brought to naught by the appearance of *flacherie* in all the cages with which we had to do. In each instance, however, the mortality was more immediate, and at first much greater, among the lots treated with the bacterial cultures than among those not purposely infected ; but the results arrived at are not insisted on, and no detailed account of these experiments is deemed advisable.

THE WALNUT CATERPILLAR.
(*Datana angusi*, G. & R.)

I have to report under this species a series of observations, cultures, and experiments, the longest which I attempted. Although these failed, in part, of their original purpose, they brought out incidental and unintended results of considerable interest, and seem to me worthy of somewhat detailed description.

On the 14th of August, 1883, a lot of the larvæ of *Datana angusi* were collected from a black walnut tree *(Juglans nigra)* in the university grounds at Normal, and brought to the office for experiment. Seven of these were placed in a breeding cage in the further end of the Laboratory, somewhat removed from all the other experimental lots. On the 30th of August one of these was found dead in the cage, having certainly perished since the preceding day. The body of this individual was very limp and flaccid and considerably shrunken, and no food occurred in the alimentary canal. Mounted slides of the blood show vast numbers of the short, broad Bacillus, with rounded or sub-truncate ends and pale central area, which I have distinguished as *Bacillus intrapallens.* The blood was, in fact, a nearly or quite pure culture of this organism, only some smaller and apparently undeveloped forms being possibly micrococci, but more probably the above Bacillus in its earlier stages. These bacilli measured upon an average 1.25 μ by 2.5 μ, and occurred singly and in doubles, the doubles with truncate opposed ends and broadly rounded free extremities. Besides the above, the intestinal contents presented spherical micrococci, usually single, but occasionally in process of division, .8 μ to 1 μ in diameter. I strongly suspect that these apparent micrococci also were the above *Bacillus intrapallens,* undeveloped.

The next morning a second larva of this lot was found dead, having apparently succumbed several hours previously. The intestinal fluids contained a great variety of bacteria, including Bacterium, and multitudes of minute spherical micrococci; but no slides or precise descriptions were prepared.

On September 2 another larva died which had been ailing for two or three days. But very few bacteria were found in the blood, while the intestinal fluids were full of double ovals, not flagellate. Mounted slides show numerous spherical or slightly quadrate micrococci, with many single and double ovals. The spherical form is .75 μ to 1 μ in diameter, some of the single ovals attaining a length of 1.5 μ. The usual length of the latter is, however, about 1.25 μ.

Another larva of this lot died during the night of the 3d September, and was examined on the following morning. Its intestinal contents were brown and nearly solid, requiring to be moistened for examination. They were noted as "full of single and double micrococci," but the slide prepared is so excessively poor that nothing satisfactory can be determined from it.

From this last larva a culture was made as follows : On September 1 freshly prepared strong beef broth was filtered, while hot, through sterilized filter paper into a four-ounce flask which had just been heated for an hour in an oven at 275°-300° Fahrenheit. This was stopped at once with a three-inch plug of raw cotton, freshly sterilized by several hours' heating as above, and was boiled with the plug inserted. This flask was left undisturbed until the 4th September, when it remained perfectly clear. It was then boiled five minutes without removing the plug and left to cool. A particle of the alimentary contents of the above larva, about as large as the head of a pin, was now taken up on the point of a recently heated needle. The plug of the flask was removed, the infection material introduced, and the flask plugged again with fresh sterilized cotton still hot from the oven. A check flask was set aside at the same time.

On the 5th September the fluid was evidently turbid throughout, but especially so at the edges, and a slight film was apparent upon the surface. The plug was loosened, and a droplet of the fluid was obtained upon a freshly heated glass rod. The mounted slide of this material was, unfortunately, worthless, but, from notes made at the time, it appears that the bacteria occurring were rather large "double ovals," nearly all motionless, but with an occasional flagellate individual. Compared with the original infection material, there was no question of the identity of the two.

On the 6th September these fluids were milky, and a film had formed on the glass at the edges, where the fluid had a somewhat ropy appearance when shaken. The check flask was perfectly clear. On September 8 the infected infusion was covered with a thick white surface-scum and the whole mass of the fluid was strongly turbid. A droplet of the liquid contents was now drawn out for examination, with a freshly-made capillary tube pushed down through the plug. The thin film upon the slide was milky with bacteria, which presented, under the microscope, an appearance of double ovals with occasional small clusters or patches of the same object, and occasional strings of three. No other form was seen among myriads passed under the eye, and no flagellate motion was detected ; this was, consequently, an unmistakably pure culture of this single organism. Admirable slides of this material, prepared at this time, further illustrate the purity of the culture, and show that many, perhaps all, of the so-called "double ovals" of my notes were immature *Bacillus intrapallens,* in most of which the pale center was but just beginning to show. On the 13th of September a number of additional slides were made from this same flask, the contents of which were now extremely turbid, the lower half thick with a whitish sediment, and the surface and the flask about the edges covered with a scum. These slides contain only the above Bacillus, somewhat increased in size, and showing the characteristic pale center more distinctly. Considering the frequency with which this form occurred in the dead Datana larvæ of this lot, I have no doubt that this was a successful culture of this particular Bacillus.

On the 17th September these fluids were selected for an experiment intended to test the possibility of preserving throughout the winter the bacteria contained in them, and a number of films were spread upon glass slides previously sterilized by heating, dried immediately with moderate warmth, and laid away for preservation. At the same time small glass tubing was taken, heated thoroughly in the flame of a lamp, and divided by melting, while still almost red hot, into short tubes closed at both ends. As soon as cooled, these partially exhausted tubes were first filled with the bacterial culture by breaking off beneath the fluid, with sterilized forceps, the tip of the tube, which then filled by atmospheric pressure; and were then immediately re-sealed by heat and laid away in cotton for the winter. Several of them were opened in the spring and summer of 1884, at various dates, and found always to contain only a pure culture of the original Bacillus, the results of the first examination, made April 4, not differing in this respect in any particular from the last, made July 30. These bacteria stained much less freely than those in the fresh culture, — a fact probably to be accounted for by their dormant condition. Occasionally a spherical or subquadrate form, 1 μ to 1.25 μ, is distinguishable in the field by a deeper stain,— possibly a spore of the preceding.

Next came a culture in beef broth made by the usual method from the contents of these tubes on the 23d of June, 1884. Two days later this was slightly turbid, decidedly so on the 26th, and on the 27th, when slides were made and the material was used for an infection experiment, they were almost milky. The contained bacteria now consisted of two forms: that frequently mentioned above as *Bacillus intrapallens*, and a spherical form indistinguishable from rather large micrococci. The bacilli occurred singly, doubly, and in strings, were 1 μ by 3 μ in typical specimens, but varied considerably, especially in transverse diameter, reaching sometimes a width of 1.5 μ. The spherules, on the other hand, averaged about 1 μ in transverse dimension. These occurred in various arrangement, but especially in long chaplets. Many of them presented a slightly quadrate outline and in a great number of instances strings of these were continuous with shorter filaments of the bacilli. Occasionally I satisfied myself that two or three of these spherical forms were contained within the Bacillus cells; that they were probably, indeed, to be considered as spores of the cells or, as seems to me more consistent with the facts, as an alternate form of the Bacillus. They seemed not to be developed by the transformation of the contents of an entire Bacillus filament, but rather to be separated off from the end of such a filament by a transformation of the protoplasm in the thickened ends of the cells.

Numerous other cultures were made from this same material. One commenced July 30 was found, August 1, to be decidedly turbid, and on the 2d to have formed a thin transparent pellicle over the whole surface. On the 3d this tube was opened. The fluid was covered with a rather thick film made up wholly of the above *Bacillus intrapallens,* as determined at the time and as shown by beautifully stained and well-mounted slides which I have studied recently Many of these were in long filaments, but none showed any sign whatever of flagellate motion. This culture, like the preceding, was subsequently used for an infection experiment.

Similar cultures from the same material were made April 21 and 24, three tubes being inoculated on the latter date. From all these was obtained the same bacillar form, having occasionally associated with it the sphericals already mentioned, and in a single instance containing also a small Micrococcus about .5 μ in diameter.

The general results of these cultures unquestionably establish the possibility of preserving through several months the bacterial form here dealt with, and afterwards cultivating it successfully in beef broth.

I have next to describe the infection experiments with this Bacillus, showing the possibility of instituting disease in healthy larvæ by means of it, and of procuring its multiplication within their bodies for some days subsequent to the infection.

THE ZEBRA CATERPILLAR.

(*Mamestra picta,* Harris.)

A small colony of zebra caterpillars found on cabbage near Bloomington was brought to the Laboratory June 1, for infection experiments with one of the above cultures,— that begun June 23 and found to contain the *Bacillus intrapallens* and the spherical Micrococcus, as detailed above. A quantity of this fluid was poured into a dish June 27, and a single cabbage leaf was soaked in it for an hour and then fed to the larvæ. These ate freely of it, and were thereafter fed daily with fresh cabbage and carefully attended, this first infection, being the only one purposely made. A check lot of the same brood was placed in a separate cage, but unfortunately removed only a few yards from that infected.

On the next day a single larva of the first lot was found almost dead, and, being isolated, died during. the night.

Examined June 29, at nine o'clock a. m., the fluid obtained by snipping off a proleg was found swarming with large bacilli, motionless at first, but beginning to move actively in all directions when exposed to the air under the cover. These bacilli measured from 2.5 μ to 5 μ in length, one apparently undivided reaching a length of 8 μ, with a transverse diameter of 1.5 μ. These presented no appearance of spores; the ends were broadly rounded, the sides parallel. Small numbers of micrococci occurred in the same slides, about .7 μ in diameter, strictly spherical, in singles and doubles. An examination of carefully stained slides leaves little room for question of the identity of these bacilli with some of those introduced with the food, but the interval was too short to make it certain that they had multiplied since ingestion. Their occurrence, however, in such vast numbers in the blood so soon after death, makes it very unlikely that they merely represented an escape of the intestinal fluids, especially as we shall soon see that the same bacilli occurred abundantly in the blood of larvæ not yet dead. The intestinal contents were full of the above Bacillus and the usual Micrococcus, 1 μ in diameter, in singles, doubles, and patches. The food contents were partially digested.

Besides the above bacteria, the blood was yellow with masses of cells with granular contents, many with a large nucleus each. These cells were apparently derived from the fatty bodies, which seemed to be in process of disorganization, but differed from the usual mulberry bodies which result from pupal histolysis, by the fact that there was no appearance of the division of the cell contents into mulberry granules.

Another larva observed this day, June 29, evidently torpid and apparently sick, seemed to have moulted imperfectly, fragments of the skin still clinging to the shrunken posterior segments. The body was flaccid, but not discolored. A proleg being snipped off, no flow of blood followed, but the fluid pressed out contained a moderate number of the above bacilli, no micrococci, but many well-defined mulberry cells and granules. Each of the cells contained from ten to fifteen or twenty of the latter. The alimentary contents contained micrococci with an occasional Bacillus, but none of the mulberry granules, both forms of bacteria being in this larva much less abundant than was usually the case with individuals so seriously affected. The epithelial cells of the intestine contained granular masses, seemingly of the micrococci, and the fluid bathing them was thick with the same objects. Occasionally patches or clusters of the micrococci occurred in contact with the food. The stained and mounted slides of the blood show chiefly mulberry granules, spherical or somewhat angular in outline, 1.5 μ to 3 μ in diameter. A small number of spherical micrococci also occurred, many of them minute, ranging from .6 μ to .8 μ. These appear in all the usual forms of aggregation, including doubles, short chaplets, and patches of considerable size. Bacilli also occasionally occur, with parallel sides and rounded ends, from 1.25 μ to 1.5 μ in transverse diameter, and from 3 μ to 4 μ in length. A single *Bacillus intrapallens* was noticed in process of development, measuring 1.75 μ by 2 μ.

On the 30th June still another larva died, the grayish fluids of which contained immense numbers of the spherical micrococci, single and double, with vast quantities of the bacilli above described, — motionless at first, but soon, near the edges of the cover or in the vicinity of a bubble, commencing active flagellate movements. The body of the next larva to die, (July 1,) was flaccid, and contained little fluid. Immense numbers of spherical micrococci, 1 μ in diameter, occurred in the blood, mostly in doubles, together with many ovals about 1.5 μ long. Neither Bacterium nor Bacillus were detected in this specimen.

On the 2d, a caterpillar, evidently diseased, shrunken, and shortened, but with colors yet bright, was found lying upon the floor of the cage, able to right itself when turned over, but making no effort to escape. Blood from a foot of this larva contained a great number of unsegmented cells, similar to blood corspucles, but of variable size and shape, some with and some without nuclei. A few hours later, when the blood was examined again, besides these cells were found a considerable number of segmented bodies and mulberry cells, the latter evidently due to dissolution of the former. The next day this segmentation of the cells in question had gone still farther in this larva, and very many mulberry cells were distinguishable, together with others but partly segmented. .

Now killing the larva, I found the fluids full of mulberry cells and granules, together, with a great number of spherical micrococci,—so determined by staining coagulated films.

On the evening of July 1 a number of larvæ in this cage were curiously affected, the prolegs, except the anal pair, being enlarged and swollen, with a slight reddish discoloration. These larvæ were evidently greatly annoyed by their condition, and dragged themselves clumsily about as if half paralyzed. One was seen to turn violently upon itself, and bite the swollen prolegs, as if in pain, so that the blood flowed from them freely. On the following morning one of these caterpillars was crawling about with the abdomen twisted and the prolegs turned almost upwards. Carefully snipping one of these swollen legs, I found in the blood an extraordinary number of lymphoid corpuscles, and a very considerable number of mulberry cells, but little, if any, larger than corpuscles of the blood, varying from circular to oval in optical section. Frequently a nucleus was visible in the midst of the mulberry granules, but no cell walls were distinguishable. The unstable character of the segmentation of these cells was unexpectedly demonstrated by the effect of a little carbolized water run under the cover. As a consequence, the segmentation entirely disappeared, the mulberry cells being all re-converted into simple nucleated corpuscles with granular contents. In fact, I happened to witness this retrogression of a mulberry cell,—a mass of distinct granules with a nucleus dimly seen among them, converted, with a curious internal commotion, into a common lymphoid corpuscle, of rather large size, with clearly distinguishable nucleus. In this condition the cells were indistinguishable from dead blood corpuscles. No bacteria were visible in these fluids.

On the 3d July one of these larvæ died. The body contained but little fluid, but this was loaded with cells, some unsegmented nucleated sphericals of various sizes, without trace of cell wall, staining deeply with aniline; and others well-developed mulberry cells, but so similar to the foregoing as to have been apparently derived from them. On the mounted slides of this material are also great numbers of separate mulberry granules and the usual spherical micrococci, the latter averaging 1 μ in diameter, with an occasional Bacillus like those already several times mentioned. Micrococci and bacilli were, however, less abundant in these fluids than is commonly the case with larvæ destroyed by bacterial disease.

In a peculiar larva which died July 2, a small specimen that had scarcely grown since it was first placed in the cage, a few micrococci were found, and a considerable quantity of the mulberry granules, although this individual caterpillar must have been far from the pupal stage of development.

In another larva examined at the same time, likewise dwarfed, although larger than the preceding, the blood was gray with the usual Micrococcus, both free and in masses, and contained likewise great numbers of mulberry cells and granules. On the 12th July a larva died in whose blood no bacteria were detected, save a few of the usual bacilli. Its fluids contained, however, an immense number of mulberry cells and granules.

From the 12th to the 14th July eight more larvæ died in this lot with symptoms and microscopic characters like those already described,— the body usually somewhat shrunken and flaccid and the colors unchanged. The blood was occasionally gray with micrococci, but more commonly differed in appearance from that of healthy larvæ, only by the slightly yellowish or whitish tinge. The original Bacillus found in the earlier specimens occurred but once in these, and then in trifling quantity. The ordinary Micrococcus was more commonly present, sometimes, indeed, profusely abundant, but at other times in relatively trivial numbers. The unvarying and characteristic feature was the number of free cells in the blood, of variable form and size, some of them being altered blood corpuscles and others evidently derived from the fatty bodies. These occurred in all stages of segmentation, from a mere trace of commencing subdivision to a complete separation of the entire contents of the cell into more or less equal granules. The absence of an enclosing wall was unquestionably evident, granular masses being occasionally found from which a single one of the mulberry granules had broken away, leaving the remainder undisturbed. When the segmentation of these cells was incomplete or indefinite, they readily reverted to nucleated cells with granular protoplasm, if treated with alcohol or carbolized water. In many of the mulberry cells the nucleus persisted, surrounded and obscured by completely formed granules, but in others this seemed likewise to have participated in the metamorphosis of the body of the cell. The number of granules in a single cluster varied from three or five to fifteen or twenty in an optical section of the mass. The few remaining larvæ of this lot were now transferred to alcohol and glycerine for histological study.

In the meantime matters had taken a somewhat unfortunate course in the so-called check lot, these larvæ commencing to die mysteriously on the 30th of June. The first victims were two dwarfed specimens which had evidently moulted very imperfectly, being still covered with fragments of the old integument. An examination of the fluids of these specimens afforded no explanation of their death, as they contained neither bacteria in any appreciable number nor any cellular bodies. Another affected larvæ proved to have been parasitized.

Next two larvæ were found dead upon the morning of July 3, the fluids of which were grayish in hue. These contained no recognizable bacteria whatever, but were loaded with segmented mulberry cells.

On the 10th of July a larva died whose blood contained a moderate number of micrococci in doubles and chains, concerning which no further notes were made at the time and the slides illustrating which were lost.

A larva evidently diseased on the evening of this day was noticed the next morning with several spherical masses of excrement clinging to the vent, connected with each other by a delicate film. This film was dissected off, stained and mounted, and found to consist of an exceedingly delicate, structureless, but rather firm, membrane (doubtless the cuticle of the intestine) through which were dispersed great numbers of micrococci,—unquestionably a pure culture. These were mostly collected in patches, some compact and well defined, others more or less diffused. The compact clusters varied in outline from nearly circular to elongate oval. One of the latter was 35 μ long by 8 μ wide; others were respectively 18 μ by 20 μ, 16 μ by 16 μ, and 12 μ by 20 μ. The micrococci composing them were 1 μ in average diameter, slightly oval to the eye, though not measurably so.

On the morning of the 12th two other larvæ were dead. The blood of one contained only immense numbers of mulberry granules with a moderate number of possible spherical micrococci, — not positively distinguishable in our slides, however, from the smallest mulberry granules. The blood of the other larva was in a similar condition, heavily loaded with mulberry cells and the results of their disintegration, but contained, likewise, a small number of various bacteria,—rarely a short, broad Bacillus, apparently identical with that first used in the experiment ; more abundantly a small spherical Micrococcus, differing in appearance from the usual form ; also a double oval Micrococcus, and an occasional patch of the true spherical so abundant in these experiments. These last were sometimes associated on the slides with patches of unsegmented cells, which evidently had their origin in the fatty bodies.

The third larva dead this day was soft, shrunken, and nearly dry. The scanty fluids were full of micrococci and thick with mulberry cells and granules. The effect of carbolized water upon the cells was, in this case, to cause separation into their constituent particles.

The results of all the above observations and experiments upon the zebra caterpillar may be summarized as follows : At least one of the bacillar forms occurring in the culture used in this infection was conveyed to the larvæ under experiment with fatal effect, and probably multiplied there successfully. This Bacillus almost wholly disaapeared, however, in the later stages of the experiment, and so is not certainly a true pathogenic form. Associated with this in the fluids of the larvæ treated were the usual spherical micrococci of this disease, clearly identical with those applied to the food, and certainly multiplying freely in the bodies of the larvæ. These presented, consequently, the characteristics of a pathogenic microbe. A curious change was observed in the phenomena of the disease in the experimental lot. Death seemed at first occasioned by the immediate action of the bacteria ingested or cultivated in the blood and alimentary fluids ; but at a later period after the infection, these bacterial forms became less abundant, and the blood was loaded with the products of histolysis, partly, in all probability, of the blood cells and partly of the fatty bodies. There seems to have been in general an inverse relation between the abundance of the bacteria and the abundance of these histolysis products, the former becoming less numerous with lapse of time and the latter more so. These facts have an interesting application to those observed in the silkworm, as detailed on previous pages, the condition of the later examples of the zebra caterpillar being, in fact, almost precisely similar, so far as microscopic appearances go, to that of silkworms supposed to be suffering from jaundice.

I have, consequently, to suggest a similar explanation of these phenomena ; viz., that in the case of the latter larva the bacterial affection largely lost its power, but still retained sufficient energy to overthrow the physiological balance as the larvæ approached the age of pupation, death resulting from the premature histolysis of certain of the larval structures, — notably the fatty bodies.

The history of the check lot gives no evidence of serious bacterial infection, but rather of that modified form of it which produces premature pupal histolysis. Reviewing the entire series of slides and cultures, I have no doubt that these indicate the successful preservation through the winter and transference to the bodies of the zebra caterpillars of certain of the forms characteristic of *flacherie* in the walnut caterpillar, *Datana angusi.*

THE EUROPEAN CABBAGE WORM.

(*Pieris rapæ*, L.)

A second infection experiment was begun with the same fluids as the foregoing upon fifty cabbage worms, twenty-five of which were selected for treatment, and an equal number isolated as a check.

On the 6th August, four days after the infection, a larva was found dead upon the bottom of the cage. On puncturing the back a clear, greenish fluid exuded, which was swarming with a large and very active Bacillus, occurring usually in doubles. Stained slides of this exhibit the same characteristics as those made directly from the culture used for the infection, but nothing else is evident.

On the same day another larva was found dead and blackened, clinging to the side of the cage, in quite different condition, however, from cabbage worms affected by their own peculiar disease. The body contained but little fluid, and that was of a paste-like consistence, full of the above bacilli, which the mounted slides show to be an absolutely pure culture.

Another larva, which died the following day, August 7, was found to present precisely the same microscopic characters, only large bacilli occurring in the slide. By the 10th ten of the specimens under experiment had either pupated or were evidently making preparations for that change. But two were apparently diseased. One of these last perished on the 12th, its body soft, pale, blackened posteriorly, but not deliquescent. The blood contained a multitude of minute spherical granules, some Bacillus-like structures, more slender than those previously occurring, and also floating cells of the fatty bodies containing mulberry granules, irregular in size, and sometimes showing also a central nucleus. With these were many large micrococci, 1 μ in diameter, circular, or sometimes slightly oval, commonly in singles or doubles, with rarely a chaplet of four. This larva soon became deliquescent, as if affected by the original *flacherie* of the cabbage-worm ; its condition, in fact, indicating a mingling of two diseases,—that conveyed by the infection to the larvæ, just described, and the one native to the species. It will be noted that one of the effects of the original infection seemed already to have waned, and that the development of the mulberry cells and granules characteristic of this condition had already occurred, — a phenomenon especially significant, since in the native disease of these cabbage caterpillars no similar condition of the fluids was ever seen. Another larva, dead this day, presented appearances so precisely similar to the preceding that no special description of it was made. The check lot, in the meantime, had progressed without injury. August 14 this experiment was interrupted, owing to a discovery of the fact that, through some oversight of the attendant, the full number of the larvæ placed in the breeding cages could not be accounted for, several having, apparently, been allowed to escape as the food was changed. This partial experiment can, consequently, only be held to verify the conclusion drawn from the one just previously described, to the effect that the Bacillus used for infection may be at least temporarily propagated in healthy larvæ with destructive effect. It is proper to add that in the remnants of both the infected and check lots, the common *flacherie* of the cabbage worm afterwards broke out, showing that these insects had been exposed to this disease before they were brought to the office for the experiment.

MUSCARDINE.

This disease, long well known in the silkworm, is not a bacterial affection, but is due to an invasion of the body of the insect by the filaments of a "thread fungus" (Hyphomycetes), whose spores germinate on the surface. These send thread-like processes through the skin which at first bud off from their free ends, within the body, short cells (sometimes called "conidia") with which the blood of the diseased insect speedily becomes loaded. These multiply by division, and finally result in a thread-mycelium which makes its appearance on the surface of the insect, and bears vast numbers of spores, white or green, with which the body becomes covered as with a fine dust. An affected larva is commonly flaccid and shrunken at death, but finally, as a consequence of the *post mortem* development of the fungus, becomes filled with threads and spores, and distended to its original size, drying without shrinkage into a hard and brittle mummy.

These later stages of the development of the fungus are greatly affected by the weather, a drouth preventing the conspicuous external appearance of the mycelium and the development of spores, and thus limiting the spread of the disease.

Every experienced collector finds occasional examples of this disease in the field in the form of stiff and mummified insects, often covered with a dense white or greenish bloom; but few observations of any wholesale destruction of a superabundant species by it have been recorded, — none for America

as far as I am aware. The following observations on the history
of a tremendous outburst, in southern Illinois, of a species of
caterpillar, one of the most destructive insects known, and of
the means by which this irruption was apparently terminated,
will consequently be of considerable interest.

In April and May, 1883, the extreme southern part of the
state, from Cobden southward, was the scene of one of the
periodical uprisings of the forest tent caterpillar (*Clisiocampa
sylvatica*), which have doubtless occurred at intervals in that
region from time immemorial. Vast numbers of forest trees
in the southern counties of Illinois and in the adjacent parts of
Missouri and Indiana were as completely defoliated as if midwinter had suddenly burst upon them in May, and whole
orchards of many acres of apple trees were left without a single
green leaf. Oak, hickory, the black and sweet gum, and dog-
wood were the trees especially selected for destruction in the
forest, and the apple on the fruit farms, — the foliage of the
peach being scarcely touched, even when the trees were covered
with the caterpillars. Strawberry fields were likewise vigorously attacked, — young fields being occasionally nearly eaten
up.

By the 18th May, when my visit there was made, the larvæ
had nearly all attained their growth and were travelling rest-
lessly about by myriads, in every direction, in search of suitable places for pupation,—a few having, in fact, already trans-
formed along the tops of fences and under rubbish on the
ground. As I walked along the road sides my attention was
immediately caught by the great numbers of dead larvæ dried
against the boards of the fences, usually in a vertical position,
and the multitudes apparently in a diseased condition, traveling
more or less feebly, or resting motionless with the head downward. These larvæ were usually flaccid and shrunken posteriorly, but not especially discolored.

It was, unfortunately, impossible for me to make any careful examination of the disease at this time, and no other opportunity offered during the season.

Revisiting this region on the 11th July, an assistant found
that the moths had all emerged sometime previously, but that
from one half to three fourths of the cocoons had never yielded

the imago. From a few of these, parasites had evidently escaped, but in most cases there was nothing in the external appearance of the cocoon to explain the failure of its development. Returning to this region June 3d of the following year, we learned from A. J. Ayers, Esq., of Villa Ridge, that a sufficient number of larvæ hatched that spring to do considerable damage, but that when they were a little over one half an inch long they died and dried upon the leaves, sometimes whole colonies being found dead together. Occasional examples of larvæ in this condition could even then be found on the apple-trees. A few apparently healthy examples were collected at this date and brought to the Laboratory at Normal. These were carefully fed and attended, with the expectation of obtaining the imago, but all died, without exception, with symptoms precisely resembling those of the year before, as they then came under my observation.

The first of these larvæ was seen to be sick on the 27th June, ejecting from the mouth and vent a fluid which contained great numbers of oval corpuscles, not unlike those characterizing *pébrine*, but varying appreciably in size and shape. Examples were found in process of sub division, or even, in occasional instances, short strings of three not wholly separated; and other examples occurred where a spherical lobe was borne upon the end of an oval cell, as if the latter were budding endwise. All these appearances were inconsistent with the hypothesis of the presence of *pébrine*, the characteristic "corpuscles" of which develop by internal segmentation of spherical masses (Sporozoa) and are never connected in doubles nor multiply by fission. Dissections of these larvæ afforded evidence that they were attacked by muscardine. In specimens which had lain some time it was not difficult to identify a scanty mycelium in the body, although, owing probably to the dry and warm weather at this season, there was no external development of the fungus either in the form of threads or spores. These larvæ continued to die until July 5, at which time the last perished.

The individual cells found in the blood varied from 2 μ to 3.5 μ, and in length from 3.5 μ to 5 μ. They differed also in shape, some being a rather broad symmetrical oval, and others narrower towards one extremity. Nuclei about one half as long and wide as the cells containing them were visible in most. Neither cells nor nuclei stained readily with aniline.

The blood of many of the larvæ examined contained also considerable numbers of mulberry cells of rather large size, composed of granules averaging about 2 μ in diameter.

As no insects affected by muscardine had been handled by us at the time these caterpillars were received at the office, it is certain that they brought the infection with them; and as all perished, without exception, from this same disease, and this without the development of spores by which the contagion might have been conveyed from one to another, the presumption is very strong that the affection illustrated by these individuals was that which had swept away the greater part of the entire brood of the preceding year, and especially that which had caused the death of the young larvæ as reported by Mr. Ayres.

SUMMARY AND CONCLUSION.

The circumstances under which the studies above described have been made; the fact that they belong to a field of research so difficult that new comers are very properly viewed with a certain suspicion until they have clearly demonstrated their right to labor in it; and the further fact that my results have not always emerged from the cloud of experiment with perfectly clear and definite outline, have seemed to me to require in this paper a quantity of detail sometimes amounting, perhaps, to wearisome prolixity; and the following summary of the principal features and results of my research has been prepared in the hope that it may serve to make this mode of treatment less objectionable.

I have first attempted to characterize a common and highly destructive disease of the European cabbage worm *(Pieris rapæ)*, by whose ravages the injuries of these pests have received a very important check,—a disease especially marked by the whitish color of the living larvæ, amounting before death to an ashy or almost milky hue, and by a rapid *post mortem* blackening and decay. The distinguishing microscopic appearances are, first, a remarkable whiteness and opacity of the circulating fluids which are early loaded with immense numbers of very

minute spherical granules from .5 μ to .7 μ in diameter, staining
with aniline fluids, although sometimes with difficulty, and less
highly refractile than ordinary micrococci ; second, a great
degeneration of the mucous membrane of the chiliferous
stomach producing before death a marked diminution in the
thickness of the epithilial layer ; and third, the appearance in
the alimentary fluids, and usually also in the blood, of sphericals and ovals (especially the former), presenting every characteristic of unmistakable micrococci. Few if any of the blood
granules are affected by ether, and they dissolve in hot caustic
potash little, if at all, more readily than known micrococci,
bacilli, and bacillar spores,* but they are not all of them certainly to be understood as of bacterial character. The fatty
bodies are the next organs to suffer, after the alimentary canal,
and speedily undergo an immense degeneration.

That this disease is contagious is shown by its unequal distribution in the neighborhoods affected by it ; by its gradual
though rapid progression from one part of the field to another ; by its evident independence of locality, climate, and
weather ; by its apparent progress across the country from east
to west ; by the probable success of experiments made to convey it from infected regions to others at a distance, not previously invaded by it ; and, finally, by its evident bacterial character.

In 1883 and 1884, numerous cultures were attempted in
beef broth by the strictest methods of fluid culture in tubes
and flasks, the accuracy of which was attested by the fact that
the check tubes in every instance remained unchanged throughout. Not all the cultures were successful,—several careful
infections from the blood especially being without result ; in
other cases, however, such infections from the blood of still
living larvæ yielded the spherical micrococcus figured in the
plate, identical in appearance with that observed in the fluids of
the diseased larvæ, but larger in average size than the supposed

* Contrary to the statement frequently made respecting the effects
of alkalies upon bacteria, I have found that hot solutions of caustic
potash rapidly attack both the cells and spores of *Bacillus subtilis* and
the common micrococci of fermentation. Two or three times heating
to a boiling point in a strong solution is sufficient in most cases to completely destroy these microbes.

blood form. Cultures from the alimentary fluids were never without result, although occasionally impure; but the commonest forms there were micrococci like the above, and the next commonest an oval micrococcus of nearly the same size and general appearance. Specimens of Bacillus and Bacterium were frequent in these alimentary cultures, but far less constant than the micrococci. No opportunity offered for experimental infection of healthy larvæ of this or other species with the cabbage worm microbes, either native or cultivated, and consequently it must be confessed that, strictly speaking, the proof is incomplete that this affection of the cabbage worm is a germ disease, although it certainly amounts to very strong probable evidence.

More complete and conclusive studies were made of a disease of the silkworm apparently identical with that known to the French as *jaunes*, and called jaundice by English and American writers. This disease, distinguished especially to the eye by the decided yellow color and restless activity of the larvæ, by the tender skin, easily broken, and by the free flow of thin yellow blood, is microscopically characterized by an abundance, in the blood, of the spherical or polygonal granules and clusters of the same, resulting from the peculiar degeneration of the larval tissues proper to pupation, — these being in this case derived chiefly from the fatty bodies and in part also from the blood corpuscles. This disease, therefore, seems to be essentially a premature pupal histolysis of the fatty bodies,—or, more properly, to be due to a retardation of the pupation of the larva which takes unequal effect on the different tissues, the fatty bodies breaking down before the muscles and membranes are ready for pupal transformation.

Spherical micrococci .75 μ to 1 μ in diameter occur in the walls of the alimentary canal as accompaniments of this disease, and are believed to be one, at least, of the exciting causes of it, although it seems not impossible that other retarding influences may produce a similar effect in overthrowing the normal physiological balance as pupation approaches.

That this supposed jaundice was contagious, was shown by the phenomena of its occurrence at Champaign, and that the bacteria accompanying it were capable of exciting disease in other larvæ was proven by first cultivating them repeatedly in beef broth and then producing in cabbage worms *(Pieris rapæ)* a similar disease by moistening their food with the culture fluids containing the bacteria. While this disease, artificially induced, in some cases came so near that of the native cabbage worm as to suggest that the bacterial treatment served only to excite the natural disease of the larvæ, in other cases it was clearly different from the above and presented characters so clearly like those of the silk worm *jaundice* that there could be little doubt of an actual transference of the original disease, especially when the blood of the sick cabbage worms was found loaded with the mulberry cells and granules of pupal histolysis.

I have next reported at length on a breeding-cage disease attacking the YELLOW-NECKED APPLE CATERPILLAR *(Datana ministra)* and the WALNUT CATERPILLAR *(Datana angusi)*, so similar to the well-know *flacherie* of the silkworm that I have not hesitated to call it by that name. Its principal symptoms are those indicating a gradual weakening of the larvæ, usually accompanied by brownish fluid discharges from the vent and a consequent shrinking and softening of the body. The alimentary canal contains always great numbers of microbes, commonly of considerable variety,—including bacilli, bacteria, and micrococci, the most abundant and characteristic being oval and spherical micrococci not distinguishable from those mentioned above. The method of the appearance and spread of the disease in our breeding room indicated a contagious character; and this conclusion was verified by culture of some of the bacterial forms encountered and their successful use as an experimental virus.

The cultures (in beef broth and on thin gelatine films) related to both micrococci and bacilli, and both were preserved over winter in plugged test tubes and in small sealed tubes, cultivated the following season, and applied to the food of another species of larva,—the ZEBRA CATERPILLAR *(Mamestra picta)*. The first result of this treatment was the destruction of several of the larvæ, in from two to six days, with a disease marked by the appearance in their intestines of great numbers of bacilli (in the specimens first to succumb) and micrococci (later). The affection seemed then to change its character to one resembling jaundice of the silkworm, the characteristic histolysis granules commencing to appear in the blood of slightly affected larvæ as early as the fourth day after infection. Caterpillars thus attacked did not commence to die until the sixth day, and most lived until the 15th.   As in the case of the silkworm jaundice with which this is compared, the bacterial affection was less evident than in more rapid and pronounced cases of disease, but the usual intestinal micrococci were always present in varying numbers.

The last infection experiment I had to report, began August 2, 1884, with the same fluid, applied to the food of the European cabbage worm, was abandoned August 14 because the assistant in charge was unable to account for all the larvæ,—some having evidently been allowed to escape when the food was changed. As far as carried, it tended to confirm the indications of the preceding experiment, the blood of those dying up to the 7th August being full of a large active Bacillus only, similar to that used in the infection, and those perishing later containing chiefly large micrococci together with mulberry cells and granules.   Later the common *flacherie* of the cabbage worm appeared in the remnants of both the infected and check lots.

Finally in a note on muscardine I have attributed largely to this affection the disappearance of a vast host of the forest tent caterpillar *(Clisiocampa sylvatica)* which devastated the forests and orchards of a part of southern Illinois in 1883, basing this conclusion upon the observed phenomena of the disease appearing among them as compared with those accompanying the death of larvæ of this species from the same localities, perishing in our breeding cages the following year of demonstrated muscardine.

There now remains to me only the pleasing duty of acknowledging my grateful obligations for aid in this work to my first assistant, Mr. W. H. Garman, to whose faithful care and unimpeachable accuracy of manipulation the larger part of the bacterial cultures were due ; to Prof. T. J. Burrill, who has had the kindness to examine many of my slides, giving me the benefit of his extensive acquaintance with the bacteria ; and to Dr. H. J. Detmers, now of the State University of Ohio, to whom I owe, among many other favors of this character, the excellent photographs of micrococci reproduced in the plate.

THE LAKE AS A MICROCOSM

Stephen Alfred Forbes

ARTICLE IX.—*The Lake as a Microcosm**. BY STEPHEN A. FORBES.

A lake is to the naturalist a chapter out of the history of a primeval time, for the conditions of life there are primitive, the forms of life are, as a whole, relatively low and ancient, and the system of organic inter-actions by which they influence and control each other has remained substantially unchanged from a remote geological period.

The animals of such a body of water are, as a whole, remarkably isolated—closely related among themselves in all their interests, but so far independent of the land about them that if every terrestrial animal were suddenly annihilated it would doubtless be long before the general multitude of the inhabitants of the lake would feel the effects of this event in any important way. It is an islet of older, lower life in the midst of the higher, more recent life of the surounding region. It forms a little world within itself—a microcosm within which all the elemental forces are at work and the play of life goes on in full, but on so small a scale as to bring it easily within the mental grasp.

Nowhere can one see more clearly illustrated what may be called the *sensibility* of such an organic complex, expressed by the fact that whatever affects any species belonging to it, must have its influence of some sort upon the whole assemblage. He will thus be made to see the impossibility of studying completely any form out of relation to the other forms; the necessity for taking a comprehensive survey of the whole as a condition to a satisfactory understanding of any part. If one wishes to become acquainted with the black bass, for example, he will learn but little if he limits himself to that species. He must evidently study also the species upon which it depends for its existence, and the various conditions upon which *these* depend. He must likewise study the species with which it comes in competition, and the entire system of conditions affecting their prosperity; and by the time he has studied all these sufficiently he will find that he has run through the whole compli-cated mechanism of the aquatic life of the locality, both animal and vege-table, of which his species forms but a single element.

It is under the influence of these general ideas that I propose to examine briefly to-night the lacustrine life of Illinois, drawing my data

*This paper, originally read February 25, 1887, to the Peoria Scientific Associa-tion (now extinct), and published in their Bulletin, was reprinted many years ago by the Illinois State Laboratory of Natural History in an edition which has long been out of print. A single copy remaining in the library of the Natural History Survey is used every year by classes in the University of Illinois, and a professor of zoology in a Canadian university borrows a copy regularly from a Peoria library for use in his own classes. In view of this long-continued demand and in the hope that the paper may still be found useful elsewhere, it is again reprinted. with trivial emendations, and with no attempt to supply its deficiencies or to bring it down to date.

from collections and observations made during recent years by myself and my assistants of the State Laboratory of Natural History.

The lakes of Illinois are of two kinds, fluviatile and water-shed. The fluviatile lakes, which are much the more numerous and important, are appendages of the river systems of the state, being situated in the river bottoms and connected with the adjacent streams by periodical overflows. Their fauna is therefore substantially that of the rivers themselves, and the two should, of course, be studied together.

They are probably in all cases either parts of former river channels, which have been cut off and abandoned by the current as the river changed its course, or else are tracts of the high-water beds of streams over which, for one reason or another, the periodical deposit of sediment has gone on less rapidly than over the surrounding area, and which have thus come to form depressions in the surface which retain the waters of overflow longer than the higher lands adjacent. Most of the numerous "horseshoe lakes" belong to the first of these varieties, and the "bluff-lakes," situated along the borders of the bottoms, are many of them examples of the second.

These fluviatile lakes are most important breeding grounds and reservoirs of life, especially as they are protected from the filth and poison of towns and manufactories by which the running waters of the state are yearly more deeply defiled.

The amount and variety of animal life contained in them as well as in the streams related to them is extremely variable, depending chiefly on the frequency, extent, and duration of the spring and summer overflows. This is, in fact, the characteristic and peculiar feature of life in these waters. There is perhaps no better illustration of the methods by which the flexible system of organic life adapts itself, without injury, to widely and rapidly fluctuating conditions. Whenever the waters of the river remain for a long time far beyond their banks, the breeding grounds of fishes and other animals are immensely extended, and their food supplies increased to a corresponding degree. The slow or stagnant backwaters of such an overflow afford the best situations possible for the development of myriads of Entomostraca, which furnish, in turn, abundant food for young fishes of all descriptions. There thus results an outpouring of life—an extraordinary multiplication of nearly every species, most prompt and rapid, generally speaking, in such as have the highest reproductive rate, that is to say, in those which produce the largest average number of eggs and young for each adult.

The first to feel this tremendous impulse are the protophytes and Protozoa, upon which most of the Entomostraca and certain minute insect larvæ depend for food. This sudden development of their food resources causes, of course, a corresponding increase in the numbers of the latter classes, and, through them, of all sorts of fishes. The first fishes to feel the force of this tidal wave of life are the rapidly-breeding, non-predaceous kinds; and the last, the game fishes, which derive from the others their principal food supplies. Evidently each of these classes must act as a check upon the one preceding it. The development of animalcules is arrested and soon sent back below its highest point by the consequent development of Entomostraca; the latter, again, are met, checked, and reduced in number by the innumerable shoals of fishes with which the water speedily swarms. In this way a general adjustment of numbers to the new conditions would finally be reached spontaneously; but long before any such settled balance can be established, often of course before the full effect of this upward influence has been exhibited, a new cause of disturbance intervenes in the *disappearance of the overflow*. As the waters retire, the lakes are again defined; the teeming life which they contain is restricted within daily narrower bounds, and a fearful slaughter follows; the lower and more defenceless animals are penned up more and more closely with their predaceous enemies, and these thrive for a time to an extraordinary degree. To trace the further consequences of this oscillation would take me too far. Enough has been said to illustrate the general idea that the life of waters subject to periodical expansions of considerable duration, is peculiarly unstable and fluctuating; that each species swings, pendulum-like but irregularly, between a highest and a lowest point, and that this fluctuation affects the different classes successively, in the order of their dependence upon each other for food.

Where a water-shed is a nearly level plateau with slight irregularities of the surface many of these will probably be imperfectly drained, and the accumulating waters will form either marshes or lakes according to the depth of the depressions. Highland marshes of this character are seen in Ford, Livingston, and adjacent counties,* between the head-waters of the Illinois and Wabash systems; and an area of water-shed lakes occurs in Lake and McHenry counties, in northern Illinois.

The latter region is everywhere broken by low, irregular ridges of glacial drift, with no rock but boulders anywhere in sight. The intervening hollows are of every variety, from mere sink-holes, either dry or occupied by ponds, to expanses of several square miles, forming marshes or lakes.

This is, in fact, the southern end of a broad lake belt which borders Lakes Michigan and Superior on the west and south, extending through eastern and northern Wisconsin and northwestern Minnesota, and occupying the plateau which separates the headwaters of the St. Lawrence from those of the Mississippi. These lakes are of glacial origin, some filling beds excavated in the solid rock, and others collecting the surface waters in hollows of the drift. The latter class, to which all the Illinois lakes belong, may lie either parallel to the line of glacial action, occupying valleys between adjacent lateral moraines, or transverse to that line and bounded by terminal moraines. Those of our own state

---

*All now drained and brought under cultivation.

all drain at present into the Illinois through the Des Plaines and Fox; but as the terraces around their borders indicate a former water-level considerably higher than the present one it is likely that some of them once emptied eastward into Lake Michigan. Several of these lakes are clear and beautiful sheets of water, with sandy or gravelly beaches, and shores bold and broken enough to relieve them from monotony. Sportsmen long ago discovered their advantages and club-houses and places of summer resort are numerous on the borders of the most attractive and easily accessible. They offer also an unusually rich field to the naturalist, and their zoology and botany should be better known.

The conditions of aquatic life are here in marked contrast to those afforded by the fluviatile lakes already mentioned. Connected with each other or with adjacent streams only by slender rivulets, varying but little in level with the change of the season and scarcely at all from year to year, they are characterized by an isolation, independence, and uniformity which can be found nowhere else within our limits.

Among these Illinois lakes I did considerable work during October of two successive years, using the sounding line, deep-sea thermometer, towing net, dredge, and trawl in six lakes of northern Illinois, and in Geneva Lake, Wisconsin, just across the line. Upon one of these Illinois lakes I spent a week in October, and an assistant, Prof. H. Garman, now of the University, spent two more, making as thorough a physical and zoölogical survey of this lake as was possible at that season of the year.

I now propose to give you in this paper a brief general account of the physical characters and the fauna of these lakes, and of the relations of the one to the other; to compare, in a general way, the animal assemblages which they contain with those of Lake Michigan —where also I did some weeks of active aquatic work in 1881—and with those of the fluviatile lakes of central Illinois; to make some similar comparisons with the lakes of Europe; and, finally, to reach the subject which has given the title to this paper—to study the system of natural interactions by which this mere collocation of plants and animals has been organized as a stable and prosperous community.

First let us endeavor to form the mental picture. To make this more graphic and true to the facts, I will describe to you some typical lakes among those in which we worked; and will then do what I can to furnish you the materials for a picture of the life that swims and creeps and crawls and burrows and climbs through the water, in and on the bottom, and among the feathery water-plants with which large areas of these lakes are filled.

Fox Lake, in the western border of Lake county, lies in the form of a broad irregular crescent, truncate at the ends, and with the concavity of the crescent to the northwest. The northern end is broadest and communicates with Petite Lake. Two points projecting inward from the southern shore form three broad bays. The western end opens into Nippisink Lake, Crab Island separating the two. Fox River enters the lake from the north, just eastward of this island, and flows directly through the Nippisink. The length of a curved line extending through the central part of this lake, from end to end, is very nearly three miles, and the width of the widest part is about a mile and a quarter. The shores are bold, broken, and wooded, except to the north, where they are marshy and flat. All the northern and eastern part of the lake was visibly shallow—covered with weeds and feeding water-fowl, and I made no soundings there. The water there was probably nowhere more than two fathoms in depth, and over most of that area was doubtless under one and a half. In the western part, five lines of soundings were run, four of them radiating from Lippincott's Point, and the fifth crossing three of these nearly at right angles. The deepest water was found in the middle of the mouth of the western bay, where a small area of five fathoms occurs. On the line running northeast from the Point, not more than one and three fourths fathoms is found. The bottom at a short distance from the shores was everywhere a soft, deep mud. Four hauls of the dredge were made in the western bay, and the surface net was dragged about a mile.

Long Lake differs from this especially in its isolation, and in its smaller size. It is about a mile and a half in length by a mile in breadth. Its banks are all bold except at the western end, where a marshy valley traversed by a small creek connects it with Fox Lake, at a distance of about two miles. The deepest sounding made was six and a half fathoms, while the average depth of the deepest part of the bed was about five fathoms.

Cedar Lake, upon which we spent a fortnight, is a pretty sheet of water, the head of a chain of six lakes which open finally into the Fox. It is about a mile in greatest diameter in each direction, with a small but charming island bank near the center, covered with bushes and vines—a favorite home of birds and wild flowers. The shores vary from rolling to bluffy except for a narrow strip of marsh through which the outlet passes, and the bottoms and margins are gravel, sand, and mud in different parts of its area. Much of the lake is shallow and full of water plants; but the southern part reaches a depth of fifty feet a short distance from the eastern bluff.

Deep Lake, the second of this chain, is of similar character, with a greatest depth of fifty-seven feet—the deepest sounding we made in these smaller lakes of Illinois. In these two lakes several temperatures were taken with a differential thermometer. In Deep Lake, for example, at fifty-seven feet I found the bottom temperature $53\frac{1}{2}°$— about that of ordinary well-water—when the air was 63°; and in Cedar Lake, at forty-eight feet, the bottom was 58° when the air was 61°.

Geneva Lake, Wisconsin, is a clear and beautiful body of water about eight miles long by one and a quarter in greatest width. The banks are all high, rolling, and wooded, except at the eastern end, where its outlet rises. Its deepest water is found in its western third, where it reaches a depth of twenty-three fathoms. I made here, early in November, twelve hauls of the dredge and three of the trawl, aggregating about three miles in length, so distributed in distance and depth as to give a good idea of the invertebrate life of the lake at that season.

And now if you will kindly let this suffice for the background or setting of the picture of lacustrine life which I have undertaken to give you, I will next endeavor—not to paint in the picture; for that I have not the artistic skill. I will confine myself to the humble and safer task of supplying you the pigments, leaving it to your own constructive imaginations to put them on the canvas.

When one sees acres of the shallower water black with water-fowl, and so clogged with weeds that a boat can scarcely be pushed through the mass; when, lifting a handful of the latter, he finds them covered with shells and alive with small crustaceans; and then, dragging a towing net for a few minutes, finds it lined with myriads of diatoms and other microscopic algæ, and with multitudes of Entomostraca, he is likely to infer that these waters are everywhere swarming with life, from top to bottom and from shore to shore. If, however, he will haul a dredge for an hour or so in the deepest water he can find, he will invariably discover an area singularly barren of both plant and animal life, yielding scarcely anything but a small bivalve mollusk, a few low worms, and red larvæ of gnats. These inhabit a black, deep, and almost impalpable mud or ooze, too soft and unstable to afford foothold to plants even if the lake is shallow enough to admit a sufficient quantity of light to its bottom to support vegetation. It is doubtless to this character of the bottom that the barrenness of the interior parts of these lakes is due; and this again is caused by the selective influence of gravity upon the mud and detritus washed down by rains. The heaviest and coarsest of this material necessarily settles nearest the margin, and only the finest silt reaches the remotest parts of the lakes, which, filling most slowly, remain, of course, the deepest. This ooze consists very largely, also, of a fine organic *debris*. The superficial part of it contains scarcely any sand, but has a greasy feel and rubs away, almost to nothing, between the fingers. The largest lakes are not therefore, as a rule, by any means the most prolific of life, but this shades inward rapidly from the shore, and becomes at no great distance almost as simple and scanty as that of a desert.

Among the weeds and lily-pads upon the shallows and around the margin—the Potamogeton, Myriophyllum, Ceratophyllum, Anacharis, and Chara, and the common Nelumbium,—among these the fishes chiefly swim or lurk, by far the commonest being the barbaric bream[1] or "pumpkin-seed" of northern Illinois, splendid with its green and scarlet and purple and orange. Little less abundant is the common perch (*Perca lutea*) in the larger lakes—in the largest out-numbering the bream itself. The whole sunfish family, to which the latter belongs, is in fact the dominant group in these lakes. Of the one hundred and thirty-two fishes of Illinois only thirty-seven are found in these waters—about twenty-

---

[1] Lepomis gibbosus.

eight per cent.—while eight out of our seventeen sunfishes (*Centrar-chinae*) have been taken t.1ere. Next, perhaps, one searching the pebbly beaches or scanning the weedy tracts will be struck by the small number of minnows or cyprinoids which catch the eye or come out in the net. Of our thirty-three Illinois cyprinoids, only six occur there— about eighteen per cent.—and only three of these are common. These are in part replaced by shoals of the beautiful little silversides (*Labides-thes sicculus*), a spiny-finned fish, bright, slender, active, and voracious —as well supplied with teeth as a perch, and far better equipped for self-defense than the soft-bodied and toothless cyprinoids. Next we note that of our twelve catfishes (*Siluridae*) only two have been taken in these lakes—one the common bullhead (*Ictalurus nebulosus*), which occurs everywhere, and the other an insignificant stone cat, not as long as one's thumb. The suckers, also, are much less abundant in this region than farther south, the buffalo fishes[1] not appearing at all in our collections. Their family is represented by worthless carp[2] by two redhorse[3], by the chub sucker[4] and the common sucker (*Catostomus teres*), and by one other species. Even the hickory shad[5]—an ich-thyological weed in the Illinois—we have not found in these lakes at all. The sheepshead[6], so common here, is also conspicuous there by its ab-sence. The yellow bass[7], not rare in this river, we should not expect in these lakes because it is, rather, a southern species; but why the white bass[8], abundant here, in Lake Michigan, and in the Wisconsin lakes, should be wholly absent from the lakes of the Illinois plateau, I am unable to imagine. If it occurs there at all, it must be rare, as I could neither find nor hear of it.

A characteristic, abundant, and attractive little fish is the log perch (*Percina caprodes*)—the largest of the darters, slender, active, barred like a zebra, spending much of its time in chase of Entomostraca among the water plants, or prying curiously about among the stones for minute insect larvæ. Six darters in all (*Ethcostomatinae*), out of the eighteen from the state, are on our list from these lakes. The two black bass[9] are the most popular game fishes—the large-mouthed species being much the most abundant. The pickerels[10], gar[11], and dogfish[12] are there about as here; but the shovel-fish[13] does not occur.

Of the peculiar fish fauna of Lake Michigan—the burbot[14], white fish,[15] trout,[16] lake herring or cisco,[17] etc., not one species occurs in these smaller lakes, and all attempts to transfer any of them have failed completely. The cisco is a notable fish of Geneva Lake, Wisconsin, but does not reach Illinois except in Lake Michigan. It is useless to at-tempt to introduce it, because the deeper areas of the interior lakes are too limited to give it sufficient range of cool water in midsummer.

In short, the fishes of these lakes are substantially those of their

[1]Ictiobus bubalus. [2]Ictiobus cyprinus. [3]Moxostoma aureolum and M. macro-lepidotum. [4]Erimyzon sucetta. [5]Dorosoma cepedianum. [6]Haploidonotus. [7]Roccus interruptus. [8]Roccus chrysops. [9]Micropterus. [10]Esox. [11]Lepidosteus. [12]Amia. [13]Polyodon. [14]Lota. [15]Coregonus clupeiformis. [16]Salvelinus namaycush. [17]Coregonus artedi.

region—excluding the Lake Michigan series (for which the lakes are too small and warm) and those peculiar to creeks and rivers. Possibly the relative scarcity of catfishes (*Siluridae*) is due to the comparative clearness and cleanness of these waters. I see no good reason why minnows should be so few, unless it be the abundance of pike and Chicago sportsmen.

Concerning the molluscan fauna, I will only say that it is poor in bivalves—as far as our observations go—and rich in univalves. Our collections have been but partly determined, but they give us three species of Valvata, seven of Planorbis, four Amnicolas, a Melantho, two Physas, six Limnæas, and an Ancylus among the Gastropoda, and two Unios, an Anodonta, a Sphærium, and a Pisidium among the Lamellibranchiates. *Pisiduim variabile* is by far the most abundant mollusk in the oozy bottom in the deeper parts of the lakes; and crawling over the weeds are multitudes of small Amnicolas and Valvatas.

The entomology of these lakes I can merely touch upon, mentioning only the most important and abundant insect larvæ. Hiding under stones and driftwood, well aware, no doubt, what enticing morsels they are to a great variety of fishes, we find a number of species of ephemerid larvæ whose specific determination we have not yet attempted. Among the weeds are the usual larvæ of dragon-flies—Agrionina and Libellulina, familiar to every one; swimming in open water the predaceous larvæ of Corethra; wriggling through the water or buried in the mud the larvæ of Chironomus—the shallow water species white, and those from the deeper ooze of the central parts of the lakes blood-red and larger. Among Chara on the sandy bottom are a great number and variety of interesting case-worms—larvæ of Phryganeidæ—most of them inhabiting tubes of a slender conical form made of a viscid secretion exuded from the mouth and strengthened and thickened by grains of sand, fine or coarse. One of these cases, nearly naked, but usually thinly covered with diatoms, is especially worthy of note, as it has been reported nowhere in this country except in our collections, and was indeed recently described from Brazil as new. Its generic name is Lagenopsyche, but its species undetermined. These larvæ are also eaten by fishes.

Among the worms we have of course a number of species of leeches and of planarians,—in the mud minute Anguillulidæ, like vinegar eels, and a slender Lumbriculus which makes a tubular mud burrow for itself in the deepest water, and also the curious *Nais probiscidea,* notable for its capacity of multiplication by transverse division.

The crustacean fauna of these lakes is more varied than any other group. About forty species were noted in all. Crawfishes were not especially abundant, and most belonged to a single species, *Cambarus virilis.* Two amphipods occurred frequently in our collections; one, less common here but very abundant farther south—*Crangonyx gracilis*—and one, *Allorchestes dentata,* probably the commonest animal in these waters, crawling and swimming everywhere in myriads among the submerged water-plants. An occasional *Gammarus fasciatus* was also taken in the dredge. A few isopod Crustacea occur, belonging to *Mancasellus tenax*—a species not previously found in the state.

I have reserved for the last the Entomostraca—minute crustaceans of a surprising number and variety, and of a beauty often truly exquisite. They belong wholly, in our waters, to the three orders, Copedoda, Ostracoda, and Cladocera—the first two predaceous upon still smaller organisms and upon each other, and the last chiefly vegetarian. Twenty-one species of Cladocera have been recognized in our collections, representing sixteen genera. It is an interesting fact that twelve of these species are found also in the fresh waters of Europe. Five cyprids have been detected, two of them common to Europe, and also an abundant Diaptomus, a variety of a European species. Several Cyclops species were collected which have not yet been determined.

These Entomostraca swarm in microscopic myriads among the weeds along the shore, some swimming freely, and others creeping in the mud or climbing over the leaves of plants. Some prefer the open water, in which they throng locally like shoals of fishes, coming to the surface preferably by night, or on dark days, and sinking to the bottom usually by day to avoid the sunshine. These pelagic forms, as they are called, are often exquisitely transparent, and hence almost invisible in their native element—a charming device of Nature to protect them against their enemies in the open lake, where there is no chance of shelter or escape. Then with an ingenuity in which one may almost detect the flavor of sarcastic humor, Nature has turned upon these favored children and endowed their most deadly enemies with a like transparency, so that wherever the towing net brings to light a host of these crystalline Cladocera, there it discovers also swimming, invisible, among them, a lovely pair of robbers and beasts of prey—the delicate Leptodora and the Corethra larva.

These slight, transparent, pelagic forms are much more numerous in Lake Michigan than in any of the smaller lakes, and peculiar forms occur there commonly which are rare in the larger lakes of Illinois and entirely wanting in the smallest. The transparent species are also much more abundant in the isolated smaller lakes than in those more directly connected with the rivers.

The vertical range of the animals of Geneva Lake showed clearly that the barrenness of the interiors of these small bodies of water was not due to the greater depth alone. While there were a few species of crustaceans and case-worms which occurred there abundantly near shore but rarely or not at all at depths greater than four fathoms, and may hence be called littoral species, there was, on the whole, little diminution either in quantity or variety of animal life until about fifteen fathoms had been reached. Dredging at four or five fathoms were nearly or quite as fruitful as any made. On the other hand, the barrenness of the bottom at twenty to twenty-three fathoms was very remarkable. The total product of four hauls of the dredge and one of the trawl at that depth, aggregating fully a mile and a half of continuous dragging, would easily go into a two-dram vial, and represents only nine animal species—not counting dead shells, and fragments which had probably floated in from shallower waters. The greater part of this little collection was composed of specimens of Lumbriculus and larvæ of Chironomus. There were a few Corethra larvæ, a single Gammarus, three small leeches, and some sixteen mollusks, all but four of which belonged to Pisidium. The others were two Sphæriums, a *Valvata carinata*, and a *V. sincera*. None of the species taken here are peculiar, but all were of the kinds found in the smaller lakes, and all occurred also in shallower water. It is evident that these interior regions of the lakes must be as destitute of fishes as they are of plants and lower animals.

While none of the deep-water animals of the Great Lakes were found in Geneva Lake, other evidences of zoölogical affinity were detected. The towing net yielded almost precisely the assemblage of species of Entomostraca found in Lake Michigan, including many specimens of *Limnocalanus macrurus* Sars; and peculiar long, smooth leeches, common in Lake Michigan but not occurring in the small Illinois lakes, were also found in Geneva. Many *Valvata tri-carinata* lacked the middle carina, as in Long Lake and other *isolated* lakes of this region.

Comparing the Daphnias of Lake Michigan with those of Geneva Lake, Wis. (nine miles long and twenty-three fathoms in depth), those of Long Lake, Ill. (one and a half miles long and six fathoms deep), and those of other, still smaller, lakes of that region, and the swamps and smaller ponds as well, we shall be struck by the inferior development of the Entomostraca of the larger bodies of water in numbers, in size and robustness, and in reproductive power. Their smaller numbers and size are doubtless due to the relative scarcity of food. The system of aquatic animal life rests essentially upon the vegetable world, although perhaps less strictly than does the terrestrial system, and in a large and deep lake vegetation is much less abundant than in a narrower and shallower one, not only relatively to the amount of water but also to the area of the bottom. From this deficiency of plant life results a deficiency of food for Entomostraca, whether of algæ, of Protozoa, or of higher forms, and hence, of course, a smaller number of the Entomostraca themselves, and these with more slender bodies, suitable for more rapid locomotion and wider range.

The difference of reproductive energy, as shown by the much smaller egg-masses borne by the species of the larger lakes, depends upon the vastly greater destruction to which the paludal Crustacea are subjected. Many of the latter occupy waters liable to be exhausted by drought, with a consequent enormous waste of entomostracan life. The opportunity for reproduction is here greatly limited—in some situations to early spring alone—and the chances for destruction of the summer eggs in the dry and often dusty soil are so numerous that only the most prolific species can maintain themselves.

Further, the marshes and shallower lakes are the favorite breeding grounds of fishes, which migrate to them in spawning time if possible, and it is from the Entomostraca found here that most young fishes get their earliest food supplies—a danger from which the deep-water species are measurably free. Not only is a high reproductive rate rendered unnecessary among the latter by their freedom from many dangers to which the shallow-water species are exposed, but in view of the relatively small amount of food available for them, a high rate of multiplication would be a positive injury, and could result only in wholesale starvation.

All these lakes of Illinois and Wisconsin, together with the much larger Lake Mendota at Madison (in which also I have done much work with dredge, trawl, and seine), differ in one notable particular both from Lake Michigan and from the larger lakes of Europe. In the latter the bottoms in the deeper parts yield a peculiar assemblage of animal forms which range but rarely into the littoral region, while in our inland lakes no such deep water fauna occurs, with the exception of the cisco and the large red Chironomus larva. At Grand Traverse Bay, in Lake Michigan, I found at a depth of one hundred fathoms a very odd fish of the sculpin family (*Triglopsis thompsoni* Gir.) which, until I collected it, had been known only from the stomachs of fishes; and there also was an abundant crustacean, Mysis—the "opossum shrimp", as it is sometimes called—the principal food of these deep lake sculpins. Two remarkable amphipod crustaceans also belong in a peculiar way to this deep water. In the European lakes the same Mysis occurs in the deepest part, with several other forms not represented in our collections, two of these being blind crustaceans related to those which in this country occur in caves and wells.

Comparing the other features of our lake fauna with that of Europe, we find a surprising number of Entomostraca identical; but this is a general phenomenon, as many of the more abundant Cladocera and Copepoda of our small wayside pools are either European species, or differ from them so slightly that it is doubtful if they ought to be called distinct.

It would be quite impossible, within reasonable limits, to go into details respecting the organic relations of the animals of these waters, and I will content myself with two or three illustrations. As one example of the varied and far-reaching relations into which the animals of a lake are brought in the general struggle for life, I take the common black bass. In the dietary of this fish I find, at different ages of the individual, fishes of great variety, representing all the important orders of that class; insects in considerable number, especially the various water-bugs and larvæ of day-flies; fresh-water shrimps; and a great multitude of Entomostraca of many species and genera. The fish is therefore directly dependent upon all these classes for its existence. Next, looking to the food of the species which the bass has eaten, and upon which it is therefore indirectly dependent, I find that one kind of the fishes taken feeds upon mud, algæ, and Entomostraca, and another upon nearly every animal substance in the water, including mollusks and decomposing organic matter. The insects taken by the bass, themselves take other insects and small Crustacea. The crawfishes are nearly omnivorous, and of the other crustaceans some eat Entomostraca and some algæ and Protoza. At only the second step, therefore, we find our bass brought into dependence upon nearly every class of animals in the water.

And now, if we search for its competitors we shall find these also extremely numerous. In the first place, I have found that all our young fishes except the Catostomidæ feed at first almost wholly on Entomostraca, so that the little bass finds himself at the very beginning of his life engaged in a scramble for food with all the other little fishes in the lake. In fact, not only young fishes but a multitude of other animals as well, especially insects and the larger Crustacea, feed upon these Entomostraca, so that the competitors of the bass are not confined to members of its own class. Even mollusks, while they do not directly compete with it do so indirectly, for they appropriate myriads of the microscopic forms upon which the Entomostraca largely depend for food. But the enemies of the bass do not all attack it by appropriating its food supplies, for many devour the little fish itself. A great variety of predaceous fishes, turtles, water-snakes, wading and diving birds, and even bugs of gigantic dimensions destroy it on the slightest opportunity. It is in fact hardly too much to say that fishes which reach maturity are relatively as rare as centenarians among human kind.

As an illustration of the remote and unsuspected rivalries which reveal themselves on a careful study of such a situation, we may take the relations of fishes to the bladderwort[1]—a flowering plant which fills many acres of the water in the shallow lakes of northern Illinois. Upon the leaves of this species are found little bladders—several hundred to each plant—which when closely examined are seen to be tiny traps for the capture of Entomostraca and other minute animals. The plant usually has no roots, but lives entirely upon the animal food obtained through these little bladders. Ten of these sacs which I took at random from a mature plant contained no less than ninety-three animals (more than nine to a bladder), belonging to twenty-eight different species. Seventy-six of these were Entomostraca, and eight others were minute insect larvæ. When we estimate the myriads of small insects and Crustacea which these plants must appropriate during a year to their own support, and consider the fact that these are of the kinds most useful as food for young fishes of nearly all descriptions, we must conclude that the bladderworts compete with fishes for food, and tend to keep down their number by diminishing the food resources of the young. The plants even have a certain advantage in this competition, since they are not strictly dependent on Entomostraca, as the fishes are, but sometimes take root, developing then but very few leaves and bladders. This probably happens under conditions unfavorable to their support by the other

---

[1]Utricularia.

method. These simple instances will suffice to illustrate the intimate way in which the living forms of a lake are united.

Perhaps no phenomenon of life in such a situation is more remarkable than the steady balance of organic nature, which holds each species within the limits of a uniform average number, year after year, although each one is always doing its best to break across boundaries on every side. The reproductive rate is usually enormous and the struggle for existence is correspondingly severe. Every animal within these bounds has its enemies, and Nature seems to have taxed her skill and ingenuity to the utmost to furnish these enemies with contrivances for the destruction of their prey in myriads. For every defensive device with which she has armed an animal, she has invented a still more effective apparatus of destruction and bestowed it upon some foe, thus striving with unending pertinacity to outwit herself; and yet life does not perish in the lake, nor even oscillate to any considerable degree, but on the contrary the little community secluded here is as prosperous as if its state were one of profound and perpetual peace. Although every species has to fight its way inch by inch from the egg to maturity, yet no species is exterminated, but each is maintained at a regular average number which we shall find good reason to believe is the greatest for which there is, year after year, a sufficient supply of food.

I will bring this paper to a close, already too long postponed, by endeavoring to show how this beneficent order is maintained in the midst of a conflict seemingly so lawless.

It is a self-evident proposition that a species can not maintain itself continuously, year after year, unless its birth-rate at least equals its death-rate. If it is preyed upon by another species, it must produce regularly an excess of individuals for destruction, or else it must certainly dwindle and disappear. On the other hand, the dependent species evidently must not appropriate, on an average, any more than the surplus and excess of individuals upon which it preys, for if it does so it will continuously diminish its own food supply, and thus indirectly but surely exterminate itself. The interests of both parties will therefore be best served by an adjustment of their respective rates of multiplication such that the species devoured shall furnish an excess of numbers to supply the wants of the devourer, and that the latter shall confine its appropriations to the excess thus furnished. We thus see that there is really a close *community of interest* between these two seemingly deadly foes.

And next we note that this common interest is promoted by the process of natural selection; for it is the great office of this process to eliminate the unfit. If two species standing to each other in the relation of hunter and prey are or become badly adjusted in respect to their rates of increase, so that the one preyed upon is kept very far below the normal number which might find food, even if they do not presently obliterate each other the pair are placed at a disadvantage in the battle for life, and must suffer accordingly. Just as certainly as the thrifty business man who lives within his income will finally dispossess his shiftless competitor who can never pay his debts, the well-adjusted aquatic animal will in time crowd out its poorly-adjusted competitors for food and for the various goods of life. Consequently we may believe that in the long run and as a general rule those species which have survived, are those which have reached a fairly close adjustment in this particular.[1]

Two ideas are thus seen to be sufficient to explain the order evolved from this seeming chaos; the first that of a general community of interests among all the classes of organic beings here assembled, and the second that of the beneficent power of natural selection which compels such adjustments of the rates of destruction and of multiplication of the various species as shall best promote this common interest.

Have these facts and ideas, derived from a study of our aquatic microcosm, any general application on a higher plane? We have here an example of the triumphant beneficence of the laws of life applied to conditions seemingly the most unfavorable possible for any mutually helpful adjustment. In this lake, where competitions are fierce and continuous beyond any parallel in the worst periods of human history; where they take hold, not on goods of life merely, but always upon life itself; where mercy and charity and sympathy and magnanimity and all the virtues are utterly unknown; where robbery and murder and the deadly tyranny of strength over weakness are the unvarying rule; where what we call wrong-doing is always triumphant, and what we call goodness would be immediately fatal to its possessor,—even here, out of these hard conditions, an order has been evolved which is the best conceivable without a total change in the conditions themselves; an equilibrium has been reached and is steadily maintained that actually accomplishes for all the parties involved the greatest good which the circumstances will at all permit. In a system where life is the universal good, but the destruction of life the well-nigh universal occupation, an order has spontaneously arisen which constantly tends to maintain life at the highest limit—a limit far higher, in fact, with respect to both quality and quantity, than would be possible in the absence of this destructive conflict. Is there not, in this reflection, solid ground for a belief in the final beneficence of the laws of organic nature? If the system of life is such that a harmonious balance of conflicting interests has been reached where every element is either hostile or indifferent to every other, may we not trust much to the outcome where, as in human affairs, the spontaneous adjustments of nature are aided by intelligent effort, by sympathy, and by self-sacrifice?

[1]For a fuller statement of this argument, see Bul. Ill. State Lab. Nat. Hist. Vol. I. No. 3. pages 5 to 10.

STUDIES OF THE FOOD
OF FRESH-WATER FISHES

Stephen Alfred Forbes

ARTICLE VII. — *Studies of the Food of Fresh-Water Fishes.** By S. A. FORBES.

FAMILY GADIDÆ.

LOTA MACULOSA, Pennant. BURBOT; LAWYER.

The cod family is represented in Illinois by only a single species, the burbot (*Lota maculosa*), occurring in the interior of Lake Michigan, and making its way at irregular intervals to the shallow waters within the reach of ordinary fishing operations. Since the opening of canals between the Great Lakes and the river systems of the State, occasional specimens have been taken in the Illinois and Mississippi.†

Its predaceous character is too well known to make special description of its alimentary structures necessary. It is reported by Mr. G. Brown Goode‡ to feed upon various small fishes and Crustacea which frequent the bottom, devouring more particularly fishes with habits like its own. It is extremely

---

* This article is to be considered as a continuation of the studies reported in Volume I. of the Bulletin of the Illinois State Laboratory of Natural History, Nos. 3 and 6, the first published in 1880, and the second in 1883.
The data here presented relate to the fishes of the State of Illinois, and most of them the lower families of the series. They are derived from collections made by my assistants and myself in various parts of the State at intervals from 1876 to 1887, for the special purpose of ascertaining the principal characters of the food, and the feeding habits of the fishes of our native fauna.

†I have seen a specimen taken from the Mississippi at Canton, Mo., in 1887, and sent to Mr. S. P. Bartlett, one of the State Fish Commissioners of Illinois. One occurred some years ago at Naples, on the Illinois River, and in a letter dated April 10, 1886, Prof. J. Lindahl, of Augustana College, Rock Island, says that three specimens have been taken from the Mississippi River within his knowledge, all small, the largest hardly a foot in length.

‡ "The Fishery Industries of the United States." p. 239.

voracious, with a wonderfully distensible stomach; and not only captures the most active fishes, such as the pike, but will eat carrion, and may even swallow stones. It is reported to be nocturnal in habit, and often to secure its prey by stealth.

It is illustrated in our collection by ten examples; five taken in spring and five in November. All but one had eaten fishes, these making eighty-three per cent. of the food of the entire group. One of the spring specimens had taken crayfishes only—*Cambarus propinquus*, the species commonest in the lake. Two others of this lot had likewise eaten crayfishes, fifty per cent. of the food of one and fifteen per cent. that of the other consisting of this same species. The fishes taken, with the exception of one young white-fish (*Coregonus clupeiformis*) and a small unrecognizable residue, were the common perch of the lakes, *Perca lutea*.

FAMILY ESOCIDÆ.

This family is represented within our limits by the European species, *Esox lucius* (the so-called common "pickerel" of the streams and smaller lakes of Illinois), by the noble muskallunge, *Esox nobilior* of Lake Michigan, and by the small grass pickerel, *Esox umbrosus*. No fishes of our waters, unless it be the gars, have become so strictly adapted to a predaceous life, —an adaptation which probably limits them, *nolens volens*, to a living prey.

Esox lucius, Linn. Pike; Pickerel.

Our specimens of this species, thirty-seven in number, of nine different lots, were from various parts of the Illinois River, except a single one from Fourth Lake in northern Illinois.

One had eaten larvæ of dragon flies (twenty per cent.), but the entire food of the remainder consisted only of fishes, these making, consequently, ninety-nine per cent. of the whole. Nine per cent. were not otherwise recognizable. Twenty-one per cent. were sunfishes and black bass—one of the latter the small-mouthed species—and nine per cent. were croppie (Pomoxys),—eaten however by only one of the specimens. Twenty of the thirty-seven pike had taken gizzard shad *(Dorosoma)*, which made forty-six per cent. of the entire food of the species. Cyprinidæ (chiefly *Notropis hudsonius*) were found in two, and three had eaten buffalo fish (*Ictiobus cyprinellus* and *I. bubalus*).

ESOX VERMICULATUS, LeS. BROOK PICKEREL.

This fish — so far as its food structures are concerned a miniature of the preceding—is abundant throughout the State in ponds and lakes and along the borders of streams, especially by the weedy margins of rivers. I have studied the food of eighteen examples, and found it to differ from that of the larger species only as was to be expected from the smaller size of this pickerel, (which rarely reaches a foot in length), and from the character of its favorite haunts.

The specimens selected for examination were from various localities in northern, central, and southern Illinois; represented lakes, rivers, and smaller ponds; and were collected in June, July, and October of different years.

Two had eaten the tadpoles of frogs, and eight had captured fishes,—which made about half of the food of the entire group. Only three of these were recognizable; one a cyprinoid, one a sunfish, and the other *(Gambusia patruelis)* a common top minnow of the southern part of the State.

Aquatic insects formed the next most important element of the food, reaching thirty-five per cent., and eaten by nine of the specimens. The greater part of these were larvæ of Agrion and larger Odonata, only four per cent. being Hexagenia larvæ. One specimen had taken an isopod (Asellus), but no other crustaceans occurred.

The food of this group may consequently be generalized as consisting of the larger aquatic insect larvæ and the smaller fishes in nearly equal ratio, with occasional larvæ of Batrachia.*

---

*Five additional specimens of this species, too large to be reckoned examples of the young and yet too small to class as adults had eaten, like the full-grown examples, chiefly fishes and neuropterous larvæ. A specimen only an inch and a half in length had swallowed a fish ; one three and a fourth inches long had likewise taken only a

FAMILY SALMONIDÆ.

The common lake trout, the white-fish, and the lake herring, are the representatives of this great family in the waters of Illinois, and occur there only in Lake Michigan. None of the smaller lakes of the State contain the herring, or so-called "cisco," as do some of those of Indiana and Wisconsin.

The food of the trout and adult white-fish having been already studied by the assistants of the United States Fish Commissioner,* I have given them no special attention.†

CoREGONUS ARTEDI, LeS. LAKE HERRING.

Gills long, deeply arched; gill chamber consequently capacious but narrow. Gill-rakers rather long and slender, allowing considerable separation of the gills. Only one row on the anterior arch, about thirty-eight in number, projecting almost directly forward, at least equal in length to the corresponding filaments of the gill. The anterior row on the second gill are as stout as those of the first, but only half as long; the second row represented by about ten triangular rudiments at the lower end of the arch. Anterior row gradually shorter on succeeding gills, posterior row longer; the secon drow on the fourth gill opposing a similar series on the pharyngeal arch. Each filament with a double row of fine teeth along the inner edge. No pharyngeal teeth ; pharynx with numerous fine longitudinal ridges which are covered with minute recurved spines. Intestine short and straight, anterior part provided with an immense number of small cœca. Alimentary canal a little shorter than the head and body without the tail.

My specimens of this species available for a study of their food were only five in number, obtained at South Chicago in

small fish ; and a third, five inches in length, had eaten a young centrarchid. The two others, respectively two and three fourths and four inches long, had filled themselves with larvæ of Agrion and small libellulid larvæ. One had taken, in addition, a minute larval Corixa and a small univalve mollusk.

* See "The Fishery Industries of the United States," pp. 490, 513.
† For a discussion of the first food of the common white-fish, see Bulletin Ill. St. Lab., Nat. Hist., Vol. I., No. 6, pp. 95–109.

October, 1881, and at Chicago in 1885. Numbers of others were examined, but without result, as they had been kept until the food was all digested.

These five specimens had taken only animal food, one of them only Entomostraca—ninety per cent. of these being the common Daphnia of the lakes (*D. hyalina*), and the remainder consisting of a few specimens of Bosmina, *Chydorus sphericus*, and Cyclops. The food of the remaining four was altogether insects of terrestrial origin. In one were recognized great quantities of winged ants (Myrmicidæ), another had eaten only Lepidoptera, and still another winged tipulids (craneflies). In the food of one, numerous specimens of the common squash beetle (*Diabrotica vittata*) were recognized, and a large quantity of undetermined Homoptera. An example of the homopterous insect *Diedrocephala mollipes* was detected in another.

Two small specimens of this species, hardly to be classed as young, respectively two and six inches long, had fed, like most of the adults examined, chiefly upon terrestrial insects, the shortest specimen upon small Diptera (ninety per cent.)and the homopterous insect Typhlocyba. The other example was taken from the stomach of a lake catfish (*Ictalurus lacustris*) from Lake Michigan. It had eaten a variety of terrestrial species, including an ant, several minute Homoptera, *Coriscus ferus*, a species of Amnestus, and examples of the families Staphylinidæ and Anthicidæ.

FAMILY DOROSOMATIDÆ.

Dorosoma cepedianum, LeS. Gizzard shad; Hickory Shad; Mud Shad; Thread Herring.

This remarkable fish occurs everywhere in the larger streams and in the ponds connected with them, but not in isolated lakes. It is marine in origin, swarming in the coast waters from Delaware to Mexico.

The mouth is toothless except in youth.* The gills are remarkably disposed within a rather small gill chamber. The

---

* For its juvenile characters and an account of the food of the young, see Bulletin Ill. St. Lab. Nat. Hist., Vol. I., No. 3, pp. 68-70.

dorsal portion of each gill projects far forward in the palatal region, and then turns abruptly backward, forming an acute angle in the roof of the mouth. This course of the arches is necessitated by the large accessory organ upon the fourth branchial arch.* The arches are all provided with numerous short rakers projecting horizontally upon either side, and forming an unusually effective straining apparatus. The intestine is very long and slender and much convoluted, the œsophagus small and long, and the stomach very short and muscular, like the gizzard of a granivorous bird. The small intestine is beset with a multitude of slender cœca, and its mucous surface is everywhere remarkably villose.

The species was represented in our collections by many specimens, but the food was so uniform in character that a prolonged study of it seemed unnecessary, especially as the critical analysis of such large quantities of material, minutely divided and thoroughly intermingled, was a very tedious and time-consuming process.

The adult specimens examined were eleven in number; ten from the Illinois river between Havana and Ottawa, and one from the Pecatonica, in northern Illinois. Eight dates and five localities are represented by them, the former ranging from April to October.

The species has, in general, the habit of swallowing quantities of fine mud, containing, on an average, about twenty per cent. of vegetable *débris*. Occasionally, in the vicinity of distilleries, it feeds, like the buffalo fish, on distillery slops, and sometimes a greater percentage of vegetation occurs mingled with the mud. Traces of animal food were common; but the ratio in most of my specimens was insignificant, averaging only four per cent. of the whole; although in one shad taken in spring in northern Illinois one fourth of the food consisted of Entomostraca (Cypris). Univalve mollusks occurred in one, fragments of Coleoptera in another, and young Corixa in still another; and spiders and water mites were also noted. Five specimens, in all, had taken Entomostraca — four of them

---

* This accessory organ is correlated by Sagemehl with the limophagous habit of the fishes in which it occurs.—*Morphologisches Jahrbuch*, XII., p. 318.

Cypris, one Cyclops, and two Alona. The vegetable food of the group amounted to thirty-two per cent., eaten by all the specimens. Beside the distillery slops already mentioned, Lemna, Wolffia, various diatoms and other unicellular plants, and occasionally filamentous Algæ, were noted in the food. It is probable that in some situations and at some seasons of the year, Entomostraca would be found a more important element; otherwise one can hardly see the advantage of the excellent branchial strainer borne by this species. The great length of the intestine and the unusual development of the mucous surface are seemingly correlated here, as among the cyprinoids, with the limophagous habit.

In five specimens, two and a half inches in length, the food was intermediate in character between that of the adult and that of the young, about sixty per cent. of it being Algæ, mixed with an abundance of dirt, and the remainder Cladocera (twenty-two per cent.) and insect larvæ — about half of them Chironomus.

A single specimen, five and a fourth inches long, had fed principally on Entomostraca (Bosmina, Daphnia, and Cyclops), with a very few Chironomus larvæ.

FAMILY CLUPEIDÆ.

Only a single species of the herring family occurs in this State — the golden shad, *Clupea chrysochloris*, Raf.-- and this not by any means commonly with us. It seems to be strictly predaceous, the three specimens taken by me at Pekin and Peoria in September and October of three different years having eaten only fishes — two of them the gizzard shad (Dorosoma) and the third some undetermined kind. A single small specimen, two and a fourth inches long, had fed wholly upon terrestrial insects, among which were noticed *Triphleps insidiosus*, a species of Typhlocyba, a chalcid (Eurytoma), small Diptera (including Culicidæ and Muscidæ), and some small spiders.

FAMILY HYODONTIDÆ.

HYODON TERGISUS, LeS.   MOON EYE; TOOTHED HERRING.

This species, not common in our collections, is represented in these studies by only five specimens, obtained from the Illinois River at Peoria and Havana, on four dates in August and October of two different years (1878 and 1887). Their food consisted wholly of insects (two thirds of them terrestrial) with the exception of a trace of univalve Mollusca. A single one, two and seven eighths inches long, had derived its food about equally from terrestrial and aquatic insects, including Orthoptera, Chironomus larvæ, and *Corixa tumida.*

FAMILY CATOSTOMATIDÆ.

One of the most striking characteristics of the fish fauna of Illinois, and indeed of the Mississippi Valley, is the prominence of the sucker family, which includes within our limits six genera and fifteen recognized species. Several of these are among the most abundant of our larger fishes, and most are very generally distributed.

With reference to the essential characteristics of their food, I find them dividing into three tolerably distinct groups. The first includes the cylindrical suckers (Moxostoma, Catostomus, and the like), in which the pharyngeal bones are heavy, the lower teeth thick and strong, usually with a well-developed grinding surface, and the gill-rakers short, thick, and few. In the second are the deep-bodied suckers, in which the pharyngeal jaws and teeth are well developed, although not as strong as in the cylindrical group, while the gill-rakers are of moderate length and number. The third contains the still deeper-bodied and thinner species, with light pharyngeal jaws and teeth, and long, slender, and more numerous gill-rakers. To this group belong the species commonly placed in the genus Carpiodes. Or, if we arrange the genera in a series, with reference to their food structures, we shall find Placopharynx at one extreme and Carpiodes at the other, the change consisting in a gradual increase in number, length, and effectiveness of the gill-rakers, correlated with an increase in length of the pharyngeal bones and in the number of their teeth, and a converse diminution in the size and strength of these structures. The intestine also becomes longer and smaller as one passes from the cylindrical suckers to the deep-bodied buffalo and carp.

The data concerning the food of this family here presented are drawn from a study of the alimentary contents of one hundred and nine specimens, collected chiefly from the Illinois and Mississippi Rivers and their immediate tributaries. They indicate, in general, that about one tenth of the food consists of vegetation, taken chiefly by the buffalo fishes (Ictiobus), and in them largely composed of distillery slops. Mollusks and insects appear in nearly equal ratio in the food of the family at large, the former taken much the more generally by the cylindrical suckers, with heavy pharyngeal jaws and solid teeth, and the latter about equally by all, with the single exception of the stone roller (Hypentelium), whose peculiar haunts and feeding habits explain its departure from the average. On the other hand, the ten per cent. of Entomostraca were eaten chiefly by the deeper-bodied species.

PLACOPHARYNX CARINATUS, Cope.

This species has the general appearance of one of the red horse (Moxostoma), and has possibly been commonly overlooked in our collections, as we have noted it very rarely.

Its branchial apparatus is not noticeably different from that of the following genus, the gill-rakers being short and few, and effective only on the upper part of the arch, the lower arm being, like that of Moxostoma, covered by a ridged pad.

The fish is very remarkably distinguished, however, by the heavy pharyngeal jaws and the thick and strong pharyngeal teeth with conspicuous grinding surface. The latter number about thirty on each pharyngeal, the upper ones minute and useless rudiments, the lower ten very large, occupying about two thirds the length of the arch,—the lower six, in fact, one half of it. It is altogether likely that this apparatus is related to a preference for molluscan food, but the number of specimens available for my examination was too small to verify this supposition.

Two large examples taken from the Illinois at Havana in October, 1887, were found to have eaten similar food. In one, sixty per cent. consisted of small univalve Mollusca (*Valvata carinata* and Amnicola), the remainder being almost wholly insects — chiefly larvæ of water beetles (Hydrophilidæ) and larval Ephemeridæ (largely Cænis). About five per cent. of Lemna occurred in this fish, — probably taken by accident, as the river was covered with a film of duckweed at the time. A few Chironomus larvæ and an Allorchestes were also noted. In the other specimen only five per cent. of the food consisted of mollusks (the same species as before, together with a small. Sphærium). Larval Hydrophilidæ made eighty per cent. of the contents of the intestine, and Ephemeridæ (Cænis) more than ten per cent. Chironomus and other dipterous larvæ, Plumatella, and a little Wolffia, were likewise recorded.

In a third example, only five and a half inches long, the locality of which is not known, the food was chiefly Plumatella, the only other elements being small case-flies (Phryganeidæ), a minute univalve shell (Strepomatidæ), and a few small Chironomus larvæ.

MOXOSTOMA MACROLEPIDOTUM, LeS.   COMMON RED HORSE ;
    WHITE SUCKER.

The genus Moxostoma, the commonest and most typical of the cylindrical suckers, is represented in Illinois by three species, two of which, *aureolum* and *macrolepidotum*, occur everywhere in lakes, rivers, and smaller streams. We have encountered *M. carpio* but rarely, and my studies relate only to the two former species.

In *macrolepidotum* the gill-rakers of the anterior row are twenty-five to twenty-seven in number, the upper twenty to twenty-two being elongate, triangular, stout, and crenate within, about three fourths as long as the filaments of the gill; while the lower five or six of this series, all of the second series of the anterior arch, and all of the other rakers of the gills, including the row upon the pharyngeals, have the form of transverse leaf-like plates with crenate edges, projecting in triangular outline a little beyond the margin of the thick gill arch. The gills seem but slightly separable, and the branchial apparatus is coarse and ineffective.

Pharyngeals moderately heavy, the teeth about forty-five on each side, the lower ten thickened and broadened, with smooth terminal edges, but alternately higher and lower in the specimen examined. The other teeth are hooked at the anterior angle, and irregularly crenate on the cutting edge. The intestine is small, one and a fourth times the length of the head and body.

The salient features of the food of *Moxostoma macrolepidotum*, as exhibited by twelve specimens examined, are the abundance of univalve Mollusca and the bivalve Sphærium, the insignificance of the vegetable element, and the absence of Crustacea and the larger and more active insect larvæ. The insect food consisted almost wholly of larvæ of Chironomus and other small mud-inhabiting species.

The molluscan food, taken by eleven of the twelve specimens, amounted to more than half the total, the principal forms represented being Vivipara and Melantho (twenty-two per cent.), Somatogyrus and Amnicola (six per cent.), and the following pulmonates,— Limnea, Physa, and Planorbis. Three of the specimens had eaten Sphærium, but the Unionidæ were only doubtfully represented. The insects — about one third the food — were practically all aquatic, and nearly all dipterous larvæ. Two specimens, however, had taken a small quantity of hydrophilid larvæ, one an Agrion larva, and two others larvæ of Ephemeridæ. The Entomostraca recognized belonged to Alona and Cyclops. The vegetable food consisted of distillery slops, eaten by one of the specimens, with a little Wolffia, Chara, filamentous Algæ, and some miscellaneous matter.

This group of specimens was taken from the Illinois River at Henry, Peoria, Pekin, and Havana, and from Crystal Lake in northern Illinois, at dates ranging from May to November of four different years.

Five additional examples of this genus, the species of which was not determined but which almost certainly belonged to *macrolepidotum*, had eaten a still larger ratio of Mollusca than the preceding group, these making now three fourths of their food, — the greater part Sphærium. Melantho, and Amnicola also occurred, the former making one fourth of the food of the five.

Moxostoma aureolum, LeS. Red Horse.

This species, less abundant in central Illinois than the preceding, takes almost identical food, so far as one may judge from the six specimens examined by me from Pekin, Peoria, and Crystal Lake in northern Illinois. The food was practically all animal, about one half of it Mollusca — largely Vivipara and Sphærium. The insects were, as before, mostly Chironomus larvæ and pupæ, the only other form worthy of note being smooth, slender, distinctly segmented, footless larvæ with elongate brown heads — very common in the food of fishes, but not yet identified.*

Minytrema melanops, Raf. Striped Sucker ; Spotted Mullet.

In this species, not uncommon throughout the State in suitable streams, the alimentary structures are not essentially different from those of Moxostoma, the pharyngeal teeth being, however, more numerous and more closely set,— about fifty-five in the series, the lower five to ten enlarged, but less so than in Moxostoma, and with the grinding surface less distinctly defined, most of even these largest teeth still presenting a somewhat crenate margin.

So far as indicated by the four specimens examined, the food of this species is similar to that of the preceding, being nearly all Mollusca,— differing, however, in the fact that the thin-shelled bivalve Sphærium had been taken in preference to the thick-shelled univalves. A Cyclops and a larger percentage of Cypris represented the Entomostraca. The small ratio of insects noticed were all Chironomus larvæ.

Catostomus teres, Mitch. Common Sucker; White Sucker ; Brook Sucker ; Fine-scaled Sucker.

Abundant northward, occurring rarely in the Illinois as far south as Peoria, and still more rarely in the extreme southern part of the State. Wherever abundant, it inhabits nearly all waters, both lakes and flowing streams. It is common in Lake Michigan.

* This larva has the superficial characters of the Mycetophilidæ, and was doubtfully assigned to that group by Dr. Williston, in a recent letter to me.

Pharyngeal jaws strong, thick, nearly twice as wide as high ; teeth about thirty-five in number, the lower four or five much thickened, occupying about one fourth the length of the jaw. The crown is expanded transversely to the axis of the jaw, rounded, not crenate or hooked. The crowns of the teeth above the sixth or seventh are hooked and slightly crenate, but less so than in Moxostoma. Compared with that genus, both teeth and jaws constitute a more effective crushing and grinding apparatus.

The system of gill-rakers is similar to that of Moxostoma, but is less effective as a strainer, the anterior row of the first gill being less numerous, shorter, and thicker. These divide into two sets of about equal length, the upper series projecting forward, rather short, triangular, about one third the length of the corresponding filaments, fifteen or sixteen in number, the lower series, five or six, in the form of low lamellar ridges. Rakers of the other gills thick, lamellar, with tubercles on the free edges ; corresponding lamellæ on anterior margin of the pharyngeal jaw.

Alimentary canal about two and a half times the length of the head and body. The alimentary structures in general indicate better adaptation to molluscan food than those of the stone roller, and inferior adaptation to Entomostraca.

The number of specimens examined was too small to make it worth while to report their food, especially as they were evidently under size. The branchial and pharyngeal structures and known habits of the species indicate that its food is not especially different from that of Moxostoma, just discussed, and it will probably be found to consist chiefly of Mollusca and insect larvæ, the former in larger ratio than in Moxostoma, and in smaller ratio than in the species next to follow.

HYPENTELIUM NIGRICANS, LeS. STONE ROLLER; HAMMER-
HEAD.

This curious fish, distinguished both in form and habit from its allies of the family, occurs usually in rapid shallows of clear streams, commonest to the northward. It is taken rarely in lakes.

The square, strong head of this species is related to its mode of life, but the cylindrical body, the large rounded pectoral fin, and relatively high coloration, give the fish the aspect of a darter among the suckers; and its habit of searching for its food among the stones in swift and shallow waters is another point of affinity with that interesting group. Curiously different as are the food and feeding habits of this species when compared with its nearest ally, *Catostomus teres*, the alimentary structures are not remarkably unlike. The pharyngeals are somewhat lighter, the pharyngeal teeth more slender and more prominently cuspidate, and the gill-rakers somewhat stouter, possibly affording a better apparatus for the separation of the relatively large insect larvæ upon which this species chiefly feeds. Its alimentary structures are extremely different, however, from those of the Etheostomatidæ, whose food, haunts, and habits it copies so closely. It is, in short, a molluscan feeder, which has become especially adapted to the search for insect larvæ occuring in rapid water under stones.

The pharyngeals bear about forty teeth on each side, which are unusually high, thin, and acute, all the upper ones with an uncommonly prominent hook or cusp at the internal angle. The six lower teeth are cultrate, without hook or distinct grinding surface, but only two or three are noticeably thickened.

The anterior gill-rakers are short and stout, twenty-five in number, six of them on the horizontal part of the arch. Those of the upper series are thin plates with the base about half the length, and are one third to one half as long as the corresponding filaments. The lower rakers of the series, more prominent than those of *C. teres*, are much like the upper, but shorter, the height scarcely equal to the base. There are five or six tubercles on the upper edge of each. The remaining gill-rakers, similar to those just mentioned, interlock by their tips, which are much more prominent and more tuberculate than those of Moxostoma. The stouter filaments of the strainer are probably related to the larger and more active insect larvæ on which this species feeds.

The intestine is small, considerably convoluted, and about twice the length of the head and body.

The food of six specimens taken in the Fox River and Mackinaw Creek contained no vegetation and but a small ratio of mollusks (Sphærium), but was nearly all aquatic insect larvæ (ninety-two per cent.). The great majority of these were Ephemeridæ, more than half the food consisting of a single form, abundant under stones, belonging to the genus Cænis. A few Chironomus larvæ, taken by all the specimens, some larvæ of Coleoptera, and traces of terrestrial insects were the only other elements.

ERIMYZON SUCETTA, Lac.   CREEK FISH; CHUB SUCKER.

Everywhere abundant in streams and lakes, ascending creeks in spring. Occurs in our collections from McHenry to Union county. Rarely taken by us, however, and not repre-sented in the material used for these studies.

Pharyngeal jaws moderately heavy, short for the size of the fish, bearing about sixty teeth, the lower ten filling the lower third of the arch, these moderately enlarged, with incon-spicuous grinding surface, the terminal edges being irreg-ularly rounded. The remaining teeth are hooked, the upper ones of the series crenate on the cutting edge.

Anterior gill-rakers thirty-four in number, upper twenty-one short and thick, about one third the length of the gill filaments; tips of the lower members of the series laterally flat-tened to a paddle shape. About eight of the lower gill-rakers of the anterior series fuse to form a thick ridged pad. Rakers of the remaining arches similar to those of Moxostoma, but more prominent, the tips of the transverse plates projecting further beyond the surface of the arch.

This species presents an ovoid thickening of the palatal region upon either side, which fills the greater part of the branchial chamber, but is less conspicuous than in Ictiobus.

Two young specimens, one and three fourths and three inches respectively, differed but little, in food, from those men-tioned on page seventy-two of Bulletin 3 (Vol. I.) of the Illi-nois State Laboratory of Natural History. The larger one had eaten chiefly the smallest of our Entomostraca (Canthocamptus), with a trace of Chironomus larvæ. The smaller had taken a moderate ratio of Entomostraca (Cypris, Cyclops, and undetermined Lynceidæ), a much larger proportion of Protozoa (especially Difflugia and Arcella), a few Squamella and other rotifers, and unicellular Algæ, including Protococcus, Chroöcoccus, Closterium, and Cosmarium.

GENUS ICTIOBUS.   BUFFALO AND RIVER CARP.

In this genus are included only the deeper-bodied suckers with light pharyngeal jaws and relatively long gill-rakers. The species differ, however, in these particulars, and may be arranged in a series exhibiting a progressive lengthening of the gill structures, a lightening of the pharyngeal jaws, and an increase in number and a decrease in size of the pharyngeal teeth. Related to these differences of structure are the inferior importance of mollusks in the food (especially of the thick-shelled univalves), the greater number of insects, the appearance of Entomostraca as an important element, and the considerable percentage of vegetation taken. The insects eaten are well distributed instead of being essentially limited, as in Moxostoma, to dipterous larvæ.   In short, correlatively with the greater number and smaller size of the pharyngeal teeth, the weaker jaws, and the greater development of the straining apparatus, in Ictiobus we find the food generalized, and drawn from numerous sources; while in Moxostoma the food and the food prehensile structures are specialized in the direction of a rather close dependence on the smaller mollusks.

The feeding habits of these fishes, like those of all species inhabiting the muddy waters of central Illinois, are very difficult of determination, but several fishermen, and others with unusual opportunities for observation, have reported to me that one or more species of this genus have the peculiar habit of whirling around in shallow water or plowing steadily along, with their heads buried in the mud, and their tails occasionally showing above the surface. These operations have nothing to do with spawning, and it is likely that fishes thus engaged are burrowing for small mollusks and for mud-inhabiting larvæ.

ICTIOBUS BUBALUS, Raf.   QUILL-BACK ; SMALL-MOUTHED BUFFALO.

This is a very abundant fish in the larger streams and in the lakes and river bottoms, being one of the three species most

commonly shipped from the Illinois and Mississippi under the
name of buffalo fish. They all sell as "coarse fish," but from
their abundance and their fair character as food, are, on the
whole, the most important commercial fishes in our streams.

The gills of this species are very compactly disposed in a
rather small branchial chamber, the upper ends of the arches
being decurved and the lower elevated so that each gill forms
about three fourths of a circle. Ten of the lower rakers of the
anterior series are reduced to thickened ridges which extend
obliquely across the horizontal portion of the arch. The remainder of this series, thirty-five in number, are flattened,
minutely toothed, the central ones about as long as the corresponding filaments of the gill, the others regularly shortened
above and below. The other rakers are similar to those of
Moxostoma, having the form of toothed triangular plates, with
their apices slightly projecting beyond the opposed surfaces of
the arches. The interlocking tips are a little more prominent
than in Moxostoma, and the whole apparatus is somewhat better developed.

The pharyngeal bones are moderately heavy, triangular in
section, about as thick as high; and the teeth, about one hundred and thirty upon each jaw, project directly backwards and
act, as in Moxostoma, against a semi-circular rim of cartilage.
They are compressed, and more or less crenate on the cutting
margin, the upper ones minute, the others gradually thickening downwards so that the lower twelve occupy about one
fourth of the length of the arch. The edges of these lower
teeth are rounded, not acute.

Seventeen specimens of this species, distributed in seven
lots, collected from the central course of the Illinois River and
from the Mississippi at Quincy in the years 1880, 1882, and
1887, and in various months from April to October, give the
following general view of the food.

In decided contrast to the preceding members of the family,
about one fifth of the food consisted of vegetation — taken by
sixteen of the fishes—nearly all aquatic, but with an occasional
admixture of terrestrial rubbish. The principal vegetable element was a small duckweed (Wolffia) especially abundant in
fishes taken from the Illinois during the autumn of 1887,

2

when it made in some cases as high as ninety-five per cent.
The larger duckweed (Lemna), fragments of Ceratophyllum,
diatoms, and other unicellular Algæ, are also worthy of mention.

The animal food (eighty per cent.) was fairly well divided
between Mollusca, insects, and Crustacea, respectively thirty,
twenty-nine, and twenty per cent. Only occasional traces of
univalves were noticed (Vivipara and Planorbis); but the thin-
shelled bivalve Sphærium was a very important element, taken
by seven of the fishes, and reckoned at thirty per cent. of the
food of the group. Several individuals had eaten nothing else.

Insect larvæ were very generally taken, and, in fact, occurred in the food of every specimen examined. Chironomus
larvæ were reckoned at nearly a fifth of the food, and were
found in fourteen out of the seventeen fishes. Neuroptera
larvæ, on the other hand, occurred in relatively insignificant
number, most of them Ephemeridæ; although a small number of case-worms (Leptocerus) and of dragon-fly larvæ
(Agrion) were also noticed. Hydrachnida occurred in the food
of one, and Crustacea were eaten by thirteen specimens,—all
Entomostraca with the exception of a single small crayfish
and an amphipod.

Curiously, the entomostracan eaten most freely by these
large fishes was the smallest of the Copepoda — Canthocamptus.
In the food of ten specimens taken at Peoria April 16, 1880,
and October 6, 1887, this made nineteen per cent. of the food
of the entire group. Specimens of Cyclops, Cypris, Pleuroxus,
Iliocryptus, Bosmina, and Simocephalus occurred in numbers
too small to figure in the ratios. Fresh-water Vermes were
almost wholly wanting, only a few Anguillulidæ occurring in
the food of one. Eight had eaten Polyzoa, including both
Plumatella and Pectinatella. The latter was recognized by
its statoblasts only, detected in seven specimens collected in
October, 1887, in situations where the gigantic colonies formed
by this polyzoan had been earlier very abundant. It is probable, consequently, that these statoblasts, widely dispersed with
the death and decay of the translucent mass in which they are
developed, had been picked up by accident with the other food.*

---

* Some notes on the young of this genus, published in the Bulletin
of this Laboratory, Vol. I, No. 3, page 73, show that specimens varying

ICTIOBUS CYPRINELLUS, C. & V.   RED-MOUTH BUFFALO.

The statements made concerning the abundance, distribution, and commercial value of the preceding species will apply equally well to this. The fishermen report, however, that the quill-back frequents deeper water than the red-mouth. The structures of food prehension differ from those of *bubalus* in the lighter pharyngeal jaws, the greater number and smaller size of the teeth, and the more efficient branchial apparatus. The pharyngeal jaws are relatively thin, the thickness being about one fourth the height. The teeth are about seventy-five in number on each jaw, minute above, gradually but not greatly thickened below, the ten lowest occupying nearly one fifth the length of the jaw. These largest teeth have the cutting edges obtuse, and are slightly hooked within. The remaining teeth are more or less crenate on the cutting edge, each with conspicuous hook or cusp at the inner angle. The posterior edges are also acute.

The gill-rakers are similar to those of the quill-back, but more efficient as a straining apparatus. The longer rakers of the anterior row (seventy-five in number) are fully equal in length to the corresponding filaments, and are armed within with a double row of clusters of minute teeth. Eight or ten of the lower rakers are fused in the form of thick oblique ridges, The tips of the rakers of the other rows project beyond the borders of the arches a distance about equal to the line of attachment to the arch. The pharyngeal enlargements are very conspicuous and thick, nearly filling the pharyngeal cavity.

---

in length from seven eighths of an inch to two inches, fed largely upon unicellular Algæ and rotifers, the remainder of their food being chiefly the smallest Entomostraca. I add here the details from two additional specimens, taken in June, from the Illinois River, at Pekin, one three fourths of an inch in length and the other eight tenths. The greater part of the food of these consisted of rotifers, Protozoa, and gelatinous and other unicellular Algæ, a single Bosmina in each being the only entomostracan form determined. The rotifers included Brachionus and Anurea ; and among the Protozoa were Actinosphærium, *Arcella vulgaris*, and *A. discoidea.* Closterium was noticed among the Algæ, with numerous gelatinous Algæ related to Protococcus, and a filament of Oscillatoria. Spores of fungi were found in both, and a fragment of vegetation penetrated by a fungus mycelium occurred in one.

This species seems to differ in food from the preceding, especially in the inferior amount of mollusks and the larger ratio of vegetation. The animal food of seventeen specimens collected in seven lots from the Illinois and Mississippi Rivers and the northern lakes in various months from April to October of five different years, was about two thirds the whole, the remaining third consisting largely of Algæ, unicellular and filamentous, and otherwise chiefly of distillery slops (taken by Illinois River specimens) and miscellaneous vegetation of terrestrial origin. This last was occasionally found in quantities sufficient to show that it had been intentionally swallowed, making in one instance the greater part of the food. The molluscan food of these specimens amounted to only three per cent., nearly all Sphærium; the insect food to thirty-three per cent., practically all aquatic, and very largely larvæ of Chironomus (twenty per cent.). The Neuroptera were chiefly Hexagenia larvæ (nine per cent.). Except a single Crangonyx, the Crustacea were all Entomostraca. These occurred in much greater variety than in *cyprinellus*, among them being representatives of Daphnella, Simocephalus, Bosmina, Chydorus, Pleuroxus, Alona, Cypris, Cyclops, and Canthocamptus. Fragments of Plumatella were noticed in a single specimen, Difflugia in two.

ICTIOBUS URUS, Ag.   BLACK BUFFALO ; MONGREL BUFFALO; BIG-MOUTHED BUFFALO; CHUCKLE-HEAD.

This species occurs commonly with the preceding, but less abundantly. Said by fishermen to frequent shallower water.

With respect to food, it closely resembles *cyprinellus*, our seventeen specimens, well distributed as to date and place, having taken almost identical ratios of animal and vegetable food — sixty-seven per cent. and thirty-three per cent. respectively. Twelve per cent. were mollusks,— nearly all Sphærium, as before. The large ratio of insect food (about forty-two per cent.) was more than half Chironomus larvæ, most of the remainder being Hexagenia larvæ, taken, however, by only one of the specimens. The Crustacea (thirteen per cent.) were practically all Entomostraca, fragments of a young crayfish appearing in only a single specimen.

Among the vegetable elements, distillery slops (eaten by three of the specimens) were the most important (twenty-one per cent.).   The rather insignificant amount of aquatic vegeta-tion (six per cent.) was distributed as usual among a number of the lower plants, chiefly duckweeds and the unicellular Algæ.

ICTIOBUS CYPRINUS, LeS.   RIVER CARP; CARP SUCKER.

Under this specific head I include, for the purposes of this paper, all the so-called species of river carp sometimes separated under the genus Carpiodes, and hitherto described under some eight specific names. This form is abundant in the great rivers of the State and in their larger tributaries, and also in Lake Michigan and the smaller lakes of northern Illinois. It is extremely common in the lakes and ponds of the river bottoms, but occurs in running water in smaller numbers than the other species of its genus.

In its structures of food prehension it exhibits an extreme development and a correlative degradation of branchial appa-ratus and pharyngeal structures respectively. The gills are re-markably compacted, the upper and lower ends nearly meeting when the mouth is closed. The pharyngeal protuberances are enormous, almost filling the branchial cavity. Anterior gill-rakers in two series, as usual, the upper about sixty-seven in number on three fourths of the arch, the longest a little longer than the corresponding filaments. The lower part of the gill with about ten thick, papillar, coherent ridges extending down-ward a distance equal to the length of the filaments of the same vicinity. The longer rakers have each two closely alter-nating rows of tubercles on the inner edge, roughened with extremely minute denticles. Inner surface of the arch with transverse tuberculate ridges springing from the bases of the rakers of the gill, and terminating inwardly in slight projec-tions representing the posterior row of rakers. The other arches are similarly tuberculate and ridged, and the whole ap-paratus closely embraces the pharyngeal thickenings. Pharyn-geal bones very thin and brittle, less than a millimeter thick in a fish ten inches long, the thickness one seventh the height to the base of the teeth. The latter about two hundred, minute above, gradually increasing downwards, but not much thickened or elongate, about thirty on the lower fourth of the arch. Crowns emarginate or doubly emarginate, with the inner angle similarly produced, forming a hook or cusp. Intestine very slender, four times as long as head and body in the specimen examined.

Nineteen examples of the species, representing thirteen dates and localities, from April to October, and from 1877 to 1887, collected from Crystal Lake in northern Illinois, from the lakes of the Ohio near Cairo, and from the Illinois River at Ottawa, Peoria, and Havana, show that the native carp differs from the other species of Ictiobus chiefly in the inferior amount of vegetation eaten, in the greater quantity of mud mingled with the food, in the absence of the larger insect larvæ, and in the lack of univalve Mollusca. It resembles closely *Ictiobus cyprinellus*, but from this differs also with respect to the vegetation taken, and in its filthy feeding habits. The vegetable food was only eight per cent., mostly Wolffia, and that eaten by only two of the specimens. A few diatoms were mingled with the mud in three, and miscellaneous aquatic vegetation occurred in five. Mollusks made about a fourth of the food,— all the thin-shelled Sphærium. Insects averaged about one third, the greater part Chironomus larvæ. Neuroptera were eaten by only four of the specimens, and contributed only two per cent. to the food, case-worms (Phryganeidæ) being the only forms identified. Entomostraca made nearly a fourth, distributed through a considerable list, which included *Simocephalus americanus*, Bosmina, Chydorus, Alona, Cypris, Cyclops, and Canthocamptus. No Vermes or Polyzoa were observed, but occasional Protozoa were noticed, especially Centropyxis and Difflugia.

Looking now at the food of the family, as exhibited by the one hundred and seven specimens discussed, representing, as they do, five genera and eleven species, we conclude that the sucker family is essentially carnivorous, the vegetable food amounting to only eight per cent. of the whole, and no element of this being especially prominent. The smaller mollusks are the most important single class, the ratio of these being forty-one per cent., about three fourths of them Sphærium. The large quantity of aquatic insects (one third of them Chironomus and a fourth ephemerid larvæ), the relative insignificance of Crustacea (about ten per cent.,— nearly all Entomostraca), and the practical absence of Vermes and Protozoa are the remaining salient features of the food characters of this family.

FAMILY SILURIDÆ.

The family of catfishes taken together is nearly omnivorous in habit, and their alimentary structures have a correspondingly generalized character. The capacious mouth, wide œsophagus, and short broad stomach, admit objects of relatively large size and of nearly every shape; the jaws, each armed with a broad pad of fine sharp teeth, are well calculated to grasp and hold soft bodies as well as hard; the gill-rakers are of average number and development; and the pharyngeal jaws — broad, stout arches below and oval pads above, with thin opposed surfaces covered with minute, pointed denticles — serve fairly well to crush the crusts of insects and the shells of the smaller mollusks and to squeeze and grind the vegetable objects which appear in the food. The use made of the jaws in tearing mollusks from their shells, as described further on, is probably the most peculiar feeding practice of these animals; and the indifference of several of the species to the past history or the present condition of their food, distinguishes them as the only habitual scavengers among our common fishes.

The family is a very abundant and characteristic one in this region. It ranges in size from the smaller species of Noturus, only an inch or two in length, to monsters more than two hundred pounds in weight; and inhabits every kind of water from the greatest rivers of the continent to small temporary ponds of surface water, where its presence is the standing wonder of the fisherman and the naturalist.

In Illinois we have three genera and twelve species of these fishes, as at present classified, none of them unfit for food except the smallest ones, and two or three of them the equals of any river fish.

My studies of their food were based upon one hundred and twenty specimens, belonging mostly to five species of Ictalurus and Noturus. The data are especially deficient with respect to the food of the largest lake and river species.

ICTALURUS FURCATUS, C. & V.   CHANNEL CAT; FORK-TAILED CAT; WHITE FULTON.

This is the catfish *par excellence*, and is the best food fish of its family. It occurs only in the deeper water of the larger streams. It is common in the Mississippi and the Ohio, although much less so than the following species, but is taken rather rarely in the Illinois, where it is often called the "Mississippi cat." It is never found in lakes and ponds, and feeds, according to the reports of fishermen, almost exclusively upon other fishes. A single specimen taken at Quincy Oct. 25, 1887, had eaten fishes only.

The gill apparatus is better developed than in Amiurus, but is nevertheless very incomplete. The anterior arch has only one row of rakers, eleven in number below the angle, four or five above. These are longest near the upper end of the lower part of the gill, where they are about half the length of the corresponding filaments. The other gills have similar but shorter rakers, the third and fourth a double row of about equal length. None of the rakers are toothed or tuberculate. The pharyngeals, both superior and inferior, are similar to those of Amiurus, but relatively smaller.

ICTALURUS PUNCTATUS, Raf.   BLUE FULTON; SPOTTED CAT; FIDDLER; SWITCH TAIL.

An abundant species in the larger rivers, much commoner than the preceding, but not quite so good for food, smaller, ranging more freely, and clearly a more general feeder, although its alimentary structures are not noticeably different.

The gill-rakers of the anterior arch are a trifle shorter, the longer ones being about one third the length of the corresponding filaments, and the pharyngeal structures seemingly a little heavier.

Forty-three specimens of this species were taken from the Illinois River at Peoria, Pekin, and Havana, and from the Mississippi River, near Quincy. Their dates of capture represent the spring, summer, and autumn months of the years 1878, 1880, and 1887.

About a fourth of the food consisted of vegetable matter, much of it miscellaneous and accidental, but chiefly Algæ — Cladophora being the most abundant form. This and other filamentous Algæ made a large part of the food of several fishes taken in October, 1878 and 1887, three having eaten nothing else. Fragments of Potamogeton were taken by other October specimens, making twenty per cent. of the food of three. The fact that the floating Lemna occurred but rarely, and then in the smallest quantity, is evidence that these cat-fishes are strictly bottom feeders. A single specimen had fed on still-house slops, as shown by the considerable amount of meal in its alimentary contents.

A dead rat, pieces of ham, and other animal *débris* attest the easy-going appetite of this thrifty species.

Fragments of fishes were found in eleven examples of this group,— commonly, however, in pieces so large as to make it certain that they were derived from those already dead. Occasionally, as in examples taken in August, 1887, from the Mississippi River, fishes probably taken alive composed the whole of the food. The species were not identifiable.

Molluscan food was a decidedly important element, being found in fifteen of the fishes and amounting to fifteen per cent. of the whole. Several specimens had taken little or nothing else, —notably six secured at Havana in September, 1887, and one at Peoria in October of the same year. The Mollusca were about equally divided between gasteropods and lamelli-branchs, the former largely Melantho and Vivipara, the latter usually Unio or Anodonta.

Notwithstanding the number of bivalves eaten by these fishes, no fragment of a shell was ever found in their stomachs, but the bodies of the animals had invariably been torn from the shell while yet living — as shown both by the fresh condition of the recently ingested specimens and likewise by the fact that the adductor muscles were scarcely ever present in the frag-

ments. Indeed in only a single instance had the posterior adductor been torn loose. The Unionidæ were usually large and
thin — probably in most cases Anodonta.

I have been repeatedly assured by fishermen that the catfish seizes the foot of the mollusk while the latter is extended
from the shell, and tears the animal loose by vigorously jerking
and rubbing it about. One intelligent fisherman informed me
that he was often first notified of the presence of catfishes in
his seine, in making a haul, by seeing the fragments of clams
floating on the surface, disgorged by the struggling captives.

Still more interesting and curious was the fact that the
univalve Mollusca found in the stomachs of these fishes were
almost invariably naked, the more or less mutilated bodies
having only the opercles attached. How these fishes manage
to separate mollusks like Melantho and Vivipara from the shell,
I am scarcely able to imagine, unless they have the power
to crack the shells in their jaws as a boy would nuts, and then
to pick out the body afterward. Certainly the shells are not
swallowed, either whole or broken.

The number of mollusks sometimes taken by a single catfish is surprising. As high as one hundred and twenty bodies
and opercles of Melantho and Vivipara were counted in a spotted catfish taken at Havana in September of last year.

Insects were, however, the principal food of the specimens
studied, making forty-four per cent. of all, eaten by twenty-
eight of the specimens; five, in fact, had eaten nothing else,
and nine others had taken ninety per cent. or more of insects.
These were mostly aquatic, although now and then a fish had
filled itself with terrestrial specimens. About half the insects were Neuroptera, nearly equally dragon-fly larvæ and
larvæ of Ephemeridæ; but Hexagenia larvæ were rarely recognized. Chironomus larvæ made thirteen per cent. of the food,
and were so frequently taken with the sand tubes they inhabit
as to make it certain that they were commonly obtained from
the bottom. Leeches appeared in the food of three of the specimens, and Gordius in one. Fragments of Plumatella were noticed in two, and a fresh water sponge likewise in two.

Four immature examples of this species, ranging from two
and a half to four inches in length, had fed almost wholly

upon insects, a few specimens of *Allorchestes dentata* and Daphnia being the only other items. Eggs and young of Hexagenia and other ephemerids composed the greater part of the food, Chironomus larvæ amounting to about one half as much.*

ICTALURUS NATALIS, LeS. YELLOW CAT.

This species occurs everywhere throughout Illinois, but less abundantly than *nebulosus*, and usually in larger streams. It has not been taken by us from ponds and lakes except where these were immediately connected with rivers subject to overflow.

The alimentary structures of this species closely resemble those of *I. nebulosus*, described under the next head. Twelve specimens were collected from the Illinois River at Peoria, the Fox River at McHenry, and from one of the smaller lakes in northern Illinois, in the months of May, August, October, and November of 1878, 1880, and 1887.

The food was wholly animal with the exception of a trace of duckweeds (Lemna and Wolffia) taken by a single specimen. The scavenger habit of the species was shown by the food of the Fox River specimen, three fourths of which consisted of the remnants of a dead cat. Fishes made a larger ratio of the food than in the preceding species, amounting to about one third, most of them apparently taken alive. One, however, a sucker, was represented only by the stomach and intestines, doubtless picked up near a fish boat. The gizzard shad, certain Cyprinidæ, and undetermined suckers (Catostomatidæ) were recognized, four of the twelve specimens having fed wholly or almost wholly upon them. The molluscan food of these specimens was insignificant, no bivalve mollusks having been taken by them and only a few Vivipara and Melantho, amounting in all to five per cent. While insects had been eaten by four of the specimens and reached a ratio of thirty per cent., they were practically all Hexagenia larvæ, taken in October, 1878 and 1887. On the other hand, seventeen per cent. of the food was catfishes, taken by four of the specimens in May and August.

---

* A hint of the winter food is given by six specimens received from the Illinois River at Havana, February, 1888, all of which had fed only upon Chironomus larvæ or larvæ of Agrion.

Seven immature examples, from two to three and a half inches long, had fed chiefly upon Entomostraca, which made about one half the food. Among these, Daphnia, *Simocephalus americanus*, Acroperus, *Macrothrix laticornis*, Cyclops, and Cypris were determined. One fourth the food consisted of the univalve Physa, and one fifth of it of insect larvæ, chiefly ephemerids and Chironomus. A little Wolffia and other aquatic vegetation likewise occurred.

ICTALURUS NEBULOSUS, LeS.   BULL-HEAD ; HORNED-POUT.

This superabundant species occurs in all waters and in all parts of the state, but frequents by preference ponds and muddy streams. It grows to a larger size in the rivers than elsewhere, and has many marked varieties. Its feeding habits are apparently essentially the same in all situations.

Gill-rakers fourteen in number on the anterior gill, in one row, thick, stout, not toothed, at the angle of the arch about half as long as the filaments, shortening rapidly above and below. Second gill also with a single row, shorter than those of the first; succeeding gill with two rows each of still shorter rakers, the posterior row shorter than the anterior ; a smaller row upon the pharyngeal arch. The upper pharyngeals are large and broad oval pads, with convex surfaces paved with close-set, minute, sharp teeth, and act against the broad lower pharyngeals, which are similarly armed. Intestine to head and body as 1.2 to 1.

Thirty-six specimens were collected for a study of the food, — at Normal, Peoria, Pekin, and Havana, in Central Illinois ; and from the Fox River and several of the small lakes in the northern part of the state. The collections were made in May, July, August, September, and October, of four different years.

The vegetable food nearly equaled that of *I. punctatus*, and was taken by seven of the specimens. One had eaten distillery slops, and in the food of the others were found Ceratophyllum, Potamogeton, Chara, and various Algæ.

Fishes made one fifth of the food,— taken however by only two of the specimens, which had eaten nothing else. One of the fishes was a perch and the other a sunfish (Centrarchidæ).

Mollusks made one fifth of the entire amount of the food, — more than one half of them Sphærium. This genus made nearly all the food of a large group taken from the Illinois River at Pekin in September, 1882, and also of two other specimens taken in the Illinois River at Peoria in October, 1887. Univalves were rarely present, amounting to only two per cent. of the food, taken however by eight of the specimens. These included the usual forms —Valvata, Melantho, and Amnicola, together with two or three specimens of Physa. Examples of Pisidium were rarely noted, and two had eaten Unios.

Nearly a fourth of the food was insects, mostly aquatic, and the larger part of them larvæ of Diptera — especially Chironomus and Corethra. Seven per cent. of Neuroptera larvæ (Hexagenia, Libellulidæ, and Phryganeidæ), together with a miscellaneous assortment of terrestrial species, complete the account of the insect food.

The Crustaceans (thirteen per cent.) were nearly all crayfish, traces of Diaptomus, Leptodora, Chydorus, etc., appearing, however, in here and there a specimen, and the little amphipod *Allorchestes dentata*, appearing in three. A leech and a nematoid worm occurred, each in one.

It will be seen that the food of this species was very widely distributed, being composed about equally of fishes, mollusks, aquatic insects, and vegetable structures, with a very considerable ratio (thirteen per cent.) of crustaceans added.

Two smaller specimens, two and three and a half inches respectively, had fed chiefly on ephemerid and Chironomus larvæ, small crayfish, and Asellus. To these were added *Corixa tumida*, Cyclops, Daphnia, filaments of Spongilla, Chydorus, *Scapholeberis mucronatus*, a few Diatoms, and traces of filamentous Algæ.

ICTALURUS MARMORATUS, Holbrook. MARBLED CAT.

This species is scarcely more than a deep-water variety of the common bull-head (*I. nebulosus*), distinguished only by the color. It occurs in the larger rivers of the State and their immediate tributaries, but nowhere, so far as I know, in stagnant waters. Our thirteen specimens were all from Peoria and Havana, taken in August, October, and November of 1878 and 1887.

The food of this species as represented by these thirteen specimens, is unusually simple for a catfish, consisting chiefly of bivalve mollusks, larvæ of Chironomus and Hexagenia, distillery slops, and accidental rubbish. Fishes are conspicuous by their absence, only a single specimen exhibiting any trace of them.

Sphærium and Unio made about a fourth of the food, and aquatic insect larvæ amounted to one half (Hexagenia thirty-five per cent. and Chironomus fourteen). A hydrophilid and a few terrestrial insects, a few specimens of Vivipara and a Physa, sialid larvæ (taken by two), slender leeches eaten by five, and a trace of Potamogeton in one, are the minor elements of this record. One of the specimens, taken in November, had eaten eighteen leeches, which made one fourth of its food. It will be noticed that three fourths of the food consisted of bivalve mollusks and insect larvæ.

LEPTOPS OLIVARIS, Raf.   MUD CAT; YELLOW CAT; MORGAN CAT.

Common in the deeper waters of the larger streams. Obtained by us only from the Illinois, Wabash, and Ohio.

This is one of the largest of the river catfishes, repulsive in appearance, but above the average as food. It is reported by fishermen to feed only upon animal food — chiefly fishes — and such was the case with the two specimens examined from collections made at Quincy in August, 1887. These had fed upon the common river sunfish (Lepomis), several cyprinoids, and an Amiurus four inches long.

NOTURUS GYRINUS, Mitch.

This little catfish, the most abundant of the small species of the family, occurs throughout Illinois, but has been confined in our collections mostly to lakes, rivers, and large creeks. It is not by any means restricted to rocky situations, but seems rather to prefer the muddy parts of both the rivers and lakes in which it occurs,

Thirteen specimens were secured at Pekin and Peoria, from Clear Lake in Kentucky, and from the Fox River in McHenry county. Their food was wholly animal, with the exception of a trace of Algæ found in two. This group had eaten practically nothing but Crustacea, nearly all Amphipoda (Allorchestes) and Isopoda (Asellus), the former eaten by nine, and the latter by two — both together making forty-seven per cent. of the entire food. As might be supposed from the small size of these specimens, Entomostraca were apparent in the food, although in moderate numbers (five per cent.). The forms recoguized were Simocephalus, Chydorus, Pleuroxus, Alona, Cypris, Candona, Cyclops, and Canthocamptus. A planarian worm was noted in one, and specimens of Difflugia in another. A single example had eaten a small fish. Most of the insects were Chironomus larvæ (twenty-five per cent.), case-worms, and larvæ of day flies (twelve per cent.).

Comparing the principal genera of this family, as represented by the one hundred and twenty specimens examined, we find that the larger deep-water species from the great rivers of the State are apparently ichthyophagous ; that the relatively minute stone cats feed on the smaller insect larvæ and the medium sized Crustacea ; that the spotted cat is essentially insectivorous ; that among the bull-heads the yellow cat eats the largest percentage of fishes and the marbled cat the smallest ; that the latter feeds more generally upon Unio than any of the other species; and that mollusks at large make about one sixth of the food of the group of species which feeds upon them.

FAMILY AMIIDÆ.

AMIA CALVA, Linn.   DOG FISH; MUD FISH; GRINDLE.

This species is very abundant throughout the State in the lakes and larger streams, and also common in ponds of southern Illinois. Not commonly eaten, but often caught for sport.

The food of twenty-one specimens taken from northern, central, and southern Illinois, in April, May, June, August, September, and October, was wholly animal, about one third of it fishes, among which were recognized some undetermined cyprinoids and a small buffalo fish (Ictiobus). The other important elements were mollusks — about one fourth — and crustaceans (forty per cent.), insects being represented by an insignificant ratio (two per cent.). Even the usually abundant Chironomus and ephemerid larvæ had been eaten by only one or two specimens each.   The mollusks were more than two thirds Sphærium, the remainder being Vivipara and Planorbis.   The Crustacea were chiefly crayfish, among them *Cambarus virilis* and *obesus.*   Besides these, I noticed Crangonyx, Allorchestes, and Asellus, Cyclops and a few Cladocera *(Simocephalus americanus, Scapholeberis mucronatus,* and Chydorus).

FAMILY LEPIDOSTEIDÆ.

A half dozen of the river gars *Lepidosteus platystomus* and *L. osseus* had eaten nothing but fishes, including the hickory shad (Dorosoma), black bass (Micropterus), and some minnows (Cyprinidæ).

FAMILY POLYODONTIDÆ.

Polyodon spathula, Wall.   Shovel - fish; Paddle - fish; Spoon - bill Cat; Duck - bill Cat.

This remarkable and most interesting fish, the most notable inhabitant of our waters, occurs abundantly in the Illinois, Mississippi, and Ohio, but not elsewhere within our limits.   It has a more or less distinct habit of migration, being much the most abundant in spring, although taken sparingly throughout the remainder of the year.   It is a gigantic species, reaching a weight of thirty pounds and upwards, and a length of six feet or more, including the paddle.   It is now quite generally dressed for the market, and sold at the same rate as catfish.

It has an alimentary apparatus not less remarkable than its other characters.   The broad blade-like snout, the enormous mouth and equally large gill slits, the efficient branchial strainer, and the peculiar structure of the intestine,— all indicate a peculiar alimentary regimen and unusual feeding habits.   Both the upper and lower jaws of the young are provided with small, acute teeth—the upper with a band upon the vomer and palatines, besides a row on the maxillaries, and the lower with a longitudinal row extending nearly its full length—but the jaws of the adult are toothless and smooth.

This fish depends, therefore, entirely upon the very remarkable straining apparatus borne by the gills, the immense oral opening, and the equally free provison for the exit of water from the gill chamber, enabling it to pass vast quantities of water through its branchial apparatus. The gills are very elongate, each having the form, when the mouth is closed, of a slender U with the sides parallel and closely approximated, the lower arm, however, extending somewhat further forward than the upper. Each gill bears throughout its whole length a double series of very long, fine, numerous, and slender rakers, the two rows separated by a membranous partition borne upon the anterior surface of the arch,— this partition a little higher than the rows of rakers, and slightly thickened on the internal edge, so as to enclose the tips of the rakers when the parts of the apparatus are approximated. These rakers average fully twice the length of the corresponding gill filaments, and numbered, on the first gill of a specimen about one and a half feet long, five hundred and sixty rakers in the anterior series. A half row of similar rakers is borne by the fifth branchial arch, corresponding to the inferior pharyngeal bones of most fishes. The individual rakers are toothless, smooth, cartilaginous, and nearly naked, the filaments covered by a thin epithelium, thickened at the tip. Interlocking as these do when the branchial apparatus is extended, they form a strainer, sufficient to arrest the smallest living forms above the Protozoa. There are no pharyngeal jaws or teeth, nor is there any apparatus of mastication elsewhere.

In the absence of any raptatorial teeth or crushing apparatus in its large and feeble jaws or in its throat, it is certain that this species cannot feed upon fishes or mollusks; and the character of the intestine makes it very probable that it never purposely swallows mud or takes a large percentage of vegetable food. On the other hand, its enormous mouth, and the remarkable straining apparatus in its branchial cavity give it access to the immense stores of minute insect and crustacean life most commonly reserved for young fishes ; while its structures are likewise evidently adapted to the larger soft-bodied insects and insect larvæ.

The use of the paddle-like snout is as yet a matter of conjecture, slightly assisted, perhaps, by a knowledge of the principal features of the food. The relatively minute size of the objects on which it feeds, the absence of mud from its intestine, and its seemingly positive preference for animal food, indicate that it is not only able to gather large quantities of very minute objects among the weeds and from the muddy bottom without filling itself with mud, but that it can separate the Entomostraca from the Algæ among which they swim. I cannot see how this is done unless its paddle be used to stir up the weeds in its advance, as it swims along, thus driving up the animal forms within reach of its branchial strainer, while the mud and vegetation settle out of its way.

What is the meaning of the minute and evanescent teeth on the jaws of Polyodon, I am unable to surmise, but judge that they can only be accounted for by reference to primitive conditions of life of which the present habits of the fish give us no hint.

Eight specimens obtained from Peoria, Pekin, and Henry on the Illinois, from the Ohio River at Cairo, and from the Mississippi at Quincy, in six different years, will probably suffice to give a fair general idea of the food, taken in connection with suggestions made above, based on a study of the structures of alimentation.

The vegetable elements of the food were eaten by four of the specimens, and amounted to only seven per cent. It is to be noted, however, that one of the specimens taken at Quincy had derived thirty per cent. of its food from a species of Nostoc, while another, taken at Peoria in May, had found about one fifth of its food among vegetable objects. A little Potamogeton, some filamentous algæ and diatoms, together with a small amount of terrestrial rubbish, were the elements recorded.

Fishes and mollusks were without representation in the alimentary contents of these specimens; while insects and crustaceans made by far the larger part of the food,— the former taken by all the specimens, and in nearly twice the ratio of the latter. The minor items of this class were Corethra larvæ (twelve per cent.) and Chironomus larvæ (five per cent.). Larvæ of Neuroptera made one half the food, and were eaten by six of the specimens,— Hexagenia larvæ alone amounting to forty-seven per cent. A few case-worms (Phryganeidæ), dragon-fly larvæ (Libellulidæ and Agrion), and Cænis larvæ, with a few Corisas, aquatic beetles (Coptotomus), and chance terrestrial insects, were the remaining items of this class.

The crustaceans were all Entomostraca, with the exception of the amphipod *Allorchestes dentata*, noted in two specimens. Five of the specimens had eaten Entomostraca, one of them ninety per cent., and another eighty, — the remaining ratios being thirty-five, thirty, and twenty. Water mites (Hydrachnida) were noticed in a single specimen, leeches also in one, and Plumatella in another. The smaller Crustacea were so numerous that no attempt was made to exhaust the possible determinations ; but in some cursory examination of this material the following forms were observed : *Daphnia pulex*, Bosmina, Chydorus, Eurycercus, Leptodora, Cypris, Cyclops, and Canthocamptus.

To the comparative anatomist, Polyodon is peculiarly notable as among the oldest of fishes, distinguished, when compared with higher species, by the persistence of juvenile characters ; and similarly we find that the most remarkable feature of its food is one which it shares with young fishes in general. This is, however, a simulated correlation, the food habit not being due to a persistence of youthful structures of alimentation, but to a remarkable specialization of the apparatus of food prehension. It must consequently be correlated with a superabundant supply of minute animal life when and where these structures originated, or, at least, when they took their present form ; and taken together with the great size of this fish and its out-worn dental furniture, seemingly indicates a radical change in the feeding habits of the species, and a capacity for adaptation to new circumstances which possibly accounts for its long survival.

FOOD OF SMALLER FAMILIES.*

KINDS OF FOOD. NUMBER OF SPECIMENS EXAMINED.	Lota maculosa	Esox lucius	Esox vermiculatus	Summary of Esox	Coregonus artedi	Dorosoma cepedianum	Clupea chrysochloris	Hyodon tergisus	Amia calva	Lepidosteus platystomus	Lepidosteus osseus	Summary of Lepidosteus	Polyodon spathula
	10	37	19	56	5	11	4	8	12	2	4	6	8
RATIOS IN WHICH EACH ELEMENT OF FOOD WAS FOUND.													
ANIMAL FOOD	1.00	1.00	1.00	1.00	1.00	.04	1.00	1.00	1.00	1.00	1.00	1.00	.93
I. BATRACHIA (tadpoles)			.13	.07									
II. FISHES	.83	.98	.52	.75			1.00		.33	.67	1.00	.83	
Cycloid		.09	.11	.10									
Acanthopteri	.62	.23	.11	.17					.09	+		+	
Perca lutea	.62												
Centrarchinæ		.21	.11	.16									
Micropterus		.03		.02									
Pomoxys		.69		.04									
Gambusia patruelis			.04	.02									
Coregonus	.10												
Dorosoma cepedianum		.46		.23			.67			.50	.25	.38	
Hyodon		.05		.03									
Cyprinidæ		.03	.09	.06									
Catostomatidæ		.03		.01					.04	+	.20	.10	
III. MOLLUSCA			+	+		+		+	.24				—
1. *Univalves*			+	+		+		+	.07				—
Vivipara									.06				
2. *Bivalves* (Sphærium)						+			.17				
IV. INSECTA		.02	.35	.18	.50	+	+	1.00	.02				.59
Terrestrial					.50	+		.64					.01
Aquatic		.02	.29	.15	+			.36	.02				.58
1. *Hymenoptera*					.11			+					
Myrmicidæ					.11								
2. *Lepidoptera*					.13			+					
3. *Diptera*					.13			+	.01				.18
Terrestrial					.13			+					+
Tipulidæ					.13			+					
Aquatic larvæ									.01				.18
Corethra													.12
Chironomidæ									.01				.05
4. *Coleoptera*					.02	+		+					.02
Terrestrial					.02	+		+					+
Chrysomelidæ					.02			+					
Aquatic								+					.01
Dytiscidæ								+					.01
Hydrophilidæ								+					
5. *Hemiptera*			+	+	.11	+		+	+				+

* The sign + indicates a ratio not estimated.

FOOD OF SMALLER FAMILIES.—*Continued.*

	Lota maculosa	Esox lucius	Esox vermiculatus	Summary of Esox	Coregonus artedi	Dorosoma cepedianum	Clupea chrysochloris	Hyodon tergisus	Amia calva	Lepidosteus platystomus	Lepidosteus osseus	Summary of Lepidosteus	Polyodon spathula
NUMBER OF SPECIMENS EXAMINED.	10	37	19	56	5	11	4	8	12	2	4	6	8
KINDS OF FOOD.	RATIOS IN WHICH EACH ELEMENT OF FOOD WAS FOUND.												
Terrestrial						.11			+				
Homoptera						.11							
Aquatic			+	+		+		+	+				+
Corisa			+	+		+		+	+				+
6. *Orthoptera* (Tettix)								+					
7. *Neuroptera* (larvæ)		.02	.29	.15					.66	.01			.51
Phryganeidæ													.02
Odonata		.02	.25	.13						.01			.02
Libellulinæ		.02	.02	.02						.01			+
Agrion			.15	.07									
Ephemeridæ			.04	.02					.66	+			.47
Hexagenia			.04	.02					.33				.47
V. ARACHNIDA						+			+				.01
VI. CRUSTACEA	.17		+	+	.50	.04		+		.41	.33	.17	.33
1. *Decapoda* (Cambarus)										.38	.33	.17	
2. *Amphipoda*										.01			+
3. *Isopoda* (Asellus)			+	+					+				
ENTOMOSTRACA					.50	.04		+		.02			.33
4. *Cladocera*						+	+	+	+				+
5. *Ostracoda* (Cypris)							.04						+
6. *Copepoda*						+	+			.02			.08
VII. VERMES (leech)													+
VIII. BRYOZOA (Plumatella)													+
IX. PROTOZOA (Difflugia)						+							
VEGETABLE FOOD			+	+		.32		+	+				.07
Miscellaneous						.16		+					+
Terrestrial			+	+									.01
Aquatic						+				+			.06
Lemna						+							
Wolffia						+							
Algæ						+							.05
Nostoc													.05
Distillery Slops						.16							
MUD AND SAND						.64							

FOOD OF CATOSTOMATIDÆ.

	Placopharynx carinatus	Moxostoma aureolum	Moxostoma macrolepidotum	Summary of Moxostoma*	Minytrema melanops	Hypentelium nigricans	Catostomus teres	Carpiodes cyprinus	Ictiobus bubalus	Ictiobus urus	Ictiobus cyprinella	Summary of Ictiobus	Summary of Catostomatidæ
NUMBER OF SPECIMENS EXAMINED.	2	6	12	23*	4	5	3	19	17	17	17	51	107
KINDS OF FOOD.	RATIOS IN WHICH EACH ELEMENT OF FOOD WAS FOUND.												
ANIMAL FOOD	.95	.97	.87	.95	1.00	1.00	.94	.83	.80	.67	.65	.71	.90
I. FISHES						+							+
II. MOLLUSCA	.32	.49	.55	.60	.87	.10	.42	.24	.30	.12	.03	.15	.41
1. *Univalves*	+	.16	.40	.27	.01		.12	+	+	+	+	+	.06
Viviparidæ		.13	.22	.20						+		+	.03
Somatogyrus			.06	.02									.01
Limnæa			.10	.03									.01
2. *Bivalves*	+	.33	.15	.33	.86	.10	.30	.24	.30	.14	.02	.15	.30
Sphærium	+	.17	.15	.28	.86	.10	.30	.24	.30	.14	.02	.15	.29
Unionidæ			+	+									+
III. INSECTA	.63	.48	.32	.35	.10	.90	.03	.32	.29	.42	.33	.35	.37
Terrestrial			+	+		+			.01	+	+	+	+
Aquatic	.63	.46	.31	.34	.10	.90	.03	.32	.28	.42	.36	.37	.37
1. *Diptera*	.03	.46	.30	.33	.10	.04	.03	.30	.19	.25	.24	.23	.17
Terrestrial			+	+						+		+	+
Aquatic larvæ	.03	.46	.30	.33	.10	.04	.03	.30	.19	.25	.24	.24	.17
Chironomidæ	.03	.30	.15	.15	.10	.04	.03	.22	.19	.23	.20	.22	.12
2. *Coleoptera*	.50		+	+		.13			.02	+	+	+	.10
Terrestrial			+	+		.01			.01	+		+	+
Aquatic larvæ	.50		+	+		.12			.01		+	+	.10
Hydrophilidæ	.50		+	+					.01			+	.09
3. *Hemiptera*									.01	.01	.01		+
Corisa									.01	.01	.01		+
4. *Neuroptera*	.10		.01	.01		.72		.02	.08	.16	.11	.11	.10
Terrestrial										+		+	+
Aquatic larvæ	.10		.01	.01		.72		.02	.08	.16	.11	.11	.10
Phryganeidæ									.01	.01			+
Sialidæ												.01	+
Odonata			+	+					.01	.01		.01	+
Ephemeridæ	.10		.01	.01		.72				.06	.15	.10	.10
Cænis						.53							.05
Hexagenia						.12				.14	.09	.08	.02
IV. ARACHNIDA			+	+			.02		+	+	+	+	+
V. CRUSTACEA	+		+	+	.03		.30	.23	.20	.13	.29	.21	.11

* Includes five specimens of undetermined species.

FOOD OF CATOSTOMATIDÆ.—*Continued.*

KINDS OF FOOD.	Placopharynx carinatus	Moxostoma aureolum	Moxostoma macrolepidotum	Summary of Moxostoma*	Minytrema melanops	Hypentelium nigricans	Catostomus teres	Carpiodes cyprinus	Ictiobus bubalus	Ictiobus urus	Ictiobus cyprinella	Summary of Ictiobus	Summary of Catostomatidæ
NUMBER OF SPECIMENS EXAMINED	2	6	12	23*	4	5	3	19	17	17	17	51	107
RATIOS IN WHICH EACH ELEMENT OF FOOD WAS FOUND.													
1. Decapoda (Cambarus)								+	.01	+		+	+
2. Amphipoda	+							+	.01		+	01	.01
ENTOMOSTRACA			+	+	.03		.30	.23	.19	.13	.29	.20	.10
3. Cladocera			+	+			.30	.03	+	.05	.08	.05	.04
Daphnella											+	+	+
Daphniidæ								.03	+	.05	+	.02	.01
Lynceidæ			+	+			.30	+	+	.01	.15	.05	.03
4. Ostracoda (Cypridæ)					.03		+	.04	+	.02	.05	.03	.02
5. Copepoda			+	+	+		+	.08	.19	.04	+	.08	.03
VI. VERMES							.02	+	+			+	+
Rotifera							+						+
VII. POLYZOA	+							.01				+	+
VIII. PROTOZOA (Rhizopoda)							.15	+	+			+	.01
VEGETABLE FOOD	.05	.03	.10	.04		+	.06	.08	.20	.33	.35	.29	.08
Seeds						+	.02		.02	+	.09	.04	.01
Aquatic	.05	.01	.07	.02		+	.04	.08	.15	.06	.17	.12	.05
Lemna	.03								+	+	+	+	.01
Wolffia	.01		.10	+				.05	.11	+		.03	.01
Algæ			+	+			.04	+	+	.01	.15	.05	.01
Filamentous			+	+			.02	+	+		.05	.02	.01
Diatoms							+	+	+		+	+	+
Distillery Slops			.03	.01						.21	.09	.10	.02
MUD		+	.03	.01	+			.09				+	.02

* Includes five specimens of undetermined species.

FOOD OF SILURIDÆ.

	Ictalurus furcatus	Ictalurus punctatus	Amiurus natalis	Amiurus nebulosus	Amiurus marmoratus	Summary of Amiurus	Leptops olivaris	Noturus gyrinus	Summary of Siluridæ
NUMBER OF SPECIMENS EXAMINED.	1	43	12	36	13	61	2	13	120
KINDS OF FOOD.	RATIOS IN WHICH EACH ELEMENT OF FOOD WAS FOUND.								
ANIMAL FOOD	1.00	.75	1.00	.77	.93	.90	1.00	.98	.96
Dead animal matter	....	.02	.13	....	.14	.09	....	....	.02
I. FISHES	1.00	.10	.34	.20	.01	.18	1.00	+	.44
Percidæ	....	....	....	.20	....	.06	.43	....	.16
Lepomis	....	....	....	....	....	....	.43	....	.14
Cyprinidæ	....	....	.04	....	....	.01	.30	....	.10
Catostomatidæ	....	....	.17	....	....	.06	....	....	.01
Amiurus	....	....	....	....	....	....	.20	....	.07
II. MOLLUSCA	....	.15	.05	.19	.26	.16	....	....	.04
1. *Univalves*	....	.08	.05	.02	.01	.02	....	....	.01
Vivipara	....	.01	.04	....	+	.01	....	....	.01
Melantho	....	.03	+	.01	....	.01	....	....	+
2. *Bivalves*	....	.07	....	.17	.25	.14	....	....	.03
Sphærium	....	....	....	.13	.12	.08	....	....	.02
Unionidæ	....	....	....	.04	.13	.06	....	....	.01
III. INSECTA	....	.44	.30	.28	.50	.36	....	.46	.26
Terrestrial	....	.07	....	+	+	+	....	.01	.01
Aquatic	....	.36	.30	.27	.50	.36	....	.41	.24
1. *Hymenoptera*	....	+	....	....	....	....	....	....	+
2. *Lepidoptera*	....	+	....	....	....	....	....	....	+
3. *Diptera*	....	.13	....	.20	.14	.11	....	.27	.12
Terrestrial	....	+	....	....	....	....	....	.01	+
Aquatic	....	.13	....	.20	.14	.11	....	.26	.12
Corethra	....	....	....	.09	+	.03	....	....	.01
Chironomus	....	.13	....	.10	.14	.08	....	.25	.11
4. *Coleoptera*	....	+	....	+	.01	.01	....	.02	.01
Terrestrial	....	+	....	+	+	+	....	....	+
Carabidæ	....	+	....	....	+	+	....	....	+
Staphylinidæ	....	+	....	....	....	....	....	....	+
Aquatic larvæ	....	+	....	+	.01	.01	....	.02	.01
Dytiscidæ	....	+	....	+	....	+	....	....	+
Hydrophilidæ	....	....	....	....	.01	.01	....	....	+
5. *Hemiptera*	....	.01	....	+	+	....	....	....	+
Terrestrial	....	+	....	....	....	....	....	....	+
Aquatic	....	.01	....	+	+	....	....	....	+
Corixa	....	+	....	+	+	....	....	....	+
6. *Orthoptera*	....	.05	....	....	+	+	....	....	.01

FOOD OF SILURIDÆ.—*Continued.*

NUMBER OF SPECIMENS EXAMINED.	Ictalurus furcatus	Ictalurus punctatus	Amiurus natalis	Amiurus nebulosus	Amiurus marmoratus	Summary of Amiurus	Leptops olivaris	Noturus gyrinus	Summary of Siluridæ
	1	43	12	36	13	61	2	13	120
KINDS OF FOOD.	RATIOS IN WHICH EACH ELEMENT OF FOOD WAS FOUND.								
Acrididæ	....	.04	....	....	....	....	....	....	.01
7. *Neuroptera* (larvæ).........	....	.23	.30	.07	.35	.24	....	.12	.10
Phryganeidæ	....	.01	....	+	....	+	....	.02	.01
Sialidæ	....	+	....	.01	+	+	....	....	+
Odonata	....	.08	....	.02	....	.01	....	....	.01
Libellulinæ...............	....	.07	....	.02	....	.01	....	....	.01
Agrioninæ	....	.01	....	....	....	....	....	....	+
Ephemeridæ	....	.09	.30	.04	.35	.23	....	.10	.08
Hexagenia...............	....	.03	.30	.04	.35	.23	....	....	.05
IV. ARACHNIDA...............	....	+	.01	....	+	+	....	....	+
V. CRUSTACEA	....	.03	.17	.13	....	.10	....	.52	.19
1. *Decapoda* (Cambarus)......	....	.03	.17	.12	....	.10	....	....	.02
2. *Amphipoda* (Allorchestes).	....	+	....	.01	....	+	....	.25	.08
3. *Isopoda* (Asellus).........	....	....	....	....	....	....	....	.22	.07
ENTOMOSTRACA...............	....	....	....	+	....	+	....	.05	.02
4. *Cladocera*	....	....	....	+	....	+	....	.02	.01
5. *Ostracoda*	....	....	....	....	....	....	....	+	+
6. *Copepoda*	....	....	....	+	....	+	....	.03	.01
VI. VERMES	....	.01	....	.01	.02	.01	....	+	.01
Hirudinei	....	.01	....	.01	.02	.01	....	....	.01
Nematodes	....	+	....	+	....	+	....	....	+
VII. BRYOZOA (Plumatella)..	....	+	....	....	....	....	....	....	+
VIII. PORIFERA (Spongilla)..	....	+	....	....	....	....	....	....	+
IX. PROTOZOA (Difflugia)	....	....	....	....	....	....	....	+	+
VEGETABLE FOOD	....	.25	+	.23	.07	.10	....	.02	.04
Miscellaneous	....	.01	....	.05	+	.02	....	.01	+
Aquatic	....	.23	+	.16	+	.06	....	.01	.03
Lemna...............	....	+	+	....	....	....	....	....	+
Wolffia...............	....	....	+	....	....	+	....	....	+
Potamogeton	....	.06	....	.01	+	.01	....	....	+
Algæ...............	....	.16	....	.02	....	.01	....	.01	.02
Distillery slops...............	....	.01	....	+	.07	.02	....	....	.01
MUD	....	....	....	+	....	+	....	....	+

ON THE FOOD RELATIONS
OF FRESH-WATER FISHES

Stephen Alfred Forbes

ARTICLE VIII.—*On the Food Relations of Fresh-Water Fishes: a Summary and Discussion.*— By S. A. FORBES.

The principal object of the research reported in the series of papers* of which this is the concluding number, is to determine more precisely than has hitherto been done the relations to nature of the various genera and families of the fishes of an interior region. This purpose has led especially to a study of the *food relations* of the groups, for through these, chiefly, fishes exert their influence on the outer world, and are themselves impressed in turn; and thus have appeared a number of subordinate considerations having a bearing, more or less direct, on the main intention of the study.

An examination of the special relations of their food and feeding structures gives us clues, not only to the present significance of fishes, but also to their past effect on life at large, showing how they must have modified the course of evolution; and the occasional occurrence in a fish of food prehensile structures out of present relation to its feeding habits, may throw light on the history of its group, indicating conditions of existence once normal to it but now outgrown. Evidence of similar application may also be obtained by a comparison of the food of the young and of the adult.

The feeding apparatus exhibits some of the most significant examples of correlation of structure, important to an acquaintance with the course of development in fishes, but not comprehensible without a knowledge of the food for whose appropria-

---

* Published at intervals from 1877 to 1888, in the first and second volumes of the Bulletin of this Laboratory, as follows: "The Food of Illinois Fishes" (Vol. I., No. 2, pp. 71-89), "The Food of Fishes" (No. 3, pp. 18-65), "On the Food of Young Fishes" (No. 3, pp. 66-79), "The Food of the Smaller Fresh-Water Fishes" (No. 6, pp. 65-94), The First Food of the Common White-fish, (No. 6, pp. 95-109), and "Studies of the Food of Fresh-Water Fishes" (Vol II., Art. VII., pp. 433-473).

tion it is adapted. I need hardly recall the fact that the defensive apparatus of one species may have its explanation only in the raptatorial structures of another.

We shall find also in a study of the food evidence of the indirect but powerful action of a number of external conditions which take effect only through the food relation, and are incomprehensible or perhaps unnoticed unless this is understood —conditions of climate, season, locality, and the like; and especially may we hope for this when we remember that the distribution and abundance of a species may be determined, not so much by ordinary conditions, as by those prevailing at critical intervals, periods of stress, when a slight advantage or a trivial disability may have prolonged and multiplied effects. As the range of a plant is often limited, not by the average temperature of the year, but by the extremes of cold or heat, so the existence of an animal may be decided by the presence or absence of some structural modification adapted to carry it safely through a single brief period of unusual scarcity or of extraordinary competition.

That the study here set forth should give us details not to be otherwise obtained of the struggle for existence among fishes themselves, goes without saying; and that it may thus explain some peculiarities of distribution, seems also probable. I have thought it not impossible that by taking into account all the data collected, and the mass of related facts, structural, biological, and other, that materials might be found bearing on the interesting question of the precedence in time and the relative evolutionary importance of desire and effort on the one hand and structural aptitudes on the other.

Among the purely practical results to be anticipated, are a more accurate knowledge of the conditions favorable to the growth and multiplication of the more important species; the ability to judge intelligently of the fitness of any body of water to sustain a greater number or a more profitable assemblage of fishes than those occurring there spontaneously; guidance as to the new elements of food and circumstance which it will be necessary to supply to insure the successful introduction into any lake or stream of a fish not native there; and a clear recognition of the fact that intelligent fish culture must take into account the necessities of the species whose increase is desired, through all ages and all stages of their growth, at every season of the year, and under all varieties of condition likely to arise. We should derive, in short, from these and similar researches, a body of full, precise, and significant knowledge to take the place of the guess-work and empiricism upon which we must otherwise depend as the basis of our efforts to maintain the supply of food and the incitement to healthful recreation afforded by the waters of the State.

As a contribution to the general subject, I present herewith a summary account of the food of twelve hundred and twenty-one fishes obtained from the waters of Illinois at intervals from 1876 to 1887, and in various months from April to November. These fishes belonged to eighty-seven species of sixty-three genera and twenty-five families. They were derived from waters of every description, ranging from Lake Michigan to weedy stagnant ponds and temporary pools, and from the Mississippi and Ohio Rivers to the muddy prairie creeks, and the rocky rivulets of the hilly portions of the State. Nine hundred and fourteen of the examples studied were practically adult, so far as the purposes of this investigation are concerned, the remaining three hundred and seven being young, in the first stage of their food and feeding habits. More than half these young belonged to a single species,— the common lake white-fish,— but the remainder were well distributed.

I have arranged the matter under the following general heads: (1) a summary statement of the food, so made as to exhibit (*a*) the kinds and relative importance of the principal competitions among fishes and (*b*) the relative value to the principal species of fishes of the major elements of their food; (2) a brief account of the food of the young; (3) an examination of the permanency and definiteness of distinctions with respect to food, between different species, and also between higher groups; (4) a review of the structures of fishes related to food prehension and to their feeding habits; and, finally, (5) a classified list of the objects detected in the food of fishes, with a statement, against each object, of the species feeding on it and the number of specimens in which it was. found.

THE FOOD OF ADULT FISHES.

An analysis of our facts made with reference to the kinds of fishes eating each of the principal articles in the dietary of the class and showing the relative importance of these elements in the food of the various species, will exhibit the competitions of fishes for food more clearly and precisely than my earlier discussions, and also the nature and the energy of the restraints imposed by fishes on the multiplication of their principal food species.

PISCIVOROUS FISHES.

The principal fish-eaters among our species — those whose average food in the adult stage consists of seventy-five per cent. or more of fishes — are the burbot[1], the pike-perch[2] or wall-eyed pike, the common pike[3] or "pickerel," the large-mouthed black bass,[4] the channel cat,[5] the mud cat,[6] and the gars.[7] Possibly also the golden shad[8] will be found strictly ichthyophagous, this being the case with the four specimens which I studied.

Those which take fishes in moderate amount — the ratios ranging in my specimens from twenty-five to sixty-five per cent. — are the war-mouth (Chænobryttus), the blue-cheeked sunfish,[9] the grass pickerel,[10] the dog-fish,[11] the spotted cat,[12] and the small miller's thumb[13]. The white[14] and striped bass,[15] the common perch,[16] the remaining sunfishes (those with smaller mouths), the rock bass,[17] and the croppie,[18] take but few fishes, these making, according to my observations, not less than five nor more than twenty-five per cent of their food.

Those which capture living fishes, to a trivial extent, at most, are the white perch or sheepshead,[19] the gizzard

[1] Lota maculosa. [2] Stizostedion vitreum. [3] Esox lucius. [4] Micropterus salmoides. [5] Ictalurus furcatus. [6] Leptops olivaris. [7] Lepidosteus. [8] Clupea chrysochloris. [9] Lepomis cyanellus. [10] Esox vermiculatus. [11] Amia calva. [12] Ictalurus punctatus. [13] Uranidea richardsonii. [14] Roccus chrysops. [15] Roccus interruptus. [16] Perca lutea. [17] Ambloplites rupestris. [18] Pomoxys. [19] Aplodinotus.

shad,[1] the suckers,[2] and the shovel fish[3] among the larger species; the darters,[4] the brook silversides,[5] the stickleback,[6] the mud minnows,[7] the top minnows,[8] the stonecats,[9] and the common minnows[10] generally, among the smaller kinds.

Our eight specimens of the toothed herring[11] had taken no fishes whatever; while our nineteen examples of the pirate perch[12] had eaten only two per cent.

Rough-scaled fishes with spiny fins (Acanthopteri) were eaten by the miller's thumb, the common pike, the wall-eyed pike, the large-mouthed black bass, the croppies, the dog-fish, the common perch, the burbot, the bull-head,[13] the common sun-fish *(Lepomis pallidus)*, the small-mouthed black bass,[14] the grass pickerel, the gar, and the mud cat (Leptops). Among these, the common perch and the sunfishes[15] were most frequently taken — doubtless owing to their greater relative abundance — the perch occuring in the food of the burbot, the large-mouthed black bass, and the bull-head; and sunfishes in both species of the wall-eyed pike, the common pike, the gars, pickerel, bull-heads, and mud cat. Black bass were taken from the common pike (Esox), the wall-eyed pike (Stizostedion), and the gar. Croppie and rock bass I recognized only in the pike. Even the catfishes (Siluridæ) with their stout, sharp, and poisoned spines, were more frequently eaten than would be expected,— taken, according to my notes, by the wall-eyed pike, both black bass, and the mud-cat (the latter a fellow species of the family).

The soft-finned fishes were not very much more abundant, on the whole, in the stomachs of other species than were those with ctenoid scales, spiny fins, and other defensive structures,— an unexpected circumstance which I cannot at present explain, because I do not know whether it expresses a normal and fixed relation, or whether it may not be due to human interference. It will be shown, however, under another head, that even when the primitive order of nature prevails, the relative numbers of soft-finned and predaceous fishes vary greatly from year to year under the influence of varying circumstances.

---

[1] Dorosoma cepedianum. [2] Catostomatidæ. [3] Polyodon spathula. [4] Etheostomatinæ. [5] Labidesthes sicculus. [6] Eucalia inconstans. [7] Umbra limi. [8] Zygonectes. [9] Noturus. [10] Cyprinidæ. [11] Hyodon tergisus. [12] Aphredoderus sayanus. [13] Amiurus nebulosus. [14] Micropterus dolomiei. [15] Centrarchidæ.

Only the catfishes seem to have acquired defensive structures equal to their protection, the predatory apparatus of the
carnivorous fishes having elsewhere outrun in development the
protective equipment of the best-defended species.

Among the soft-finned fishes the most valuable as food
for other kinds is the gizzard shad (Dorosoma),— this single
fish being about twice as common in adults as all the minnow
family taken together. It made forty per cent. of the food of
the wall-eyed pike; a third that of the black bass; nearly
half that of the common pike or "pickerel"; two thirds that
of the four specimens of golden shad examined; and a third of
the food of the gars. The only other fishes in whose stomachs
it was recognized were the yellow cat *(Amiurus natalis)* and
the young white bass *(Roccus).* It thus seems to be the especial
food of the large game fishes and other particularly predaceous
kinds.

The minnow family (Cyprinidæ) are in our waters especially
appropriated to the support of half-grown game fishes, and
the smaller carnivorous species. They were found in the wall-
eyed pike, the perch, the black bass, the blue-cheeked sunfish,
the croppie, the pirate perch, the pike, the little pickerel,[1]
the chub minnow,[2] the yellow cat, the mud cat, the dog-fish,
and the gar.

Suckers (Catostomatidæ) I determined only from the pike,
the sheepshead, the blue-cheeked sunfish, the yellow cat,
and the dog-fish (Amia). Buffalo[3] and carp[4] occurred in the
pike, the dog-fish, and the above sunfish.

MOLLUSK EATERS.

The ponds and muddy streams of the Mississippi Valley are
the native home of mollusks in remarkable variety and number, and these form a feature of the fauna of the region not
less conspicuous and important than its characteristic and leading groups of fishes. We might, therefore, reasonably expect to
find these dominant groups connected by the food relation; and
consistently with this expectation, we observe that the sheepshead, the cat-fishes, the suckers, and the dog-fish find an impor-

---

[1] Esox vermiculatus. [2] Semotilus. [3] Ictiobus. [4] Carpiodes.

tant part of their food in the molluscan forms abundant in the waters which they themselves most frequent. The class as a whole makes about one fourth of the food of the dog-fish and the sheepshead,— taking the latter as they come, half-grown and adults together,—about half that of the cylindrical suckers,—rising to sixty per cent. in the red horse,[1]— and a considerable ratio (fourteen to sixteen per cent.) of the food of the perch, the common catfishes (Amiurus and Ictalurus), the small-mouthed sunfishes, the top minnows, and the shiner (Notemigonus). Notwithstanding the abundance of the fresh water clams or river mussels (Unio and Anodonta), only a single river fish is especially adapted to their destruction, viz., the white perch or sheepshead; and this species derives, on the whole, a larger part of its food from univalve than from bivalve mollusks, the former being eaten especially by half-grown specimens, and the latter being the chief dependence of the adults.

The ability of the catfishes to tear the less powerful clams from their shells has been especially discussed in another paper* containing the details of the food of the family. Even the very young Unios were rarely encountered in the food of fishes, my notes recording their presence in only three sunfishes, a brook silversides, and a perch. Large clams were eaten freely by the full-grown sheepshead — whose enormous and powerful pharyngeal jaws with their solid pavement teeth are adapted to crushing the shells of mollusks — and by the bull-heads (Amiurus), especially the marbled cat.[2] The small and thin-shelled Sphæriums are much more frequent objects in the food of mollusk-eating fishes than are the Unios. This genus alone made twenty-nine per cent. of the food of our one hundred and seven specimens of the sucker family, and nineteen per cent. of that of a dozen dog-fishes. Among the suckers it was eaten greedily by both the cylindrical and the deep-bodied species, although somewhat more freely by the former. Even the river carp,[3] with its weak pharyngeal jaws and delicate teeth, finds these sufficient to crush the shells of Sphærium, and our nineteen specimens had obtained about

---

* Bull. Ill. St. Lab. Nat. Hist., Vol. II., pp. 457, 458.
[1] Moxostoma. [2] Amiurus marmoratus. [3] Carpiodes.

one fourth of their food from this genus. Besides the above families, smaller quantities of the bivalve mollusks occurred in the food of one of the sunfishes *(Lepomis pallidus)* and — doubtless by accident only — in the gizzard shad.

The gasteropod mollusks (snails of various descriptions) were more abundant than bivalve forms in the sheepshead and the sunfishes and all the smaller fishes which feed upon Mollusca, but less abundant in the suckers and the catfishes. In the sheepshead they made one fifth of the food of the twenty-five specimens examined, but the greater part of these had not yet passed the insectivorous stage, this being much longer continued in the sheepshead than in many other fishes. A few of these univalve Mollusca occurred in the food of the common perch and in certain species of sunfishes, especially in the superabundant bream or pumpkin-seed. They made fifteen per cent. of the food of the minute top minnows, and occurred in smaller quantities among the darters, the grass pickerel, the mud minnows, and the cyprinoids. The heavier river snails, Vivipara and Melantho, were eaten especially by the cylindrical suckers, and the catfishes. The delicate pond snails (Succinea, Limnæa, and Physa) were taken chiefly by the smaller mollusk-eating fishes,— a few of them also by the catfishes and the suckers.

Further particulars concerning the molluscan food may be obtained by the interested reader from the list of food elements at the end of this article.

INSECTIVOROUS SPECIES.

It is from the class of insects that adult fishes derive the most important portion of their food, this class furnishing, for example, forty per cent. of the food of all the adults which I examined.

The principal insectivorous fishes are the smaller species, whose size and food structures, when adult, unfit them for the capture of Entomostraca, and yet do not bring them within reach of fishes or Mollusca. Some of these fishes have peculiar habits which render them especially dependent upon insect life,— the little minnow Phenacobius, for example, which, according to my studies, makes nearly all its food from insects (ninety-eight per cent.) found under stones in running water. Next are the pirate perch, Aphredoderus (ninety-one per cent.), then the darters (eighty-seven per cent.), the croppies (seventy-three per cent.), half-grown sheepshead (seventy-one per cent.), the shovel fish (fifty-nine per cent.), the chub minnow (fifty-six per cent.), the black warrior sunfish (Chænobryttus) and the brook silversides (each fifty-four per cent.), and the rock bass and the cyprinoid genus Notropis, (each fifty-two per cent.)

Those which take few insects or none are mostly the mud-feeders and the ichthyophagous species, Amia (the dog-fish) being the only exception noted to this general statement. Thus we find insects wholly or nearly absent from the adult dietary of the burbot, the pike, the gar, the black bass, the wall-eyed pike, and the great river catfish, and from that of the hickory shad[1] and the mud-eating minnows (the shiner, the fat-head,[2] etc.). It is to be noted, however, that the larger fishes all go through an insectivorous stage, whether their food when adult be almost wholly other fishes, as with the gar and the pike, or mollusks, as with the sheepshead. The mud-feeders, however, seem not to pass through this stage, but to adopt the limophagous habit as soon as they cease to depend upon Entomostraca.

Terrestrial insects, dropping into the water accidentally or swept in by rains, are evidently diligently sought and largely depended upon by several species, such as the pirate perch, the brook minnow, the top minnows or killifishes (cyprinodonts), the toothed herring and several cyprinoids (Semotilus, Pimephales, and Notropis).

Among aquatic insects, minute slender dipterous larvæ, belonging mostly to Chironomus, Corethra, and allied genera, are of remarkable importance, making, in fact, nearly one tenth of the food of all the fishes studied. They are most abundant in Phenacobius and Etheostoma, which genera have become especially adapted to the search for these insect forms in shallow rocky streams. Next I found them most generally in the pirate perch, the brook silversides, and the stickleback, in which they averaged forty-five per cent. They amounted to about one third the food of fishes as large and important as the red

---

[1] Dorosoma. [2] Pimephales.

horse and the river carp, and made nearly one fourth that of fifty-one buffalo fishes. They appear further in considerable quantity in the food of a number of the minnow family (Notropis, Pimephales, etc.), which habitually frequent the swift water of stony streams, but were curiously deficient in the small collection of miller's thumbs (Cottidæ) which hunt for food in similar situations. The sunfishes eat but few of this important group, the average of the family being only six per cent.

Larvæ of aquatic beetles, notwithstanding the abundance of some of the forms, occurred in only insignificant ratios, but were taken by fifty-six specimens, belonging to nineteen of the species,— more frequently by the sunfishes than by any other group. The kinds most commonly captured were larvæ of Gyrinidæ and Hydrophilidæ; whereas the adult surface beetles themselves (Gyrinus, Dineutes, etc.) — whose zigzag-darting swarms no one can have failed to notice — were not once encountered in my studies.

The almost equally well-known slender water-skippers (Hygrotrechus) seem also completely protected by their habits and activity from capture by fishes, only a single specimen occurring in the food of all my specimens. Indeed, the true water bugs (Hemiptera) were generally rare, with the exception of the small soft-bodied genus, Corisa, which was taken by one hundred and ten specimens, belonging to twenty-seven species, — most abundantly by the sunfishes and top minnows.

From the order Neuroptera fishes draw a larger part of their food than from any other single group. In fact, nearly a fifth of the entire amount of food consumed by all the adult fishes examined by me consisted of aquatic larvæ of this order, the greater part of them larvæ of day flies (Ephemeridæ), principally of the genus Hexagenia.* These neuropterous larvæ were eaten especially by the miller's thumb, the sheepshead, the white and striped bass, the common perch, thirteen species of the darters, both the black bass, seven of the sunfishes, the rock bass and the croppies, the pirate perch, the brook silversides, the sticklebacks, the mud minnow, the top min_

---

* The winged adults of this and related genera are often called "river flies" in Illinois.

nows, the gizzard shad, the toothed herring, twelve species each of the true minnow family and of the suckers and buffalo, five catfishes, the dog-fish, and the shovel fish,—seventy species out of the eighty-seven which I have studied.

Among the above, I found them the most important food of the white bass, the toothed herring, the shovel fish (fifty-one per cent.), and the croppies; while they made a fourth or more of the alimentary contents of the sheepshead (forty-six per cent.), the darters, the pirate perch, the common sunfishes (Lepomis and Chænobryttus), the rock bass, the little pickerel, and the common sucker (thirty-six per cent).

Ephemerid larvæ were eaten by two hundred and thirteen specimens of forty-eight species — not counting young. The larvæ of Hexagenia, one of the commonest of the "river flies," was by far the most important insect of this group, this alone amounting to about half of all the Neuroptera eaten. They made nearly one half of the food of the shovel fish, more than one tenth that of the sunfishes, and the principal food resource of half-grown sheepshead; but were rarely taken by the sucker family, and made only five per cent. of the food of the catfish group.

The various larvæ of the dragon flies, on the other hand, were much less frequently encountered. They seemed to be most abundant in the food of the grass pickerel, (twenty-five per cent.), and next to that, in the croppie, the pirate perch, and the common perch (ten to thirteen per cent.).

Case-worms (Phryganeidæ) were somewhat rarely found, rising to fifteen per cent. in the rock bass and twelve per cent. in the minnows of the Hybopsis group, but otherwise averaging from one to six per cent. in less than half of the species.

THE CRUSTACEAN ELEMENT.

Of the four principal classes of the animal food of fishes; viz., fishes, mollusks, insects, and Crustacea, the latter stand third in importance according to my observations, mollusks alone being inferior to them. That insect larvæ should be more abundant in the food of fresh-water fishes than are crustaceans, is a somewhat unexpected fact, but while the former made about twenty-five per cent. of the food of our entire collection, the crustaceans amounted to only fourteen per cent. These divide conveniently into crayfishes, the medium-sized, sessile-eyed crustaceans (Isopoda and Amphipoda), and Entomostraca. The so-called fresh-water shrimps (Palæmon and Palæmonetes) appeared so rarely in the food that they need scarcely be taken into the account.

Crayfishes made about a sixth of the food of the burbot; about a tenth that of the common perch, a fourth that of half a dozen gars, not far from a third that of the black bass, * the dog-fish, and our four rock bass. Young crayfishes appeared quite frequently in some of the larger minnows (Semotilus and Hybopsis), and also in catfishes, especially the pond and river bull-heads, averaging nearly fifteen per cent. of the entire food of the two most abundant species.

The small, sessile-eyed crustaceans eaten by fishes were nearly all of four species; viz., *Allorchestes dentata,*— excessively abundant in the northern part of the State,— a species of Gammarus not uncommon in running streams, and two representatives of the isopod genera Asellus and Mancasellus. To fishes at large, this group is of little importance; but the perch of northern Illinois finds about one third of its food among them, and the common sunfishes (Lepomis) eat a considerable ratio (eleven per cent.). The miller's thumb of southern Illinois seems also to search for them among the stones.

The little Allorchestes mentioned above I found in a single white bass, in eleven of the common perch, in one of the largest darters, in five young black bass, in seventeen sunfishes of various species, in the rock bass, the pirate perch, a single grass pickerel and six top minnows, in only two of the true minnow family, in two only of the sucker tribe, in seventeen catfishes,— mostly young or of the smallest species,— in a single dog-fish, and in a single spoon-bill.[1] The common

* Our specimens — especially of the small-mouthed black bass — were too few in number to make this average reliable.

[1] Polyodon.

Asellus, or water wood louse, was less generally eaten; by only two of the miller's thumb, a single sheepshead, a white bass, four perch, two young black bass, eight sunfishes (Lepomis), two pirate perch, a grass pickerel, three small catfishes, and a dog-fish.

The minute crustaceans commonly grouped as Entomostraca are a much more important element. Among full-grown fishes, I find them especially important in the shovel fish,— where they made one third the food of the specimens studied, — in the common lake herring,[1] in the brook silversides (forty per cent.), in the stickleback (thirty per cent.), in the darter family (eleven per cent.), and in the mud minnows (ten per cent.). The perch had taken scarcely a trace of them. Among the sunfishes at large they were present in only insignificant ratio; but two genera (Pomóxys and Centrarchus), distinguished by long and numerous rakers on the anterior gill, had derived about one tenth of their food from these minute crustaceans. In the early spring especially, when the backwaters of the streams are filled with Entomostraca, the stomachs of these fishes are often distended with the commonest forms of Cladocera.

Notemigonus and Notropis among the minnows, represented in my collections by one hundred and twenty-five and one hundred specimens respectively, had obtained about a sixth of their food from Entomostraca.

Ten per cent. of the food of the sucker family consisted of them, mostly taken by the deep-bodied species Carpiodes and Ictiobus, in which they made a fourth or a fifth of the entire food. This fact is explained, it will be remembered, by the relatively long, slender, and numerous gill-rakers of these fishes. Large river-buffalo were occasionally crammed with the smallest of these Entomostraca,— the minute Canthocamptus, only a twenty-fifth of an inch in length.

I have several times remarked the peculiar importance of Entomostraca to the shovel fish,— one of the largest of our fresh-water animals,— a fact accounted for by the remarkable branchial strainer of this species, probably the most efficient apparatus of its kind known to the ichthyologist. Here,

---

[1] Coregonus artedi.

again, the smallest forms were the most abundant. Generally, however, the Cladocera were more common than the other orders, the bivalve Cypris (most frequent in the mud) being much less abundant in the food. I have shown elsewhere,* at length, that Entomostraca compose by far the greater part of the food of young fishes of all descriptions,— with the partial exception of the sucker family, the young of which feed largely on still more minute organic forms,— and present an abstract of these facts in this article under another head.†

Particulars concerning the use of this abundant and varied group as food for fishes, are so numerous as to make them difficult to summarize, and the interested reader is again referred to the detailed list accompanying this paper.

VERMES AS FOOD FOR FISHES.

Probably to those accustomed to the abundance of true worms (Vermes) in marine situations, no feature of the poverty of fresh-water life will be more striking than the small number of this subkingdom occurring in the course of miscellaneous aquatic collections in the interior. Similarly we notice that in the food of fishes the occurrence of Vermes is so rarely noticed that they might be left out of account entirely without appreciably affecting any of the important ratios.

The minnows (cyprinoids) had eaten more of them than any other family,— three per cent. of the food of twenty-two specimens of Semotilus being credited to them, and one per cent. of that of thirteen specimens of Pimephales, besides a trace in the food of Notropis. More precisely analyzed, we find that a single Nais, a Lumbriculus, two examples of Gordius (doubtless taken as insect parasites) and several minute rotifers (wheel-animalcules) are the forms upon which this estimate is based.

A trace of Vermes likewise appears in the food of suckers, — mostly a polyzoan species (Plumatella) and minute rotifers sucked up with the mud.

* Bull. Ill. St. Lab. Nat. Hist., Vol. I., No. 3, pp. 75, 76.
† See pp. 495 and 496.

Catfishes alone seem purposely to eat leeches, these occurring in nine specimens of three different species of this family, and also in one common sucker and in a single shovel fish. This leech last mentioned and a small quantity of Plumatella were the only Vermes eaten by the shovel fishes which I examined.

A planarian worm occurred in one small stone cat, while rotifers were recognized in a common minnow, eight young red-horse, six young chub suckers,[1] five of the common sucker,[2] a single Carpiodes (young), and seven young buffalo. Polyzoa were noted, in addition to the instances above mentioned, in four common sunfishes, the croppie, and seven buffalo.

SPONGES AND PROTOZOA.

One of the fresh water sponges (Spongilla) had been eaten in considerable quantities by two examples of the spotted cat taken in September, but this element was not encountered elsewhere in my studies.

That the minutest and simplest of all the animal forms, far too small for the eye of a fish to see without a microscope, should have been recognized in the food of seventeen species of fishes is, of course, to be explained only as an incident of the feeding habit. It is possible, however, that these Protozoa, where especially abundant, may be recognized in the mass by the delicate sensory structures of the fish; and they seem in most cases to have been taken with mud and slime rich in organic substances. As most of them are extremely perishable, and can scarcely leave a trace a few seconds after immersion in the gastric juices of the fish, it is probable that they contribute much more generally than our observations indicate to the food of some fishes, especially to those which feed upon the bottom.

Young suckers under six inches in length clearly take them purposely, substituting them in great part for the Entomostraca taken by other fishes of their size and age.

I detected Protozoa in the food of several genera of Cyprinidæ, in the young of buffalo, the river carp, the chub sucker, the red horse, the stone roller,[3] in the common sucker,

---

[1]Erimyzon sucetta. [2]Catostomus teres. [3]Hypentelium.

in a single gizzard shad, in a stone cat, and in a top minnow. The commonest forms, as would be supposed, were those protected by permanent shells; viz., Difflugia, Centropyxis, Arcella, and the like; but occasionally specimens of Actinosphærium, Euglena, and Dinobryon were present and recognized.

SCAVENGERS.

The only scavenger fishes of our collection were three species of the common catfishes; the spotted cat, the yellow cat, and the marbled cat,—all of which had eaten dead animal matter, including pieces of fish, ham, mice, kittens, and the like. A single large-mouthed black bass had likewise eaten food of this description.

VEGETABLE FEEDERS.

Considering the wealth of vegetation accessible to aquatic animals, and the fact that few other strictly aquatic kinds have the vegetarian habit, it is indeed remarkable that the plant food of fishes is an unimportant part of their diet. Taking our nine hundred specimens together, the vegetation eaten by them certainly would have amounted to less than ten per cent. of their entire food, and excluding vegetable objects apparently taken by chance, it probably would not reach five per cent.

The greatest vegetarians are among the minnow family, largely in the genera Hybopsis, Notemigonus, and Semotilus, thirteen specimens of the first and twenty-five of the second having taken about half their food from vegetable objects. One hundred and twelve Notropis, twenty-two Semotilus, eighteen Hybognathus, and nine Campostoma, had found in the vegetable kingdom a fourth or fifth of their food. Counting each genus as a unit, I find that the family as a whole obtained from plants about twenty-three per cent. of its food. The little Phenacobius, already reported as strictly insectivorous, was the only one studied in which vegetation can scarcely be said to occur.

The mud minnows (Umbridæ) are also largely vegetarian (forty-one per cent.); and likewise the cyprinodonts, the vegetable average in the food of thirty-three specimens being seventeen per cent. Plant structures made about one fourth the food of seven sticklebacks.

Certain of the sunfishes evidently take plant food purposely, on occasion, this making, for example, nearly a tenth of the food of forty-seven specimens of Lepomis. Among the larger fishes, the principal vegetarian is the gizzard shad, in which this element was reckoned at about a third,— taken, however, not separately, but with quantities of mud. A considerable part of it was distillery slops obtained near towns.

The buffalo fishes are likewise largely vegetarian, more than a fourth of their food coming from plants,— about a third of this in our specimens, refuse from distilleries. Vegetation made a tenth of the food of the larger genera of catfishes (Amiurus and Ictalurus),— some of it distillery refuse,— and nearly as large a ratio of that of the great Polyodon.

Not infrequently, terrestrial vegetable rubbish — seeds of grasses, leaves of plants, and similar matter — was taken in quantity to make it certain that its appropriation was not accidental.

Besides a great variety of Algæ, both filamentous and unicellular, including considerable quantities of diatoms, the principal plant forms found in the food of fishes were the duckweeds Lemna and Wolffia. The deep-bodied suckers, especially, occasionally take quantities of these little plants during the autumnal months.

MUD.

The principal mud-eating fishes are the gizzard shad, the common shiner, and the genera of minnows belonging to the groups with elongate intestines and cultrate pharyngeal teeth; viz., Pimephales, Hybognathus, Chrosomus, and Campostoma. Much mud was taken also by the cylindrical members of the sucker family, but apparently as an incident to their search for mollusks.

SUMMARY OF THE FOOD OF THE YOUNG.*

By an examination of three hundred and seven specimens, representing twenty-seven species, twenty-six genera, and twelve families of Illinois fishes, I learn that the food of many species differs greatly according to age, and that, in fact, the life of most of our fishes divides into at least two periods, and that of many into three, with respect to the kinds of food chiefly taken.

In the first of these periods a remarkable similarity of food was noticed among species whose later feeding habits are widely different. The full grown black bass, for example, feeds principally on fishes and crayfishes, the sheepshead on mollusks, and the gizzard shad on mud and Algæ, while the catfishes are nearly omnivorous; yet all these agree so closely in food when very small, that one could not possibly tell from the contents of the stomachs which group he was dealing with.

I will now summarize the facts concerning the earliest food of the principal species, taken *seriatim*.

The food of six common perch *(Perca lutea)* from an inch to an inch and a quarter long, consisted wholly of Entomostraca (ninety-two per cent.) and minute larvæ of Chironomus. No very small white bass (Labracidæ) were found, the youngest being an inch and a quarter long. Half the food of this consisted of Entomostraca, and the other half of minute gizzard shad. Forty-three sunfishes (Centrarchidæ) from five eighths of an inch to two inches long, had made ninety-six per cent. of their food of Entomostraca and the small larvæ of gnats (Chironomus) already mentioned, seventy per cent. of the first and twenty-six of the second. This group comprised five specimens of black bass under three quarters of an inch in length, two rock bass of similar size, two of the large-mouthed sunfish (Chænobryttus) from seven eighths of an inch to an inch long, nineteen of the commoner sunfishes (Lepomis) ranging in length from an inch to two inches, five of the genus Centrarchus, one inch and under, four croppies

---

* For detailed treatment of this topic see Bull. Ill. St. Lab. Nat. Hist., Vol. I., No. 3, p. 66, and No. 6, p. 95.

(Pomoxys) from three quarters of an inch to an inch and a half, and six indeterminable specimens, probably Lepomis, from seven sixteenths to five eighths of an inch long. A single sheepshead an inch and an eighth in length had eaten Chironomus larvæ (seventy-five per cent.) and larvæ of the "river fly" (Hexagenia). A single grass pickerel about an inch and a quarter long had taken about sixty per cent. of its food from Entomostraca and young Amphipoda, the remainder consisting of little fishes.

The first food of the common white-fish was determined experimentally, the breeding habits of this species making direct observation impossible. Three hundred and forty very young fry fed with fragments of the brook shrimp, Gammarus, in a hatching house, were examined in January, 1881, and thirty-five of them, which had apparently taken food, were dissected. Minute fragments of Gammarus were found in but eighteen of these, while five contained minute insect larvæ, four, Entomostraca, and eight, small particles of vegetation,— objects accidentally conveyed to them in the water of the hatching house. In two hundred and forty-two others, confined in spring water; only eight were found to have eaten anything, and these had taken only Algæ and vegetable fragments. In February of the same year, fourteen specimens, confined in a small aquarium and supplied with living objects, plant and animal, from stagnant pools, were proven to feed freely upon the smallest Entomostraca presented to them,— chiefly Cyclops and Canthocamptus, ten of the fourteen eating Cyclops, three Canthocamptus, and one a specimen of each.

A little later, a more extensive experiment was conducted by means of a large aquarium, in which there were placed several hundred fry, kept constantly supplied with all the living objects which a fine gauze net would separate from the waters of Lake Michigan. Of one hundred and six of these, dissected within the following fortnight, sixty-three had taken food consisting almost wholly of the smallest Entomostraca occurring in the Lake (a minute Cyclops and a slender Diaptomus). The other objects encountered were rotifers, and diatoms and other unicellular Algæ, appearing, however, in such trivial quantity as to contribute nothing of importance to the support of the fry.

A dozen specimens of small gizzard shad, ranging in length from four fifths of an inch to nearly two inches, had eaten about ninety per cent. of Entomostraca, two per cent. of Chironomus larvæ, and, for the remainder, Algæ.

The true minnows (Cyprinidæ) seem to agree with the suckers in the more minute character of their early food. Six examples — three eights to three fourths of an inch long — too small to determine, but apparently belonging to the genera Minnilus, had eaten Entomostraca, Chironomus larvæ, many Protozoa, and unicellular Algæ, a few filamentous Algæ and minute fungi and fungus spores, a water mite, and a few accidental insects. In several specimens of the common chub minnow (Semotilus), from five eighths of an inch to an inch in length, seven per cent. of the food was Entomostraca, and the remainder consisted of filamentous Algæ. It should be noted, however, that twenty per cent. of that of the smallest specimen, which was five eighths of an inch long, was Cyclops, and it may be that Semotilus lives wholly on Entomostraca at first, merely changing its habit earlier than most of its allies. Two other minnows of the genus Notropis, an inch and a half in length, had eaten nothing but Entomostraca. The Cyprinidæ, like the sucker family, are toothless when young.

Thirty young suckers were studied, representing five genera of their family. The very smallest were found feeding on Entomostraca only, and it is possible that these usually form the first food of the family; but later they resort to elements still more minute: viz., rotifers, Protozoa, and unicellular Algæ, quantities of which were found in the intestines of young suckers six inches or more in length. Young stone rollers (Hypentelium) not more than an inch and a half long, had taken chiefly larvæ of Chironomus (ninety per cent.), the remaining tenth being principally Entomostraca. A single small black sucker (Minytrema) had eaten little but Cyclops. Four chub suckers (Erimyzon), two three quarters of an inch, and two an inch and a quarter long, had eaten only Entomostraca and a trace of water mites. In two larger specimens, however, still minuter forms were the leading feature of the food, including rotifers, Protozoa, and unicellular Algæ. Another example, three inches long, had eaten a trace of

Chironomus larvæ, but for all the rest, one of the smallest of the Entomostraca (Canthocamptus). Ten young red horse (Moxostoma), varying in length from an inch to two and three fourths inches, had fed largely upon Protozoa,— especially the largest of the specimens,— but the smallest of them had taken a considerable amount of Entomostraca,—notably the bivalve cyprids occurring on the bottom. Two of the commonest buffalo fish (Ictiobus), seven eighths of an inch long, had eaten most freely of unicellular Algæ (sixty-three per cent.), the remainder of the food consisting of rotifers and Entomostraca. Four of the river carp (Carpiodes), seven eighths of an inch to two inches long, had fed like the preceding, except that the Entomostraca amounted to nearly half the food, while the rotifers were comparatively few.

Young catfishes, only three eighths of an inch in length, belonging to the genus Amiurus, but quite too small to be specifically determinable, were filled with various Entomostraca and Chironomus larvæ. Other examples of this genus, making thirteen in all, none longer than an inch and five eighths, had fed almost wholly on Entomostraca and larvæ of Chironomus, the latter, however, composing seventy-four per cent. of the food of all, and the former eighteen per cent. Six small stone cats (Noturus), varying in length from seven eighths of an inch to one and a half inches, had taken more Chironomus larvæ and scarcely any Entomostraca.

A single dog-fish (Amia), one and three fourths inches long, had eaten seventy per cent. of Entomostraca, a few larvæ of Chironomus, some small crustaceans, and aquatic insects. Others of the species, under an inch in length, had the intestine packed with Entomostraca. Of the common river gars one, an inch and a quarter long, had filled itself with minute Entomostraca, while two other specimens had eaten only the smallest fry of fishes.

To recapitulate, I find that, taking together the young of all the genera studied, considering each genus as a unit, and combining the minute dipterous larvæ with the Entomostraca as having essentially the same relation, about seventy-five per cent. of the food taken by young fishes of all descriptions is made up of these elements.

From the above it is clear that young fishes in general depend at first on Entomostraca and certain small insect larvæ (chiefly those of two genera of gnats), beginning with the smallest of these forms, or with those especially exposed to their attack. One-celled plants and animals are also eaten freely by the young of two of the largest families.

Correlated with these facts, I find that two at least of the genera, which are toothless when adult, have minute raptatorial teeth in this early stage; viz., Coregonus and Dorosoma. Otherwise young fishes have no apparatus specially adapted to the capture of their minute prey, but this is brought within their reach merely by their own small size and the corresponding minuteness of their structures of food prehension. Later, as the larger species grow, this apparatus becomes too coarse to retain objects so minute, but other food resources are made available, usually through some adaptive modification of the fishes themselves.

In other words, one-celled organisms and Entomostraca are the natural, and practically the only, food of an undifferentiated small fish; and to be at liberty to grow, the fish must either change its food (as is usually done) or must develop a special apparatus (commonly a set of fine long gill-rakers) for the separation of Entomostraca from the waters in which they swim.

Of the fishes which emerge from this earliest stage, through increase in size with failure to develop alimentary structures especially fitted to the appropriation of minute animal forms, some become mud-eaters, like Campostoma and the gizzard shad; a few apparently become vegetarians at once; but most pass into or through an insectivorous stage. After this a few become nearly omnivorous, like the bull-heads; others learn to depend chiefly on molluscan food, — the sheepshead and the red horse species, — but many become essentially carnivorous. In fact, unless the gars are an exception, as they now seem to be, (attacking young fishes almost as soon as they can swallow,) all our specially carnivorous fishes make a progress of three steps, marked, respectively by the predominance of Entomostraca, of insects, and of fishes, in their food; and the same is true of those strictly fitted for a molluscan diet.

While small fishes of all sorts are evidently competitors for food, this competition is relieved to some extent by differences of breeding season, the species dropping in successively to the banquet, some commencing in very early spring, or even, like the white-fish, depositing their eggs in fall, that their young may be the first at the board, while others delay until June or July. The most active breeding period coincides, however, with that of the greatest evolution of Entomostraca in the backwaters of our streams; that is, the early spring.

That large adult fishes, with fine and numerous rakers on the gills — like the shovel-fish and the river carp — may compete directly with the young of all other species, and tend to keep their numbers down by diminishing their food supply — especially in times of scarcity — is very probable, but is not certainly true; for these larger fishes have other food resources also, and may resort to Entomostraca only when these are superabundant, thus appropriating the mere excess above what are required for the young of other groups.

ON THE DEFINITENESS AND PERMANENCY OF THE FOOD HABITS OF FISHES.

It is always posssible that the seemingly specific differences of food exhibited by data derived from miscellaneous collections not strictly comparable as to dates and localities, are really due to differences of circumstance affecting the representatives of the species, and not to differences in the food habits or the regimen of the species in general. Date, locality, and other circumstantial conditions, may have more to do with the distinctions of food detected than structure and specific habit. It is true that the probability of such errors of inference is reduced to a minimum where alimentary peculiarities can be clearly correlated with peculiarities of structure, as has usually been done in my discussions; but to test still further the distinctness of species and genera with respect to food habits and preferences, I have assorted my observations according to dates and localities of the collections on which they were made and have compared species with species as occurring under the same general conditions and at the same time. If perch and catfishes caught in the same haul of the seine show more marked differences in food between the two groups than those exhibited by the individuals of each group among themselves, the probability is considerable that the differences are specific instead of accidental; and such probability becomes greater the greater the number of species found to present corresponding differences under corresponding circumstances. Although it was rarely the case that examples enough of two or more species comparable as to size and range had been taken at the same time and place to afford a tolerable average of the food under local conditions, yet a sufficient number of such cases was found to give a considerable amount of evidence on this point.

Thus three specimens of the marbled cat, *Amiurus marmoratus*, taken at Peoria, Nov. 1, 1878, had derived nine tenths of their food from Hexagenia larvæ, the remainder consisting of leeches and a few spiders; while eight specimens of the large-mouthed black bass, *Micropterus salmoides*, taken at the same time and place, had eaten nothing but the young gizzard shad (Dorosoma).

Comparing the food of four examples of the channel cat (*Ictalurus punctatus*) with seven croppies (Pomoxys), both taken at Peoria, Apr. 10, 1878, I found that aquatic insects made ninety-eight per cent. of the food of the latter, seventy per cent. being Hexagenia larvæ, while only sixty-two per cent. of the food of the catfishes consisted of insects (ephemerid larvæ twenty-eight per cent.), the remainder consisting of vegetation and scraps of dead fishes.

A contrast equally decided is shown by three specimens of the gizzard shad (Dorosoma) and four of the rock bass (*Ambloplites rupestris*), all obtained at Ottawa, July 8, 1879. The former had swallowed large quantities of fine mud containing about twenty per cent. of minutely divided vegetable *débris*, while the latter had fed wholly upon insects, fishes, and crayfishes,— the first chiefly aquatic larvæ.

Even in the shallow muddy pools left behind in the retreating overflow of the Mississippi in southern Illinois, fishes of the same size but differing widely in alimentary structures exhibit corresponding differences in the selections made from the meager food resources of their localities. Two of the common blunt-jawed minnows *(Hybognathus nuchalis)* had fed here almost wholly upon mud mixed with Algæ and miscellaneous vegetation; while three of the little pirate perch (Aphredoderus) had eaten little but Chironomus larvæ, half the food of one of the specimens being wholly small fishes, and insignificant quantities of Entomostraca occurring in the stomachs of the others.

A small collection, made from the Little Fox River, in White county, in southern Illinois, Oct. 5, 1882, of four specimens each of Labidesthes and *Zygonectes notatus* enables us to bring into comparison the food of two extremely different species taken together from the same pools in a running stream. The Labidesthes, although predaceous in habit and feeding most commonly upon Entomostraca, was here giving its attention wholly to terrestrial insects,— more than two thirds of them winged Chironomus; while the Zygonectes had eaten in addition to thirty-seven per cent. of terrestrial insects (scarcely any of them Chironomus imagos), about thirty per cent. of aquatic vegetation, nine per cent. of Entomostraca, eleven per cent. of aquatic insects, and fourteen per cent. of mollusks. These differences in food have no apparent relation to the essential structural differences of the species, but must be considered an illustration of the 'various effect of like conditions when applied to different species.

On the other hand, three bull-heads *(Amiurus nebulosus)* and six common perch (Perca) taken from Fox River, at McHenry, May 9, 1880, did not differ remarkably in food, both groups having eaten crayfishes, mollusks, aquatic insects, and vegetation. One of the catfishes had taken another fish, and one had eaten leeches. It is to be noted, however, that these species are both bottom feeders, and that both lots of these specimens had taken about the average food of their kind.*

The above are examples of the food relations of fishes widely separated from each other in the classification and decidedly different in alimentary structures and in feeding habits. Illustrations of the differences in food apparent in

---

* See Bull. Ill. St. Lab. Nat. Hist., Vol. I., No. 3, p. 35.

species allied in classification but differing with respect to the structures concerned in the appropriation of food are given by the following examples.

Two species of minnows, *Chrosomus erythrogaster* and *Semotilus atromaculatus*— the first represented by fourteen specimens, and the second by six, all collected from a small tributary of the Fox, near Plano, Sept. 8, 1882 — were brought into comparison with reference to their food, with the result that the characteristic differences of the species, as shown in the general discussion of the group published in our Bulletin 6, Vol. I., were clearly manifested by this small number. In the former lot seventy-five per cent. of the food was mud, the remainder being indiscriminate vegetable *débris*; while in the latter the entire mass consisted of insects (chiefly terrestrial) except a single insect parasite (Gordius).

From one of the permanent ponds or so-called lakes of southern Illinois, covered in September with a film of Wolffia and other vegetation, three specimens of *Gambusia patruelis* and five of *Umbra limi* were examined. The former had eaten little but Wolffia, which amounted to more than ninety per cent. of the food, the remainder consisting of Entomostraca, mollusks, and aquatic insect larvæ, while the Wolffia made less than sixty per cent. of the food of the Umbra,— about one fourth consisting of Entomostraca, and the remainder of unrecognized insects.

Two minnows of similar range (*Phenacobius mirabilis* and *Notropis whipplei*) agree essentially in gill structure and pharyngeal teeth, and differ but little in the relative length of intestine; and they have consequently been placed by me in the same alimentary group.* They are unlike, however, in the form of the mouth and in their haunts and feeding habits. This difference is reflected in the food of a small collection made in the Galena River, in April, 1880, three specimens of Phenacobius having eaten only aquatic larvæ and pupæ (nearly all chironomid), while the food of the Notropis, represented by six specimens, was of a varied character, containing few aquatic larvæ (only one per cent. of Chironomus), but consisting chiefly of miscellaneous collections of terrestrial insects,

seeds and anthers of terrestrial plants, and other accidental
rubbish.

From a collection made at Henry, Illinois, Nov. 1, 1887,
four specimens of croppie (*Pomoxys nigromaculatus*) are
comparable with five sunfishes (*Lepomis pallidus*), and
three large-mouthed black bass (*Micropterus salmoides*) may
be compared with three striped bass (*Roccus chrysops*). Eighty-
four per cent. of the food of the Pomoxys consisted of Hexagenia larvæ, an additional six per cent. being other aquatic
larvæ, and the remaining ten per cent. consisting of fishes;
while the Lepomis had eaten but twelve per cent. of Hexagenia larvæ, eight per cent. of other aquatic insects, and no
fishes at all, — the remaining elements being terrestrial insects
(about one fourth), worms (Nais and Lumbriculus, fifteen per
cent.), and mollusks (thirty-seven per cent.).

The black bass had eaten chiefly fishes and a mouse,
together with a few aquatic insects; while the food of the
striped bass was nearly all ephemerid larvæ with only a trace
of fishes.

A collection of small fishes, made from Mackinaw Creek,
in Woodford county, August 20, 1879, affords an interesting
opportunity to compare the food of a number of the smaller
species (cyprinoids, darters, etc.). About half that of four
specimens of *Notropis megalops* collected there, consisted of
insects, the remainder being terrestrial and aquatic vegetation;
and substantially the same statement may be made with respect
to six specimens of *Notropis whipplei*, — these two species
belonging respectively to the third and fourth groups of my
paper on the "Food of the Smaller Fresh Water Fishes."[*]

Two specimens of *Hybopsis biguttatus*, on the other
hand, had eaten only aquatic vegetation; and two examples of
Phenacobius — a species extremely darter-like in its haunts and
habits — had taken only Chironomus larvæ.

The darters were represented by four examples of Boleosoma and six of Hadropterus, the former and smaller species
having eaten mostly Chironomus larvæ and Entomostraca, —
eighty-nine per cent. and eleven per cent. respectively, —
while the larger had taken only aquatic larvæ, — nearly all
ephemerids.

[*] Bull. Ill. St. Lab. Nat. Hist., Vol. I., No. 6, p. 76.

Finally, eight of the slender, active, and wholly predaceous little brook silversides (*Labidesthes sicculus*) had eaten a single fish, fourteen per cent. of Entomostraca, and about eighty per cent. of insects — somewhat more than half of aquatic origin. In brief, the structures of Labidesthes, the habits of Phenacobius and the darters, and the differences in size of the species of Boleosoma and Hadropterus were all reflected in the food of this little group.

The obverse fact of the unifying effect of similarity of alimentary structures is apparently shown by a small collection of minnows, all belonging to the first two groups of the paper cited above*, made from an extremely muddy little creek in Jersey county, which contained no visible vegetation and few, if any, Entomostraca. Twelve of these fishes, representing the genera Campostoma, Pimephales, Hyborhynchus, Hybognathus, and Notemigonus, agreed in food almost precisely, all having swallowed the fine mud of the creek bottom, with a slightly varying admixture of unicellular Algæ and vegetable *débris.*

As an example of a contrast between two species agreeing in alimentary structures, but differing in size and somewhat, also, in habitual range, we may take three examples of *Notropis heterodon* and three of *Notropis megalops,* captured at McHenry, May 8, 1880. More than half the food of the latter group consisted of vegetation, and of the former only ten per cent. The remaining ninety per cent. of the food of *heterodon* was Entomostraca; but these were not represented at all in the *megalops,* the remaining food of these specimens consisting of insects and amphipod Crustacea.

Sensible and even conspicuous differences in food often appear between groups which are neither widely separate in classification nor yet distinguished by marked differences in alimentary structures, as between species of the same genus. Sometimes these are apparently due to differences in habit with respect to the search for food; but sometimes seem dependent upon distinction of habit or preferences even more obscure.

Six specimens of the channel cat (*Ictalurus punctatus*), taken at Peoria, October 6, 1887, had eaten insects, mollusks, and vegetation at the rate of forty-one, nineteen, and forty per cent. respectively, the vegetation being nearly all Cladoph-

---

* See the preceding page.

ora and Potamogeton; while the same number of bull-heads (*Amiurus nebulosus*) had derived thirty-seven per cent. of their food from insects, and sixty-three per cent. from mollusks. The difference here was substantially a larger ratio of mollusks for Amiurus, replacing the vegetable food of the Ictalurus group. By a comparison of these differences with those detected between the species at large, as explained on pages 456–461, it will be seen that the former do not represent the specific differences in food, but simply give evidence that two species may be differently affected by the same conditions.

Other specific differences in the same genus are shown by the collections made Oct. 27, 1875, from Peoria Lake. Eight examples of the wall-eyed pike *(Stizostedion vitreum)* had eaten only soft-finned fishes, — excepting one small sunfish, — while four of ten specimens of the related species *S. canadense*, had eaten spiny-finned fishes, and in only three were the fishes recognizable as belonging to the soft-finned species. Three specimens of Micropterus taken with the above had eaten crayfishes and fishes (including a catfish).

Among my specimens of the sucker family (Catostomatidæ), a lot obtained at Quincy, Aug. 25, 1887, are comparable for the present purpose. Four examples each of *Ictiobus urus* and *I. cyprinella* presented a decided contrast with respect to the elements of their food, that of *I. urus* consisting almost wholly of Chironomus larvæ, with large quantities of dirt, while three of the specimens of *I. cyprinella* had eaten scarcely anything but Algæ, ninety per cent. of the food of the fourth being Chironomus larvæ, and the remainder, larvæ of Neuroptera,— Hexagenia and Corydalis.

On the other hand, two small collections of the same species made at Peoria, Oct. 9, 1878 — four of *I. urus* and five of *I. cyprinella* — exhibit similar food, composed chiefly of Entomostraca, Chironomus larvæ, distillery waste (meal, etc.), and aquatic vegetation. The *urus* group alone had eaten Entomostraca, these being replaced in the other by a larger quantity of meal.

The facts above recited are evidence that fishes are not mere animated eating-machines, taking indiscriminately and indifferently whatever their structures fit them to capture, to strain from the waters, or to separate from the mud, but that psychological preferences as well as physical capabilities have something to do with their choice of food.

THE STRUCTURES OF ALIMENTATION.

A brief review of the principal facts respecting the structures of alimentation in fishes will be necessary to exhibit clearly the relation of habit and organization in this particular.

These structures may be conveniently divided into those of *search*, of *prehension*, of *mastication*, and of *digestion*. Means of defence and escape may also properly be mentioned, as belonging to the obverse side of the food relation.

Structural peculiarities relating to the methods and situation of the search for food are illustrated by the barbels of the catfishes and the sturgeons, the shovel of Polyodon, the square head of the stone roller, the flat heads of the top minnows, and the pointed snouts of the darters,— which fit them for prying about between and under stones in running water. Similarly related, are the bare breasts of many darters and the large pectoral fins of the stone roller and Phenacobius.

The structures of food prehension are the lips, the jaws, the teeth, and the gill-rakers, with which should be considered, perhaps, the gill slit or branchial opening. The sucking lips of the Catostomatidæ, organs of touch as well as of prehension, are of course related to the mud-searching habit of these fishes, the protractile jaws aiding in this use. The stout wide jaws of the catfishes, with their wide bands of minute, pointed teeth, are probably to be understood as an apparatus for seizing, holding, and pulling about relatively large objects, whether hard or soft, and are perhaps most useful in feeding upon mollusks. The very large but weak jaw of the shovel fish is explained by the minute character of its food, which offers no resistance, but necessitates the passage of large quantities of water through the mouth; while the long and slender jaws of the long-nosed gar (Lepidosteus) armed with several rows of acute raptatorial teeth, are the best apparatus in our waters for the destruction of a relatively small but active living prey.

The teeth of our fresh-water fishes are always pointed and acute, there being no examples of pavement teeth or cutting incisors among them, such as are found in several marine forms, nor are there any instances of either jaw being toothed and the other not. The evanescent teeth of the young of several species which become toothless when mature, are sometimes to be understood as rudiments, as in the shovel fish, and sometimes as related to the early food, as in the white-fish and the gizzard shad.

The gill-rakers of fishes vary widely in number, length, and usefulness, but are as important and significant as any other part of the feeding apparatus. As they oppose the only obstacle to the escape through the gill slit, of objects which enter the mouth with the water of respiration, they set the minimum of size for objects of the fishes' food, the only exception to this rule being afforded by the few fishes which swallow mud with little or no discrimination.

They are usually arranged in two rows on each gill arch, with frequently one also on the pharyngeal, behind the last gill slit. Occasionally only one row is developed on each gill (lake " herring "), and commonly the second row, if present, is less prominent than the first. The shovel fishes are, however, an exception to this latter statement, for in them both rows are equally and remarkably developed. As the anterior rakers guard the relatively large passage-way between the foremost gill and the opercle, while the other rows merely prevent the escape of objects between the several pairs of gills, the anterior row is almost invariably longer than the remaining series. The shovel fish and the gizzard shad are exceptions. The rakers of this row are commonly longest in the middle of the arch, shortening toward each end; but the particulars of this disposition depend on the length and shape of the arch and the concavity of the inner surface of the opercle. In the gizzard shad, however, the short but very numerous and fine gill-rakers project in a nearly horizontal direction.

The gill-rakers, when short and ineffective, are often armed with minute denticles, variously arranged, but are never branched or pinnate. In several of the sucker family, the rakers of the lower horizontal arm of the arch are represented

by a thick, broad pad, transversely ridged (the ridges representing the separate rakers) so that when approximated these
structures form a continuous floor for the sides of the buccal
cavity. The rakers may vary in number in different species
from ten or twelve in a series, as in some sunfishes, to more
than five hundred, as in the shovel fish; and in length from
mere tubercles, to two or three times the length of the corresponding filaments of the gill. Rarely they are completely
wanting, as in the pike. The anterior row is commonly so set
upon the arch as to be obliquely divaricated by the separation
of the branchial structures, being thus automatically adapted
to the respiratory movements.

They are little developed in young fishes, the small branchial arches and the narrow slits between them serving to separate from the water the minute objects of the earliest food.
Their development with the growth of the fish simply enables
it to retain as elements of its dietary, objects which the coarse-
ness of its branchial structures would otherwise compel it to
forego.

Concerning their relations to food prehension, we may say
in general that if numerous, long, and fine, they indicate the
importance of Entomostraca to the fish. If less numerous,
but moderately long and stout, in a fish of medium size, we
may presume that insects form a considerable ratio of the food.
If wanting, or rather short and strong, the presumption is
(except for the smaller fishes) that the species is either piscivorous or feeds largely upon mollusks, the dental and pharyngeal
apparatus easily showing which.

The pike-perch (Stizostedion) is somewhat remarkable in
the fact that although strictly piscivorous when adult, it has
long and strong gill-rakers, much longer in fact than in the
less piscivorous related species, the common perch. In this
case the rakers seem to have been retained, and even further
developed, as a basis of attachment for several rather large
recurved teeth borne on their inner surfaces, useful in preventing
the escape of a living prey.

The masticatory apparatus of fishes (sometimes wanting)
comprises always a pair of pharyngeal bones,—the lower pharyngeal jaws, a pair of modified branchial arches. These are

commonly opposed by superior pharyngeals, which most frequently consist of osseous and cuticular thickenings of the upper ends of the gill arches, — sometimes of only one or two, as in the catfish family, sometimes of all, as in the sunfishes. In the cyprinoids, the upper pharyngeal is a quadrate or triangular pad, rarely, if ever, toothed, borne upon an oblique, expanded process of the basioccipital. In the sucker family the sickle-shaped lower pharyngeals act against a more or less indurated palatal arch supported by the same cranial process, the firmness and width of this hardened band varying with the development of the lower arches of the apparatus. In most of the Acanthopteri and in the catfish family the lower pharyngeals have a fusiform outline, varying in width according to the food, the upper surface set with minute denticles, sharppointed in the insectivorous species, more or less blunt and conical in those which take a considerable percentage of molluscan food. The immense development of these structures in the sheepshead (Aplodinotus), as a crushing apparatus for Mollusca, is too well known to require description. In the Catostomatidæ the number of teeth may vary from thirty or less to two hundred or more, reduction in number going with increase in size (especially in the lower part of the arch,) both being related to an increased importance of molluscan food.

In the cyprinoids or minnow family, this is practically an insectivorous apparatus, except in some of the species with very long intestine and the limophagous habit, where it seems useful chiefly as a means of grinding up the mud ingested.

In the piscivorous species, and in those with highly developed gill-rakers, the lower pharyngeals are commonly slight and insignificant; but in the former group the upper pharyngeals may be preserved and enlarged as a basis for the insertion of hooked teeth, to aid in the retention of their struggling prey.

Concerning the digestive structures, I will only remark that the fishes with the longest intestine are mud-feeders, as a rule, and that in one of them, — the gizzard shad, a mud lover, *par excellence*, — the pharyngeal jaws (which in the mud-eating cyprinoids are evidently used to grind the food) are functionally replaced by a bulbous, muscular stomach, the pharyngeals themselves being reduced to thin and delicate plates, scarcely better than rudiments.

In this connection the adult size of the fish ought always to be mentioned, since this has, perhaps, at least as much to do with the food as any structural endowment, and frequently, in fact, has had a determining influence on the latter. Many fishes can enjoy the advantages of large size only on condition that they acquire some new capacity of food prehension, adapting them to new food relations. Simple and symmetrical growth of a small fish would render it incapable of straining out Entomostraca without fitting it for the appropriation of any other food, except, perhaps, the larger Crustacea and some aquatic insects; and beyond this insectivorous stage nothing is possible without new adaptations.

CORRELATIONS OF ALIMENTARY ORGANS.

Correlations of structure may be either mediate or immediate, in the latter case modification of one organ being directly dependent on modification of another, and in the former both parties to the correlation being modified by a common cause. The immediate class of correlations are relatively few and simple in the alimentary structures of fishes, while several of the mediate class are less obvious and more suggestive. That a fish with canine teeth has a strong jaw is a less interesting fact than the weakness of the jaw in one with long and numerous gill-rakers, or the incompatibility of canine teeth and heavy lower pharyngeals. The first is an immediate adaptive adjustment which a child might foresee, while the others are to be understood only when the peculiarities of the food are known to which both owe their character. The weak jaw of the shovel fish and the slight lower pharyngeals of the pike-perch illustrate the law of disuse (especially when we take into account the teeth of the young in the former and the large pharyngeals of the common perch), and the branchial apparatus of the shovel fish and the canine teeth of the pike-perch are examples of special adaptation to particular kinds of food.

Some mediate correlations are inverse, others coincident, the related structures varying oppositely or in the same direction. An interesting inverse correlation is exhibited by the gill-rakers and the pharyngeals in the suckers; as the former lengthen and multiply, the latter become weaker and bear smaller and more numerous teeth. The cause of this correlation is seen in the food, the species with heavy pharyngeals, few and large pharyngeal teeth, and few and short gill-rakers being mollusk feeders, and the other group depending largely on insects and crustaceans and using mollusks sparely, and then only the small and thin-shelled sorts. A similar inverse relation is seen between the large mouths and the weak pharyngeals of many piscivorous fishes; between the weak pharyngeals and the muscular stomach of the gizzard shad; and between the long gill-rakers and the rudimentary pharyngeals of the shovel fish. Such correlations are often evidence of a specialization and corresponding limitation of the feeding habit, — the increased efficiency of one structure corresponding to the increased importance to the fish of the related kind of food, and the defective development of the correlated structure indicating an abandonment of the food for whose appropriation it was especially fitted. On the other hand, the absence of these inverse correlations marks an omnivorous habit, — as in the cat-fishes, whose jaws, teeth, gill-rakers, and pharyngeals are all moderately developed, while the food is correspondingly indiscriminate.

DETAILED RECAPITULATION OF DATA.*

ANIMAL FOOD.

Dead animal matter: 1 Micropterus salmoides, Nov.; 6 Ictalurus punctatus, Mar., Apr., June, Aug.; 2 Amiurus natalis, May; 1 A. marmoratus, Oct.

Tadpoles: 2 Esox vermiculatus, June, July.

FISHES.

Ctenoid fishes: 1 Uranidea richardsoni, Aug.; 1 Esox lucius, Sept.

Cycloid fishes: 1 Stizostedion canadense, Nov.; 1 Esox lucius, May, Nov.; 2 E. vermiculatus, July; 1 Ictalurus punctatus, Aug.

ACANTHOPTERI.

Undetermined: 11 Stizostedion canadense, June; 1 Micropterus salmoides, Nov.; 3 Pomoxys, Oct., Nov.; 1 Esox lucius, Sept.; 1 Amia calva, Oct.

Aplodinotus grunniens: 2 Stizostedion canadense, Oct.

Percidæ: 1 Perca lutea, May.

Perca lutea: 8 Lota maculosa, Nov.; 1 Micropterus salmoides, May; 1 Amiurus nebulosus, May.

Etheostomatinæ: 1 Lepomis pallidus, Nov.

Etheostoma: 1 Perca lutea, Oct.

Percina caprodes: 1 Micropterus dolomiei, June.

---

* The figures in the following lists show the number of examples of the species of fish in which the given food element was detected.

Where a family or other general name above that of a species occurs in the body of the list, the data placed against it are to be understood as relating only to specimens of the group not further determined; the species names, for example, placed against the family names Percidæ, Cyprinidæ, and the like, indicate the species and specimens in whose food *undetermined* examples of those families were noted—the more precise determinations being given lower down.

Boleosoma maculatum: 1 Pomoxys, **Mar.**
Centrarchinæ: 1 Stizostedion canadense, Nov.; 1 S. vitreum,
Oct.; 4 Esox lucius, Sept., Oct.; 1 E. vermiculatus, 5 in.,
Oct.; 1 Amiurus nebulosus, **Aug.**
Micropterus: 1 Esox lucius, Nov.; 1 Lepidosteus platystomus,
June.
M. dolomiei: 1 Stizostedion canadense, Nov.; 1 Esox lucius,
Nov.
Lepomis: 1 Leptops olivaris, **Aug.**
Ambloplites rupestris: 1 Esox lucius, **Nov.**
Pomoxys: 1 Esox lucius, **Sept.**

HAPLOMI.

Gambusia patruelis: 1 Esox vermiculatus, **July.**

ISOSPONDYLI.

Coregonus artedi: 1 Lota maculosa, **Nov.**
C. clupeiformis: 1 Lota maculosa, **Nov.**
Dorosoma cepedianum: 2 Roccus interruptus, yg.; 4 Stizoste-
dion canadense, Oct., Nov.; 7 S. vitreum, Apr., Oct.; 8
Micropterus salmoides, Nov.; 16 Esox lucius, Sept., Oct.;
2 Clupea chrysochloris, Sept., Oct.; 1 Amiurus natalis,
Oct.; 1 Lepidosteus platystomus, Sept.; 2 L. osseus, July.
Hyodon: 1 Esox lucius.

EVENTOGNATHI.

Cyprinidæ: 2 Stizostedion vitreum, Oct.; 4 Perca lutea, May,
Oct.; 1 Micropterus dolomiei, yg.; 1 Lepomis cyanellus;
1 Pomoxys, Oct.; 1 Aphredoderus sayanus, July; 3 Esox
lucius, Nov.; 2 E. vermiculatus, July, Oct.; 1 Semotilus
atromaculatus, July; 1 Amiurus natalis, Aug; 1 Leptops
olivaris, Aug.; 2 Amia calva, May; 1 Lepidosteus platys-
tomus, June; 1, $1\frac{3}{4}$ in., June; 1 L. osseus, July; 1, 2 in.,
July.
Semotilus atromaculatus: 1 Stizostedion vitreum, Oct.
Notropis: 1 Pomoxys, **Mar.**
N. hudsonius: 1 Esox lucius, **Nov.**
Campostoma anomalum: 1 Micropterus salmoides, **Nov.**
Catostomatidæ: 1 Aplodinotus grunniens, Sept.; 1 Esox lucius,
Sept.; 1 Amiurus natalis, **Aug.**

Ictiobus: 1 Lepomis cyanellus, July; 2 Esox lucius, Nov.; 1
Amia calva.
I. bubalus: 1 Esox lucius, Sept.
Carpiodes: 1 Esox lucius, Nov.

NEMATOGNATHI.

Siluridæ: 1 Stizostedion canadense, Nov.; 1 Micropterus sal-
moides, Oct.
Amiurus: 1 Stizostedion canadense, Oct.; 1 Leptops olivaris,
Aug.
Noturus flavus: 1 Micropterus dolomiei; June.

MOLLUSCA.

GASTEROPODA.

Pleurocera: 1 Ictalurus punctatus, Sept.
Amnicola: 4 Lepomis gibbosus, May, July, Aug.; 1 L. notatus,
Sept.; 1 L. pallidus, Oct.; 2 Placopharynx carinatus, Oct.;
1 Moxostoma, Nov.; 1 M. macrolepidotum, Sept.; 1 Miny-
trema melanops, Oct.; 1 Ictalurus punctatus, Oct.; 3
Amiurus nebulosus, May, Aug., Oct.
Somatogyrus: 3 Moxostoma macrolepidotum, Sept.
Valvata tricarinata: 1 Perca lutea, May; 2 Lepomis gibbosus,
May; 2 Notemigonus chrysoleucus, May; 1 Placopharynx
carinatus, Oct.; 2 Moxostoma macrolepidotum, Sept; 1
Ictiobus urus, Aug.; 1 Amiurus nebulosus, July.
V. sincera: 1 Gambusia patruelis, Sept.
Vivipara: 2 Lepomis pallidus, July, Nov.; 1 Moxostoma aure-
olum, June; 3 M. macrolepidotum, Sept., Oct.; 1 Ictiobus
bubalus, Oct.; 7 Ictalurus punctatus, Apr., Sept., Oct.; 2
Amiurus natalis, Oct.; 1 A. marmoratus, Oct.; 1 Amia
calva, Aug.
Melantho: 1 Moxostoma, Nov.; 3 M. macrolepidotum, Oct.;
7 Ictalurus punctatus, Sept., Oct.; 1 Amiurus natalis, Oct.;
1 A. nebulosus, Oct.
M. decisa: 2 Aplodinotus grunniens, Oct.
Lioplax subcarinata: 2 Ictalurus punctatus, Sept.
Succinea: Perca lutea, Aug.

Limnæa: 1 Notropis whipplei, Apr.; 1 Moxostoma macrolepidotum, May.

Physa: 1 Lepomis gibbosus, yg.; 2 L. pallidus, Nov.; 1 Umbra limi, Sept.; 3 Gambusia patruelis, Sept., Oct.; 1 Zygonectes dispar, July; 3 Z. notatus, Sept., Oct.; 1 Moxostoma macrolepidotum, May; 2 Amiurus natalis, 2⅛ in., July; 3 A. nebulosus, Aug., Sept.; 1 A. marmoratus, Aug.

P. heterostropha: 2 Perca lutea, May; 1 Amiurus nebulosus, Oct.

Planorbis: 1 Aplodinotus grunniens, June; 1 Lepomis gibbosus, July; 1 L. notatus, Sept.; 1 L. pallidus, Nov.; 1 Umbra limi, July; 1 Gambusia patruelis, Sept.; 1 Zygonectes dispar, July; 2 Fundulus diaphanus, June, Oct.; 2 Moxostoma macrolepidotum, May, Sept.; 2 Ictiobus bubalus, Oct.; 1 Ictalurus punctatus, Oct.; 1 Amia calva, Aug.

P. deflectus, yg.: 1 Notemigonus chrysoleucus, May.

Ancylus: 1 Percina caprodes, Aug.

LAMELLIBRANCHIATA.

Sphærium: 2 Aplodinotus grunniens, June, Oct.; 1 Perca lutea, Oct.; 1 Lepomis pallidus, Oct.; 1 Dorosoma cepedianum; 1 Placopharynx carinatus, Oct.; 4 Moxostoma, June; 1 M. aureolum, June; 3 M. macrolepidotum, June, Sept.; Minytrema melanops, Sept., Oct.; 2 Hypentelium nigricans, Aug.; 1 Catostomus teres, Oct.; 3 Ictiobus velifer, Aug., Oct.; 7 I. bubalus, Aug., Oct.; 4 I. urus, Aug., Nov.; 2 I. cyprinella, June, Oct.; 13 Amiurus nebulosus, May, Sept., Oct.; 1 A. marmoratus, Oct.; 2 Amia calva, Sept.

S. sulcatum: 1 Ictiobus bubalus, Oct.; 4 Amiurus nebulosus, Sept.; 1 A. marmoratus, Aug.

Pisidium: 1 Fundulus diaphanus, June; 1 Amiurus nebulosus, Sept.

Unionidæ: 1 Lepomis notatus, Sept.; 1 Labidesthes sicculus, Oct.; 1 Ictiobus urus, Apr.; 2 Ictalurus punctatus, Sept.; 2 Amiurus nebulosus, May, Oct.; 1 A. marmoratus, Oct.

Unio: 1 Aplodinotus grunniens, June; 1 Perca lutea, May; 2 Lepomis gibbosus, yg.; 1 Moxostoma macrolepidotum, May; 1 Catostomus teres, June.

Anodonta: 2 Aplodinotus grunniens, June; 1 Lepomis megalotis, June; 2 Ictalurus punctatus, Aug., Oct.

INSECTA.

Eggs: 4 Lepomis gibbosus, yg.; 1 L. pallidus, July; 1 Hyodon tergisus, Oct.; 1 Notropis hudsonius, June; 1 N. stramineus, Apr.; 1 Amiurus natalis, Nov.

Pupæ: 1 Perca lutea, Oct.; 1 Hadropterus phoxocephalus, Apr.; 1 Notropis megalops, June; 1 N. whipplei, Aug.

Larvæ: 1 Uranidea richardsoni, Aug.; 1 Micropterus salmoides, yg.; 1 Lepomis pallidus, yg.; 2 L. megalotis, June; 1 L. cyanellus, yg.; 1 Ambloplites rupestris, July; 1 yg.; 1 Umbra limi, Sept.; 1 Zygonectes notatus, Sept.; 1 Dorosoma cepedianum. 2½ in., July; 1 Semotilus atromaculatus, July; 1 Notropis megalops, June; 1 N. whipplei, June; 1 Moxostoma aureolum, June; 1 M. macrolepidotum, May; 1 Ictiobus urus, Aug.; 2 I. cyprinella, July; 2 Ictalurus punctatus, Apr., June; 2 Amiurus natalis, 2–2½ in., July.

Terrestrial: 1 Dorosoma cepedianum, July; 3 Hyodon tergisus, May, June, Aug.; 3 Notropis megalops, May, July, Aug.; 1 N. whipplei, June; 1 Ictiobus bubalus, Oct.; 1 I. cyprinella, Oct.

Terrestrial pupæ: 1 Notropis analostanus, Oct.

Aquatic: 1 Notropis whipplei, June; 2 Ictiobus bubalus, Oct.; 1 Ictalurus punctatus, Sept.; 1 Polyodon spathula, Aug.

Aquatic larvæ: 3 Uranidea richardsoni, Aug.; 1 Lepomis gibbosus, yg.; 1 L. notatus, Sept.; 1 L. pallidus, July; 3 Aphredoderus sayanus, Sept.; 1 Semotilus atromaculatus, May; 1 Hybopsis biguttatus, Aug.; 3 Notropis megalops, May, June; 3 N. whipplei, Apr., June; 1 N. lutrensis, July; 1 N. heterodon; 1 Hypentelium nigricans, Aug.; 1 Ictiobus velifer, Oct.; 1 I. urus, Aug.; 1 I. cyprinella, Oct.; 1 Ictalurus punctatus, Apr.; 1, 2½ in., Sept.; 2 Noturus gyrinus, June; 1 Polyodon spathula, June.

HYMENOPTERA.

Undetermined: 2 Lepomis pallidus, Nov.; 1 Pomoxys, May; 1 Labidesthes sicculus, Oct.; 4 Zygonectes notatus, July, Sept., Oct.; 1 Semotilus atromaculatus, June; 1 Notropis atherinoides, Apr.; 1 N. megalops, June; 1 N. whipplei, Apr.

Apis mellifica: 1 Hyodon tergisus, May; 1 Ictalurus punctatus, Oct.

Sphegidæ: 1 Hyodon tergisus, Oct.

Larrada montana: 1 Hyodon tergisus, Oct.

Formicidæ: 1 Lepomis pallidus, Nov.; 1 Centrarchus macropterus, July; 1 Coregonus artedi, 6 in., Aug.; 3 Semotilus atromaculatus, July, Aug.; 1 Notropis megalops, July; 3 N. whipplei, Aug.; 1 Ictalurus punctatus, Oct.; 1, 3½ in., Sept.

Myrmicidæ: 1 Gambusia patruelis, Sept.; 1 Zygonectes notatus, Sept.; 1 Coregonus artedi, Oct.; 1 Ictalurus punctatus, Oct.

Solenopsis: 1 Zygonectes notatus, July.

Chalcididæ: 2 Labidesthes sicculus, Oct.; 1 Fundulus diaphanus, Oct.

Eurytominæ: 1 Clupea chrysochloris, 2¼ in., Sept.

Ichneumonidæ: 1 Lepomis pallidus, July.

Amblyteles subrufus: 1 Hyodon tergisus, Oct.

LEPIDOPTERA.

Undetermined: 1 Coregonus artedi, Oct.; 2 Hyodon tergisus, Oct.; 1 Hybopsis biguttatus, Nov.; 3 Notropis megalops, July, Aug.; 1 N. whipplei, June.

Larvæ: 3 Lepomis pallidus, May, July, Nov.; 1 Ambloplites rupestris, yg.; 1 Semotilus atromaculatus, July; 1 Hybopsis biguttatus, June; 1 Notropis atherinoides, Apr.; 2 N. whipplei, Aug.; 1 Pimephales promelas, May; 1 Ictalurus punctatus, Oct.

Heterocera: 1 Notemigonus chrysoleucus, July.

DIPTERA.

Terrestrial: 2 Lepomis cyanellus, yg.; 6 Labidesthes sicculus, June, Aug., Oct.; 1 Coregonus artedi, 2 in., Aug.; 1 Clupea chrysochloris, 2¼ in., Sept.; 1 Hyodon tergisus, Oct.; 2 Notemigonus chrysoleucus, Sept.; 3 Semotilus atromaculatus, June, Sept.; 2 Notropis atherinoides, July; 2 N. whipplei, Aug.; 2 N. heterodon, May, July; 1 Moxostoma macrolepidotum, Sept.; 2 Ictalurus punctatus, Apr.; 1 Amiurus nebulosus, 2 in., Aug.; 1 Noturus flavus, Oct.; 1 Polyodon spathula, Aug.

Aquatic larvæ: 2 Uranidea richardsoni, Aug.; 2 Aplodinotus grunniens, June, Sept.; 1 Roccus chrysops, Nov.; 1 Etheostoma fusiforme, July; 3 E. cœruleum, June, July; 2 E. zonale, June; 1 Hadropterus phoxocephalus, Apr.; 1 Percina caprodes, Aug.; 1 Boleosoma maculatum, July; 1 Ammocrypta pellucida, June; 2 Lepomis gibbosus, yg.; 1 L. megalotis, June; 1 L. cyanellus, yg.; 2 Chænobryttus gulosus, yg., 1 Ambloplites rupestris, July; 3 Pomoxys, Apr.; 5 Aphredoderus sayanus, Sept.; 4 Eucalia inconstans, Oct.; 1 Umbra limi, July; 1 Gambusia patruelis, Sept.; 4 Zygonectes notatus, June, Sept., Oct.; 7 Fundulus diaphanus, June, Oct.; 1 Dorosoma cepedianum, $2\frac{1}{2}$ in., July; 1 yg.; 1 Hybopsis biguttatus, Sept.; 1 Notropis atherinoides, Aug.; 2 N. megalops, May; 4 N. whipplei, June; 1 N. stramineus, Apr.; 4 N. heterodon, May, July; 1 Pimephales promelas, May; 1 Placopharynx carinatus, Oct.; 1 Moxostoma, Nov.; 1 M. aureolum, June; 5 M. macrolepidotum, June, Sept., Nov.; 1, 2 in., July. 1 Minytrema melanops, Oct.; 1 Erimyzon sucetta, $3\frac{1}{4}$ in.; 2 Hypentelium nigricans, Aug.; 8 Ictiobus velifer, Apr., July, Oct.; 1 I. bubalus, Oct.; 7 I. urus, Apr., June, July, Aug., Oct., Nov.; 4 I. cyprinella, Apr., June, July, Oct.; 12 Ictalurus punctatus, Apr., May, June, Aug., Oct.; 1, $2\frac{1}{2}$ in., Sept.; 4 Amiurus, yg.; 1 A. natalis $2\frac{1}{2}$ in., July; 2 A. nebulosus, Oct.; 3 Noturus gyrinus, May, Oct.; 2 Polyodon spathula, May, June.

Nemocera: 2 Notropis atherinoides, Aug.; 4 N. whipplei, May; 1 N. heterodon, May.

Brachycera: 4 Labidesthes sicculus, Oct.; 1 Zygonectes notatus, Sept.; 5 Notropis whipplei, May, June.

Simulium, larvæ: 1 Eucalia inconstans, June; 1 Notropis atherinoides, Apr.; 2 N. whipplei, June.

Bibio albipennis: 1 Notropis atherinoides, May.

Culicidæ: 1 Alvarius punctulatus, May; 1 Lepomis cyanellus, yg.; 1 Aphredoderus sayanus, July; 1 Zygonectes notatus, June; 1 Clupea chrysochloris, $2\frac{1}{4}$ in., Sept.; 2 Notropis atherinoides, July; 1 Noturus gyrinus, Sept.

Culicidæ, larvæ: 2 Micropterus dolomiei, yg.; 1 M. salmoides, yg.; 1 Polyodon spathula, May.

Corethra, larvæ: 1 Pomoxys, Oct.; 4 yg.; 7 Aphredoderus sayanus, Aug., Sept.; 1 Dorosoma cepedianum, yg.; 2 Amiurus nebulosus, Aug.; 1 A. marmoratus, Oct.; 3 Polyodon spathula, May, Aug.

Chironomidæ: 3 Aplodinotus grunniens, yg.; 1 Roccus interruptus, yg.; 17 Labidesthes sicculus, June, Aug., Oct.; 2 Zygonectes notatus, Oct.; 1 Fundulus diaphanus, Oct.; 1 Phenacobius mirabilis, Oct.; 1 Notropis heterodon, May; 2 Ictiobus urus, Aug.

Chironomidæ, larvæ and pupæ: 3 Aplodinotus grunniens, yg.; 1 Roccus interruptus, yg.; 1 Roccus chrysops, Nov.; 2 yg.; 1 Perca lutea, May; 6 yg.; 8 Alvarius punctulatus, May, June; 3 Etheostoma fusiforme, July; 2 E. jessiæ, Sept.; 6 E. cœruleum, June, July, Aug.; 5 E. lineolatum, Apr., May, June, July; 2 E. zonale, June; 3 Hadropterus aspro, Aug.; 2 H. phoxocephalus, Apr., Aug.; 9 Percina caprodes, Apr., Aug., Sept.; 1 Boleosoma camurum; 9 B. maculatum, Apr., July, Aug.; 2 Crystallaria asprella, June; 3 Ammocrypta pellucida, June; 2 Micropterus dolomiei, yg.; 2 M. salmoides, yg.; 3 Lepomis gibbosus, May, July, Aug.; 13 yg.; 1 L. notatus, Sept.; 2 L. pallidus, July, Nov.; 11 yg.; 4 L. megalotis, June; 5 L. cyanellus, yg.; 4 Chænobryttus gulosus, yg.; 3 Ambloplites rupestris, yg.; 7 Pomoxys, Apr., May, Nov.; 13 yg.; 2 Centrarchus macropterus, July; 4 yg.; 8 Aphredoderus sayanus., Aug., Sept.; 1 Labidesthes sicculus, July; 5 Eucalia inconstans, June; 2 Fundulus diaphanus, Oct.; 2 Dorosoma cepedianum, 2½–5¼ in., July, Oct.; 1 yg.; 1 Hyodon tergisus, 2⅞ in., June; 1 Cyprinidæ, yg.; 1 Notemigonus chrysoleucus, July; 3 Hybopsis biguttatus, June, Sept.; 9 Phenacobius mirabilis, Apr., Aug., Sept., Oct.; 3 Notropis, yg.; 5 N. whipplei, Apr., June, July; 1 N. hudsonius, May; 4 N. heterodon, Apr., July; 1 Campostoma anomalum, Sept.; 2 Placopharynx carinatus, Oct.; 1 Moxostoma, June; 1 yg.; 3 M. aureolum, Apr.; 7 M. macrolepidotum, May, Aug., Sept., Nov.; 3, 1¼–2¾ in., July, Aug.; 2 Minytrema melanops, Oct.; 2 Erimyzon sucetta, 1¾–3 in., Oct.; 5 Hypentelium nigricans, Aug.; 10 yg.; 1 Catostomus teres, Oct.; 11 Ictiobus velifer, Mar., June, July,

Aug., Oct.; 14 I. bubalus, Apr., Aug., Sept., Oct.; 10 I.
urus, July, Aug., Oct.; 6 I. cyprinella, June, July, Aug.,
Oct.; 12 Ictalurus punctatus, Apr., Aug., Sept.; 3, 2½–4 in.,
June, Sept.; 9 Amiurus, yg.; 5 A. natalis, 2–3½ in., July,
Oct.; 3 A. nebulosus, Sept., Oct.; 2, 2½–3½ in., June, Aug.;
4 A. marmoratus, Oct.; 2 Noturus, yg.; 11 N. gyrinus,
May, Aug., Sept., Oct.; 2 Amia calva, June, Sept.; 1 yg.;
5 Polyodon spathula, Aug., Sept., Nov.

Tipulidæ: 1 Coregonus artedi, Oct.; 1 Hyodon tergisus, Oct.;
1 Notropis atherinoides, Apr.

Tipulidæ, larvæ: 1 Notropis atherinoides, Apr.

Tipulidæ, eggs: 1 Coregonus artedi, Oct.; 1 Hyodon tergisus,
Oct.

Tabanus, larvæ: 1 Ictalurus punctatus, Apr.; 1 Amiurus nebulosus, May.

Muscidæ: 1 Clupea chrysochloris, 2½ in., Sept.; 1 Hyodon tergisus, Oct.

Muscidæ, larvæ: 1 Micropterus dolomiei, yg.

COLEOPTERA.

Larvæ: 1 Roccus chrysops, Nov.; 1 Micropterus dolomiei, yg.;
1 Lepomis pallidus, July; 1 Ambloplites rupestris, yg.; 2
Pomoxys, Apr., May; 1 Notropis whipplei, June; 1 Ictiobus urus, July; 1 Noturus, yg.

Terrestrial: 3 Lepomis pallidus, July, Nov.; 1 Dorosoma cepedianum, July; 2 Hyodon tergisus, Oct.; 1 Semotilus atromaculatus, July; 1 Hybopsis biguttatus, Sept.; 1 Notropis
atherinoides, May; 1 N. megalops, July; 1 N. whipplei,
Aug.; 1 Moxostoma macrolepidotum, Aug.; 3 Hypentelium
nigricans, Aug.; 1 Ictiobus bubalus, Oct.; 1 Amiurus marmoratus, Aug.; 1 Polyodon spathula, Nov.

Aquatic: 1 Hyodon tergisus, May.

Aquatic larvæ: 1 Aphredoderus sayanus, Oct.; 2 Hypentelium
nigricans, Aug.; 1 Ictiobus cyprinella, July; 1 Noturus
gyrinus, May.

Cicindelidæ: 1 Hyodon tergisus, Oct.

Carabidæ: 1 Lepomis pallidus, Nov.; 1 Notropis atherinoides,
Apr.; 1 Ictalurus punctatus, Apr.; 1 Amiurus nebulosus,
July.

Carabidæ, larvæ: 1 Ictiobus urus, July.
Clivina: 1 Hyodon tergisus, Aug.
Bembidium: 1 Notropis atherinoides, May.
Pterostichus sayi: 1 Hyodon tergisus, Oct.
Harpalini: 2 Lepomis pallidus, July, Nov.
Agonoderus pallipes: 1 Ambloplites rupestris, July; 1 Hyodon
tergisus, Aug.; 1 Notropis atherinoides, Apr.; 1 Ictalurus
punctatus, May.
Harpalus: 1 Semotilus atromaculatus, Sept.
Stenolophus: 1 Hyodon tergisus, Aug.
Anisodactylus discoideus: 1 Lepomis pallidus, Nov.; 1 Hyodon
tergisus, Oct.
Haliplus: 1 Lepomis notatus, Sept.; 1 Semotilus atromaculatus,
Aug.
Cnemidotus 12-punctatus: 1 Lepomis notatus, Sept.
Dytiscidæ: 1 Lepomis cyanellus, yg.; 1 Ambloplites rupestris,
July.
Dytiscidæ, larvæ: 2 Micropterus dolomiei, yg.; 1 Lepomis pal-
lidus, July; 1 Pomoxys, Apr.; 1 Ictalurus punctatus, Apr.;
1 Amiurus nebulosus, Oct.
Hydroporus undulatus: 1 Ambloplites, yg.
H. hybridus: 1 Lepomis notatus, Sept.
Coptotomus interrogatus: 1 Lepomis pallidus, July; 1 Polyodon
spathula, May.
Cybister fimbriolatus: 1 Hyodon tergisus, Aug.
Gyrinidæ: 1 Amiurus nebulosus, Aug.
Gyrinidæ, larvæ: 1 Aplodinotus grunniens, Oct.; 3 yg.; 2
Lepomis pallidus, July, Nov.; 1 yg.; 1 L. megalotis, June;
4 Pomoxys, Apr., Nov.; 1 Notropis megalotis, Apr.; 1
Moxostoma macrolepidotum, Sept.
Hydrophilidæ: 2 Lepomis gibbosus, May, Aug.; 1 yg.; 1
Lepomis pallidus, May, July; 1 Fundulus diaphanus, Oct.;
1 Semotilus atromaculatus, Sept.
Hydrophilidæ, larvæ: 2 Aplodinotus grunniens, yg.; 1 Microp-
terus dolomiei, yg.; 1 Lepomis pallidus, July, Oct.; 2 Placo-
pharynx carinatus, Oct.; 2 Moxostoma macrolepidotum;
Sept.; 2 Ictiobus bubalus, Apr., Oct.; 1 Amiurus marmor-
atus, Aug.
Hydrophilus: 1 Lepomis cyanellus.

H. nimbatus: 1 Lepomis pallidus, July; 1 Ambloplites rupestris, July; 1 Hyodon tergisus, Oct.

H. glaber: 1 Hyodon tergisus, Oct.

Berosus striatus: 1 Hyodon tergisus, Oct.

Philhydrus: 3 Zygonectes notatus, Sept., Oct.

Silvanus: 1 Notropis atherinoides, Aug.

Histeridæ: 1 Hyodon tergisus, Oct.

Heterocerus: 1 Coregonus artedi, Oct.

H. undatus: 1 Ictiobus urus, July.

Staphylinidæ: 1 Lepomis cyanellus, yg.; 2 Zygonectes notatus, Oct.; 1 Fundulus diaphanus, Oct.; 1 Coregonus artedi, 6 in., Aug.; 1 Ictalurus punctatus, May.

Staphylinus tomentosus: 1 Ictalurus punctatus, Oct.

Elateridæ: 1 Lepomis pallidus, July; 1 Zygonectes notatus, July.

Drasterius elegans: 1 Hyodon tergisus, Aug.

Lampyridæ: 1 Hyodon tergisus, Oct.

Scarabæidæ: 1 Lepomis pallidus, Nov.; 1 Semotilus atromaculatus, July; 1 Notropis atherinoides, Apr.; 1 N. megalops, July.

Aphodius fimetarius: 1 Lepomis pallidus, May; 1 Hyodon tergisus, Oct.

A. inquinatus: 1 Lepomis pallidus, Nov.; 1 Notropis atherinoides, Oct.; 1 Polyodon spathula, Nov.

Melolonthinæ: 1 Notemigonus chrysoleucus, July.

Anomala binotata: 1 Chænobryttus gulosus.

Chalepus trachypygus: 2 Hyodon tergisus, Oct.

Tetramera: 1 Dorosoma cepedianum, July.

Chrysomelidæ: 1 Lepomis pallidus, Nov.; 1 L. megalotis, June; 1 Semotilus atromaculatus, July.

Cryptocephalus 4-maculatus: 1 Lepomis pallidus, Nov.

Colaspis brunnea: 1 Hyodon tergisus, Aug.

Doryphora 10-lineata: 1 Lepomis pallidus, Nov.

Diabrotica 12-punctata: 2 Lepomis pallidus, Nov.

D. vittata: 1 Coregonus artedi, Oct.

D. longicornis: 1 Hyodon tergisus, Aug.

Halticini: 1 Lepomis pallidus, Nov.

Disonycha limbicollis: 1 Hyodon tergisus, Oct.

Anthicidæ: 1 Coregonus artedi, 6 in., Aug.

Rhynchophora: 1 Notropis megalops, July; 2 N. hudsonius, May.
R. brevirostres: 1 Notropis hudsonius, May.
Curculionidæ: 1 Lepomis pallidus, Nov.
Macrops: 2 Hyodon tergisus, Aug., Oct.
Sphenophorus ochreus: 1 Hyodon tergisus, Oct.

HEMIPTERA.

Terrestrial: 1 Coregonus artedi, Oct.; 2 Notropis megalops, June, Aug.; 1 Ictiobus cyprinella, Oct.
Aquatic: 1 Zygonectes notatus, Oct.; 1 Notemigonus chrysoleucus, Sept.; 1 Notropis atherinoides, July; 1 N. hudsonius, May; 1 Hypentelium nigricans, yg.
Heteroptera: 1 Micropterus salmoides, yg.; 1 Zygonectes notatus, Sept.; 1 Fundulus diaphanus, Oct.; 1 Notropis atherinoides, May.
Terrestrial Heteroptera: 1 Fundulus diaphanus, Oct.; 1 Hyodon tergisus, Oct.
Amnestus: 1 Coregonus artedi, 6 in., Aug.
Pentatomidæ: 1 Lepomis pallidus, Nov.; 1 Hyodon tergisus, Oct.; 1 Ictalurus punctatus, May.
Podisus: 1 Ictalurus punctatus, Apr.
Euschistus: 1 Ictalurus punctatus, Oct.
Coreidæ: 1 Pomoxys, June.
Lygæidæ: 1 Gambusia patruelis, Sept.
Lygus pratensis: 1 Coregonus artedi, Oct.
Triphleps insidiosus: 1 Clupea chrysochloris, 2¼ in., Sept.
Tingitidæ: 1 Zygonectes notatus, Sept.
Piesma: 1 Notropis whipplei, Aug.
Tingis: 2 Zygonectes notatus, Sept.
Coriscus ferus: 1 Zygonectes notatus, Sept.; 1 Hyodon tergisus, Aug.
Melanolestes picipes: 1 Hyodon tergisus, Oct.
Hygrotrechus: 1 Ambloplites rupestris, yg.
Zaitha fluminea: 1 Micropterus salmoides, Nov.; 2 Hyodon tergisus, Oct.
Nepa: 1 Lepomis pallidus, May.
Ranatra: 1 Lepomis pallidus, July.
Notonecta: 1 Micropterus salmoides, yg.

Plea: 1 Gambusia patruelis, Sept.; 1 Ictalurus punctatus, **May.**

Corisa: 2 Perca lutea, yg.; 2 Hadropterus aspro, Aug.; 2 Percina caprodes, July, Sept.; 6 Micropterus dolomiei, yg.; 5 M. salmoides, yg.; 4 Lepomis pallidus, June, July, Nov.; 4 yg.; 1 Lepomis megalotis, June; 5 L. cyanellus, yg.; 1 Chænobryttus gulosus, Oct.; 1 yg.; 1 Ambloplites rupestris, yg.; 4 Pomoxys, Apr., May.; 4 yg.; 1 Centrarchus irideus, yg.; 3 Aphredoderus sayanus, July, Sept.; 1 Esox vermiculatus, 4 in., June; 1 Zygonectes dispar, July; 2 Z. notatus, Sept., Oct.; 1 Dorosoma cepedianum July; 3 Semotilus atromaculatus, July, Sept.; 1 Notropis megalops, Aug.; 1 N. whipplei, July; 2 Ictiobus urus, July, Aug.; 2 I. cyprinella, July; 1 Ictalurus punctatus, Apr.; 1 Amiurus nebulosus, Oct.; 1 Amia calva, June; 1 yg.; 1 Polyodon spathula, Aug.

C. alternata: 1 Perca lutea, yg.; 3 Micropterus salmoides, **yg.**; 1 Pomoxys, Apr.; 3 Zygonectes notatus, Sept.; 1 Ictalurus punctatus, Apr.; 1 Polyodon spathula, **May.**

C. signata: 4 Micropterus dolomiei, **yg.**

C. tumida: 2 Perca lutea, yg.; 1 Hadropterus aspro, Aug.; 8 Micropterus dolomiei, yg.; 3 M. salmoides, yg.; 1 Lepomis pallidus, Nov.; 1 L. megalotis; 1 L. cyanellus; 1 yg.; 1 Chænobryttus gulosus; 2 yg.; 3 Ambloplites rupestris, **yg.**; 1 Pomoxys, July; 1 yg.; 1 Centrarchus irideus, July; 1 Hyodon tergisus, 2⅞ in., June; 1 Amiurus nebulosus, 3½ in., June.

Homoptera: 1 Gambusia patruelis, Sept.; 1 Coregonus artedi, Oct.; 1, 6 in., Aug.; 1 Hyodon tergisus, Oct.; 2 Notropis whipplei, Apr., Aug.; 1 Ictalurus punctatus, **May.**

Tettigoninæ: 1 Labidesthes sicculus, Oct.; 1 Zygonectes notatus, **Oct.**

Diedrocephala mollipes: 1 Coregonus artedi, Oct.

Typhlocyba: 1 Coregonus artedi, 2 in., Aug.; 1 Clupea chrysochloris, 2¼ in., Sept.

Aphididæ: 1 Gambusia patruelis, Sept.; 3 Zygonectes notatus, Sept., Oct.; 1 Notropis, yg.; 3 N. atherinoides, July, Aug.

Aphis: 1 Labidesthes sicculus, Oct.; 1 Zygonectes notatus, **Oct.**

Thrips: 1 Labidesthes sicculus, Oct.; 1 Zygonectes notatus, Oct.; 1 Fundulus diaphanus, Oct.; 1 Moxostoma, **yg.**

ORTHOPTERA.

Undetermined: 1 Hyodon tergisus, 2⅞ in., June; 1 Amiurus marmoratus, Aug.

Acridida: 1 Roccus interruptus, May; 3 Semotilus atromaculatus, Sept.; 2 Ictalurus punctatus, Oct.

Tettigina: 1 Ictalurus punctatus, June.

Tettix: 1 Hyodon tergisus, Oct.; 1 Ictalurus punctatus, Oct.

Tettigidea: 1 Lepomis pallidus, June, Nov.

Locustida: 1 Lepomis pallidus, May; 2 Semotilus atromaculatus, Sept.

Phaneroptera curvicauda: 1 Lepomis pallidus, Nov.

Nemobius vittatus: 1 Lepomis pallidus, Nov.

Blatta: 1 Ictalurus punctatus, June.

NEUROPTERA.

Larva: 2 Roccus chrysops, yg.; 1 Lepomis gibbosus, June; 1 L. pallidus, yg.; 1 Chænobryttus gulosus, yg.; 1 Ambloplites rupestris, yg.; 1 Aphredoderus sayanus, July; 1 Semotilus atromaculatus, May; 2 Hybopsis biguttatus, Aug., Sept.; 1 Phenacobius mirabilis, Oct.; 5 Notropis megalops, May, June; 2 N. whipplei, Apr., July; 1 Moxostoma macrolepidotum, Sept.; 1 Ictiobus velifer; 1 I. cyprinella, July.

Terrestrial: 1 Ictiobus urus, Aug.

Phryganeida: 1 Lepomis pallidus, July; 1 Ambloplites rupestris, July; 1 Ictiobus bubalus, Oct.; 1 I. urus, Aug.

Phryganeida, larva: 2 Perca lutea, May; 2 Etheostoma cœruleum, June; 1 Percina caprodes, Apr.; 2 Lepomis gibbosus, May; 1 L. megalotis, July; 4 Hybopsis biguttatus, Aug., Sept.; 1 Phenacobius mirabilis, Apr.; 3 Notropis atherinoides, July, Aug.; 4 N. megalops, Apr. June; 1 N. stramineus, Apr.; 3 Ictiobus velifer, Aug., Oct.; 5 I. bubalus, Aug., Oct.; 7 Ictalurus punctatus, Apr., May, Aug.; 1 Amiurus nebulosus, May; 1 Noturus gyrinus, May; 3 Polyodon spathula, June, Aug., Sept.

Leptoceridæ, larvæ: 1 Gambusia patruelis, Sept.; 1 Ictalurus punctatus, Oct.

Leptocerus, larvæ: 1 Lepomis gibbosus, July; 1 Ictiobus bubalus, Oct.

Sialidæ, larvæ: 1 Ictiobus cyprinella, July; 1 Ictalurus punctatus, Aug.; 1 Amiurus nebulosus, Oct.; 2 A. mamoratus, Oct.

Sialis infumata: 3 Lepomis pallidus, May, Aug.

Corydalis, larvæ: 1 Lepomis cyanellus, Apr.; 1 Ictiobus cyanellus, Aug.

Corydalis cornutus, larvæ: 1 Pomoxys, Oct.

Odonata, larvæ: 1 Labidesthes sicculus, July; 4 Esox vermiculatus, June, Oct.; 1, 4 in., June; 1 Polyodon spathula, May.

Libellulinæ, larvæ: 4 Aplodinotus grunniens, Sept.; 5 Perca lutea, Mar., May; 2 Lepomis gibbosus, May; 5 L. pallidus, May, Oct.; 1 L. cyanellus, Apr.; 1 yg.; 1 Ambloplites rupestris, July; 1 Pomoxys, May; 1 Aphredoderus sayanus, Oct.; 1 Esox lucius, Aug.; 1 E. vermiculatus, 2¾ in., June; 2 Ictiobus bubalus, Aug., Oct.; 1 I. urus, Aug.; 6 Ictalurus punctatus, Mar., Apr., May, Sept.; 2 Amiurus nebulosus, May; 2 Amia calva, May, Aug.

Agrioninæ, larvæ: 3 Perca lutea, Mar., May; 1 yg.; 1 Hadropterus aspro, Aug.; 2 Micropterus dolomiei, yg.; 1 M. salmoides, Nov.; 2 Lepomis gibbosus, yg.; 3 L. pallidus, May, June, July; 1 yg.; 3 Chænobryttus gulosus, yg.; 1 Ambloplites rupestris, July; 9 Pomoxys, Mar., April, May; 1 Erimyzon sucetta, yg,; 2 Ictalurus punctatus, Mar., Apr.

Agrion, larvæ: 1 Roccus interruptus, May; 1 Labidesthes sicculus, June; 3 Esox vermiculatus, June, July; 1, 2¾ in., June; 1 Zygonectes notatus, Sept.; 1 Moxostoma, June; 1 M. macrolepidotum, Aug.; 1 Ictiobus bubalus, Oct.; 2 Ictalurus punctatus, Apr.; 1 Polyodon spathula, May.

Ephemeridæ: 1 Roccus interruptus, May.

Ephemeridæ, larvæ: 1 Aplodinotus grunniens, yg.; 5 Roccus chrysops, Nov.; 2 Perca lutea, May; 3 yg.; 1 Alvarius punctulatus, May; 3 Etheostoma fusiforme, July; 2 E. jessiæ, Sept.; 2 E. cœruleum, July, Aug.; 4 E. lineolatum, Apr., June; 6 Hadropterus aspro, July, Aug.; 5 H. phoxocephalus, Apr., Aug.; 1 Percina caprodes, July, Aug.; 1 Boleosoma camurum; 1 B. maculatum,

Aug.; 2 Ammocrypta pellucida, June; 3 Micropterus dolomiei, yg.; 3 M. salmoides, yg.; 1 Lepomis gibbosus, Aug.;
5 yg.; 2 L. pallidus, July, Aug.; 3 yg.; 1 L. megalotis,
July; 3 L. cyanellus, Apr.; 5 yg.; 2 Chænobryttus gulosus,
yg.; 2 Ambloplites rupestris, July; 3 Pomoxys, Mar., Apr.;
2 yg.; 2 Centrarchus irideus, July; 1 yg.; 3 Aphredoderus
sayanus, July, Sept., Oct.; 2 Fundulus diaphanus, Oct.; 2
Hyodon tergisus, June; 1 Hybopsis biguttatus, June; 3
Notropis atherinoides, Apr., Aug., Oct.; 1 N. megalops,
July; 2 N. whipplei, June; 1 N. hudsonius, Aug.; 2 N.
stramineus, July; 1 N. heterodon, Sept.; 2 Moxostoma
macrolepidotum, Aug., Sept.; 2 Hypentelium nigricans,
Aug.; 1 yg.; 3 Ictiobus bubalus, Apr., Oct.; 3 I. urus,
June, Aug.; 1 I. cyprinella, July; 13 Ictalurus punc-
tatus, Mar., Apr., May, Aug.; 1, $2\frac{3}{4}$ in., Oct.; 1 Amiurus
natalis, $3\frac{1}{2}$ in., Oct.; 1 Amiurus nebulosus, 2 in., Aug.; 1
Noturus, yg.; 6 N. gyrinus, May, Oct.; 1 Amia calva,
June; 4 Polyodon spathula, May, June, Aug., Sept.

Cænis, larvæ: 2 Placopharynx carinatus, Oct.; 5 Hypentelium
nigricans, Aug.; 1 Polyodon spathula, May.

Baëtis, larvæ: 1 Lepomis pallidus, June; 1 Ambloplites rupestris, July.

Hexagenia, larvæ: 15 Aplodinotus grunniens, June, Sept., Oct.;
4 yg.; 1 Roccus interruptus, Oct.; 2 R. chrysops, Sept.,
Oct.; 1 Perca lutea, Oct.; 4 yg.; 1 Hadropterus phoxocephalus, Aug.; 1 Lepomis gibbosus, May; 2 L. pallidus,
Nov.; 1 L. cyanellus, Apr.; 2 Chænobryttus gulosus, Oct.;
18 Pomoxys, Mar., Apr., June, July, Oct., Nov.; 3 Aphredoderus sayanus, Sept.; 1 Esox vermiculatus, July; 3
Hyodon tergisus, Aug., Oct.; 1 Notropis megalops, Aug.;
2 N. hudsonius, June, Aug.; 1 Hypentelium nigricans,
Aug.; 1 Ictiobus urus, Nov.; 2 I. cyprinella, Aug.; 7
Ictalurus punctatus, Apr., Oct.; 1, 4 in., Sept.; 4 Amiurus
natalis, Oct., Nov.; 4 A. nebulosus, May, Oct.; 7 A. marmoratus, Oct., Nov.; 4 Polyodon spathula, May, June,
Sept., Nov.

THYSANURA.

Podura: 3 Labidesthes sicculus, Aug., Oct.

ARACHNIDA.

Undetermined: 1 Hyodon tergisus, Oct.

Araneina: 4 Lepomis pallidus, Oct., Nov.; 2 Labidesthes sicculus, June, Oct.; 1 Eucalia inconstans: 3 Zygonectes notatus, June, Oct.; 1 Fundulus diaphanus, Oct.; 1 Dorosoma cepedianum, July; 1 Clupea chrysochloris, $2\frac{1}{4}$ in., Sept.; 1 Hyodon tergisus, Oct.; 1 Notropis atherinoides, Apr.; 1 N. whipplei, Apr.; 1 Ictalurus punctatus, Oct.; 1 Amiurus natalis, Oct.; 2 A. marmoratus, Nov.

Terrestrial Araneina: 3 Labidesthes sicculus, Aug., Oct.

Acarina: 2 Umbra limi, July; 1 Fundulus diaphanus, Oct.; 1 Notemigonus chrysoleucus, July; 1 Notropis megalops, Aug.; 2 N. heterodon, May, July; 1 Ictiobus urus, Aug.; 1 I. cyprinella, Oct.

Hydrachnidæ: 1 Lepomis pallidus, July, Nov.; 5 yg.; 1 Centrarchus irideus, yg.; 1 Labidesthes sicculus, June; 1 Umbra limi, July; 1 Dorosoma cepedianum, July; 2 Moxostoma macrolepidotum, $2\frac{1}{2}$-$2\frac{3}{4}$ in., Aug.; 2 Erimyzon sucetta, yg.; 1 Hypentelium nigricans, yg.; 1 Catostomus teres, Aug.; 1 Ictiobus bubalus, Oct.; 2 I. cyprinella, July; 1 Polyodon spathula, Aug.

Hydrachna: 1 Ambloplites rupestris, July.

Atax: 1 Lepomis pallidus, yg.

CRUSTACEA.

DECAPODA.

Cambarus: 1 Perca lutea, May; 1 Lepomis pallidus, Nov.; 1 L. cyanellus; 2 Ambloplites rupestris, July; 3 Semotilus atromaculatus; 2 Hybopsis biguttatus, Sept., Nov.; 1 Ictiobus urus, Aug.; 3 Ictalurus punctatus, Apr., May, June; 4 Amiurus natalis, May, Aug.; 6 A. nebulosus, May, Aug.; 1, $3\frac{1}{2}$ in., June; 11 Amia calva, Apr., May, June; 1 Lepidosteus platystomus, Apr.

C. virilis: 2 Perca lutea, May; 1 Anguilla rostrata, Aug.; 4 Amia calva, May.

C. propinquus: 3 Lota maculosa, Nov.; 2 Micropterus dolomiei, June.

C. immunis: 1 Micropterus salmoides, Oct.

C. obesus: 1 Amia calva, Apr.

Palæmonetes exilipes: 1 Perca lutea; 1 Lepomis cyanellus; 1 Amiurus natalis, 2 in., July.

AMPHIPODA.

Gammarus, yg.: 2 Alvarius punctulatus.

Gammarus fasciatus: 1 Micropterus dolomiei, yg.

Crangonyx: 1 Alvarius punctulatus, June; 1 Ictiobus cyprinella, July; 1 Amia calva, June.

C. gracilis: 1 Umbra limi, Sept.; 1 Gambusia patruelis, Sept.; 1 Zygonectes notatus, June.

Allorchestes dentata: 1 Roccus interruptus, May; 7 Perca lutea, Mar., May, Aug.; 4 yg,; 1 Percina caprodes, Aug.; 5 Micropterus dolomiei, yg.; 2 Lepomis gibbosus, Aug.; 3 yg.; 8 L. pallidus, May, June, July, Aug.; 1 yg.; 2 L. megalops, June; 1 Ambloplites rupestris, yg.; 1 Centrarchus irideus, yg.; 1 Aphredoderus sayanus, Oct.; 1 Esox vermiculatus, yg.; 6 Fundulus diaphanus, June, Oct.; 1 Notropis megalops, May; 1 N. heterodon, May; 1 Placopharynx carinatus, Oct.; 1 Ictiobus velifer, Oct.; 1 Ictalurus punctatus, May; 1, 4 in., June; 2 Amiurus, yg.; 1 A. natalis, 2½ in., July; 3 A. nebulosus, May; 9 Noturus gyrinus, May, Aug., Oct.; 1 Amia calva, June; 1 Polyodon spathula, May.

ISOPODA.

Asellus: 2 Uranidea richardsoni, Aug.; 1 Aplodinotus grunniens, Apr.; 1 Roccus chrysops; 3 Perca lutea, Mar., Aug.; 1 yg.; 2 Micropterus dolomiei, yg.; 2 Lepomis gibbosus, May, Aug.; 4 L. pallidus, May, Aug.; 1 L. megalotis, June; 1 L. cyanellus, yg.; 2 Aphredoderus sayanus, July, Aug.; 1 Esox vermiculatus, July; 1 Amiurus nebulosus, 3½ in., June; 2 Noturus gyrinus, June, Aug.; 1 Amia calva, June.

Mancasellus tenax: 3 Perca lutea, Mar.; 1 yg.

ENTOMOSTRACA.

Eggs: 1 Dorosoma cepedianum, Oct.

CLADOCERA.

Daphnella: 1 Percina caprodes, Sept.; 1 Pomoxys, yg.; 1 Notropis heterodon, July; 1 Ictiobus cyprinella, July.

Daphniidæ: 1 Roccus interruptus; yg.; 1 Stizostedion vitreum, yg.; 5 Perca lutea, yg.; 2 Percina caprodes, Aug.; 4 Centrarchinæ, yg.; 1 Micropterus dolomiei, yg.; 4 M. salmoides, yg.; 2 Lepomis gibbosus, yg.; 2 L. pallidus, yg.; 3 L. cyanellus, yg.; 1 Ambloplites rupestris, yg.; 2 Pomoxys, Mar.; 4 yg.; 1 Eucalia inconstans, June; 2 Zygonectes notatus, June; 1 Dorosoma cepedianum, June; 1 Notemigonus chrysoleucus, Sept.; 1 Notropis atherinoides, Oct.; 1 N. whipplei, Aug.; 1 Hypentelium nigricans, yg.; 1 Ictiobus urus, Aug.; 1 Amiurus, yg.; 2 Polyodon spathula, Aug.

Daphniidæ, eggs: 1 Ictiobus urus, Aug.; 1 Polyodon spathula, Aug.

Daphnia: 3 Perca lutea, yg.; 1 Percina caprodes, Aug.; 1 Centrarchinæ, yg.; 2 Chænobryttus gulosus, yg.; 1 Zygonectes notatus, Sept.; 1 Dorosoma cepedianum, 5¼ in., Oct.; 3 yg.; 1 Ictiobus velifer, yg.; 1 Ictalurus punctatus, 4 in., June; 1 Amiurus natalis, 2¾ in., Oct.; 1 A. nebulosus, 3½ in., June.

Daphnia, eggs: 1 Coregonus artedi.

D. pulex: 1 Perca lutea, yg.; 1 Lepomis pallidus, yg.; 1 Labidesthes sicculus, Aug.; 1 Dorosoma cepedianum, yg.; 2 Notemigonus chrysoleucus, July; 1 Polyodon spathula, June.

D. hyalina: 4 Labidesthes sicculus, June, Aug.

D. retrocurva: 3 Labidesthes sicculus, June.

Simocephalus: 3 Lepomis gibbosus, yg.; 3 L. pallidus, yg.; 3 Chænobryttus gulosus, yg.; 4 Pomoxys, Mar., Apr.; 3 yg.; 2 Notropis heterodon, May, July; 1 Ictiobus velifer, Mar.; 3 Ictiobus urus, Apr., Aug.; 2 I. cyprinella, Apr.; 3 Amiurus, yg.; 2 Notropis gyrinus, Oct.

Simocephalus, eggs: 1 Ictiobus urus, Aug.

S. vetulus: 1 Pomoxys, Mar.

S. americanus: 1 Perca lutea, yg.; 1 Alvarius punctulatus, May; 3 Micropterus salmoides, yg.; 1 Lepomis cyanellus, yg.; 4 Pomoxys, yg.; 1 Centrarchus irideus, yg.; 1 Labidesthes sicculus, Aug.; 1 Esox vermiculatus, yg.; 1 Dorosoma cepedianum, yg.; 1 Ictiobus velifer, Apr.; 2 Amiurus, yg.; 1 A. natalis, $3\frac{1}{2}$ in., Oct.; 1 Amia calva, June; 1 yg.

Ceriodaphnia: 1 Ictiobus urus, Aug.; 1 Amiurus, yg.

C. dentata: 1 Dorosoma cepedianum, yg.

Scapholeberis: 1 Amia calva, yg.

S. mucronatus: 1 Erimyzon sucetta, yg.; 1 Ictiobus velifer, yg.; 2 Amiurus, yg.; 1 A. nebulosus, $3\frac{1}{2}$ in., June; 1 Amia calva, June; 1 Lepidosteus platystomus, yg.

Macrothrix laticornis: 1 Boleosoma maculatum, July; 3 Amiurus, yg.; 2 A. natalis, $2\frac{1}{2}$ in., July.

Bosmina: 2 Perca lutea, yg.; 1 Centrarchinæ, yg.; 4 Lepomis pallidus, May; 2 Chænobryttus gulosus, yg.; 1 Pomoxys, June; 1 yg.; 5 Labidesthes sicculus. June, Aug. Oct.; 1 Eucalia inconstans, June; 1 Coregonus artedi; 1 Dorosoma cepedianum, $5\frac{1}{4}$ in., Oct.; 8 yg.; 1 Cyprinidæ, yg.; 2 Notemigonus chrysoleucus, Sept.; 2 Notropis atherinoides, Nov.; 2 Ictiobus, yg.; 3 I. velifer, Mar., Sept., Oct.; 2 I. bubalus, Oct.; 2 I. urus, Oct.; 4 I. cyprinella, Apr., May, Oct.; 5 Polyodon spathula, May, June, Aug.

B. longirostris: 4 Micropterus salmoides, yg.; 2 Dorosoma cepedianum, yg.; 2 Notropis atherinoides, Oct.

Iliocryptus: 1 Notropis heterodon; July; 1 Ictiobus bubalus, Sept.

Lynceidæ: 1 Perca lutea, yg.; 1 Alvarius punctulatus, May; 1 Chænobryttus gulosus, yg.; 1 Labidesthes sicculus, June; 1 Umbra limi, Sept.; 1 Zygonectes dispar, July; 4 Z. notatus, June, Sept., Oct.; 1 Moxostoma, yg.; 1 M. macrolepidotum, Sept.; 1 Erimyzon sucetta, $1\frac{3}{4}$ in.; 1 Ictiobus velifer, Oct.; 2 I. bubalus, Sept., Oct.; 3 I. cyprinella, Apr.

Chydorus: 3 Perca lutea, yg.; 4 Alvarius punctulatus, May; 1 Centrarchinæ, yg.; 1 Micropterus salmoides, yg.; 5 Lepomis gibbosus, yg.; 3 L. cyanellus, yg.; 1 Chænobryttus gibbosus; 1 yg.; 8 Pomoxys, yg.; 1 Labidesthes sicculus, Oct.; 6 Eucalia inconstans, June; 1 Esox vermiculatus, yg.; 1 Umbra limi, Sept.; 2 Zygonectes notatus, Sept., Oct.;

2 Fundulus diaphanus, Oct.; 2 Dorosoma cepedianum, yg.;
1 Notemigonus chrysoleucus, Sept.; 2 Notropis atherinoides, Oct.; 1 N. megalops, May; 1 N. whipplei; 5 N. hudsonius, June, July; 10 N. heterodon, May, July, Sept.;
2 Erimyzon sucetta, yg.; 1 Carpiodes, Apr.; 1 Ictiobus
urus, Aug.; 2 I. cyprinella, July; 2 Amiurus, yg.; 1 A.
nebulosus, May; 1, 2 in., Aug.; 3 Noturus gyrinus, Oct.;
1 Amia calva, June; 1 yg.; 1 Polyodon spathula, May.

C. denticulatus: 1 Pomoxys, yg.

C. sphericus: 1 Coregonus artedi.

Pleuroxus: 1 Perca lutea, yg.; 2 Micropterus salmoides, yg.; 1
Lepomis gibbosus, yg.; 2 L. pallidus, yg.; 2 L. cyanellus,
yg.; 1 Chænobryttus gulosus, yg.; 1 Ambloplites rupestris,
yg.; 2 Pomoxys, yg.; 1 Labidesthes sicculus, Oct.; 1 Zygonectes notatus, Sept.; 1 Notropis heterodon, Sept.; 4 Moxostoma macrolepidotum, $2\frac{1}{8}$-$2\frac{3}{4}$ in., Aug.; 2 Erimyzon
sucetta, yg.; 1 Ictiobus cyprinella, July; 1 Amiurus, yg.;
3 Noturus gyrinus, Oct.

P. dentatus: 1 Lepomis pallidus, yg.; 1 Notropis heterodon,
July; 2 Amiurus, yg.

Alona: 1 Lepomis pallidus, Aug.; 3 yg.; 1 Pomoxys, yg.; 1 Centrarchus irideus, yg.; 1 Labidesthes sicculus, Oct.; 3 Umbra
limi, Sept.; 3 Fundulus diaphanus, Oct.. 2 Dorosoma
cepedianum, July; 1 yg.; 1 Notropis hudsonius, June; 4
N. heterodon, May, July; 1 Moxostoma, yg.; 1 M. macrolepidotum, Sept.; 1 Erimyzon sucetta, yg.; 7 Hypentelium
nigricans, yg.; 2 Catostomus teres, June, Aug.; 3 Carpiodes, Apr., July, Oct.; 1 Ictiobus bubalus, Apr.; 2 I .cypri-
nella, July; 3 Amiurus yg.; 1 Noturus gyrinus, Oct.

Acroperus: 1 Notropis heterodon, May.; 2 Ictiobus cyprinella,
July; 3 Amiurus natalis, 2-2$\frac{1}{2}$in., July.

A. leucocephalus: 1 Zygonectes notatus, Oct.; 1 Fundulus
diaphanus, Oct.; 1 Notropis megalops, Aug.; 1 N. heterodon, May.

Camptocercus macrurus: 2 Fundulus diaphanus, Oct.

Eurycercus: 1 Pomoxys, yg.; 1 Labidesthes sicculus, June; 1
Fundulus diaphanus, June; 1 Polyodon spathula, May.

Eurycercus lamellatus: 1 Percina caprodes, Aug.: 1 Micropterus salmoides, yg.; 2 Lepomis pallidus, yg.; 1 Amiurus, yg.

Leptodora: 1 Roccus interruptus, yg.; 1 Micropterus salmoides, yg.; 1 Dorosoma cepedianum, yg.; 1 Hyodon tergisus, June; 1 Amiurus nebulosus, Aug.; 1 Polyodon spathula, **Aug.**

OSTRACODA.

Cypridæ: 1 Stizostedion vitreum, yg.; 1 Alvarius punctulatus; 1 Percina caprodes, Aug.; 1 Centrarchinæ, yg.; 8 Lepomis gibbosus, yg.; 2 L. pallidus, July, Aug.; 2 yg.; 1 L. cyanellus, yg.; 3 Centrarchus irideus, yg.; 1 Notropis heterodon, July; 3 Moxostoma, yg.; 2 M. macrolepidotum, $1\frac{1}{4}$-$2\frac{3}{4}$ in., July, Aug.; 1 Erimyzon sucetta, $1\frac{3}{4}$ in.; 1 yg.; 1 Hypentelium nigricans, yg.; 4 Carpiodes, Mar., Apr., Aug.; 2 yg.; 2 Ictiobus cyprinella, July; 2 Amiurus, yg.; 3 A. natalis, 2-$2\frac{1}{2}$ in., July.

Cypris: 3 Perca lutea, yg.; 1 Percina caprodes, Aug.; 1 Lepomis pallidus, May; 1 yg.; 1 L. cyanellus, yg.; 1 Pomoxys, Apr.; 1 yg.; 1 Centrarchus irideus, yg.; 2 Aphredoderus sayanus, Sept.; 1 Eucalia inconstans, Oct.; 6 Umbra limi, Sept.; 2 Zygonectes notatus, Sept., Oct.; 3 Fundulus diaphanus, Oct.; 4 Dorosoma cepedianum, Apr., July, Oct.; 1 yg.; 1 Notemigonus chrysoleucus, Sept.; 1 Notropis megalops, Aug.; 4 N. heterodon, May, July, Sept.; 1 Pimephales notatus, Sept.; 1 Moxostoma, yg.; 1 M. macrolepidotum, 2 in., July; 1 Minytrema melanops, Oct.; 1 Catostomus teres, Aug.; 1 Carpiodes, June; 5 Ictiobus bubalus, Apr., Oct.; 3 I. urus, Aug., Oct.; 3 I. cyprinella, Apr., June; 1 Amiurus, yg.; 1 A. natalis, $2\frac{1}{2}$ in., July; 3 Noturus gyrinus, May, Aug.; 2 Polyodon spathula, June, **Aug.**

C. vidua: 1 Eucalia inconstans; 1 Fundulus diaphanus, Oct.; 2 Notropis hudsonius, July.

Candona: 1 Fundulus diaphanus, Oct.; 1 Noturus gyrinus.

C. bifasciata: 2 Amiurus, yg.

COPEPODA.

Nauplius: 1 Erimyzon sucetta, yg.

Cyclops: 1 Aplodinotus grunniens, yg.; 2 Roccus interruptus, yg.; 7 Perca lutea, yg.; 8 Alvarius punctulatus, May, June;

1 Etheostoma lineolatum, July: 2 Hadropterus aspro, Aug,; 1 Percina caprodes, July; 3 Boleosoma maculatum, July, Aug.; 4 Centrarchinæ, yg.; 1 Micropterus dolomiei, yg.; 8 M. salmoides, yg.; 13 Lepomis pallidus, yg.; 5 L. cyanellus, yg.; 4 Chænobryttus gulosus, yg.; 2 Ambloplites rupestris, yg.; 3 Pomoxys, Apr., June; 15 yg.; 2 Centrarchus irideus, July; 4 yg.; 3 Aphredoderus sayanus, Aug., Sept.; 3 Eucalia inconstans; 1 Gambusia patruelis, Sept.; 2 Zygonectes notatus, Sept., Oct.; 1 Coregonus artedi; 1 Dorosoma cepedianum, July; 1, 5½ in., Oct.; 10 yg.; 4 Notemigonus chrysoleucus, July, Sept.; 2 Semotilus atromaculatus, July; 1 Phenacobius mirabilis, Sept.; 2 Notropis whipplei, June; 1 N. stramineus, Apr.; 12 N. heterodon, Apr., May, July, Sept.; 1 Moxostoma, yg.; 1 M. macrolepidotum, Sept.; 2, 2½–2¾ in., Aug.; 1 Minytrema melanops, Oct.; 1 Erimyzon sucetta, 1¾ in.; 3 Hypentelium nigricans, yg.; 1 Catostomus teres, June; 8 Carpiodes, Mar., Apr., July, Aug., Oct.; 2 yg.; 10 Ictiobus bubalus, Apr., Sept., Oct.; 1 yg.; 3 I. urus, Apr., Aug., Oct.; 4 I. cyprinella, Apr., June, July; 11 Amiurus, yg.; 2 A. natalis, 2–2⅛ in., July; 2 A. nebulosus, 2–3½ in., June, Aug.; 2 Noturus, yg.; 6 N. gyrinus, Oct.; 1 Amia calva, June; 1 yg.; 3 Polyodon spathula, June, Aug.

C. thomasi: 1 Labidesthes sicculus, Aug.

Canthocamptus: 1 Labidesthes sicculus, Oct.; 1 Notropis stramineus, Apr.; 1 N. heterodon, May; 1 Erimyzon sucetta, 3 in., Oct.; 1 Hypentelium nigricans, yg.; 6 Carpiodes, Mar., Apr., June, Oct.; 1 yg.; 10 Ictiobus bubalus, Apr., Oct.; 1 I. urus, Oct.; 1 I. cyprinella, Oct.; 2 Noturus gyrinus, Oct.; 1 Polyodon spathula, May.

Diaptomus: 1 Perca lutea, yg.; 1 Labidesthes sicculus, Aug.; 1 Notropis atherinoides, Nov.; 1 N. heterodon, July; 1 Amiurus nebulosus, Aug.

Epischura lacustris: 3 Labidesthes sicculus, June, Aug.

Limnocalanus: 1 Labidesthes sicculus, Aug.

VERMES.

Polyzoa: 3 Lepomis pallidus, May, Aug., Oct.; 1 Pomoxys, yg.

Pectinatella magnifica: 1 Lepomis pallidus, yg.; 7 Ictiobus bubalus, Oct.

Plumatella: 1 Placopharynx carinatus, Oct.; 3 Ictiobus bubalus, Oct.; 1 I. cyprinella, Oct.; 2 Ictalurus punctatus, Sept.; 1 Polyodon spathula, May.

Hirudinei: 1 Catostomus teres, Oct.; 3 Ictalurus punctatus, Apr., June; 1 Amiurus nebulosus, May; 5 A. marmoratus, Oct., Nov.; 1 Polyodon spathula, May.

Chætopoda: 1 Aphredoderus sayanus, Sept.

Naididæ: 1 Pimephales promelas, Aug.; 1 Moxostoma macrolepidotum, 2 in., July.

Lumbriculus: 1 Notropis megalops, June.

Lumbricus: 1 Lepomis pallidus, Nov.; 1 yg.

Nematoda: 1 Amiurus nebulosus, Aug.

Gordius: 2 Semotilus atromaculatus, Sept.; 1 Ictalurus punctatus, Oct.

Anguillulidæ: 1 Ictiobus bubalus, Apr.; 1 I. cyprinella, June.

Rotifera: 1 Notropis heterodon, July; 1 Moxostoma, yg.; 3 M. macrolepidotum, $2\frac{1}{4}$–$2\frac{3}{4}$ in., Aug.; 1 Erimyzon sucetta, $1\frac{3}{4}$ in.; 1 yg.; 2 Catostomus teres, June, Aug.; 1 Carpiodes, yg.; 1 Ictiobus, yg.; 1 I. bubalus, yg.

Anurœa: 2 Erimyzon sucetta, yg.; 2 Ictiobus, yg.; 2 I. bubalus, yg.

Brachionus: 1 Ictiobus, yg.

Metopidea: 1 Moxostoma, yg.; 3 M. macrolepidotum, $2\frac{1}{8}$–$2\frac{3}{4}$ in., Aug.; 1 Erimyzon sucetta, $1\frac{3}{4}$ in.; 1 yg.; 2 Catostomus teres, June, Aug.

Rotifer vulgaris: 1 Catostomus teres, June.

Planaria: 1 Noturus gyrinus, Oct.

PORIFERA.

Spongilla: 2 Ictalurus punctatus, Sept.

PROTOZOA.

Dinobryon: 1 Ictiobus, **yg.**
Euglena viridis: 4 Notropis, **yg.**
E. acus: 3 Notropis, **yg.**
Actinosphærium: 2 Ictiobus, **yg.**
Centropyxis: 1 Carpiodes, Apr.
C. ecornis: 1 Notropis heterodon, July.
Arcella: 1 Erimyzon sucetta, 1¾ in.; 2 yg.; 1 Carpiodes, yg.; 1 Ictiobus, **yg.**
A. discoides: 1 Ictiobus, **yg.**
A. vulgaris: 1 Ictiobus, **yg.**
Difflugia: 1 Dorosoma cepedianum, July; 1 Notropis, yg.; 3 N. heterodon, May, July; 1 Pimephales notatus, Aug.; 1 Campostoma anomalum, Aug.; 3 Moxostoma, yg.; 5 **M.** macrolepidotum, 1¼-2¾ in., July, Aug.; 1 Erimyzon sucetta, 1¾ in.; 2 yg.; 4 Hypentelium nigricans, yg.; 2 Catostomus teres, June, Aug.; 4 Carpiodes, Apr., Oct.; 1 yg.; 2 Ictiobus urus, Aug., Oct.; 2 I. cyprinella, July; 1 Noturus gyrinus, Oct.
D. globulosa: 1 Gambusia patruelis, Sept.

VEGETABLE FOOD.

Seeds: 3 Fundulus diaphanus, Oct.; 1 Semotilus atromaculatus, July; 1 Notropis atherinoides, July; 2 N. megalops, Apr., Aug.; 6 N. whipplei, Apr., June, Aug.; 3 N. heterodon, May; 1 Moxostoma, yg.; 1 Ictiobus bubalus, **Apr.**
Corn meal (distillery slops): 1 Dorosoma cepedianum, July; 1 N. whipplei, Aug.; 1 Moxostoma macrolepidotum, Sept.; 3 Ictiobus urus, Oct.; 4 I. cyprinella, Oct.; 2 Ictalurus punctatus, Aug., Oct.; 1 Amiurus nebulosus, Sept.; 1 A. marmoratus, Oct.
Exogenæ: 6 Notropis megalops, June; 1 N. hudsonius, June; 1 Ictalurus punctatus, Apr.
Endogenæ: 1 Micropterus dolomiei, yg.; 3 Notropis megalops, June, July.

Fungi: 3 Notropis, yg.; 1 N. megalops, Aug.; 1 Hybognathus nuchalis, Aug.; 1 Chrosomus erythrogaster, Sept.; 2 Ictiobus, yg.

Terrestrial vegetation: 1 Centrarchinæ, yg.; 1 Pomoxys, Apr.; 1 Esox vermiculatus, June; 1 Hybopsis biguttatus, Sept.; 3 Notropis atherinoides, Apr., May; 4 N. megalops, Apr. Aug.; 3 N. whipplei, Apr., Aug.; 1 N. heterodon, Sept.; 1 Pimephales notatus, Aug.; 1 Ictiobus bubalus, Sept.; 1 I. urus, July; 1 I. cyprinella, June; 4 Ictalurus punctatus, Mar., Apr., Aug.; 1 Amiurus nebulosus, May; 1 Polyodon spathula, May.

Graminew, seeds: 2 Notemigonus chrysoleucus, May, Aug.; 7 Hybopsis biguttatus, June, Aug., Sept.; 1 Notropis whipplei, Apr.; 3 N. stramineus, Apr., July; 2 I. bubalus, Apr., Oct.; 1 I. urus, July.

Setaria, seeds: 1 Catostomus teres, Oct.

Aquatic vegetation: 2 Notemigonus chrysoleucus, Aug.; 4 Notropis megalops, Apr., July, Aug.; 3 N. hudsonius, June; 1 Chrosomus erythrogaster, June; 1 Campostoma anomalum, Sept.; 1 Placopharynx carinatus, Oct.; 1 Moxostoma aureolum, Apr.; 2 M. macrolepidotum, May, Sept.; 1 Hypentelium nigricans, Aug.; 5 Carpiodes, July, Oct.; 8 Ictiobus bubalus, Apr., Aug., Oct.; 3 I. urus, July, Oct.; 4 I. cyprinella, Oct.; 4 Ictalurus punctatus, Aug.; 1 Amiurus natalis, 3½ in., Oct.; 1 Amia calva, Aug.; 2 Polyodon spathula, May, June.

AQUATIC PHÆNOGAMIA.

Myriophyllum: 1 Lepomis gibbosus, May.

Ceratophyllum: 1 Lepomis pallidus, May; 2 Pomoxys, Apr., May.; 1 Ictiobus bubalus, Oct.; 1 Amiurus nebulosus, May.

Lemna: 1 Umbra limi, Sept.; 1 Dorosoma cepedianum, July; 1 Placopharynx carinatus, Oct.; 2 Ictiobus bubalus, Oct.; 1 I. urus, Aug.; 1 I. cyprinella, Oct.; 4 Ictalurus punctatus, Sept., Oct.; 1 Amiurus natalis, Oct.

L. trisulca: 1 Pomoxys, May.

L. minor: 1 Ictiobus bubalus, Oct.

Wolffia : 1 Lepomis pallidus, yg.; 1 Aphredoderus sayanus, Sept.; 7 Umbra limi, Sept.; 4 Gambusia patruelis, Sept.; 1 Zygonectes notatus, Sept.; 1 Dorosoma cepedianum, July; 1, 2½ in., July; 1 Placopharynx¦ carinatus, Oct.; 1 Moxostoma macrolepidotum, Sept.; 1 Erimyzon sucetta, 1¾ in.; 2 Carpiodes, Oct.; 11 Ictiobus bubalus, Oct.; 2 I. urus, Oct.: 1 Amiurus natalis, Oct.; 2, 2–2⅛ in., July.

Naias flexilis : 3 Lepomis pallidus, May, July, Nov.

Potamogeton : 1 Ambloplites rupestris, yg.; 1 Notropis megalops, Apr.; 1 Ictiobus bubalus, Oct.; 3 Ictalurus punctatus, June, Sept.; 1 Amiurus nebulosus, May; 1 A. marmoratus, Oct.; 9 Polyodon spathula, May.

P. gramineus : 2 Ictalurus punctatus, Oct.

AQUATIC CRYPTOGAMIA.

Chara : 1 Lepomis gibbosus, July; 1 Moxostoma macrolepidotum, June; 1 Amiurus nebulosus, July.

Algæ, filamentous : 3 Percina caprodes, Aug.; 2 Lepomis gibbosus, June, Aug.; 1 yg.; 9 L. pallidus, July, Aug., Oct., Nov.; 1 Pomoxys, July; 1 Aphredoderus sayanus, Sept.; 4 Eucalia inconstans, Oct.; 1 Gambusia patruelis, Sept.; 9 Zygonectes notatus, Sept., Oct.; 1 Fundulus diaphanus, Oct.; 2 Dorosoma cepedianum, June, July; 3 Notemigonus chrysoleucus, Aug., Sept.; 4 Semotilus atromaculatus, July; 3 Hybopsis biguttatus, Aug.; 4 Notropis, yg.; 1 N. atherinoides, Aug.; 8 N. megalops, Apr., May, June; 7 N. whipplei, Apr., May, June, Aug.; 7 N. hudsonius, May, June, July; 2 N. heterodon, Apr., May; 3 Pimephales notatus,‘ July, Oct.; 1 Hybognathus nuchalis, May; 2 Chrosomus erythrogaster, Sept.; 7 Campostoma anomalum, Aug., Sept.; 2 Moxostoma, yg.; 1 M. macrolepidotum, Sept.; 1 Erimyzon sucetta, July; 2 Catostomus teres, June, Aug.; 3 Ictiobus bubalus, Aug. Oct.; 3 I. urus, Aug.; 4 I. cyprinella, June, Aug.; 3 Ictalurus punctatus, Oct.; 1 Amiurus nebulosus, May; 1, 2 in., Aug.; 1 Noturus, yg.; 2 N. gyrinus, May; 3 Polyodon spathula, May, June, Aug.

Algæ, unicellular : 1 Dorosoma cepedianum, July; 2 Notropis, yg.; 2 N. whipplei; 1 N. hudsonius, July; 1 Pimephales promelas, Aug.; 1 Moxostoma, yg.; 1 Ictiobus, yg.; 1 I. bubalus, yg.; 1 I. urus, Aug.; 1 I. cyprinella, Aug.

Cladophora: 4 Ictalurus punctatus, Oct.

C. glomerata: 1 Notropis megalops, June.

Vaucheria: 1 Ictiobus urus, Aug.; 1 Ictalurus punctatus, Aug.

Scenedesmus: 2 Ictiobus cyprinella, Aug.

Protococcus: 1 Dorosoma cepedianum, $2\frac{1}{2}$ in., July; 1 Erimyzon sucetta, $1\frac{3}{4}$ in.; 1 Ictiobus bubalus, yg.; 1 I. cyprinella, Aug.

Glæocystis: 1 Notropis whipplei, Apr.

Spirogyra: 1 Semotilus atromaculatus, July; 2 Notropis, yg.; 1 N. whipplei, Apr.

Diatomaceæ: 1 Gambusia patruelis, Sept.; 3 Dorosoma cepedianum, Apr., July; 1, $2\frac{1}{2}$ in., July; 2 Notemigonus chrysoleucus, Aug.; 3 Notropis, yg.; 2 N. megalops, May; 2 N. whipplei, May; 2 N. hudsonius, June, July; 5 N. heterodon, May, July, Sept.; 1 Pimephales notatus, July; 2 Hybognathus nuchalis, May, Sept.; 1 Chrosomus erythrogaster, Sept.; 1 Campostoma anomalum, Sept.; 1 Moxostoma, yg.; 1 Erimyzon sucetta, July; 1 yg.; 5 Hypentelium nigricans, yg.; 1 Catostomus teres, June; 3 Carpiodes, Mar., Sept., Oct.; 2 yg.; 1 Ictiobus bubalus, Aug; 4 I. urus, June, Aug.; 3 I. cyprinella, Apr., Aug.; 1 Amiurus nebulosus, 2 in., Aug; 1 Polyodon spathula, June.

Pinnularia: 1 Gambusia patruelis, Sept.

Pleurosigma: 1 Moxostoma macrolepidotum, $2\frac{3}{8}$ in., Aug.

Cymatopleura: 3 Notropis, yg.

Desmideæ: 1 Notropis megalops, May; 1 Pimephales notatus, Aug.; 2 Moxostoma, yg.; 1 M. macrolepidotum, $2\frac{3}{4}$ in., Aug.

Closterium: 3 Notemigonus chrysoleucus, Sept.; 4 Notropis, yg.; 2 Moxostoma, yg.; 5 M. macrolepidotum, $2-2\frac{1}{2}$ in., July, Aug.; 1 Erimyzon sucetta, $1\frac{3}{4}$ in.; 2 yg.; 2 Hypentelium nigricans, yg.; 2 Catostomus teres, June, Aug.; 1 Ictiobus, yg.; 1 Carpiodes, yg.; 1 Ictiobus bubalus, yg.; 1 I. cyprinella, Aug.

Cosmarium: 4 Notropis, yg.; 1 M. macrolepidotum, $2\frac{3}{4}$ in., Aug.; 1 Erimyzon sucetta, $1\frac{3}{4}$ in.; 2 yg.

Staurastrum: 1 Erimyzon sucetta, yg.; 1 Ictiobus cyprinella, Aug.

Nostoc: 2 Ictiobus cyprinella, Aug.; 1 Polyodon spathula, Aug.
Oscillaria: 1 Chrosomus erythrogaster, Sept.; 1 Ictiobus, yg.
Chroöcoccus: 1 Erimyzon sucetta, 1¾ in.; 1 yg.

Dirt: 10 Dorosoma cepedianum, Apr., June, July, Oct.; 4,
2½–5¼ in., July, Oct.; 1 Clupea chrysochloris, 1⅜ in., June;
10 Notemigonus chrysoleucus, July, Aug., Sept.; 3 Hybopsis biguttatus, Aug., Sept.; 2 Phenacobius mirabilis,
Sept., Oct.; 1 Notropis hudsonius, May; 1 N. heterodon,
Sept.: 8 Pimephales notatus, July, Aug., Sept., Oct.; 4 P.
promelas, May, Aug.; 8 Hybognathus nuchalis, May, Aug.,
Sept., Oct.; 3 Chrosomus erythrogaster, June, Sept.; 9
Campostoma anomalum, Aug., Sept.; 3 Moxostoma, June;
2 M. aureolum, June; 3 M. macrolepidotum, June, Sept.;
2, 2⅛–2¾ in., Aug.; 1 Minytrema melanops, Sept.; 1
Erimyzon sucetta, July; 6 Carpiodes, Mar., Apr., July,
Oct.; 5 Ictiobus urus, Aug.; 1 I. cyprinella, Aug.; 1
Amiurus nebulosus, Sept.

ON THE LOCAL DISTRIBUTION
OF CERTAIN ILLINOIS FISHES

Stephen Alfred Forbes

ARTICLE VIII.—*On the Local Distribution of certain Illinois Fishes: an Essay in Statistical Ecology.* By S. A. FORBES.

An animal society is composed of animals habitually occurring together in the same locality and the same class of situations. Such an association is, of course, composed of many species, variously related to their special environment, some attracted to it by one set of conditions and some by another. Although their local haunts may be virtually identical, their ecological relations, if determined in detail, may prove to be very different. A pike and a minnow may be members of the same associate group, to whose habitat, however, the pike is especially attracted by the minnow, and the minnow by the facilities which are offered there for concealment or escape from the pike.

It is usually possible to learn the contents of a local association of plants by simple inspection and enumeration; but animals come and go, elude observation, and refuse to be numbered, and the details of their associate occurrence can only be learned indirectly, by means of sample collections preserved for subsequent study. If the situations from which such collections are made are carefully chosen and correctly classified, and if the collections themselves are full enough, uniform enough, and numerous enough to be fairly representative of each situation, the essential facts concerning the assemblage of animals corresponding to any unit of environment may be readily made out. The making of such collections for such a purpose is, however, a relatively new thing, and scarcely a beginning has been made in the systematic study of animal associations by this method.

A knowledge of definitely circumscribed, or merely measurably distinct, local associations does not, however, by any means exhaust the subject of associate relations, for the animals of a region can not be wholly divided up into such definite societies, and such society groups as can be clearly recognized rarely have any precise boundaries. For a full knowledge of the intricate web of the relations to their physical environment, and through that to each other, of the animals of any composite area, it is necessary that the entire assemblage of the inhabitants of that area should be studied as a compound unit, and for this, of course, extensive and comprehensive collections must be made, such as will fairly represent the entire animal life of their region.

The possession of a miscellaneous but very large collection of Illinois fishes, obtained during various seasons of a long period of years, from all kinds of waters and in all parts of the state (see Map I.), each lot still bearing, as a rule, the original collector's data giving both the time of collection and the exact locality, has suggested to me a trial study, intended to show what may be learned with regard to the ecology of fishes by a critical analysis of the local data of such a collection.

These data may be organized and generalized for ecological study in two ways. They may be treated in one mass, without local subdivision, and in such a way as to bring out the facts concerning the association of the different species of fishes with each other, without reference, in the first instance, to the localities and situations from which the specimens have been taken; or they may be first divided and arranged according to location and surroundings, the assemblage of species from each geographical unit and from each kind of ecological situation being studied separately, as a local animal society. The first method has the advantage over the second, that it gives us much larger numbers of specimens and collections from which to generalize, and thus enables us to enter further into the details of the associate relationship without danger of error from unsafe generalization; and it also enables us to distinguish similarities and differences of ecological relationship among the species, uninfluenced by any previous discrimination or classification of ecological situations. The second method has the advantage over the first, that it attacks the problem more simply and more directly, and, if the data are sufficient, reaches results more immediately and obviously significant.

I have used both these methods in the present paper, comparing the results of the two in a way to make the one set account for and explain the other. This paper is thus to be taken as a contribution to an answer to the following questions: What Illinois fishes are habitually found in each others' society, and what is the relative frequency of their associations? How are Illinois fishes grouped and distributed according to location and situation, and in each ecological assemblage so formed what is the proportionate representation of its various constituent species? How far are the two classes of data, those of associative affiliation and of ecological relationship, comparable, and to what extent may the one be used to explain the other? An answer to these inquiries would enable us to recognize, define, and account for associate groups among our fresh-water fishes, and also to distinguish those members of each group which, being most frequently and most strictly associated, are most characteristic of it. It has, in fact, been a part of my undertaking to find a method of distinguishing clearly these central or typical members of an ecological assemblage, and to express numerically the intensity of the influence—the strength of the bond—which holds them to the local situation, as compared with the more lax or less continuous forces influencing what we may call the outlying members of the group.

Studies of this description may be expected to give us significant information, also, concerning the competitions of associated species, and concerning the evasions of competition, and the escape from its consequences, by those closely related and similarly endowed, and concerning the niceties of adaptation, psychological, physiological, and structural, exhibited by fishes inhabiting a notably uniform area.

Associative Relationships among the Etheostominæ.

For a preliminary and sample study of this description, I have chosen first a subfamily of our fishes, the *Etheostominæ*—or darters, as they are commonly called—and have endeavored to learn to what extent the species of this subfamily are ecologically affiliated, which of the species are most typical of the subfamily as an ecological group, which are to be regarded as lagging or wandering members of it, and which, if any, do not belong ecologically with their taxonomic relatives.

I shall be obliged, in these studies, to assume provisionally that my collections are large enough and numerous enough fairly to represent actual field conditions in Illinois, and that they are so numerous that they may reasonably be treated, for the present purpose, as homogeneous and similar, each collection as a unit substantially like every other, important differences among them disappearing, in aggregates and averages, by the process of mutual cancelation. In other words, I must assume provisionally, testing my supposition later by the constancy and reasonableness of the results, that these random samples of Illinois darters represent the subfamily as a whole sufficiently well to justify their use as materials for a study in statistical ecology.

The Method of the Investigation.

The species of darters which are most frequently found in each others' company are, of course, those most likely to be closely related ecologically; and the ratio of the number of collections containing both of any two species to the total number of collections containing either, may be used as a provisional measure of the ecological affinity of the two.

Furthermore, given a certain average frequency of occurrence of each of two species inhabiting a common territory, and assuming a uniform distribution of each in this territory, uninfluenced by ecological relationships, the average frequency of the joint occurrence of these species in collections may be computed; and any very marked departure, positive or negative, from this computed average will point to some ecological bond if the difference is positive, or to some cause of ecological separation if it is negative.

If, for example, it appears that several species ought to be found together, on an average, in one out of twenty of our collections, provided that they are distributed over their common area uninfluenced by causes tending to bring them together into the same situations, and if the actual average of the joint occurrences of the species is one in every five collections, then the associative bond of the species concerned may be given the value of four—a value of little significance perhaps, taken by itself, but useful, at any rate, for a comparison of the darters with other groups. And if certain of the species are associated with the other darters in an average ratio of five to one, while other species are associated with the other darters in an average ratio of only two to one, then the former species will typify the ecological group more definitely and correctly than the latter.

By this means, also, if the actual frequencies of joint occurrence of the various species of the group be compared with the computed average of such frequencies, the division of any presumably single group into two distinguishably separate ones might be made out. If it should appear, for example, that the species of darters may be divided into two groups, each of which taken separately is found to have a mutual associative ratio of six to one, while the corresponding ratio between the two groups themselves is but three to one, we may infer provisionally the division of the darters into two ecological groups, distinguishable by their predominant attraction to different sets of ecological factors in their common environment, but united in turn in one larger group by their common attraction to certain other factors.

For an analysis of the facts, we need for each species of darter a determination of the average frequency of its merely chance occurrence in collections with each of the other species, a determination of the actual frequency of these joint occurrences, and a numerical expression of the ratio of one of these frequencies to the other. Then by a systematic tabulation of these latter ratios, which may be called the *coefficients of association*, we may compare one species with another, and bring the essential data for the whole family under the eye for convenient inspection and analysis.

For the computation of these ratios, I have used, with two exceptions to be presently stated, the thousand Illinois collections most available for these studies, excluding five hundred and forty-four additional collections, which, because of imperfect data and for various other reasons, are undesirable material. I find that the species *Hadropterus aspro* has been taken in 159 of these thousand collections, which ratio of average frequency may be expressed by the fraction .159; and that the species *Hadropterus phoxocephalus* has been taken 85 times, which gives a frequency ratio of .085. That is, in any thousand similar miscellaneous collections distributed over the area inhabited by these species we may, according to these data, expect to get the first species 159 times and the second species 85 times; and the chance that any single collection will contain the first species is .159, and that it will contain the second species is .085. From this it follows that the chance that the two species will occur together in any single collection of the thousand, provided that the distribution of each is arbitrary and accidental with reference to that of the other, is the product of these fractions; and the probable number of chance joint occurrences of the two species in the thousand collections is, of course, a thousand times that product, or 13.515. As a matter of fact, however, these two species were found together in my collections 40 times instead of approximately 13.5 times, or three times as frequently as there was reason to expect provided that there had been no associative bond between the species. This number 3, indicative of the frequency of actual association as compared with the chance or accidental, is the coefficient of association for these two species. If the numbers of presumable and actual joint occurrences were equal, this coefficient would evidently be 1, in which case no associative bond would be indicated; and if it were notably less than 1, we should have some reason to suppose that the two species belonged to different ecological groups—that their ecological affinities and relationships tended to separate them instead of to bring them together.

The computation may be facilitated by the use of algebraic symbols.

Let a equal the total number of collections to be used in the computation; b, the number of collections containing the more abundant of two species to be compared with one another; c, the number of collections containing the less abundant of these species; and d, the number of collections each of which actually contains both species together. Then $\frac{b}{a}$ expresses the chance that any collection of a will contain one or more representatives of the first species; $\frac{c}{a}$, the chance that any collection will contain one or more representatives of the second species; $\frac{bc}{a^2}$, the chance that any collection will contain one or more representatives of both species at once, provided that the distribution of each is ecologically independent of that of the other; and $\frac{bc}{a}$, the probable number of chance occurrences of the first and second species together in the number of collections represented by a, the same proviso being made. Since $d =$ the actual number of such joint occurrences, $\frac{ad}{bc}$ is the formula for the ratio of actual to calculated joint occurrences—the formula, in other words, for the computation, in all cases, of our coefficients of association. For example, substituting in this formula the values already given for *Hadropterus aspro* and *Hadropterus phoxocephalus*,

$$\frac{ad}{bc} = \frac{1000 \times 40}{159 \times 85} = 2.96.$$

To determine the coefficient for any pair of species, we need only to know their separate frequencies and their joint frequencies in collections derived from the territory of their common distribution.

The above formula may be translated into the following rule for finding the coefficient of association of any two species: *Multiply the number of collections made from the common area of the species by the number containing one or more representatives of both; multiply the number of collections containing one or more representatives of one of the species by the number containing one or more of the other; and divide the first product by the second. The quotient will be the coefficient of association.*

DISCUSSION OF ASSOCIATIVE TABLES.

I have computed, by the above-described method, for thirteen species of Illinois darters—each of which was obtained more than fifteen times in my collections—the coefficients of the association of each species with each of the other twelve, and have arranged these seventy-eight coefficients (apparently one hundred and fifty-six, since each of them is entered twice) in Tables I.-V. for comparison and discussion. In computing the coefficients of two species, *Diplesion blennioides* and *Etheostoma zonale*, the first of which is found only in the eastern part of the state and the second only in the northern half, I have used as the value of a in my formula, not the entire number of collections made throughout the state, but the number made in the stream systems in which these species occur.

In Table I. the coefficients in each column are in serial order, the highest to the lowest from above downwards; and the columns for the several species are placed in the order of the average coefficients for the columns, the highest at the left.

We notice first, that the total of the one hundred and fifty-six coefficients of this table is 315.8—a general average associative coefficient of 2.02 for all the thirteen species. As the normal chance average would be but 1, we conclude, from these data, that darters were found together in my collections about twice as frequently as mere chance would indicate. This ratio of 1 to 2 is thus an approximate and provisional measure of the ecological bond in this family taken at large.

We notice next, the unlike totals and averages of the coefficients for the several species, these running from 1.22 to 2.69—an indication that the associative bond is more than 2.2 times as strong for *Hadropterus phoxocephalus* and *Etheostoma zonale* as for *Boleichthys fusiformis* and *Boleosoma camurum*. On the other hand,we find no species in which the average coefficient of association is less than 1—no indication that any of these twelve species are wholly drawn away from their family by stronger ecological affiliations with some other group. Nor do we find, in passing from the more strongly associated species to those less strongly associated, any abrupt transition in the series—a fact which may be taken as evidence that the darters of my list are a unitary group, of which certain species are ecologically more typical than others, having, that is, the darter habits and relationships more fully developed and more strongly fixed.

Typical and Non-typical Darters.

The more typical species of this list seem to be the following six, mentioned in the order of the size of their coefficients of association: *Hadropterus phoxocephalus, Etheostoma zonale, Etheostoma flabellare, Hadropterus aspro, Ammocrypta pellucida,* and *Etheostoma cœruleum*, the associative coefficients of which average 2.48. Apparently the least stringently connected with their kind by the associative relation are *Diplesion blennioides, Etheostoma jessiæ, Boleosoma camurum,* and *Boleichthys fusiformis,* the average coefficient of which is 1.36.

Furthermore, those least strictly associated with darters in general are not especially strongly associated with each other. Of the four species just mentioned, six pairs may of course be made, and the average of the coefficients of these six pairs is 1.33—less by .69 than the general average for the entire group (see Table III.). If we similarly pair the six species which I have selected as most typical, and average the fifteen coefficients of these pairs (see Table IV.), we get a general coefficient of 3.47—more by 1.5 than the average for the group. That is, those species which are laxly associated with the darters in general, are also laxly associated with each other; while those which are strongly associated with darters in general, are still more strongly associated among themselves. This last fact was to be anticipated, since in making up the special average coeffi cients of those species which exhibit strong associative affinities we omit those which have the weaker affinities, and so have a group of select associates whose average coefficient must be higher than that of the whole thirteen species, including, as this does, some with strong and others with feeble associative tendencies.

The same fact is illustrated in Table II., in which all the coefficients of the seventy-eight possible pairs of my thirteen species are arranged in the order of the magnitude of their co-efficients of association with *Hadropterus aspro* (1421). Taking the first twenty-one coefficients of the six most frequent associates of *Hadropterus aspro*, we find that they average 3.27, while the last twenty-one coefficients of the six least frequent associates of *Hadropterus aspro* average 1.4. That is, the twenty-one coefficients at the upper left angle of Table II. (above the black line) average two and a half times as much as the twenty-one coefficients at the lower right angle of that table (to the right of the black line). The most frequent associates of this species are associated with each other about two and a half times as frequently as are its least frequent associates.

It is also significant that five of the list of six most frequent mutual associates made up from Table I., are the same as those of the corresponding list made up from Table II., of *Hadropterus aspro* and its five closest associates, the two tables containing the same figures, differently arranged. We further notice that the three least frequently associated species are the same on both lists. Whether the data indicating frequency of association are arranged under each species independently, in the order of frequency, as in Table I., or with reference only to a single leading species, as in Table II., the results are nearly identical as to the darters most typical and least typical of the group.

SUFFICIENCY OF THE COLLECTIONS.

With respect to the sufficiency of the collections for the use which is here made of them, some additional evidence may be found by tabulating separately the seven species which appear least frequently in them—ranging in number of occurrences from 16 to 60, with an average of 34—and comparing the average of their coefficients of mutual association with the general average coefficient for the entire group, with its 82 occurrences to the species. From Table V. it appears that this general coefficient for the seven least frequent species taken separately is 1.85, while that for the whole group of thirteen (Table I.) is 2.02—a coincidence probably as close as could be expected in view of the fact that the former number is an average of only 21 coefficients and the latter of 78. The coefficient expressing frequency of mutual association among these least frequent species, is thus so close to the general coefficient for the entire group that even the former species may be said to occur frequently enough in the collections for the purposes of this discussion.

Relations to Physical Environment.

I have next to study the interrelations of this group of darters by means of another and widely different set of data, to be derived from an analysis of collectors' records concerning the kinds of waters and the classes of situations from which the several collections came; and to compare the conclusions thus reached concerning the physical relations of the species with those already derived from an analysis of their relations of association. For this purpose these records have been organized in a way to show the relative frequency of the occurrence of each species in our collections in each of the three sections—northern, central, and southern Illinois, as the state is commonly divided; in each of the ten stream systems, or river basins, distinguished by us; and in each kind or class of body of water—whether stream, lake, pond, or marsh—the classification made expressing differences in size, in water movement, and in the character of the bottom.

Equalization of the Data.

The data available are not equally numerous under these various heads. Those concerning the size and general character of water bodies, and the distribution by stream systems and sections of the state, are inclusive of all our collections; but in many cases data are wanting definitely *descriptive* of the waters and the situations from which the collections were made. This is owing to the fact that the present use of these materials was not foreseen in the beginning of our collection period, nor, indeed, until the greater part of the field work had been done, and the records of the earlier years are consequently often incomplete for the present purpose. Later, collectors were instructed to make full descriptive notes, from the ecological standpoint, of each body of water visited and of each location at which a haul of the seine was made, and the whole body of the data of local distribution and ecological preference is such that if used with due discretion it may be expected to throw considerable light on the associative relationships of this little group of fishes.

These data have been worked out, in the same manner as in the preceding section of this paper, in the form of percentages of frequency of the occurrence of each species in each geographic or hydrographic subdivision and in each ecological situation. As the numbers of collections made have varied widely for the several areas and situations, those from one being often many times as numerous as those from another, it was necessary to reduce the frequency ratios of the several species in each area to a common standard for comparison. These numbers have been equalized, and confusing discrepancies removed, by reducing the collection data to percentages of the same base, which, for convenience, has been made one hundred collections.

Discussion of Ecological Tables.

If equal numbers of miscellaneous collections had been made from each situation, and if the total number of collections were such that any given darter had been taken one hundred times, what number or percentage of these collections of darters would have come, according to my present data, from each of the situations represented?

The figures in Table VI. are answers to this question; and

when I say that 63 per cent. of our collections of *Hadropterus phoxocephalus* are from rivers and 26 per cent. from creeks; or that 94 per cent. of them are from waters with a bottom of rock and sand and only six per cent. from mud; this means that if miscellaneous collections of fishes of all descriptions had been made from all kinds of Illinois waters until one hundred of them contained darters of this species, then sixty-three of the hundred would have come from rivers and twenty-six of them from creeks, ninety-four of them from rock and sand, and six of them from mud.

The ratios of this table differ in significance from those of my tables of associative coefficients in the fact that while the latter exhibit various degrees of *associative* relationship between species, the former express the tendencies or preferences of the species with respect to the features of the *physical* environment. An understanding of these physical relations of a species must help us to understand and explain its associative relations, and the one set of data may be expected to serve as a test of the completeness and correctness of the other.

The Darters as an Ecological Group.

It is well known that the darters as a group are most likely to be found in comparatively swift and rocky streams, and that they are especially adapted, by their small size, their large paired fins, their pointed heads, and their habit of resting on the bottom, for maintaining themselves in swift currents, and for securing from among and under stones the insect larvæ and crustaceans on which they mainly depend for food. This fact is clearly reflected in my Table VI., of "Local Preferences of Darters", from which it appears that 70 per cent. of our collections of the thirteen species were obtained from the smaller streams, 77 per cent. from swift waters, and 82 per cent. from waters with a bottom of rock and sand. Only 12 per cent., in fact, came from lakes and ponds, and 18 per cent. from waters with a muddy bottom.

The Typical and the Non-typical Species.

A comparison, in respect to the strength of their local preferences, between the six species which, by means of an analysis of their associative ratios, I have distinguished as typical and the six less typical species, shows that the more typical group occurs in the smaller rivers and creeks in 88 per cent. of these collections, and the less typical in 47 per cent.; the first group, in swift waters in 88 per cent. of the cases, and the second in 62 per cent.; the first, in rocky or sandy streams in 91 per cent., and the second in 66 per cent. That is, the frequency of occurrence of the less typical species in small rivers and creeks is 53 per cent. of that of the more typical species; in swift waters it is 71 per cent., and on rock and sandy bottoms it is 72 per cent.,—an average of 65 per cent. for these three factors. These purely ecological ratios agree in a significant manner with the corresponding averages to be drawn from the tables of associative frequencies, as may be seen by reference to Table I. If we average separately the totals for the first six and the last six species of that table, we find the average of the latter group to be 63 per cent. of that of the former—the difference in degree of associative affiliation is essentially the same as the difference of ecological relationship, the one conclusion confirming, and likewise explaining, the other.

It is further to be noticed, of the ecological affinity of the six selected species, that no one of them has been found in upland or glacial lakes; that their occurrence in lowland lakes, ponds, and sloughs—an average of only 1 per cent.—is so rare as to be negligible; and that, omitting *Ammocrypta pellucida*, which is in some respects peculiar, the frequency ratio for the larger rivers ranges from 3 to 9 per cent., with an average of only 5.5 per cent. for these species. This uniformity of their ecological relationships, which makes of them a well defined ecological group, is the explanation, of course, of their high degree of associative affiliation. The most notable specific differences among them are the relative frequency of *Ammocrypta pellucida*, and the absence of *Diplesion blennioides*, in my two hundred and ninety-three collections from the larger rivers.

The six less typical species, on the other hand, have little in common except their difference from this more typical group. *Boleosoma nigrum*, of which we have two hundred and thirty-six collections, is an abundant and wide-ranging species, with comparatively feeble ecological preferences, as is shown by the fact that 15 per cent. of these collections are from lakes, 32 per cent. from still waters, and 11 per cent. from those with a muddy bottom. *Percina caprodes* (sixty collections) makes a similar showing, this being also a lake species in part (19 per cent.); but it differs from the preceding in the fact that it has occurred more frequently in the larger streams (10 per cent.), less frequently in still waters (7 per cent.), and not at all on muddy bottoms. *Cottogaster shumardi*, so far as may be judged from our sixteen lots of this species, is peculiar in its frequency in the larger rivers (55 per cent.) and the lowland lakes (18 per cent.), and in its avoidance of the smaller streams (only 4 per cent. in the creeks and smaller rivers). *Etheostoma jessiæ* (one hundred and fifty-eight collections) is an indifferent species, and occurs in almost equal ratios in large rivers, small rivers, creeks, and lowland lakes. *Boleichthys fusiformis*, which we have taken fifty-six times, is rare in the larger rivers, and seems to be the commonest of all our species in the upland lakes. *Boleosoma camurum* (one hundred and seven collections) is somewhat less indiscriminate in its local preferences. It is commonest in creeks (42 per cent.) and relatively rare in the larger rivers (9 per cent.). It apparently has no marked preference for swift waters over slow, nor for a hard bottom as compared with one of mud.

The ecological heterogeneity of these least typical species is reflected in their relatively feeble associative affiliations, these six species having a mutual associative ratio (derived from Table II.) of 1.4, while the corresponding ratio of the first six more typical species of Table II. is 3.28.

ASSOCIATION AND DISTRIBUTION.

The association of species may be looked upon as a consequence of their distribution. Species of wholly different general, or geographical, distribution can, of course, never be associated; and the same is true of those of wholly unlike local distribution. Those whose areas of general distribution merely overlap, will be less frequently associated, other things being equal, than those whose distribution areas are identical; and species which are equally attracted to some local situations and unequally attracted to others, will be less frequently associated than those whose local preferences are altogether similar. Furthermore, if two species which occupy the same situations in the same area have a widely unlike abundance in different parts of this area—one being much the most abundant to the north, for example, and the other to the south—these species will occur together in collections less frequently, will have a lower coefficient of association, than if the two were most abundant in the same section and least abundant in the same. The number of joint occurrences will be conditioned, in part, in each section of the common area, by the abundance there of the less abundant species. It is impossible, consequently, to distinguish, by a simple inspection of a table of coefficients, local from general factors among the determining causes of difference in associative frequency. For this purpose maps of species distribution, and tables showing the locality preferences of species (like my Table VI.) must be studied in connection with tables of associative coefficients.

The causes controlling general distribution and local distribution are alike ecological, those affecting general distribution being usually general—climatic, topographic, hydrographic, and the like—and those affecting local distribution being local. In a small area like that of Illinois, one in which there are comparatively few physical barriers to the intermingling of fishes, these two classes of causes are not widely different, but they must nevertheless be distinguished, so far as possible, if we are to have a clear and correct knowledge of ecological relationships.

COMPARATIVE STUDY OF TABLES AND MAPS.

As an example of the manner in which these factors may be separated by a comparison of my tables and maps, and of the extent to which associate relationships may be accounted for, we may take a few instances of very low, and others of very high, coefficients from Table II., and look up the facts concerning the species compared, as given in Table VI. and in the distribution maps appended to this paper. Thus far, it may be noticed, I have dealt with aggregates and average numbers only, which, owing to the heterogeneous and variable character of the data, are much more likely to be uniformly reliable than are the separate entries of the tables. The present discussion will, however, necessarily bring into comparison these separate entries, and the reasonableness and consistency of the conclusions reached by it may serve as some measure of the validity of their individual coefficients.

By reference to Table II. it will be seen that zeros appear at five points, in place of coefficients of association—an indication that representatives of the several pairs of species concerned have never been taken together by us in the same collection. This, as already pointed out, must mean either a complete difference in general distribution, so far as represented by my collections, or a very radical difference in locality preference.

Species 1443 and 1461 (*Diplesion blennioides* and *Etheostoma zonale*) are examples. A glance at the distribution maps of these species will show that each has been taken by us only in a different part of the state from the other, *blennioides* being confined to the Wabash valley, with the exception of a single collection at Chicago, and *zonale* being limited to the Illinois and Rock river systems. It seems difficult to believe that the flat and indefinite watershed separating the tributaries of the Wabash from those of the Illinois, can constitute a physical barrier sufficient to prevent the intermingling of these two species. On the other hand, it must be admitted that their ecological relations, as expressed in their preferences of situation, are, on the whole, very similar, as may be seen by a comparison of the two in Table VI.

A similar explanation is to be made in the case of *Diplesion blennioides* and *Cottogaster shumardi* (1443 and 1436). Here the areas of our collections of these two species are entirely separate, with the exception of a single collection of *Cottogaster* from the Wabash valley—to which *Diplesion* was entirely confined. Furthermore, *Cottogaster* has been taken only in the larger streams or their immediate neighborhood, as is shown by the distribution map for that species; while *Diplesion* is limited to the smaller rivers and creeks.

With respect to *Cottogaster shumardi* and *Etheostoma cœruleum* (1436 and 1477), the case is a little less clear, and it is quite possible that with a larger number of collections containing the former species, the two might have been found in company. It is true that only 4 per cent. of our collections of *Etheostoma cœruleum* have come from the larger rivers and from stagnant waters to which *Cottogaster* is confined. On the other hand, a concurrence of the locality marks on the maps of distribution of these species (Maps V. and XII.) shows that the two were taken from the same locality—although not in the same collections—in three out of nine possible cases.

The lack of any coincident occurrence of *Cottogaster shumardi* and *Etheostoma zonale* (1436 and 1461) is explained by a glance at the maps (V. and X.), as due, not to a difference of geographical distribution, which is approximately identical for the two, but to that of local preference, the former species occurring only in or near the largest streams, and the latter being limited to the smaller rivers and creeks. Indeed, the two species were not taken by us from even the same locality at any time.

Nearly the same may be said of *Diplesion blennioides* and *Boleosoma camurum* (1443 and 1448), which have come from the same locality but once, although in general distribution they are not mutually exclusive. *Blennioides*, as may be seen from Table VI., is a species of more indefinite preferences than *camurum*, and occurs in various situations from which the latter is excluded.

I take up next five pairs of species, representatives of which have been occasionally taken together by us, but the co-efficients of whose association are nevertheless very small.

Etheostoma zonale and *E. jessiœ* (1461 and 1474), for example, with an associative coefficient of only .37, show a preponderant abundance of the first in the north half of the state and of the second in central and southern Illinois. Among the twenty-nine localities from which the first of these species was taken, and the fifty-four for the second, there were but two in which both were found, and at each of these localities they occurred in only one collection. That is, in one hundred and eighty-eight separate collections of one or the other of these species from these various localities, the two were taken together but twice—a fact to be connected partly with the limitation of *Etheostoma zonale* to the northern half of the state, and partly with differences in the bodies of water in which these species habitually occur. Twenty-one per cent. of our collections of *jessiæ* came from the larger rivers, and only 3 per cent. of those of *zonale*; 19 per cent. of *jessiæ*, from the smaller rivers and creeks, and 74 per cent. of *zonale*; 24 per cent. of *jessiæ*, from lakes and ponds, and none of *zonale*.

Boleosoma camurum and *Etheostoma zonale* (1448 and 1461), whose coefficient of association is but .39, furnish an example of the relation of distribution already referred to, the area of the two species overlapping, but not coinciding throughout—that of *zonale* expanding to the northward and that of *camurum* to the southward. Partly in consequence of this fact, we have but a single joint occurrence of these species out of one hundred and thirty-eight collections containing one or the other. Their ecological relations, as shown by Table VI., are also quite unlike, *Boleosoma camurum* occurring in sluggish or stagnant waters five times as frequently as the other species, and in waters with a muddy bottom in a still greater differential ratio.

The low associative coefficient (.63) of *Hadropterus phoxocephalus* and *Boleichthys fusiformis* (1418 and 1494) is largely explained by the difference in preponderant distribution, the former being commonest in the Illinois valley and to the northward generally, while the latter is much the most abundant in the Wabash system and in extreme southern Illinois. In one hundred and thirty-eight collections containing one or the other of these species, they have occurred together but three times,—twice in branches of the Little Wabash River and once in the Saline. The ecological relationships of the species are likewise very different, *phoxocephalus* showing a much stronger tendency than *fusiformis* to the larger streams. It occurs, for example, according to our data, in rivers in 63 per cent. of the cases, as against 13 per cent. for the other species. It also prefers swift to moderate water much more strongly, if I may judge from the small number of collections for which this factor was recorded, the ratios for swift water being 87 per cent. for *phoxocephalus* and 22 per cent. for *fusiformis*. A corresponding difference is seen in respect to the character of the bottom, 66 per cent. of our collections of *fusiformis* coming from waters with a muddy bottom and only 6 per cent. of those of *phoxocephalus*.

Boleosoma nigrum and *Etheostoma jessiæ* (1446 and 1474), with their coefficient of .99, may serve as an example of species similarly distributed but essentially indifferent as associates, a coefficient of 1, it will be remembered, indicating a neutral relation. A glance at the distribution maps of the species shows at once some notable differences. *Boleosoma nigrum*, the most abundant of our darters, and taken by us in two hundred and thirty-six collections, has virtually the same geographical distribution as the other species, but it is represented in the larger rivers in very much smaller ratio. The marks of local distribution for the more abundant species are widely and rather uniformly scattered over the map, with but few on the larger streams, while those of the less abundant species are strung, like beads, along the principal rivers of the state. On the other hand, neither species is definitely excluded from either the territory or the situations of the other, as may be seen by a comparison of the figures for them given in Table VI.

Turning now to pairs of species with extraordinarily high associative coefficients, I may call attention first to *Etheostoma zonale* and *Etheostoma cæruleum* (1461 and 1477), whose coefficient reaches the remarkable figure of 8.38. The general distribution of these species is substantially the same, except that *Etheostoma cæruleum* has a greater development to the south. *Etheostoma zonale* is much less numerous than *cæruleum*, but both species have been found most frequently in the eastern part of the state. A close comparison of the distribution maps shows that both have been taken from eighteen of the thirty *localities* in which the less abundant one was found; and they have been taken together in seventeen of the one hundred and five *collections* containing either or both.

A comparison of their local preferences indicates a close agreement in ecological relationship. Each of the species was found in the larger rivers in 3 per cent. of the collections; *zonale* in 97 per cent. of those from the smaller rivers and creeks, and *cæruleum* in 89 per cent.—the remainder of the latter coming from lowland lakes and ponds (1 per cent.) and from various miscellaneous sources. Eighty-nine per cent. of the collections of *zonale* and 83 per cent. of those of *cæruleum* were from streams of swift or moderate flow; 89 per cent. of *zonale* and 92 per cent. of *cæruleum*, from rock and sandy bottom. The only notable difference between these species is the preponderant disposition of *zonale* towards the smaller rivers rather than the streams classed as creeks.

The next highest coefficient (5.69) is that of *Hadropterus phoxocephalus* and *Etheostoma zonale* (1418 and 1461), which have occurred together sixteen times in my one hundred and one collections of one or the other. Both have been taken from seventeen of the thirty localities in which we have found *zonale*. The general distribution of the two differs but little, except that *zonale* is very much less abundant than *phoxocephalus*, and has been limited much more closely to the Illinois and Rock river basins. The ecological ratios for *zonale* and *phoxocephalus* respectively are,—larger rivers, 3 per cent. and 7 per cent.; smaller rivers, 74 per cent. and 56 per cent.; creeks, 23 per cent. and 26 per cent.; lakes and ponds, 0 and 3 per cent. The ratios of preference for rapid and slow waters respectively are still more closely approximate—89 per cent. of *zonale* and 87 per cent. of *phoxocephalus* from moderate or rapid currents. The preferences of the two species for rock and sandy bottom are similarly close—89 per cent. for *zonale* and 94 per cent. for *phoxocephalus*.

The next coefficient in order of size, that of *Hadropterus phoxocephalus* and *Ammocrypta pellucida* (1418 and 1450), is 4.95. These species are virtually identical in general distribution, *pellucida* being, however, comparatively scarce. The two species have been taken in ten of the seventeen localities in which *pellucida* was found, and have occurred conjointly eight times in the ninety-six collections containing one or the other. In general ecological relationship they are very closely similar, both occurring infrequently in the larger rivers, and in smaller rivers more frequently than in creeks. *Ammocrypta [pellucida* has not been taken at all in lakes and ponds, and *phoxocephalus* only to the amount of 3 per cent. Both are rapid-water species, and strongly prefer streams flowing over rock and sand to those with muddy bottoms.

Hadropterus aspro and *Ammocrypta pellucida* (1421 and 1450), with a coefficient of 3.97 based on their twelve joint occurrences in one hundred and sixty-six collections, were taken from the same localities in ten cases of a possible seventeen. *Ammocrypta pellucida*, although much the less abundant, is distributed in general precisely like *aspro*, except that it does not show so marked a preference as does the latter species for the eastern part of the state. With respect to the character of the streams in which these species are most generally found, the ratios are unusually similar, *pellucida* occurring, however, according to our data, more commonly in the larger rivers, and *aspro* more frequently in creeks. Neither has been taken by us in lakes or ponds. The ratios of preference for waters with a clean bottom are 84 per cent. for each.

Percina caprodes and *Etheostoma zonale* (1417 and 1461) were taken together four times in the eighty-eight collections containing either or both. Their associative coefficient is 3.55. Their general distribution is different in the fact that *caprodes* is the more abundant in the central and southeastern parts of the state. They were collected from the same localities seven times out of a possible thirty. In ecological relationships they are only fairly similar. Both occur in the larger rivers, but

Percina caprodes in the larger percentage. This species was likewise frequently found in lakes and ponds, from which *zonale* was entirely absent. Their relations to slow and rapid waters seem essentially the same, but while all the collections of *caprodes* were taken from sand and rock, 11 per cent. of those of *zonale* came from a muddy bottom.

Indeed, we have, for the first time, in these last two species, a pair whose ecological records do not seem to correspond quite closely to their associative coefficients—a fact which might be due to a number of collections of these species too small to give a reliable average, or to the influence of ecological factors not covered by the classification of Table VI. *Percina caprodes* was represented by sixty collections, and *Etheostoma zonale* by thirty-two; but I have information concerning the relations of the species to the water current for only fourteen collections of the first species and eighteen of the second, and concerning their relation to the kind of bottom for only twenty of the first and nineteen of the second. On the other hand, it seems certain that the local distribution of darters must be affected by many things not referred to in Table VI. —variations in the mere instinct of segregation, in the kind of food preferred, in relations to the temperature and the chemical condition of the water, and the time of the year at which the greater part of the collections were made—involving, as this may, similarities and differences of the annual migratory movements of the species—and several other like conditions.

COLLECTIONS FOR ECOLOGICAL STUDY.

It has been the object of this paper to test the availability and the usefulness for ecological study, of the data of the careful zoological collector, by applying to them a special method of classification and analysis. At the same time, of course, the method itself has been severely tested; and it might have failed completely in this instance without being permanently discredited.

The unit of this paper is the collection; but this term as here used is highly various in its meaning, and to some extent accidental in its denotation. It usually includes everything which it was convenient or desirable to catalog under one accessions number, with a mention of the date, place, and body of water from which the collection came, and, in the majority of cases, particulars concerning the apparatus used and the more notable features of the situation. It may cover at one time the product of a single haul of a small minnow seine from a rivulet or a pond, and at another time that of a number of longer hauls with a larger seine from a great lake or from a considerable stretch of the course of a great river; and in this discussion no account has been taken of differences of condition, season, or time of day, represented by the several accessions numbers.

If each collection had been made as much like every other as practicable in respect to the apparatus used, the proportionate area covered, and the definiteness and distinctness of the unit of environment from which it was drawn; if these ecological situations had been skilfully chosen, fully described, and thoroughly "sampled" as to the contents in fishes; and if collections, of moderate size but ample in number for the territory covered, had been judiciously repeated for each situation at different seasons and under varying conditions,—we should doubtless have obtained for our tables coefficients capable of yielding a larger and more complex knowledge than I have here presented of the local distribution of fishes under the influence of their environment.

In a later paper, in course of preparation, the writer intends to discuss, in a similar manner, the local and ecological relations of all the species obtained from a limited area—that of the Wabash valley in Illinois.

ACKNOWLEDGMENTS.

For a large part of the materials of this paper, both specimens and field observations, I am indebted to a considerable series of former and present assistants of the State Laboratory of Natural History—most largely to Professor H. Garman, Professor H. A. Surface, Mr. Wallace Craig, Mr. Thomas Large, and

Mr. R. E. Richardson. I am also under special obligations to Mr. Richardson, my colleague in a study of the ichthyology of Illinois, for the compilation and primary tabulation of our data from collectors' labels and notes and from accessions catalog entries.

EXPLANATION OF TABLES AND MAPS.

In place of the names of species, the corresponding numbers of Jordan and Evermann's "Synopsis of North American Fishes" have been used in the construction of the tables, as follows:—

1417. *Percina caprodes* (Raf.).
1418. *Hadropterus phoxocephalus* (Nelson).
1421. *Hadropterus aspro* (Cope & Jordan).
1436. *Cottogaster shumardi* (Gir.).
1443. *Diplesion blennioides* (Raf.).
1446. *Boleosoma nigrum* (Raf.).
1448. *Boleosoma camurum* Forbes.
1450. *Ammocrypta pellucida* (Baird).
1461. *Etheostoma zonale* (Cope).
1474. *Etheostoma jessiæ* (Jordan & Brayton).
1477. *Etheostoma cœruleum* Storer.
1490. *Etheostoma flabellare* Raf.
1494. *Boleichthys fusiformis* (Gir.).

Table I. shows, under each species number, first, the number of collections of the species used in this study; second, the coefficients of the association of the species with each of the others of the group of thirteen represented by this table, these coefficients being arranged in order of magnitude from above downward; and, third, the totals and the averages for each column. The species columns are arranged in the order of their average coefficients. At the bottom of the table is the sum of the totals for the species and the general average of their average coefficients, the last being the general associative coefficient for the entire group.

In Table II. the species numbers are placed in like order at the top and at the side of the table, and the coefficient of any two species will be found at the point of intersection of the column for one with the line for the other. The upper right half of this table is the reversed duplicate of the lower left half, inserted for convenience in following a series of coefficients.

Table III. is constructed like Table II., but with totals and averages added, as in Table I. It contains the coefficients of mutual association of the last three species of Table I., which are distinguished by the lowest average coefficients of the whole series of thirteen.

Table IV. is constructed like Table III. It contains the coefficients of mutual association of the "typical darters"—the first six of Table I. distinguished by their high average coefficients.

Table V. contains the coefficients of mutual association of the seven species which have occurred least frequently in my collections.

Table VI. is intended to represent the relations of preference and avoidance of the various species with reference to kinds of bodies of water, to current movements, and to character of bottom, so far as these are determinable from our data. Where the ratios do not amount to 100 per cent., the difference is due to the omission of miscellaneous minor data.

The general map of the distribution of collections (Map I.) shows, by the location of the red spots, all the localities from which collections of fishes have been made by us in the work of the Natural History Survey. The distribution maps for the various species indicate in the same way all the localities from which representatives of the species have been taken. *For an accurate idea of the significance of these species maps, each should be compared with Map I.*

The following numbered list of the counties of the state corresponds to the numbers on these maps.

1. Jo Daviess.	35. Hancock.	69. Madison.
2. Stephenson.	36. McDonough.	70. Bond.
3. Winnebago.	37. Fulton.	71. Fayette.
4. Boone.	38. Mason.	72. Effingham.
5. McHenry.	39. Tazewell.	73. Jasper.
6. Lake.	40. McLean.	74. Crawford.
7. Cook.	41. Vermilion.	75. Lawrence.
8. Du Page.	42. Champaign.	76. Richland.
9. Kane.	43. Piatt.	77. Clay.
10. DeKalb.	44. Dewitt.	78. Marion.
11. Ogle.	45. Logan.	79. Clinton.
12. Lee.	46. Menard.	80. St. Clair.
13. Carroll.	47. Cass.	81. Monroe.
14. Whiteside.	48. Schuyler.	82. Randolph.
15. Rock Island.	49. Brown.	83. Washington.
16. Mercer.	50. Adams.	84. Perry.
17. Henry.	51. Pike.	85. Jefferson.
18. Bureau.	52. Scott.	86. Wayne.
19. Putnam.	53. Morgan.	87. Edwards.
20. La Salle.	54. Sangamon.	88. Wabash.
21. Kendall.	55. Christian.	89. White.
22. Grundy.	56. Macon.	90. Hamilton.
23. Will.	57. Moultrie.	91. Franklin.
24. Kankakee.	58. Douglas.	92. Jackson.
25. Iroquois.	59. Edgar.	93. Williamson.
26. Ford.	60. Clark.	94. Saline.
27. Livingston.	61. Coles.	95. Gallatin.
28. Marshall.	62. Cumberland.	96. Hardin.
29. Woodford.	63. Shelby.	97. Pope.
30. Stark.	64. Montgomery.	98. Johnson.
31. Peoria.	65. Macoupin.	99. Union.
32. Knox.	66. Greene.	100. Alexander.
33. Warren.	67. Calhoun.	101. Pulaski.
34. Henderson.	68. Jersey.	102. Massac.

TABLE I.—ASSOCIATIVE COEFFICIENTS OF THIRTEEN SPECIES OF DARTERS (ETHEOSTOMINÆ).

In order of size.

Species Numbers	1418	1461	1490	1421	1450	1477	1417	1446	1436	1443	1474	1448	1494
Collections	85	32	30	159	19	90	60	236	16	24	158	107	56
	5.69	8.38	4.44	3.97	4.51	8.38	3.55	3.25	4.17	3.48	2.77	3.34	3.34
	4.95	5.69	4.17	3.97	3.97	4.44	3.33	3.20	3.28	2.57	2.76	2.34	1.80
	3.33	3.97	3.80	3.20	3.28	2.73	3.13	2.90	3.13	2.45	1.89	1.89	1.57
	2.96	3.80	3.48	2.96	2.90	2.45	2.63	2.62	2.94	2.36	1.84	1.87	1.56
	2.94	3.55	3.25	2.73	2.63	2.36	2.45	2.45	2.77	2.14	1.69	1.32	1.52
	2.76	2.62	2.73	2.73	2.34	2.34	2.20	2.14	2.34	1.84	1.33	1.29	1.12
	2.57	1.56	1.80	2.20	2.14	1.83	1.87	1.69	1.57	1.77	1.27	.91	.99
	2.14	1.39	1.75	1.77	1.75	1.30	1.69	1.63	1.12	1.63	1.27	.62	.94
	1.83	.39	1.27	1.57	1.39	.99	1.69	1.52	.79	.24	.99	.50	.68
	1.32	.37	1.17	1.57	1.33	.77	1.30	.99	0.0	0.0	.77	.39	.63
	1.17	0.0	1.11	1.29	.94	.31	1.11	.91	0.0	0.0	.68	.31	.30
	.63	0.0	.62	1.27	.50	0.0	.30	.79	0.0	0.0	.37	0.0	.24
Totals	32.29	31.72	29.59	29.23	28.12	27.90	25.25	24.07	22.11	18.48	17.63	14.78	14.69
Averages	2.69	2.64	2.46	2.44	2.34	2.32	2.10	2.01	1.84	1.54	1.47	1.23	1.22
Probable Errors	±.28	±.43	±.25	±.18	±.24	±.26	±.15	±.16	±.20	±.16	±.14	±.16	±.16
Ratio of Prob. Errors to Averages	.104	.163	.101	.074	.103	.11	.071	.08	.109	.104	.097	.13	.131

Grand Total, 315.81.   General Average, 2.02. (±.09)

TABLE II.—ASSOCIATIVE COEFFICIENTS OF THIRTEEN SPECIES OF DARTERS (ETHEOSTOMINÆ).

In the order of the size of the coefficients of association of each species with *Hadropterus aspro* (1421).

Sp.	Collections	1421	1450	1461	1446	1418	1490	1477	1417	1443	1436	1494	1448	1474
1421	159		3.97	3.97	3.20	2.96	2.73	2.73	2.20	1.77	1.57	1.57	1.29	1.27
1450	19	3.97		1.39	2.90	4.95	1.75	2.34	2.63	2.14	3.28	.94	.50	1.33
1461	32	3.97	1.39		2.62	5.69	3.80	8.38	3.55	0.0	0.0	1.56	.39	.37
1446	236	3.20	2.90	2.62		2.14	3.25	2.45	1.69	1.63	.79	1.52	.91	.99
1418	85	2.96	4.95	5.69	2.14		1.17	1.83	3.33	2.57	2.94	.63	1.32	2.76
1490	30	2.73	1.75	3.80	3.25	1.17		4.44	1.11	3.48	4.17	1.80	.62	1.27
1477	90	2.73	2.34	8.38	2.45	1.83	4.44		1.30	2.36	0.0	.99	.31	.77
1417	60	2.20	2.63	3.55	1.69	3.33	1.11	1.30		2.45	3.13	.30	1.87	1.69
1443	24	1.77	2.14	0.0	1.63	2.57	3.48	2.36	2.45		0.0	.24	0.0	1.84
1436	16	1.57	3.28	0.0	.79	2.94	4.17	0.0	3.13	0.0		1.12	2.34	2.77
1494	56	1.57	.94	1.56	1.52	.63	1.80	.99	.30	.24	1.12		3.34	.68
1448	107	1.29	.50	.39	.91	1.32	.62	.31	1.87	0.0	2.34	3.34		1.89
1474	158	1.27	1.33	.37	.99	2.76	1.27	.77	1.69	1.84	2.77	.68	1.89	

TABLE III.— COEFFICIENT TABLE OF THE FOUR LEAST FREQUENT ASSOCIATES.

Species	1443	1448	1474	1494
1443		0.0	1.84	.24
1448	0.0		1.89	3.34
1474	1.84	1.89		.68
1494	.24	3.34	.68	
Totals	2.08	5.23	4.41	4.26

General Average, 1.33

TABLE IV.— COEFFICIENT TABLE OF THE SIX MOST FREQUENT ASSOCIATES.

Species	1418	1461	1490	1421	1450	1477
1418		5.69	1.17	2.96	4.95	1.83
1461	5.69		3.80	3.97	1.39	8.38
1490	1.17	3.80		2.73	1.75	4.44
1421	2.96	3.97	2.73		3.97	2.73
1450	4.95	1.39	1.75	3.97		2.34
1477	1.83	8.38	4.44	2.73	2.34	
Totals	16.60	23.23	13.89	16.36	14.40	19.72

General Average, 3.47.

TABLE V.— COEFFICIENT TABLE OF THE SEVEN LEAST FREQUENT DARTERS.

Species	1436	1450	1443	1490	1461	1494	1417
Collections	16	19	24	30	32	56	60
1436		3.28	0.0	4.17	0.0	1.12	3.13
1450	3.28		2.14	1.75	1.39	.94	2.63
1443	0.0	2.14		3.48	0.0	.24	2.45
1490	4.17	1.75	3.48		3.80	1.80	1.11
1461	0.0	1.39	0.0	3.80		1.56	3.55
1494	1.12	.94	.24	1.80	1.56		.30
1417	3.13	2.63	2.45	1.11	3.55	.30	
Totals	11.70	12.13	8.31	16.11	10.30	5.96	13.17

General Average, 1.85.

TABLE VI.—LOCAL PREFERENCES OF DARTERS.

303

Situations	1450	1443	1490	1418	1421	1461	1477	1446	1417	1436	1474	1494	1448	Av.
Larger rivers	.14	.00	.09	.07	.06	.03	.03	.03	.10	.55	.21	.01	.09	.11
Smaller rivers	.47	.47	.00	.56	.42	.74	.44	.25	.37	.00	.19	.12	.23	.33
Creeks	.39	.53	.87	.26	.47	.23	.45	.53	.27	.04	.16	.25	.42	.37
Lowland lakes, ponds, sloughs, etc.	.00	.00	.04	.03	.01	.00	.01	.01	.10	.18	.02	.02	.06	.06
Upland lakes, ponds, etc.	.00	.00	.00	.00	.00	.00	.00	.14	.09	.00	.00	.60	.00	.06
Moderate to rapid current	1.00	.83	1.00	.87	.69	.89	.83	.68	.93		.83	.22	.44	.77
Sluggish to stagnant	.00	.17	.00	.13	.31	.11	.17	.32	.07		.17	.78	.56	.23
Muddy bottom	.16	.00	.00	.06	.16	.11	.08	.11	.00		.23	.66	.62	.18
Rocky and sandy bottom	.84	1.00	1.00	.94	.84	.89	.92	.89	1.00		.77	.34	.38	.82

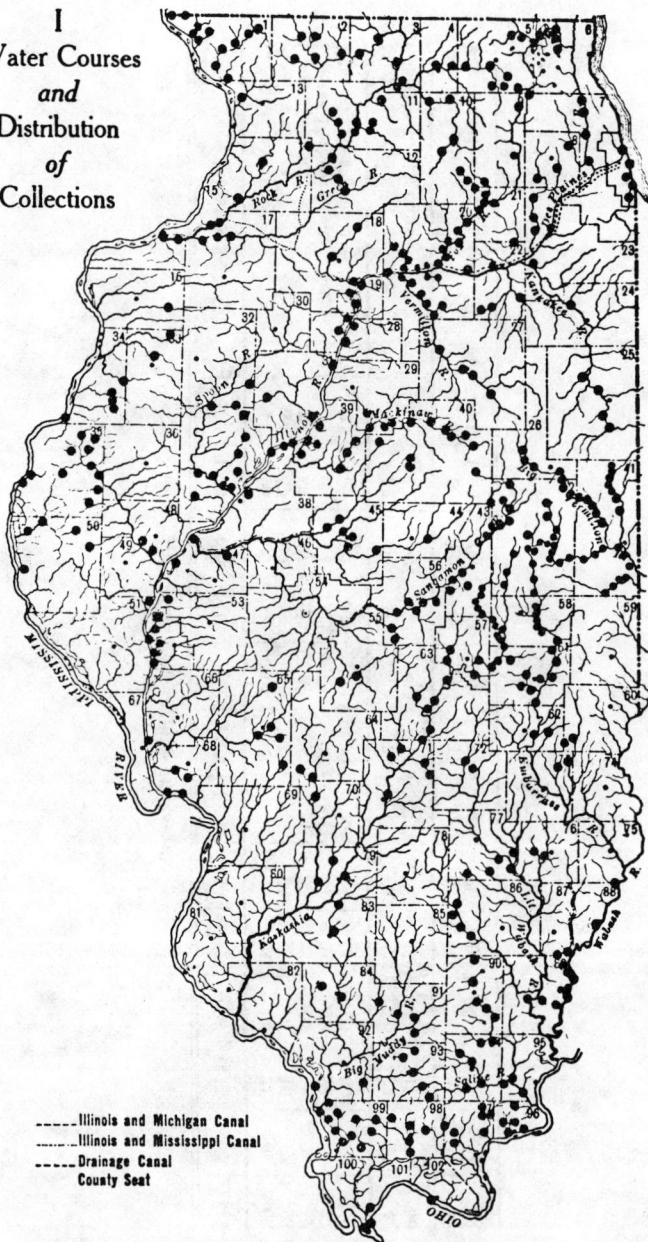

I
Water Courses
and
Distribution
of
Collections

------- Illinois and Michigan Canal
.... Illinois and Mississippi Canal
----- Drainage Canal
County Seat

II
The
Stream Systems
of
Illinois

Illinois and Michigan Canal
Illinois and Mississippi Canal
Drainage Canal
County Seat

AN ORNITHOLOGICAL CROSS-SECTION
OF ILLINOIS IN AUTUMN

Stephen Alfred Forbes

ARTICLE IX.—*An Ornithological Cross-section of Illinois in Autumn.* By S. A. FORBES.

The subject of the relations of interaction between organisms and their environment, animate and inanimate, which goes by the name of ecology, may be studied with reference to the welfare of species or to that of the general assemblage of organisms to which the species belong. The ecology of a species is special ecology; that of the assemblage is a phase or division of general ecology—more or less general according to the size and contents of the assemblage considered. In special ecology every ecological factor, every feature of the environment, is valued according to its importance to the species; in general ecology the various ecological factors are valued according to their significance in the general system of life. In special ecology the species is the all-important, dominating center; in general ecology each species takes its appropriate place—dominant, important, subordinate, or insignificant—according to its dynamic value as a part of the whole.

Precise studies in animal ecology have heretofore been made mainly in the special field, necessarily so in the beginning since a knowledge of the ecology of species must precede that of groups or assemblages of species. These special studies are, however, merely preliminary to a general study of the dynamic system of organic life as exhibited in its larger and more complex units. Without the corrective and organizing influence of such a study of the system as a whole, our ideas of that system must be badly proportioned and correspondingly inadequate or misleading—a fact readily illustrated by the state of our knowledge and opinion respecting the ecological significance of birds.

To learn what we now know of the effects of the activities of birds has required much difficult, expert, time-consuming study, especially of the details of their food, since it is mainly through the food relation that birds affect the welfare of other animals and of plants. These studies, although both qualitative and quantitative as related to the welfare of the various species of birds themselves, have been qualitative only as concerning the relation of birds to the general welfare; and we have little but vague estimate and doubtful surmise in place of a definite knowledge of the relative ecological values of the various species, and equally little knowledge, in consequence, of the total significance of birds as a class. We do know fairly well (owing, in part, to the early work of this Laboratory*, but mainly to that of the United States Biological Survey) the principal features of the food of many species of our common birds, but we can not lay these data together for an intelligent estimate of the total effect of the life of birds on their environment except on the supposition that the various species are about equally abundant wherever they occur. That this is not the fact is obvious to every one, and it must be equally obvious, consequently, that until we know how abundant, on an average, the various species are in the various parts of the country and throughout the country at large, we can make little definite application, either scientific or strictly practical, of the knowledge we now have. Our present information in this field is like a chain one of the links of which is missing and has been replaced by a piece of twine. To substitute iron for cotton at this point is the object of the studies now in progress in Illinois on the local distribution, average numbers, and ecological preferences of the various species of Illinois birds.

The Field Method.

To this end, after a preliminary quantitative study made in 1905-06 of the bird life of a single limited tract—a 400-acre stock and grain farm in central Illinois—a systematic program of field observation and statistical record was entered upon last August, with complete arrangements for its continuance through one entire year. Two acute and thoroughly reliable ornithological observers—one of whom, Mr. A. O. Gross, al-

---

* See Bull. Ill. State Lab. Nat. Hist., Nos. 3 and 6, Vol. I.

though still an undergraduate student in the University of Illinois, has had several years' experience as a collector and observer of birds—were sent into the field under instructions to traverse the state in various directions, traveling always in straight lines and always thirty yards apart, and noting and recording the species, numbers, and exact situation of all birds flushed by them on a strip fifty yards in width, including also those crossing this strip within one hundred yards to their front. No attention is paid by them, for this purpose, to any other birds.

As they are able to recognize with accuracy all species of Illinois birds at sight, and most of them by song, their movement is like that of a gigantic sweep-net 150 feet wide and 300 feet deep, so drawn across the country day by day as to capture every bird which comes in its way; with this difference, that the birds are not actually caught or even inconvenienced, and that nothing can escape the meshes of their well-trained observation.

One of these observers, Mr. H. A. Ray, also a University student, is primarily responsible for the record of distances and kinds of surface over which they travel, carrying for this purpose a pedometer whose action has been carefully tested and repeatedly checked, and a mechanical tally or "lumber-counter"—both used to make a record of the number of paces traveled over each crop or other kind of surface vegetation.

The reports of their travel made to me by Mr. Gross contain every needful detail as to date and time of day; to precise location of their line of march; to temperature, wind, and other features of the weather; to distances traveled in succession over each field or other distinguishable area; to vegetation, wild or cultivated, on each tract; and to the species and numbers of birds identified on each area and in each kind of crop.

General Results of Observations.

The present paper is a discussion of the product of one of their earlier trips, made from August 28 to October 17, 1906, across the state from east to west, from the Indiana line be-

yond Danville, Ill., to Quincy, on the Mississippi River. It has
to do with autumnal conditions in the central part of the state,
and is merely preliminary to a comprehensive report on the
whole investigation.

The entire distance covered by these observations is 191.86
miles, and the strip from which all birds were accurately determined and numbered was 150 feet in width for this whole
distance. The area thus covered was 3519 acres, or 5½ square
miles. It included every kind of surface, soil, and vegetation
traversed by the observers, with the exception of forests of too
lofty or too dense a growth for a complete and certain recognition of their bird population.

The whole number of birds identified was 4804, of which
1620 were English sparrows and 3184 were of native species.
The average number of birds seen was 25 for each mile of the
trip, which is 1.36 for each acre covered, or 874 for each square
mile. The English sparrows averaged .46, and the native species .9, per acre, or 295 per square mile for the sparrows and 579
per square mile for the native birds. The total number of
species recognized was 93; but 90 per cent. of the individual
birds seen, belonged to 20 of these species, leaving but 10 per
cent. for the other 73 species. Indeed, 15 species included 85
per cent. of the individual birds observed, leaving for the other
81 species but 728 birds—an average of 130 birds per square
mile, or one bird to each five acres.

It is evident, consequently, that the real dynamic significance of the birds of this district at this time was to be found
wholly in the fifteen most abundant species, the remainder being virtually negligible as a general ecological factor.* These
fifteen species are arranged in the order of their frequency in
the following table, which shows for each the number of indi-

* A species represented by a relatively small number of birds may have a special
ecological significance if it is concentrated in a special class of situations; and may,
indeed, be especially important ecologically if the class of situations in which it is concentrated is especially important. This aspect of the general problem must be reserved
for discussion when a larger mass and a more comprehensive variety of data are available.

viduals seen, the ratio of its numbers to the number of all the
birds observed, and the average number of the species per
square mile of the area under observation.

TABLE I.   THE FIFTEEN MOST IMPORTANT BIRDS,
INDIANA LINE TO QUINCY, AUGUST 28 TO OCTOBER 17, 1906.

SPECIES	NUMBER	PER CENT.	PER SQ. MI.
English sparrow	1620	34.	295
Crow-blackbird	517	11.	94
Meadow-lark	312	6.5	59
Crow	226	4.7	41
Cowbird	221	4.6	40
Horned lark	220	4.6	40
Mourning-dove	180	3.7	33
Swamp-sparrow	155	3.2	28
Goldfinch	134	2.7	24
Myrtle warbler	112	2.3	20
White-throated sparrow	93	1.9	17
Field-sparrow	83	1.7	15
Vesper-sparrow	72	1.5	13
Quail	69	1.4	13
Flicker	62	1.3	11
Totals	4076	85.1	743

If we exclude the English sparrow from consideration, as
an obnoxious alien whose habits should not be permitted to influence opinion concerning the ninety-two species of our native
birds, we must compute the ratios of abundance for the native
species with reference to the 3184 such birds identified on this
trip.   This is an average of 579 per square mile, instead of 874,
the former number.   To obtain 85 per cent. of all the native
birds seen we must add to the above list the next most abundant species, which are the robin, the bluebird, the killdeer, and
the blue jay.   The following table shows the ratios of abundance and the birds per square mile of the eighteen species of
this amended list.   The seventy-four native species remaining
are now represented by 499 birds—an average of 1 to about
seven acres; a proportion far too small to have any general significance.

TABLE II.   THE EIGHTEEN MOST IMPORTANT NATIVE BIRDS,
INDIANA LINE TO QUINCY.

SPECIES	NUMBER	PER CENT.	PER SQ. MI.
Crow-blackbird	517	16.	94
Meadow-lark	312	9.8	59
Crow	226	7.1	41
Cowbird	221	7.	40
Horned lark	220	7.	40
Mourning-dove	180	5.7	33
Swamp-sparrow	155	4.9	28
Goldfinch	134	4.2	24
Myrtle warbler	112	3.5	20
White-throated sparrow	93	3.	17
Field-sparrow	83	2.6	15
Vesper-sparrow	72	2.2	13
Quail	69	2.2	13
Flicker	62	2.	11
Robin	61	2.	11
Bluebird	61	2.	11
Killdeer	60	1.9	11
Blue jay	57	1.8	10

THE VEGETABLE COVERING OF THE SOIL.

As the area traversed on this trip was almost wholly under cultivation, the relation of these birds to the vegetable covering of the soil was virtually their relation to the agricultural and horticultural crops of central Illinois in autumn—almost entirely to the former, since the horticultural area is comparatively insignificant in this part of the state. Nearly all this surface was in fields of ripe corn, the stalks standing in some fields and in others cut and shocked; in blue-grass pastures; in meadows of timothy, clover, and millet, or timothy and clover mixed; in fields of stubble, mostly after a crop of oats; in fields of young wheat; in ground freshly plowed, mainly as a preparation for wheat; and in orchards, almost all of apple. Plowing for wheat was in progress when the trip began, and fields of young wheat were reported in increasing numbers after October 1. Some of the later plowing was doubtless done for corn.

The track of my observers led them also through barn-yards, and gardens of vegetables and shrubs, and occasionally across a shrubby ravine or a neglected field which had grown up to weeds With the exception of a large marshy tract in the bottoms of the Illinois River near Meredosia, there was very little waste land worth mentioning on this line.

For an analysis of the preferences of the principal species of birds with respect to the various classes of situation and kinds of food available to them at the time, it is necessary to take into account the areas in each of the crops along the line of travel. For this purpose the following table has been prepared, showing the total distance traveled through each kind of crop, and the acreage in each from which a complete count and analysis of the bird life was obtained.

TABLE III.   CROP AREAS, INDIANA LINE TO QUINCY.

CROP	MILES TRAVELED IN CROP	PER CENT. IN EACH CROP	ACRES IN 50-YARD STRIP	NUMBER OF FIELDS	ACRES IN AVERAGE FIELD*
Corn	71.87	38.	1306.64	362	32.
Stubble	37.4	19.5	680.56	205	18.2
Wheat..........	8.48	4.4	155.36	59	17.
Plowed Ground	5.76	3.	105.66	34	18.4
Pasture......... ..	50.97	26.6	926.65	345	14.
Meadow	8.4	4.5	153.18	51	21.7
Orchard....	2.5	1.3	46.7	23	9.5
Miscellaneous	6.48	2.7	134.25	36	....
Totals	191.86		3509.00	1115	

Corn, it will be seen, was the principal crop. A distance of nearly seventy-two miles was traveled through 362 corn fields of an average size of 32 acres per field, and all the birds were determined for 1306.64 acres of this crop. That is, 38 per

*Virtually all central Illinois farm-fields are rectangular, and the average form of a sufficient number is consequently that of a square. The length of one side of such an average field was found by dividing the entire distance traveled in any crop by the number of fields of that crop crossed. The square of this side is, of course, the area of this average field.

cent. of the entire journey was in fields of corn. The next largest area was in blue-grass pastures, over which my observers traveled 51 miles, determining the birds of 926.65 acres, which was 26.6 per cent. of the whole area of their observations. Thirty-seven and four tenths miles in fields of stubble, mainly oats, averaging 18.2 acres each, gave a total of 680.56 acres for the 50-yard strip, or 19.5 per cent. of its entire length. Thus the oats fields were more than one half, and the pastures more than two thirds, the area in corn, and these three crops together covered 83 per cent. of the surface. If to this we add the 4.5 per cent. of meadow-lands, we have nearly 88 per cent. of the total area in corn, oats, and grass (including in the last a small amount of clover, usually growing with timothy).

The surface in wheat is not accurately obtainable from these data, since wheat sowing had not begun and plowing for wheat was not finished when the start was made, but both were finished before the trip was ended. If virtually all the fall plowing was being done for wheat, the area in that crop was about 7 per cent., or 260 acres for the 14 miles traveled through 93 fields. About 2½ miles were traveled through 23 orchards, aggregating 1.3 per cent. of the strip, or 46.7 acres in all. The marshes, waste lands, forests, gardens, farmyards, brushy hollows, and other miscellaneous tracts examined, amount to 2.7 per cent. of the whole. An immense plain of corn, oats, and grass, the first greatly predominating, with a little wheat, less clover, and an occasional farm orchard—this is the region, quite typical for nearly all the central two thirds of Illinois, from which these data were drawn.

GENERAL DISTRIBUTION ACCORDING TO CROPS.

We have next to see how our 4800 birds, belonging to 93 species,—and especially how our 15 most abundant species, represented by 4076 birds,—had distributed themselves over the 3500 acres in these crops actually scrutinized by these observers.

This latter query admits of various answers: (1) we may simply give the number of individuals of each species observed in each kind of crop; (2) we may give the number of species on equal areas of each crop—an acre or a square mile; (3) we may give the percentage of each of the species found in each of the crops; (4) we may compare the actual numbers of each species in each crop with the number which would occur there if the species were uniformly distributed over its area, thus showing where and in what degree the species is densely or sparsely distributed above or below the average; or (5) we may compare several species one with another, and each with all the rest, in a way to show just how and how far they differ in their numerical relations to the various crop areas they inhabit. All these several forms of answer are contained in full in the following tables for our most abundant birds, and from these I will extract here and there only such data for discussion as seem adapted to a general treatment of the subject.

From Table IV., showing the distribution of all birds without distinction of species for the principal areas actually covered by this inspection, we see that 2249 of these birds were found in pastures, 955 of them in corn fields, 454 in stubble

TABLE IV. GENERAL DISTRIBUTION OF ALL BIRDS, BY CROPS, INDIANA LINE TO QUINCY.

CROP	ACRES	ACREAGE PER CENT.	BIRDS	BIRDS PER CENT.	BIRDS PER ACRE	BIRDS PER SQ. MILE
Corn	1306.64	38.	955	20.	.73	468
Stubble	680.56	19.5	454	9.	.67	429
Wheat	155.36	4.4	46	1.	.30	192
Plowed ground	105.66	3.	71	1.5	.67	430
Pasture	926.65	26.6	2249	47.	2.43	1551
Clover	79.08	2.3	51	1.	.65	416
Timothy	57.10	1.7	47	1.	.83	531
Millet	17.	.5	17	....	1.	640
Orchard	46.70	1.3	199	4.	4.23	2726
Yards	11.77	.003	121	2.5	10.28	6580·
Swamp	47.16	.013	98	2.	2.08	1331
Timber*	7.34	....	9	....	1.23	785
Miscellaneous	78.	1.1	487	....	....	....

*All forests "skipped" if high or dense.

ground, 199 in orchards, 115 in meadows, 71 on recently plowed ground, and 46 of them on young wheat. Taking into account the different acreages of these areas and computing the number of birds per square mile in each, we have 2726 per square mile in orchards, 1551 in pastures, 1587 in meadows, 468 in corn fields, 430 on plowed ground, 429 on stubble, and 192 on young wheat. A square mile of swamp land, if we may judge by the forty-seven acres examined, would have contained a population of 1331 birds; and a square mile of farmyards, 6580. Compared by percentages of all the birds in each crop, 47 per cent. were in pastures, 20 per cent. in corn, and 9 per cent. in stubble, the ratios in other crops and situations ranging from 4 per cent. down.

The above crops may be divided, from this point of view, into four classes: young wheat, with less than 200 birds to the square mile; corn, stubble, and plowed ground, with about 450 each; pastures and meadows with over 1500 each; and orchards, with 2700 birds to the like area. The fact that birds are nearly as common in old stubble fields as in corn, suggests that it is not the grain in either case which attracts them there, but rather the seeds of the weeds by which both kinds of fields are generally covered in fall. Their preference for pasture-lands is probably due to the amount of food found by them in the droppings of stock, and to the greater abundance of insect life in such a situation. Other comparative conclusions may best be postponed until the special assemblages of birds characterizing each of these principal classes of situation are more fully discussed.

THE PRINCIPAL BIRDS IN EACH CROP.

The next four tables give us the data of the distribution and abundance of the principal species of birds as related to the principal crops. In Table V. we have the numbers identified of the twelve most abundant birds in each kind of crop, without reference to differences in acreage. In Tables VI.–VIII. the list of species is reduced to nine by dropping the three passing migrants. In Table VI. the number of birds per section, or square mile, of each crop is given for each of the species; in Table

315

TABLE V. NUMBER OF PRINCIPAL BIRDS IN PRINCIPAL CROPS, INDIANA LINE TO QUINCY.

	CORN	STUBBLE	WHEAT	PASTURE	MEADOW	PLOWED GROUND	ORCHARD	YARDS	WEEDS	SHRUBBERY	FALLOW AND WASTE
English sparrow	562	38	....	530	10	9	101	119	251	...	....
Crow-blackbird	21	22	....	445	..	..	....	9	....	...	....
Meadow-lark	49	97	21	122	20	3	....	...	....	..	....
Crow	12	7	1	190	3	13	....	....	....	..	....
Cowbird	2	2	1	133	20	1	:...	....	(62*)	...	
Horned lark	6	26	7	141	1	38	....	1	.:..	....	
Mourning-dove	53	41	1	73	8	1	3	....	....		
Swamp-sparrow	14	7	....	20	15	...	7	....	....	5	87
Goldfinch	20	7	4	56	..	...	16	....	28	3	
Myrtle warbler	30	3	....	47	1	2	13	....	....	5	11
White-throated sparrow	19	8	1	15	..	...	16	....	21	13	....
Field-sparrow	11	2	1	33	1	...	8	....	11	16	....

*Sorghum.

TABLE VI. NUMBER OF BIRDS PER SQUARE MILE IN EACH CROP.

	CORN	STUBBLE	WHEAT	PASTURE	MEADOW	PLOWED GROUND	ORCHARD
English sparrow	275	36	......	366	48	55	1383
Crow-blackbird	10	21	......	307	....	....	......
Meadow-lark	24	92	86	84	98	18	......
Crow	6	7	4	131	15	79	......
Cowbird	1	2	4	89	98	6	
Horned lark	3	25	29	97	5	230	.....
Mourning-dove	26	39	4	50	39	6	42
Goldfinch	10	7	17	39		......	220
Field-sparrow	5	2	4	23	5	......	110
All birds	468	429	192	1551	553	430	2726

316

TABLE VII. PERCENTAGE OF EACH SPECIES IN EACH OF THE PRINCIPAL CROPS.*

	CORN	STUBBLE	WHEAT	PASTURE	MEADOW	PLOWED GROUND	ORCHARD	YARDS	SHRUBBERY	SORGHUM	WASTE
English sparrow	35	2	....	33	.6	.5	6	7	....	....	15
Crow-blackbird	4	4	....	90				2	....	....	....
Meadow-lark	16	31	7	38	7.	1.					
Crow	5	3	....	84	1.	6.					
Cowbird	1	1	....	60	9.	....				28	....
Horned lark	3	12	3	64	....	17.					
Mourning-dove	29	23	1	40	4.	1.	2				
Goldfinch	15	5	3	42			12		2	....	21
Field-sparrow	13	2	1	40	1.		10	....	19	....	14
All birds	20	9	1	47	2.	1.	4	3	1	1	11

*Read from left to right.

TABLE VIII. RATIO OF EACH SPECIES IN EACH CROP TO ALL BIRDS IN THAT CROP*.

	CORN	STUBBLE	WHEAT	PASTURE	MEADOW	PLOWED GROUND	ORCHARD
English sparrow	.59	.08	......	.24	.09	.13	.51
Crow-blackbird	.02	.05	......	.20			
Meadow-lark	.05	.21	.45	.05	.18	.04	
Crow	.01	.02	.02	.09	.03	.18	
Cowbird			.02	.06	.18	.01	
Horned lark		.06	.15	.06	.01	.53	
Mourning-dove	.06	.09	.02	.03	.07	.01	.01
Goldfinch	.02	.02	.09	.03			.08
Field-sparrow	.01	......	.02	.02	.01		.04

*Read from above downwards.

VII. are the percentages of each species found in the various crops; and in Table VIII. we have in each crop column, percentages showing for each species the ratio of the number of birds of that species found in that crop to the total number of all birds found in the same crop.

From Table VIII. it will be seen that the principal *corn-field* species at the times and places of this trip was the English sparrow, to which more than half the birds seen in corn fields belong, and that the mourning-dove and the meadow-lark were the species next in abundance there—6 per cent. and 5 per cent. respectively. In *stubble fields* the meadow-lark was the most abundant species, making about a fifth of all the birds seen in such fields. The next in order of abundance were the mourning-dove, the English sparrow, the horned lark, and the crow-blackbird, present in ratios ranging from 9 per cent. to 5 per cent. The meadow-lark was also much the most abundant bird on fields of *young wheat*, where it made 45 per cent. of all the birds seen; and the horned lark and the gold-finch were next to this in number, one third and one fifth as great respectively. The principal *pasture* species were the English sparrow (24 per cent.) and the crow-blackbird (20 per cent.), with the crow, the cowbird, the horned lark, and the meadow-lark following in numbers ranging from a third to about a fifth the number of the sparrows. In *meadows*, on the other hand, the meadow-lark and the cowbird were in the lead, each 18 per cent. of all the meadow birds identified, and the English sparrow and the mourning-dove were about half as numerous. On *fall plowing* more than half the birds were horned larks, and the only other abundant species were the crow (18 per cent.) and the English sparrow (13 per cent.). In the small number of *orchards* traversed the English sparrow was at this time much the most abundant bird (51 per cent.). The other common species were the goldfinch (8 per cent.), the field-sparrow (4 per cent.), and a few passing migrants—the myrtle warbler and the white-throated sparrow, for example. (See Table V.)

The Principal Species Separately.

English Sparrows.—From these tables we learn that about two thirds of the English sparrows were in corn fields and pastures, and in about equal numbers in each; that approximately half as many were found in waste weedy fields as in pastures; and that the remainder were about equally divided between barn-yards and orchards. Some 52 per cent. of this species—those in corn fields, stubble, and waste lands—were among weeds, and 40 per cent. of them were following farm stock in pastures and yards. Those in orchards (6 per cent.) were doubtless there mainly for shelter and rest. The table of numbers per square mile (Table VI.) shows that orchards were the favorite resort of the sparrows. Barn-yards, pastures, and corn fields were their principal feeding grounds, and only scattering numbers occurred in stubble, meadows, and plowed fields. Not a single one of the 1620 sparrows noted on this trip was seen in the 59 fields of young wheat. These sparrows were, in a word, barn-yard, corn-field, and pasture birds, and were doubtless feeding mainly on weed seeds and undigested fragments of grain.

Crow-blackbirds and Crows.—Blackbirds, on the other hand, were seen to be at this time essentially birds of the pasture, 90 per cent. of them occurring there, and only 4 per cent. in corn fields, 4 per cent. in stubble, and 2 per cent. in farmyards. Practically the same may be said of the crows, whose ratios of abundance are close copies of the preceding excepting for the 6 per cent. on plowed ground, the 1 per cent. in meadows, and the absence of crows from barn-yards. During this whole trip of 192 miles, only 12 crows and 21 blackbirds were seen in the 1300 acres of corn covered by these observations—an average of 6 crows and 10 blackbirds per square mile of corn. It was suggestive of a useful feature of the habits of crows that an average of 79 of these birds per square mile were seen on plowed ground, where they could have found little if any food except insect larvæ—mainly white-grubs. The record for blackbirds is disturbed by the fact that they were moving southward when the trip began, as is shown by their occurrence at the rate of 7.2 per mile of travel during the first half of the period of this trip and at only 1.1 per mile during the last half.

Meadow-larks.—That good genius of the farm, the meadowlark, was evidently at home almost everywhere on the farm premises, as is shown especially by the numbers per square mile, which are approximately equal for stubble fields, meadows, pastures, and fields of young wheat (Table VI.). These birds were about a fourth as numerous in corn fields, and a fifth as numerous on plowed ground, as in meadows and fields of stubble, and somewhat more numerous in these latter situations than in pastures and young wheat; but taking into account the actual crop areas in the country covered (Table VII.) we find meadow-larks so distributed through these crops as to be about equally common in pastures and stubble fields, and about half as common in corn, with only 7 per cent of their number in wheat and meadow-lands respectively. Their recorded numbers on plowed ground amounted to only 1 per cent. of the whole number seen. The occurrence of 86 of these birds per square mile in fields of young wheat suggests a possible economic depredation, of which, in fact, they have been sometimes accused.

Cowbirds.—The cowbird's record of occurrence for this trip would be almost exclusively that of a pasture and meadow species if it had not been for a flock of 62 seen in a field of sorghum, feeding on the seeds. Even including these in the ratios, 60 per cent. were in pastures and 9 per cent. in meadows, the remaining distribution being merely a scattering one. Tested by the number of species per square mile in each crop, as shown by Table VI., the cowbird shows no very decided choice between pastures and meadow-lands, averaging 89 per square mile for the former and 98 for the latter. The species was evidently migrating at the time, as only one example was seen during the last seventy miles of the trip.

It should be noted at this point that these generalizations concerning gregarious birds, which roost in company or feed in flocks, require a much larger body of data than those for birds of solitary habit. The averages of this paper are hence more likely to require amendment for blackbirds, cowbirds, and crows, as information accumulates, than for the other species of our list.

Horned Larks.—The birds of this species found in central

Illinois were all of the prairie variety, *praticola.* With habits much like those of the meadow-lark, they differed from that species widely in their local distribution, especially in their preference for plowed ground, on which they occurred at the rate of 230 per square mile as against 18 meadow-larks for the same area. Their next preference was for pastures, where 97 per square mile were found, the remainder occurring mostly on stubble and young wheat, 25 and 29 per square mile respectively. Nearly two thirds of their actual numbers were found in pasture-lands, 17 per cent. were on plowed ground, and 12 per cent. on stubble. The remainder were in fields of wheat and corn, 3 per cent. in each.

Mourning-doves.—Mourning-doves were mainly in pastures (40 per cent.), corn fields (29 per cent.), and stubble lands (23 per cent.), these three situations thus containing 92 per cent. of all these birds recorded. As tested by the average numbers per square mile, their preferences seem much less definite. While commonest on pasture-lands (50 to the square mile), they were almost as abundant in stubble, meadows, and orchards,— about 40 per mile in each situation,—and more than half as common in corn fields (26 to the mile). Their occurrence on plowed ground and wheat was only occasional, and their numbers there were trivial.

Goldfinches and Field-sparrows.—These little birds were at this time similarly distributed, occurring in the same situations and in nearly equal ratios in each. Both were most numerous in pastures, 42 per cent. for the goldfinches and 40 per cent. for the field-sparrows, and were otherwise rather equally scattered through corn fields and orchards and on waste patches of weeds. In birds per square mile they were about three times as common in orchards as in all the other places taken together, their next apparent preference being for pasture-lands, where, however, the sparrows averaged only 23 to the square mile and the goldfinches 39.

SUMMARY FOR PRINCIPAL SPECIES.

Summarizing now the data for all these nine species taken

together as one group, we find an average of 1755 birds to the square mile of orchard, more than three fourths of this number English sparrows; 1186 per square mile in pasture, nearly one third of them English sparrows; 394 to the square mile of plowed ground, 230 of these being horned larks; 373 to the square mile of corn, three fourths of these English sparrows; 308 to the square mile of meadow-lands, where meadow-larks and cowbirds made each about a third of the number; 231 to the square mile of stubble, about two fifths of them meadow-larks; and 148 to the square mile of young wheat, of which meadow-larks made nearly three fifths. This statement may be still further generalized and simplified by saying that the number of these birds per square mile varies in round numbers from 150 in young wheat to eight times that number in pastures, and to nearly 12 times the same number in orchards; and that the intervening ratios were 230 per square mile in stubble, 300 in meadows, 375 in corn, and 400 on plowed ground.

The wide differences of their numbers in these several situations can not be taken to demonstrate corresponding differences in the local or ecological preferences of these birds, although they do indicate something of the effects which birds may be producing on equal areas in these crops. If sparrows resort to orchards largely for resting places and for protection against the wind, they would tend to accumulate there in much greater numbers to the unit of area in a country containing only scattering small orchards than in one where many large orchards were within their reach; and if horned larks decidedly prefer bare ground to a grassy turf, there will be a larger number of them in plowed fields to the square mile when but few fields have been lately plowed than when the larger part of the agricultural area has just been broken up.

RATIOS OF FREQUENCY AND PREFERENCE.

Bearing in mind the necessity thus shown for an intelligent analysis and interpretation of certain of the facts, the following tables of *frequency ratios*, and *coefficients of preference* may

be found convenient as a compact systematic summary of my
data. The frequency ratios express the comparative densities
of population on each kind of surface, for each species tabulated and for all the birds of our list. Taking the ratio of the
number of birds found in a crop to the whole number of birds
as a dividend, and the ratio of the area in that crop to the
entire area as a divisor. the quotient is the frequency ratio for
those birds and that crop. If a species were equally distributed
over the entire area studied, this ratio would be 1 for all situations and all crops. If 40 per cent. of the area were in corn,
then 40 per cent. of the birds of that species would be in corn
fields. If, on the other hand, only 20 per cent. of the birds
were in corn, the density of population in corn fields would be
expressed by the frequency ratio of 50 per cent. All ratios
below 1 indicate a density of population less than that resulting from a uniform distribution; and all greater than 1, a
density above that limit.

The coefficients of preference are found by dividing in
succession the frequency ratios of a species for each crop by its
frequency ratios for each of the other crops. They are thus a
measure of the degree of preference of the species for one crop
or situation over another; and as arranged in my tables of
coefficients following, they enable us to see just where the
preferences lie, and how they compare one with another.
Turning, for example, to the coefficient table for the mourning-
dove (Table XI., p. 327), we find at the left of the table a list of
the crops in which this bird is found, and a like list, in the same
order, at the top. At the place of intersection of the line of
figures for one crop with the column of figures for another, will
be found the coefficient of the preference of the mourning-dove
for one of these crops as compared with the other,—the standard crop being the one whose name is at the head of the
column. Selecting, as an illustration, the column headed "corn,"
and following it to its intersection with the line for "meadows,"
we find there the coefficient 1.16,—the meaning of which is
that for every hundred mourning-doves found in a given area
of corn fields, 116 would be found, according to our data, in a

like extent of meadows.   If any number of these birds found in corn fields is multiplied by the coefficient 1.16, the product is the number which we may expect to find in meadows of the same aggregate area.

Reading upward from 1 in any column, one gets a descending series of expressions for the densities of the dove population in crops less attractive than the one named at the head of the column; and reading downward from the same point, a reverse series for crops more attractive to doves than this standard crop.   The figures on one side of the diagonal line of 1's are the reciprocals of those on the other side.

Tables of this description will be useful for a comparison of the distribution and ecology of the several species at different seasons and in different situations, and for a comparative study of the statistics of bird distribution in different parts of the state and in different states.

TABLE IX. RATIOS OF FREQUENCY, MOST ABUNDANT BIRDS, INDIANA LINE TO QUINCY.

	CORN	STUBBLE	WHEAT	PLOWED GROUND	PASTURES	MEADOWS	ORCHARDS	YARDS	SHRUBS	SWAMPS	WEEDS
English sparrow	.92	.1		.17	1.24	.13	4.61	2333			60
Crow-blackbird	.11	.21	1.59		3.42			6.66			
Meadow-lark	.42	1.59	1.59	.33	1.43	1.56					
Crow	.13	.15		2.	3.16	.22					
Cowbird	.026	.05	.68		2.26	2.					
Horned lark	.079	.61	.68	5.66	2.41		1.54				
Mourning-dove	.76	1.18	.23	.33	1.50	.88					
Goldfinch	.39	.26	.69		1.58		9.23		750		14000
Field-sparrow	.34	.10	.23		1.50	.22	7.69		3500		5500
All birds	.53	.47	.23	.33	1.76	.44	3.08	8.33		1.54	

TABLE X. COEFFICIENTS OF PREFERENCE, ALL BIRDS, INDIANA LINE TO QUINCY.

	WHEAT	PLOWED GROUND	MEADOWS	STUBBLE	CORN	SWAMP	PASTURES	ORCHARDS	YARDS
Wheat	1.	.70	.52	.49	.43	.14	.13	.075	.028
Plowed ground	1.43	1.	.75	.70	.62	.21	.19	.11	.04
Meadows	1.91	1.33	1.	.94	.83	.29	.25	.14	.053
Stubble	2.05	1.42	1.1	1.	.89	.31	.27	.15	.057
Corn	2.30	1.61	1.2	1.1	1.	.34	.30	.17	.064
Swamp	6.70	4.67	3.5	3.3	2.9	1.	.88	.5	.185
Pastures	7.65	5.33	4.	3.7	3.3	1.1	1.	.57	.21
Orchards	13.39	9.33	7.	6.6	5.8	2.	1.75	1.	.87
Yards	36.22	25.24	18.8	18.	16.	5.4	4.7	3.	1.

TABLE XI. COEFFICIENTS OF PREFERENCE, NINE MOST ABUNDANT BIRDS, INDIANA LINE TO QUINCY.

ENGLISH SPARROW	STUBBLE	MEADOWS	PLOWED GROUND	CORN	PASTURES	ORCHARDS	WEEDS	YARDS
Stubble	1.	.77	.6	.11	.08	.022	.002	.00004
Meadows	1.3	1.	.76	.14	.10	.03	.002	.00006
Pl'd ground	1.7	1.31	1.	.19	.14	.04	.003	.00007
Corn	9.2	7.08	5.41	1.	.74	.2	.015	.0004
Pastures	12.4	9.54	7.29	1.35	1.	.27	.02	.0005
Orchards	46.1	35.5	27.	5.01	3.72	1.	.08	.002
Weeds	600.	462.	353.	65.	48.4	13.	1.	.03
Yards	23330.	17946.	13724.	2536.	1881.	506.	38.9	1.

TABLE XI.—*Continued.*

CROW-BLACKBIRD	CORN	STUBBLE	PASTURES	FARMYARDS
Corn	1.	.52	.03	.02
Stubble	1.91	1.	.07	.03
Pastures	31.1	16.28	1.	.48
Farmyards	60.55	31.71	2.06	1.

MEADOW-LARK	PLOWED GROUND	CORN	PASTURES	MEADOWS	STUBBLE	WHEAT
Plowed ground	1.	.79	.23	.21	.21	.21
Corn	1.27	1.	.29	.27	.26	.26
Pastures	4.33	3.40	1.	.92	.90	.90
Meadows.......	4.73	3.71	1.09	1.	.98	.98
Stubble.........	4.82	3.78	1.11	1.02	1.	1.
Wheat..........	4.82	3.78	1.11	1.02	1.	1.

CROW	CORN	STUBBLE	MEADOWS	PLOWED GROUND	PASTURES
Corn	1.	.87	.55	.065	.04
Stubble	1.15	1.	.68	.075	.05
Meadows	1.69	1.47	1.	.11	.07
Plowed ground	15.4	13.3	9.09	1.	.63
Pastures	24.3	21.	14.4	1.58	1.

TABLE XI.—*Continued.*

COWBIRD	CORN	STUBBLE	MEADOWS	PASTURES
Corn	1.	.52	.013	.01
Stubble	1.92	1.	.025	.02
Meadows	76.9	40.	1.	.88
Pastures	80.	45.2	1.13	1.

HORNED LARK	CORN	STUBBLE	WHEAT	PASTURES	PLOWED GROUND
Corn	1.	.13	.12	.03	.01
Stubble	7.72	1.	.90	.25	.11
Wheat	8.61	1.11	1.	.28	.12
Pastures	30.51	3.95	3.54	1.	.43
Plowed ground	71.65	9.28	8.32	2.35	1.

MOURNING-DOVE	WHEAT	PLOWED GROUND	CORN	MEADOWS	STUBBLE	PASTURES	ORCHARDS
Wheat	1.	.70	.30	.26	.19	.15	.15
Plowed ground	1.43	1.	.43	.38	.28	.22	.21
Corn	3.30	2.33	1.	.86	.64	.51	.49
Meadows	3.96	2.67	1.16	1.	.75	.59	.57
Stubble	5.13	3.58	1.55	1.34	1.	.79	.77
Pastures	6.52	4.55	1.97	1.70	1.27	1.	.97
Orchards	6.70	4.67	2.03	1.74	1.31	1.03	1.

TABLE XI.—*Continued.*

FIELD-SPARROW	STUBBLE	MEADOWS	WHEAT	CORN	PASTURES	ORCHARDS	SHRUBS	WEEDS
Stubble	1.	.45	.43	.29	.07	.013	.00003	.00002
Meadows	2.2	1.	.95	.65	.15	.028	.00006	.00004
Wheat	2.3	1.05	1.	.68	.15	.03	.00007	.00004
Corn	3.4	1.55	1.48	1.	.23	.04	.0001	.00006
Pastures	15.	6.82	6.52	4.41	1.	.19	.0004	.0003
Orchards	76.9	35.	33.4	22.6	5.13	1.	.002	.001
Shrubs	35000.	15909.	15217.	10294.	2333.	455.	1.	.64
Weeds	55000.	25000.	23913.	16177.	3666.	715.	1.57	1.

GOLDFINCH	STUBBLE	CORN	WHEAT	PASTURES	ORCHARDS	SHRUBS	WEEDS
Stubble	1.	.66	.38	.17	.028	.00035	.00002
Corn	1.50	1.	.56	.25	.042	.00052	.00003
Wheat	2.65	1.77	1.	.44	.075	.00092	.00005
Pastures	6.04	4.05	2.29	1.	.171	.0021	.0001
Orchards	35.5	23.7	13.4	5.84	1.	.0123	.0006
Shrubs	2885.	1923.	1087.	475.	81.3	1.	.0534
Weeds	53846.	35897.	20290.	8861.	1517.	18.7	1.

The data of Table XI., arranged under the different species of birds, may also be classified, as in Table XII., according to the different situations, or the different kinds of crops, frequented by the birds. The one table shows us how each kind of bird is related to the various crops; and the other, how each crop is related to the various kinds of birds. Table XI. is thus essentially ornithological, showing the preferences of each kind of bird with respect to the food resources and places of resort offered it by each kind of crop or other situation. Table XII. is essentially agricultural, and shows the principal bird visitants of each kind of crop, brought into comparison with respect to their preferences for that crop alone. Referring, for example, to the section for corn, we see at the left the names of the principal birds of the corn field, arranged from above downwards in the order of their frequency in corn, the least frequent visitants uppermost. We may use this table to compare any species with another as a corn-field bird—the horned lark with the meadow-lark, for instance—by finding the place of the one species in the diagonal series of 1's and going up or down the column until the line for the other species is reached. The coefficient at the intersection of the column with the line shows the frequency relation of the one bird to the other. In this way we learn that for every hundred horned larks, 532 meadow-larks were found in corn, or, what is virtually the same thing, that for every hundred meadow-larks there were 19 horned larks on an average in corn.

It is also easy to ascertain from these tables whether there is any group of species which seem especially and strongly attracted to any special situation. We notice such a group in the horned larks, mourning-doves, and meadow-larks, considered as visitants of fields of stubble, and found there respectively about 3 times, 5 times, and 7½ times as frequently as are blackbirds; in the crows and the horned larks, considered as visitants of plowed fields, found there approximately 6 times and 17 times as frequently as are meadow-larks; and in the field-sparrows, goldfinches, meadow-larks, mourning-doves, and English sparrows in the corn fields, in which they occur from 3 to 8 times as frequently as blackbirds. The principal meadow birds, by these tables, are mourning-doves, meadow-larks, and cowbirds, since they occur in meadows 7 times, 12 times, and 15 times as commonly as English sparrows; while pastures apparently afford a common meeting ground for all the birds of this list of most important species, the coefficient of the blackbird— the most frequent pasture bird—being less than three times that of the English sparrow, the least frequent of these birds in pastures.

Numerous questions of cause, effect, and controlling condition are suggested by these data, some of them readily answerable and others doubtfully so, but the discussion of ecological problems may best be postponed until the data here presented may be brought into comparison with those obtained from other trips, made at other seasons and in other parts of the state.

TABLE XII.   COEFFICIENTS OF PREFERENCE, TABULATED BY CROPS.

CORN	COWBIRD	HORNED LARK	CROW-BLACKBIRD	CROW	FIELD-SPARROW	GOLDFINCH	MEADOW-LARK	MOURNING-DOVE	ENGLISH SPARROW
Cowbird	1.	.33	.24	.20	.08	.07	.06	.03	.03
Horned lark	3.04	1.	.72	.61	.23	.20	.19	.10	.09
Crow-blackbird	4.23	1.39	1.	.85	.32	.28	.26	.14	.12
Crow	5.	1.65	1.18	1.	.38	.33	.31	.17	.14
Field-sparrow	13.08	4.30	3.09	2.61	1.	.87	.81	.45	.37
Goldfinch	15.	4.94	3.55	3.	1.15	1.	.92	.51	.42
Meadow-lark	16.15	5.32	3.82	3.23	1.24	1.08	1.	.55	.46
Mourning-dove	29.23	9.62	6.91	5.85	2.24	1.95	1.81	1.	.83
English sparrow	35.38	11.64	8.36	7.08	2.71	2.36	2.19	1.21	1.

TABLE XII.—*Continued.*

STUBBLE	COWBIRD	FIELD-SPARROW	ENGLISH SPARROW	CROW	CROW-BLACKBIRD	GOLDFINCH	HORNED LARK	MOURNING-DOVE	MEADOW-LARK
Cowbird	1.	.5	.5	.33	.24	.19	.08	.04	.03
Field-sparrow	2.	1.	1.	.66	.48	.38	.16	.08	.06
English sparrow	2.	1.	1.	.66	.48	.38	.16	.08	.06
Crow	3.	1.5	1.5	1.	.71	.58	.24	.13	.09
Crow-blackbird	4.2	2.1	2.1	1.40	1.	.81	.34	.18	.13
Goldfinch	5.2	2.6	2.6	1.73	1.24	1.	.43	.22	.16
Horned lark	12.2	6.1	6.1	4.07	2.90	2.35	1.	.51	.38
Mourning-dove	23.6	11.8	11.8	7.87	5.62	4.54	1.93	1.	.74
Meadow-lark	31.8	15.9	15.9	10.6	7.57	6.11	2.61	1.35	1.

WHEAT	MOURNING-DOVE	FIELD-SPARROW	HORNED LARK	GOLDFINCH	MEADOW-LARK
Mourning-dove	1.	1.	.34	.33	.14
Field-sparrow	1.	1.	.34	.33	.14
Horned lark	2.96	2.96	1.	.99	.43
Goldfinch	3.	3.	1.01	1.	.43
Meadow-lark	6.91	6.91	2.34	2.30	1.

PLOWED GROUND	ENGLISH SPARROW	MEADOW-LARK	MOURNING-DOVE	CROW	HORNED LARK
English sparrow	1.	.52	.52	.08	.03
Meadow-lark	1.94	1.	1.	.16	.06
Mourning-dove	1.94	1.	1.	.16	.06
Crow	11.76	6.06	6.06	1.	.35
Horned lark	33.29	17.15	17.15	2.83	1.

PASTURES	ENGLISH SPARROW	MEADOW-LARK	MOURNING-DOVE	FIELD-SPARROW	GOLDFINCH	COWBIRD	HORNED LARK	CROW	CROW-BLACKBIRD
English sparrow..	1.	.87	.83	.83	.78	.55	.51	.39	.36
Meadow-lark	1.15	1.	.95	.95	.90	.63	.59	.45	.42
Mourning-dove...	1.21	1.05	1.	1.	.95	.66	.62	47	.44
Field-sparrow	1.21	1.05	1.	1.	.95	.66	.62	.47	.44
Goldfinch	1.27	1.10	1.05	1.05	1.	.70	.66	.50	.46
Cowbird..........	1.82	1.58	1.51	1.51	1.43	1.	.93	.71	.66
Horned lark......	1.94	1.69	1.61	1.61	1.52	1.07	1.	.76	.70
Crow.....	2.55	2.21	2.11	2.11	2.	1.40	1.31	1.	.92
Crow-blackbird ..	2.76	2.39	2.28	2.28	2.16	1.51	1.42	1.08	1.

MEADOWS	ENGLISH SPARROW	CROW	FIELD-SPARROW	MOURNING-DOVE	MEADOW LARK	COWBIRD
English sparrow..................	1.	.59	.59	.15	.08	.07
Crow...............................	1.69	1.	.1	.25	.14	.11
Field-sparrow	1.69	1.	1.	.25	.14	.11
Mourning-dove..................	6.77	4.	4.	1.	.57	.44
Meadow-lark	12.	7.09	7.09	1.77	1.	.78
Cowbird...........................	15.38	9.09	9.09	2.27	1.28	1.

Conclusion.

The circumstance that the data of this paper are summarized in numerical tables must not be permitted to obscure the fact that they merely present a fixed picture of a fleeting condition; that they are to be taken only as numerical generalizations of the observations here recorded, and do not, in themselves, warrant much by way of inference beyond their immediate contents. The view of the autumnal bird life of central Illinois which we get by their means is like a short-time photograph of a changing scene—changing so rapidly, indeed, that the effects of its transformations are noticeable even in the picture itself; for it is evident, especially from the list of species at the end of this paper, that there was some bird migration southward during the fifty days of this trip. Summer residents of central Illinois diminish in numbers, or even wholly disappear, during its course, winter residents come in, and migrants to the south, not seen in the earlier days of the journey, become abundant as they move across the line of march in the western part of the state.

Some of the effects of this migration were seen a fortnight later in the very different picture of bird life presented on a trip made by these same observers, October 31 and November 1, from Cairo, the southernmost point in Illinois, to Ullin, some twelve and a half miles north. Here, instead of the scanty average of 874 birds per square mile, as found in central Illinois, there were over 9 to the acre, or 5882 to the square mile. Two thirds of these were crow-blackbirds and robins—45 per cent. of the first and 23 per cent. of the second—and the next most abundant species was the white-throated sparrow (7 per cent.), and next to that, the quail (4 per cent.). The meadowlark was reduced to 2 per cent. of the birds observed; and, more remarkable still, the English sparrow, to a little more than 1 per cent. Into the angle formed by the meeting of the Ohio River with the Mississippi, birds from the north were dropping down by thousands as into a huge pocket, to be held there, no doubt, until cold weather or a diminution of their food supply should drive them farther south.

Definite conclusions of permanent value concerning the numbers and significance of the bird life of the state evidently can not be drawn until many such pictures as these have been assembled, compared, and adjusted in their right relations; and it has been the principal object of this paper to describe and illustrate one process, at least, by which the materials necessary to a correct general view of the ornithological ecology of the state may be brought together and made available.

LIST OF BIRDS IDENTIFIED, INDIANA LINE TO QUINCY, ILL.

CHECK-LIST NO.	SPECIES	I*	II	III	IV	V	VI
190	Botaurus lentiginosus	....	....	....	....	....	1
194	Ardea herodias	1	....	....	....	....	....
201	Butorides virescens	1	....	....	....	....	....
214	Porzana carolina	1	1	....	....	....	....
261	Bartramia longicauda	1	1	1	....	....	....
273	Oxyechus vociferus	55	....	2	3	....	....
289	Colinus virginianus	....	....	....	....	14	55
305	Tympanuchus americanus	2	7	....	....	....	....
316	Zenaidura macroura	56	22	14	42	32	14
325	Cathartes aura	4	....	2	....	....	....
331	Circus hudsonius	....	1	1	....	....	1
337	Buteo borealis	2	....	....	....	1	1
347a	Archibuteo lagopus sancti-johannis	1	....	....	1	....	....
357	Falco columbarius	....	....	....	....	....	1
360	Falco sparverius	3	1	....	....	....	....
387	Coccyzus americanus	2	1	....	....	....	....
390	Ceryle alcyon	1	....	....	....	....	....
393	Dryobates villosus	....	....	....	....	....	1
394c	Dryobates pubescens medianus	....	....	1	1	....	8
402	Sphyrapicus varius	....	....	2	....	....	....
406	Melanerpes erythrocephalus	21	....	1	....	....	....
409	Centurus carolinus	....	....	....	1	....	2
412	Colaptes auratus	23	8	14	14	2	1
420	Chordeiles virginianus	21	....	....	....	....	....
423	Chætura pelagica	2	1	....	5	1	8
444	Tyrannus tyrannus	2	....	....	....	....	1
456	Sayornis phœbe	....	....	2	....	1	3
461	Contopus virens	1	....	1	....	....	....
466	Empidonax traillii	2	....	....	....	....	....
474b	Otocoris alpestris praticola	41	61	49	17	2	50
477	Cyanocitta cristata	11	3	15	12	2	14
488	Corvus brachyrhynchos	14	5	19	20	158	10
494	Dolichonyx oryzivorus	....	....	1	....	....	....
495	Molothrus ater	60	24	63	73	....	1
498	Agelaius phœniceus	3	....	....	....	....	5
501	Sturnella magna	82	19	50	31	20	110
511b	Quiscalus quiscula æneus	309	65	11	95	....	37
517	Carpodacus purpureus	....	....	....	....	....	4
——	Passer domesticus	188	447	112	683	5	185
529	Astragalinus tristis	12	1	10	12	....	99
540	Poocætes gramineus	....	8	11	....	....	52
542a	Passerculus sandwichensis savanna	....	....	....	1	....	6
546	Coturniculus savannarum passerinus	16	11	4	1	5	7
548	Ammodramus leconteii	....	....	....	....	12	11
554	Zonotrichia leucophrys	....	....	....	2	....	6
558	Zonotrichia albicollis	....	....	....	2	....	91

* I=Indiana line to Champaign, Aug. 28-Sept. 1. II=Urbana to Decatur, Sept. 17-21. III=Decatur to Springfield, Sept. 24-29. IV=Springfield to Jacksonville, Oct. 1-4. V=Jacksonville to Meredosia, Oct. 5-8. VI=Meredosia to Quincy, Oct. 12-17.

LIST OF BIRDS IDENTIFIED.—*Continued.*

CHECK-LIST NO.	SPECIES	I	II	III	IV	V	VI
560	Spizella socialis....................		1				
563	Spizella pusilla	2	1	1	1	1	77
567	Junco hyemalis				3	1	32
581	Melospiza cinerea melodia				4	1	19
583	Melospiza lincolni				3		3
584	Melospiza georgiana..................			1	12	32	110
585	Passerella iliaca						2
587	Pipilo erythrophthalmus						14
593	Cardinalis cardinalis................			1		1	1
598	Cyanospiza cyanea...................	1					
604	Spiza americana....................	2					
611	Progne subis	4					
612	Petrochelidon lunifrons	6					
613	Hirundo erythrogaster.........	3					
614	Iridoprocne bicolor..................					9	2
619	Ampelis cedrorum..................						1
622	Lanius ludovicianus.................	3	1				1
624	Vireo olivaceus			1			
626	Vireo philadelphicus.................			1			
629	Vireo solitarius		1				
645	Helminthophila rubricapilla	1					
646	Helminthophila celata...............						1
647	Helminthophila peregrina.............		1				
655	Dendroica coronata.....		2	34	33	7	36
657	Dendroica maculosa.....		1				
667	Dendroica virens		2		1		
672	Dendroica palmarum................	2			1		
681d	Geothlypis trichas brachidactyla...			1			
687	Setophaga ruticilla		2				
697	Anthus pensilvanicus.................						25
703	Mimus polyglottos.................				1		
704	Galeoscoptes carolinensis............	2	1				
705	Toxostoma rufum...................		2	2	1		
719	Thryomanes bewickii..........			1			
721	Troglodytes aëdon..................	3			1		2
724	Cistothorus stellaris................			2	2		3
726	Certhia familiaris americana........				2		1
727	Sitta carolinensis...................						2
728	Sitta canadensis			3			1
731	Bæolophus bicolor...................						10
735	Parus atricapillus						22
736	Parus carolinenis...................				3	4	
748	Regulus satrapa					2	2
749	Regulus calendula..................						6
758a	Hylocichla ustulata swainsoni		1				
761	Merula migratoria..................	4	15	18	5		19
766	Sialia sialis..................			10	11		40
?*	....	3	1	1	2	6	10

* Indentification uncertain.

ON THE GENERAL AND INTERIOR DISTRIBUTION OF ILLINOIS FISHES

Stephen Alfred Forbes

ARTICLE III.—*On the General and Interior Distribution of Illinois Fishes.** BY S. A. FORBES.

The geography of Illinois is, in its most obvious features, so simple and so monotonous that one naturally expects a similar simplicity and monotony in the geographic distribution of its plants and animals. The plan of its hydrography is as little complicated as the geography of its land areas. Surrounded on more than two thirds of its circumference by three large rivers, the Mississippi, the Ohio, and the Wabash, with Lake Michigan covering a narrow strip at its northeast corner and draining a bordering region of scarcely greater area, its other waters flow southwestward into the Mississippi and southward into the Wabash and the Ohio, all mingling finally opposite its southernmost extremity for their journey to the Gulf. Its principal watersheds are inconspicuous ridges or slightly elevated plains, most of them originally more or less marshy, and the headwaters and tributaries of its various stream systems so approach and intermingle that in times of flood they formed an interlacing network, through which it would seem that a wandering fish might have found its way in almost any direction and to almost any place.

Its climate varies considerably, of course, within the five and a half degrees of its length from north to south, but by insensible gradations, with no lines of abrupt transition anywhere to set definite boundaries to the range of its aquatic species.

Its surface geology is more diversified than its topography, and its soils, although uniformly fertile throughout most of the state, differ notably in their origin and physical constitution, some of these differences being such as to affect more or less the surface waters and, through them, to influence the conditions of aquatic life. The extreme northwestern and the extreme southern parts of the state are bare of drift, and their soil is derived immediately from the underlying rock; but the surface of all the remainder of the state, excepting a

* This article is a reprint, with minor changes, of a chapter in the introduction to "The Fishes of Illinois," by S. A. Forbes and R. E. Richardson.

small area above the mouth of the Illinois, has been repeatedly worked over by ice in the course of the successive divisions of the glacial period. The oldest glaciated area, known as the lower Illinoisan glaciation, covers the greater part of southern Illinois and a narrow belt of the southeast part of the central section of the state. Next to this at the northwest, and immediately east of the lower half of the Illinois River, is the middle Illinoisan; above this, in the west-central part of the state, between the Illinois River and the Rock, is the upper Illinoisan; and still farther north, in the Rock River basin, are the Iowan and Preiowan glaciations, reaching northward across the Wisconsin boundary. East of the last three mentioned, and north of the southern Illinois district, the Wisconsin glaciation, the most recent of the series, covers about a fourth of the state. It is to the peculiar features of the lower Illinoisan glaciation especially that we shall presently be compelled to pay particular attention, because of their evident effect on the distribution of a considerable group of our fishes.

The topographical relations of the state to the surrounding territory are as simple and open as its own interior hydrography, and there is little to suggest the possibility of anything in the least peculiar in the general constitution or the relations of its fauna, or anything problematical or especially interesting in the details of the distribution of its native fishes. We shall find reason to believe, however, that this appearance is misleading, and that the subject, studied in detail, contains matter of unusual interest, and presents problems of considerable difficulty, a solution of which will lead us to some novel results.

It is true, however, generally speaking, that the distribution of Illinois fishes reflects, in uniformity and relative monotony, the features of the topography of the state. A few species occurring in Lake Michigan and characteristic of the Great Lakes are, in fact, the only Illinois fishes which are definitely and permanently separated from their fellows in other Illinois waters by what may be called geographical conditions, and these conditions are not physical obstacles to their passage from Lake Michigan to the Illinois River.

Excluding, for the moment, these fishes special to the Great Lakes, we find elsewhere in Illinois a general commingling and overlapping of the fish population of the surrounding territory, the limits to whose range are climatic, local, and ecological, but topographic only in a secondary sense.

THE GENERAL DISTRIBUTION

Most of the 150 species of the native fishes of Illinois range far and wide in all directions beyond its narrow boundaries, thus illustrating the breadth and the simplicity of our geographical affiliations with the surrounding territory; but a considerable number, on the other hand, coming into Illinois from one direction, do not pass beyond it in another, some part of the boundary of the general area of their distribution passing through our state. Several southern fishes go no farther north than Illinois; some northern fishes go no farther south; some eastern species find here their western limit; and a few western species range no farther east. The comparison of these geographical groups whose areas overlap by their borders here in Illinois is a matter of special interest to the student of distribution, because it is in them that we find indicated the more remote affinities of our fish fauna, and from them, if anywhere, we may glean suggestions of its various origins.

It will be convenient for a discussion of this subject to divide the general expanse over which Illinois fishes are distributed, into the following twelve districts: 1, the upper Mississippi Valley, including the Missouri and its tributaries; 2, the lower Mississippi Valley, including the Ohio and its tributaries; 3, the far North, extending northward from the headwaters of the Mississippi, east to the Lake Superior drainage, and west to the Rocky Mountains; 4, the far Northwest, separated from the preceding by the Rocky Mountains range; 5, the Great Lake region; 6, the district of Quebec and New England; 7, the Hudson River district; 8, the north Atlantic drainage, from New England to the Chesapeake Bay; 9, the south Atlantic, from the Chesapeake Bay to Florida; 10, the peninsula of Florida; 11, the east Gulf district, bounded by the Mississippi drainage on the west; and 12, the west Gulf district, bounded by the Mississippi drainage on the east, and extending west and south to include the Rio Grande and its tributaries. The following table shows the recorded distribution of our species over the territory so divided.

TABLE OF THE GENERAL DISTRIBUTION OF ILLINOIS FISHES

	Great Lake Basin	Quebec and New England	Hudson River	North Atlantic	South Atlantic	Florida Peninsula	East Gulf	Lower Miss. and Ohio	Upper Miss. and Mo.	West Gulf and Rio Grande	Far Northwest	Far North
Silvery lamprey (*Ichthyomyzon*)	+	+						+	+			
Brook lamprey (*Lampetra*)	+							+	+			
Paddle-fish (*Polyodon*)	+							+	+			
Lake sturgeon (*Acipenser*)	+	+						+	+			+
Shovel-nosed sturgeon								+	+	+		
White sturgeon (*P. albus*)								+				
Long-nosed gar	+	+		+	+	+	+	+	+	+		
Short-nosed gar	+						+	+	+	+		
Alligator-gar						+		+	+	+		
Dogfish (*Amia*)	+	+			+	+	+	+	+			
Mooneye (*alosoides*)								+	+			+
Toothed herring (*tergisus*)	+	+						+	+			+
Gizzard-shad (*Dorosoma*)	+			+	+	+	+	+	+	+		
Skipjack (*chrysochloris*)	+						+	+	+	+		
Whitefish	+	+										+
Lake herring	+	+										
Lake trout	+	+									+	+
Eel	+	+	+	+	+	+	+	+	+	+		
Black-horse (*Cycleptus*)								+	+	+		
Red-mouth buffalo (*cyprinella*)								+	+			+
Mongrel buffalo (*urus*)								+	+			
Small-mouth buffalo (*bubalus*)								+	+			

TABLE OF THE GENERAL DISTRIBUTION OF ILLINOIS FISHES—*continued*

	Great Lake Basin	Quebec and New England	Hudson River	North Atlantic	South Atlantic	Florida Peninsula	East Gulf	Lower Miss. and Ohio	Upper Miss. and Mo.	West Gulf and Rio Grande	Far Northwest	Far North
River carp (*carpio*)								+	+	+		
Blunt-nosed carp (*difformis*)								+	+			
Lake carp (*thompsoni*)	+	+							+			
Quillback carp (*velifer*)	+							+	+	+		+
Chub-sucker	+	+	+	+	+	+	+	+	+	+		
Striped sucker	+			+	+		+	+	+	+		
Common sucker (*commersonii*)	+	+	+	+	+			+	+			+
Hogsucker (*nigricans*)	+	+		+	+		+	+	+			
White-nosed sucker (*anisurum*)	+	+			+			+	+			+
Common red-horse (*aureolum*)	+	+	+					+	+			+
Short-headed red-horse (*breviceps*)	+							+	+			
Placopharynx duquesnei	+						+	+	+			
Harelipped sucker (*Lagochila*)	+							+	+			
Stone-roller (*Campostoma*)	+				+		+	+	+	+		
Red-bellied dace (*Chrosomus*)	+	+		+	+			+	+			
Silvery minnow (*H. nuchalis*)				+	+		+	+	+	+		+
Hybognathus nubila								+	+			
Black-head minnow (*P. promelas*)	+	+						+	+	+		+
Blunt-nosed minnow (*P. notatus*)	+	+		+			+	+	+			
Horned dace (*Semotilus*)	+	+	+	+	+		+	+	+			
Opsopœodus emiliæ	+				+		+	+	+			
Golden shiner (*Abramis*)	+	+	+	+	+	+	+	+	+	+		

TABLE OF THE GENERAL DISTRIBUTION OF ILLINOIS FISHES—*continued*

	Great Lake Basin	Quebec and New England	Hudson River	North Atlantic	South Atlantic	Florida Peninsula	East Gulf	Lower Miss. and Ohio	Upper Miss. and Mo.	West Gulf and Rio Grande	Far Northwest	Far North
Bullhead minnow (*Cliola vigilax*)....	+						+	+	+	+		
Notropis anogenus..............	+								+			
N. cayuga..................	+							+	+			+
N. cayuga atrocaudalis.............	+	+						+		+		
N. heterodon...................	+							+	+			
Straw-colored minnow (*N. blennius*)..	+	+						+	+	+		+
N. phenacobius...................									+			
N. gilberti..................									+			
N. illecebrosus	+							+	+			
Redfin (*N. lutrensis*)...............								+	+	+		
Spot-tailed minnow (*N. hudsonius*)...	+	+	+	+	+			+	+			
Silverfin (*N. whipplii*)..............	+	+						+	+			
Common shiner (*N. cornutus*)........	+	+	+	+	+		+	+	+			+
Notropis pilsbryi..................								+	+			
N. jejunus....................								+	+			+
Shiner (*N. atherinoides*).............	+	+						+	+			+
Notropis rubrifrons................	+	+		+				+	+			
Blackfin (*N. umbratilis atripes*)......	+			+			+	+	+			
Ericymba buccata..................	+						+	+	+			
Sucker-mouthed minnow (*Phenacobius*)								+	+	+		
Long-nosed dace (*R. cataractæ*)......	+	+		+	+			+	+	+	+	+

TABLE OF THE GENERAL DISTRIBUTION OF ILLINOIS FISHES—*continued*

	Great Lake Basin	Quebec and New England	Hudson River	North Atlantic	South Atlantic	Florida Peninsula	East Gulf	Lower Miss. and Ohio	Upper Miss. and Mo.	West Gulf and Rio Grande	Far Northwest	Far North
Black-nosed dace (*R. atronasus*)	+	+	+	+	+			+	+			
Hybopsis hyostomus							+	+	+			
Spotted shiner (*H. dissimilis*)	+							+	+			
Silver chub (*amblops*)	+						+	+	+			
Storer's chub	+							+	+			+
River chub (*kentuckiensis*)	+			+	+		+	+	+			
Flat-headed chub (*Platygobio*)								+	+			+
Blue cat (*furcatus*)								+	+	+		
Ictalurus anguilla								+		+		
Channel-cat (*punctatus*)	+					+	+	+	+	+		+
Great Lake catfish (*lacustris*)	+	+										
Yellow bullhead (*natalis*)	+				+	+	+	+	+	+		
Common bullhead (*nebulosus*)	+	+	+	+	+	+	+	+	+	+		+
Black bullhead (*melas*)	+						+	+	+			
Mud-cat (*Leptops*)							+	+	+	+		
Common stonecat (*N. flavus*)	+			+				+	+			
Tadpole cat (*S. gyrinus*)	+		+	+		+	+	+	+			
Freckled stonecat (*S. nocturnus*)								+	+	+		
Slender stonecat (*S. exilis*)	+							+	+			
Brindled stonecat (*S. miurus*)	+							+	+			
Mud-minnow	+	+		+				+	+			
Grass pike (*Esox vermiculatus*)	+						+	+	+			

TABLE OF THE GENERAL DISTRIBUTION OF ILLINOIS FISHES—*continued*

	Great Lake Basin	Quebec and New England	Hudson River	North Atlantic	South Atlantic	Florida Peninsula	East Gulf	Lower Miss. and Ohio	Upper Miss. and Mo.	West Gulf and Rio Grande	Far Northwest	Far North
Pike (*E. lucius*)	+	+	...	+				+	+	...	+	+
Muskallunge	+	+	...					+	+	...		+
Menona top-minnow (*F. diaphanus m.*)	+							+	+			...
Striped top-minnow (*F. dispar*)	+						+	+	+	...		...
Common top-minnow (*F. notatus*)	+						+	+	+	+		...
Viviparous top-minnow (*affinis*)	...		...	+	+	+	+	+	+	+	...	...
Chologaster papilliferus									+			..
Brook stickleback	+	+	...					+	+	...	...	+
Nine-spined stickleback	+	+	+								+	+
Trout-perch	+	+	...	+				+	+	...		+
Brook silverside	+	...	...		+	+	+	+	+			
Pirate-perch	+	...	...	+	+	+	+	+	+			
Pigmy sunfish (*Elassoma*)					+	...	+	+				
White crappie (*annularis*)	+			+	+		+	+	+			
Black crappie (*sparoides*)	+	+	...	+	+	+	+	+	+			
Round sunfish					+	+	+	+				
Rock bass	+	+	+	+	+	...	+	+	+	...		+
Warmouth (*Chænobryttus*)	+				+	+	+	+	+	+		
Green sunfish (*cyanellus*)	+							+	+	+		
Lepomis ischyrus									+			
L. symmetricus								+		+		...
L. euryorus	+		...						+			...

TABLE OF THE GENERAL DISTRIBUTION OF ILLINOIS FISHES—*continued*

	Great Lake Basin	Quebec and New England	Hudson River	North Atlantic	South Atlantic	Florida Peninsula	East Gulf	Lower Miss. and Ohio	Upper Miss. and Mo.	West Gulf and Rio Grande	Far Northwest	Far North
Lepomis miniatus						+	+	+	+			
Long-eared sunfish	+				+	+	+	+	+	+		
Orange-spotted sunfish (*humilis*)								+	+	+		
Bluegill (*pallidus*)	+				+	+	+	+	+	+		
Eupomotis heros							+	+		+		
Pumpkinseed (*gibbosus*)	+	+	+	+	+			+	+			
Small-mouthed black bass	+	+	+	+	+		+	+	+	+		
Large-mouthed black bass	+	+	+	+	+	+	+	+	+	+		+
Pike-perch (*S. vitreum*)	+	+		+	+		+	+	+			+
Sauger (*S. canadense griseum*)	+	+							+			+
Yellow perch	+	+	+	+	+			+	+			+
Log-perch (*P. caprodes*)	+	+		+			+	+	+	+		
Hadropterus evermanni								+	+			
H. phoxocephalus	+							+	+			
Black-sided darter (*H. aspro*)	+				+			+	+			+
Hadropterus ouachitæ								+				
H. evides	+							+	+			
H. scierus								+		+		
Cottogaster shumardi	+							+	+			
Green-sided darter (*blennioides*)	+						+	+	+			
Johnny darter (*B. nigrum*)	+	+	+	+	+			+	+			+
Boleosoma camurum							+	+	+	+		

Table of the General Distribution of Illinois Fishes—*concluded*

	Great Lake Basin	Quebec and New England	Hudson River	North Atlantic	South Atlantic	Florida Peninsula	East Gulf	Lower Miss. and Ohio	Upper Miss. and Mo.	West Gulf and Rio Grande	Far Northwest	Far North
Crystallaria asprella	+							+				
Sand darter (*Ammocrypta*)	+							+	+	+		
Banded darter (*E. zonale*)	+						+	+	+			
Blue-breasted darter (*E. camurum*)	+							+	+			
Etheostoma iowæ									+			+
E. jessiæ	+						+	+	+	+		
Rainbow darter (*E. cœruleum*)	+			+				+	+	+		
Etheostoma obeyense								+				
E. squamiceps							+	+				
Fan-tailed darter (*E. flabellare*)	+	+		+	+			+	+			
Boleichthys fusiformis	+	+		+	+	+		+	+	+		
Least darter (*Microperca*)	+			+				+	+			
White bass (*Roccus chrysops*)	+	+						+	+			
Yellow bass (*Morone*)								+	+			
Sheepshead (*Aplodinotus*)	+	+					+	+	+	+		+
Miller's thumb	+	+		+	+		+	+	+			
Cottus ricei	+											
Uranidea kumlienii	+											
Burbot (*Lota*)	+	+	+					+	+			+
Number of species	108	53	19	40	45	23	56	134	131	47	4	37

Arranged according to the number of Illinois species in each, these districts succeed each other in the following order.

Districts	No. of species	Per cent. of all Illinois species
Lower Mississippi and Ohio valleys................	134	89
Upper Mississippi and Missouri valleys..............	131	87
The Great Lake basin.............................	108	72
The east Gulf district	56	37
Quebec and New England..........................	53	36
The west Gulf and Rio Grande district	47	31
The south Atlantic district........................	45	30
The north Atlantic district........................	40	27
The far North...................................	37	25
The Florida peninsula.............................	23	15
The Hudson drainage.............................	19	13
The far Northwest...............................	4	3

Next to the two Mississippi Valley districts and the Great Lake basin, which average 124 Illinois species, our fishes are most largely represented in the east Gulf and the Quebec and New England districts, averaging 54 Illinois species—the first closely related to the lower Mississippi, and the second a continuation eastward of the Great Lake basin. Then follow the north and south Atlantic and the west Gulf districts, with an average of 43 species; the far North, the Florida peninsula, and the Hudson River districts, with 37 to 19 species; and, finally, the far Northwest, with but 4 Illinois species.

The northern and the southern affiliations of the assemblage of fishes represented in our Illinois collections may be contrasted by comparing the list of Illinois species occurring in either or both of the more northerly divisions—that is, the far North and the Quebec and New England districts—on the one hand, with a list of those found in either or all of the three most southerly districts—that is, the Florida peninsula, the east Gulf, and the west Gulf and Rio Grande—on the other hand. In this northern list of Illinois fishes there are 64 species, and in the southern list there are 77; but 25 of these species are more or less common to both north and south, leaving 39 Illinois fishes distinctively northern in their distribution and 52 distinctively southern. Northern and southern species thus mingle in our territory in unequal proportions, the southern element largely preponderating.

If we look to the further distribution of the northern and southern elements of our fish population, distinguishing northeastern from
northwestern species, and southeastern from southwestern, we find
that the southeastern species largely outnumber the southwestern
in Illinois, and that the northeastern outnumber the northwestern.
Thus there are 47 species of the west Gulf and Rio Grande region in
this state, and 58 species of the east Gulf and Florida districts.

Further, there are more species known as common to Illinois and
the far northeast than there are to Illinois and the southwestern district of the west Gulf and the Rio Grande. Notwithstanding the
much greater distance from us of the Quebec and New England
district, there are 53 of the fishes of that region known in Illinois to
47 of those of the west Gulf district. The northeastern fishes have,
however, been much more carefully collected than the southwestern, and an equal knowledge of both districts might change these
relative numbers.

THE INTERIOR DISTRIBUTION

The interior distribution of the fishes of the state may best be exhibited by treating each considerable stream-system as a unit, and
comparing the fishes of each such system with all the others. The
state may be conveniently divided into ten such hydrographic
districts, as follows:
1. The Galena district, including the streams of the northwestern unglaciated area, most of which empty into the Mississippi
through Galena, Apple, and Plum rivers. 2. The Rock River district, extending southward and westward from the northern boundary of the state to the Mississippi at the mouth of the Rock. 3. The
Illinois district, including the entire drainage of the Illinois River.
4. The Michigan district, a narrow strip along the borders of Lake
Michigan—the Lake Michigan drainage—most of which centers in
the Chicago and the Calumet rivers. 5. The Mississippi River, and
an irregular strip adjacent not included in any of the more definite
river systems and mainly drained by small streams of the bluffs and
neighboring highlands. This district is divided by the lower end
of the Illinois basin. 6. The Kaskaskia basin. 7. The Illinois
drainage of the Wabash, including that stream itself so far as it helps
to form the boundary line between Illinois and Indiana. 8. The
basin of the Big Muddy River, in the southwestern part of the state.

9. The Saline River basin, in the southeastern part of the state. 10. The Cairo district, the driftless area of extreme southern Illinois, drained by the Cache River and smaller tributaries of the Ohio. The Ohio itself is included in this last district.

The following list and table gives the details of the distribution of the species in a way to show the number of collections of each species made by us from each district. A cross opposite a species name indicates that the species occurs in the basin mentioned at the head of the column, but that it is not represented by preserved collections affording numerical data.

INTERIOR DISTRIBUTION OF ILLINOIS FISHES BY RIVER SYSTEMS
SPECIES AND NUMBER OF COLLECTIONS OF EACH

	Districts										Sections		
	Galena District	Rock River	Illinois River	Michigan Drainage	Mississippi and Creeks	Kaskaskia	Wabash	Big Muddy	Saline	Cairo District	North	Central	South
Number of species	44	92	128	57	97	69	95	42	55	101	120	123	119
Collections made	13	73	1115	20	57	41	103	10	18	95	269	1083	192
Silvery lamprey		1	12	1	+		1				+	+	+
Brook lamprey				1						1	+	0	+
Paddle-fish			8		+		+			1	0	+	+
Lake sturgeon			+	+	+					+	+	+	+
Shovel-nosed sturgeon			+		+		+			+	0	+	+
White sturgeon					4						0	+	0
Long-nosed gar		1	20	1	10	1	+			4	+	+	+

INTERIOR DISTRIBUTION OF ILLINOIS FISHES BY RIVER SYSTEMS
SPECIES AND NUMBER OF COLLECTIONS OF EACH—*continued*

| | Districts | | | | | | | | | | Sections | | |
	Galena District	Rock River	Illinois River	Michigan Drainage	Mississippi and Creeks	Kaskaskia	Wabash	Big Muddy	Saline	Cairo District	North	Central	South
Short-nosed gar		1	52	...	4	+	+			1	+	+	+
Alligator-gar			+	...	+					+	0	+	+
Dogfish			27	1	3	+	1	...	2	1	+	+	+
Mooneye			1	...	+					+	0	+	+
Toothed herring			8	1	+					7	+	+	+
Gizzard-shad	1	3	89	1	1	7	2			3	+	+	+
Skipjack	2	1	3	...	2					+	+	+	+
Whitefish				+							+	0	0
Lake herring				+							+	0	0
Lake trout				+							+	0	0
Eel			+	+	+		+			+	+	+	+
Black-horse			1	...	2						0	+	+
Red-mouth buffalo	1	1	28	...	9		2			1	+	+	+
Mongrel buffalo	1		17	1	1						+	+	+
Small-mouth buffalo	1	1	46	1	9		2			+	+	+	+
River carp		1	11		2	1	+		1	+	+	+	+
Blunt-nosed carp	1	6	54	...	8	15	21	...	3	3	+	+	+
Lake carp			10	...	1						+	+	0
Quillback carp	1	19	39	...	1	1	8	...	1	+	+	+	+

INTERIOR DISTRIBUTION OF ILLINOIS FISHES BY RIVER SYSTEMS
SPECIES AND NUMBER OF COLLECTIONS OF EACH—*continued*

				Districts								**Sections**	
	Galena District	Rock River	Illinois River	Michigan Drainage	Mississippi and Creeks	Kaskaskia	Wabash	Big Muddy	Saline	Cairo District	North	Central	South
Chub-sucker	....	4	48	...	2	21	47	6	7	10	+	+	+
Striped sucker	1	1	13	1	....	13	16	1	1	3	+	+	+
Common sucker	1	14	69	...	9	5	26	...	3	9	+	+	+
Long-nosed sucker				+							+	0	0
Hogsucker	1	11	61	...	1	9	27			1	+	+	+
White-nosed sucker	....	2	14	+	1						+	+	+
Common red-horse	2	13	90	...	5	10	25	...	1	2	+	+	+
Short-headed red-horse	....	4	39	1	3	7	2			+	+	+	+
Placopharynx duquesnei	...	1	1	...	+	....	1			+	+	+	+
Harelipped sucker							+				0	0	+
Stone-roller	1	20	99	...	14	9	36	1	1	10	+	+	+
Red-bellied dace	....	4	13	...	2					4	+	+	+
Silvery minnow	2	6	86	1	16	10	27	6	11	18	+	+	+
Hybognathus nubila	1	3	...		1					1	+	+	+
Black-head minnow	....	8	67	...	12	6	5			+	+	+	+
Blunt-nosed minnow	3	33	162	3	19	31	77	8	13	25	+	+	+
Horned dace	1	9	72	...	16	10	24	4	6	14	+	+	+
Opsopœodus emiliæ	....	3	49	1	1	1	18	3	6	4	+	+	+

INTERIOR DISTRIBUTION OF ILLINOIS FISHES BY RIVER SYSTEMS
SPECIES AND NUMBER OF COLLECTIONS OF EACH—*continued*

	Districts										Sections		
	Galena District	Rock River	Illinois River	Michigan Drainage	Mississippi and Creeks	Kaskaskia	Wabash	Big Muddy	Saline	Cairo District	North	Central	South
Golden shiner..........	1	18	183	1	8	19	50	7	10	10	+	+	+
Bullhead minnow......	1	14	110	...	5	22	38	1	3	2	+	+	+
Notropis anogenus..........			2								+	0	0
Notropis cayuga........	1	4	29	2	5	....	1	....		1	+	+	0
N. heterodon..............		5	81	1	1	....	4			3	+	+	+
Straw-colored minnow..	1	22	108	4	9	6	44	...	2	1	+	+	+
Notropis phenacobius.....			2								+	0	0
N. gilberti..............		3	15	...	10	....	2				+	+	+
N. illecebrosus..........			2	...	1	....	17	....		+	+	+	+
Spot-tailed minnow......		4	133	4	4					2	+	+	+
Redfin...................		1	142	9	16	4	.....	4	1	10	+	+	+
Silverfin..............	3	34	116	1	8	29	71	2	3	6	+	+	+
Common shiner........	1	19	105	...	11	14	22	...	1	12	+	+	+
Notropis pilsbryi.........			1								+	0	0
N. jejunus..............	1	5	21	1	10	....	5	...	2	5	+	+	+
Shiner................	3	8	82	6	8	4	19	4	6	11	+	+	+
Notropis rubrifrons.....	2	4	8	...							+	+	0
Blackfin	2	9	67	...	3	25	56	5	11	19	+	+	+
Ericymba buccata.......			4			25	58	......			0	+	+

INTERIOR DISTRIBUTION OF ILLINOIS FISHES BY RIVER SYSTEMS
SPECIES AND NUMBER OF COLLECTIONS OF EACH—*continued*

	Districts										Sections		
	Galena District	Rock River	Illinois River	Michigan Drainage	Mississippi and Creeks	Kaskaskia	Wabash	Big Muddy	Saline	Cairo District	North	Central	South
Sucker-mouthed minnow	2	15	78	...	13	17	36	1	4	8	+	+	+
Long-nosed dace........										1	0	0	+
Black-nosed-dace........		1	4							1	+	0	+
Hybopsis hyostomus....		2	1								+	+	0
Spotted shiner...........		6	3			1	1				+	+	+
Silver chub.............			2			10	37	4	2		0	+	+
Storer's chub...........		1	7	...	7		5	...	4	4	+	+	+
River chub.............	1	12	90	...	8	10	16			1	+	+	+
Flat-headed chub.......										3	0	0	+
Blue cat................			1	...	1					2	0	+	+
Ictalurus anguilla......			+	...	+					+	0	+	+
Channel-cat.............		17	108	...	7	17	26	2	1	2	+	+	+
Great Lake catfish......				+							+	0	0
Yellow bullhead.........		3	82			10	18	3	4	6	+	+	+
Common bullhead......			42	...	1	1				4	+	+	+
Black bullhead	1	11	144	...	19	15	35	4	6	10	+	+	+
Mud-cat...............	+	3	22	...	2	1	2			+	+	+	+
Stonecat...............	2	3	32	...	1	1	2			+	+	+	0
Tadpole cat.............		2	132	...	11	14	21	3	8	5	+	+	+

INTERIOR DISTRIBUTION OF ILLINOIS FISHES BY RIVER SYSTEMS
SPECIES AND NUMBER OF COLLECTIONS OF EACH—*continued*

	Districts										Sections		
	Galena District	Rock River	Illinois River	Michigan Drainage	Mississippi and Creeks	Kaskaskia	Wabash	Big Muddy	Saline	Cairo District	North	Central	South
Freckled stonecat......			5	...	1	2	+				0	+	+
Slender stonecat.......	1		1	...	2					2	+	+	+
Brindled stonecat......			1			1	26		5	1	0	+	+
Mud-minnow..........		8	18	1	1	...	4	1	1	6	+	+	+
Grass pike............		5	61	1	4	11	19	7	6	9	+	+	+
Pike.................		2	17	1	1					1	+	+	0
Muskallunge..........			+								+	0	0
Menona top-minnow....			11	7			+				+	+	0
Striped top-minnow......		1	75	1			8			5	+	+	+
Common top-minnow...	1	6	66	...	6	23	58	8	17	27	+	+	+
Viviparous top-minnow.			1	...	1	...	4	1	2	9	0	+	+
Chologaster papilliferus..										6	0	0	+
Brook stickleback......			1	2							+	0	0
Nine-spined stickleback				1							+	0	0
Trout-perch...........			14	1							+	+	0
Brook silverside........	1	6	89	2	2	1	21				+	+	+
Pirate-perch...........			54			9	11	7	11	9	+	+	+
Pigmy sunfish..........							5			1	0	0	+

INTERIOR DISTRIBUTION OF ILLINOIS FISHES BY RIVER SYSTEMS
SPECIES AND NUMBER OF COLLECTIONS OF EACH—*continued*

	Districts										Sections		
	Galena District	Rock River	Illinois River	Michigan Drainage	Mississippi and Creeks	Kaskaskia	Wabash	Big Muddy	Saline	Cairo District	North	Central	South
White crappie.........	2	9	119	2	13	6	14	3	3	6	+	+	+
Black crappie............	....	8	130	3	15	8	13	3	1	....	+	+	+
Round sunfish...........	....	....	1	...	....	....	1	1	2	8	0	0	+
Rock bass...............	...	4	35	1	3	2	1	1	...	2	+	+	+
Warmouth..............	...	3	83	...	3	5	10	6	6	11	+	+	+
Green sunfish..........	2	20	158	...	16	33	57	7	12	15	+	+	+
Lepomis ischyrus.........	....	1	3	...	....	....	....	....	....	....	+	+	0
L. symmetricus...........	....	....	2	...	....	....	3	...	....	4	0	+	+
L. euryorus.............	....	....	1	...	....	....	....	....	....	....	0	+	0
L. miniatus.............	....	....	24	...	1	....	2	...	....	....	+	+	+
Long-eared sunfish........	....	3	37	1	....	27	57	7	8	16	+	+	+
Orange-spotted sunfish.....	....	5	112	...	22	15	23	2	3	3	+	+	+
Bluegill...............	2	7	179	1	6	3	18	1	1	6	+	+	+
Eupomotis heros.........	....	....	....	....	....	....	5	...	....	1	0	0	+
Pumpkinseed............	....	4	82	4	2	....	1	...	....	1	+	+	+
Small-mouthed black bass.................	....	16	69	...	5	2	8	1	...	3	+	+	+
Large-mouthed black bass.................	....	7	135	4	13	8	33	2	4	12	+	+	+
Pike-perch.............	....	3	20	1	13	1	+	...	....	....	+	+	+

INTERIOR DISTRIBUTION OF ILLINOIS FISHES BY RIVER SYSTEMS
SPECIES AND NUMBER OF COLLECTIONS OF EACH—*concluded*

	Districts										Sections		
	Galena District	Rock River	Illinois River	Michigan Drainage	Mississippi and Creeks	Kaskaskia	Wabash	Big Muddy	Saline	Cairo District	North	Central	South
Sauger............	....	1	13	...	3	1	...	....	....	....	+	+	0
Yellow perch...........	....	+	75	3	6	...	....	....	....	....	+	+	0
Log-perch..............	....	4	35	3	5	9	8	...	1	2	+	+	+
Hadropterus evermanni..	....	....	3	...	....	....	....	....	....	....	0	+	0
H. phoxocephalus..........	....	12	58	...	3	10	6	...	2	....	+	+	+
Black-sided darter......	2	15	70	...	1	22	42	2	7	11	+	+	+
Hadropterus ouachitæ....	....	....	....	...	....	.:.	1	....	....	....	0	0	+
H. evides..............	....	1	....	...	....	....	....	....	....	....	+	0	0
H. scierus..............	....	....	1	...	....	....	1	....	....	....	0	+	0
Cottogaster shumardi.....	....	....	14	...	....	2	1	....	....	....	0	+	+
Green-sided darter......	....	....	+	...	....	....	36	....	....	....	0	+	+
Johnny darter.........	3	22	100	3	10	27	58	1	6	8	+	+	+
Boleosoma camurum......	....	1	45	2	2	12	17	7	11	10	+	+	+
Crystallaria asprella.....	1	3	....	...	2	....	1	....	....	....	+	+	+
Sand darter............	....	3	7	...	1	2	16	....	....	....	+	+	+
Banded darter.........	1	11	21	...	....	....	1	....	....	....	+	+	+
Blue-breasted darter......	....	2	6	1	....	....	....	....	....	1	0	+	0
Etheostoma iowæ..........	....	2	4	1	....	....	....	....	....	1	+	0	+
E. jessiæ..............	....	4	119	...	5	11	14	2	1	4	+	+	+

INTERIOR DISTRIBUTION OF ILLINOIS FISHES BY RIVER SYSTEMS
SPECIES AND NUMBER OF COLLECTIONS OF EACH—*concluded*

	Districts										Sections		
	Galena District	Rock River	Illinois River	Michigan Drainage	Mississippi and Creeks	Kaskaskia	Wabash	Big Muddy	Saline	Cairo District	North	Central	South
Rainbow darter........	2	9	39	...	1	2	29	1	4	13	+	+	+
Etheostoma obeyense.....	...	...	...	...	...	...	...	...	1	....	0	0	+
E. squamiceps..........	...	...	...	...	...	1	1	...	1	7	0	+	+
Fan-tailed darter......	1	6	11	...	1	1	14	...	...	3	+	+	+
Boleichthys fusiformis...	...	1	13	...	...	5	18	3	8	8	+	+	+
Least darter...........	...	1	10	...	...	...	...	...	...	1	+	0	+
White bass............	1	2	36	2	12	...	...	...	...	1	+	+	+
Yellow bass...........	...	1	95	...	5	...	...	...	...	....	+	+	+
Sheepshead...........	...	1	53	...	13	...	...	1	1	1	+	+	+
Miller's thumb	...	...	5	...	...	...	...	...	...	6	+	+	+
Cottus ricei...........	...	...	...	+	...	...	...	...	...	....	+	0	0
Uranidea kumlienii.....	...	...	...	+	...	...	...	...	...	....	+	0	0
Burbot................	...	...	3	1	...	...	...	...	...	....	+	+	0

THE ILLINOIS BASIN AND THE OTHER DISTRICTS COMPARED

The key to the distribution of Illinois fishes within the state is the species list of the Illinois basin. Covering fully one half the area of Illinois, and extending in a broad belt diagonally northeast and southwest across its northern two thirds, this basin contains nearly every variety of stream, lake, pond, and marsh to be found between the

Great Lakes on the one hand and the giant flood of the Mississippi on the other, and it is to be expected that its fish population will be highly typical of Illinois as a whole. It includes, in fact, more than four fifths of the species on our Illinois list, and the special features of the various other basins and areas may best be seen by comparing them with this characteristic central basin as a type.

The following is a list of the species of the Illinois system obtained by us in collections, arranged in the order of the frequency of their appearance in 1,115 collections made from that stream and its tributary waters.

SPECIES OF THE ILLINOIS BASIN, AND NUMBER OF COLLECTIONS CONTAINING EACH

Species	Collections	Species	Collections*
Golden shiner	183	Common red-horse	90
Bluegill	179	Gizzard-shad	89
Blunt-nosed minnow	162	Brook silverside	89
Green sunfish	158	Silvery minnow	86
Black bullhead	144	Warmouth	83
Redfin (*lutrensis*)	142	Shiner	82
Large-mouthed black bass	135	Yellow bullhead	82
Spot-tailed minnow	133	Pumpkinseed	82
Tadpole cat	132	*Notropis heterodon*	81
Black crappie	130	Sucker-mouthed minnow	78
Etheostoma jessiæ	119	Yellow perch	75
White crappie	119	Striped top-minnow	75
Silverfin	116	Horned dace	72
Orange-spotted sunfish	112	Black-sided darter	70
Bullhead minnow	110	Common sucker	69
Straw-colored minnow	108	Small-mouthed black bass	69

*A cross (+) in this column indicates the known occurrence of a species which is not represented in our collections from the Illinois basin.

403

SPECIES OF THE ILLINOIS BASIN, AND NUMBER OF COLLECTIONS
CONTAINING EACH—*continued*

Species	Collections	Species	Collections
Channel-cat............	108	Blackfin..............	67
Common shiner..........	105	Black-head minnow......	67
Johnny darter...........	100	Common top-minnow	66
Stone-roller.............	99	Hogsucker.............	61
Yellow bass.............	95	Grass pike.............	61
River chub.............	90	*Hadropterus phoxocephalus*	58
Blunt-nosed carp........	54	Pike....................	17
Pirate-perch............	54	*Notropis gilberti*.........	15
Sheepshead.............	53	White-nosed sucker......	14
Short-nosed gar..........	52	Trout-perch............	14
Opsopœodus emiliæ........	49	*Cottogaster shumardi*......	14
Chub-sucker.............	48	Striped sucker..........	13
Small-mouth buffalo	46	Red-bellied dace........	13
Boleosoma camurum.......	45	Sauger.................	13
Common bullhead........	42	*Boleichthys fusiformis*.....	13
Quillback carp...........	39	Silvery lamprey.........	12
Rainbow darter..........	39	Menona top-minnow.....	11
Short-headed red-horse....	39	Fan-tailed darter.......	11
Long-eared sunfish........	37	River carp.............	11
White bass..............	36	Least darter	10
Rock bass..............	35	Lake carp..............	10
Log-perch..............	35	Paddle-fish.............	8
Stonecat...............	32	Toothed herring........	8
Notropis cayuga...........	29	*Notropis rubrifrons*.......	8
Red-mouth buffalo.......	28	Storer's chub...........	7

SPECIES OF THE ILLINOIS BASIN, AND NUMBER OF COLLECTIONS
CONTAINING EACH—*concluded*

Species	Collections	Species	Collections
Dogfish	27	Sand darter	7
Lepomis miniatus	24	Blue-breasted darter	6
Mud-cat	22	Freckled stonecat	5
Notropis jejunus	21	Miller's thumb	5
Banded darter	21	Black-nosed dace	4
Long-nosed gar	20	*Ericymba buccata*	4
Pike-perch	20	Skipjack	3
Mud-minnow	18	Spotted shiner	3
Mongrel buffalo	17	*Lepomis ischyrus*	3
Hadropterus evermanni	3	Brindled stonecat	1
Burbot	3	Slender stonecat	1
Notropis phenacobius	2	Brook stickleback	1
Silver chub	2	Round sunfish	1
Lepomis symmetricus	2	*Lepomis euryorus*	1
Notropis anogenus	2	*Hadropterus scierus*	1
N. illecebrosus	2	Lake sturgeon	+
Viviparous top-minnow	1	Shovel-nosed sturgeon	+
Mooneye	1	Alligator-gar	+
Black-horse	1	Eel	+
Placopharynx duquesnei	1	*Ictalurus anguilla*	+
Notropis pilsbryi	1	Muskallunge	+
Hybopsis hyostomus	1	Green-sided darter	+
Blue cat	1		

Of the twenty-three Illinois species which have not been taken by us in the Illinois River or its tributaries, two are distinctively western fishes, and occur but rarely anywhere within our limits; nine are southern species, few of which have been found as far north as the mouth of the Illinois, and one other is only southern in this state; two are northern species which barely reach our borders; five are typical fishes of the Great Lakes; one has been found by us only in the main Mississippi and the Ohio; one is a subterranean fish of strictly local occurrence; and the two remaining species are very rare in this state.

Further particulars as to the species of these various geographical groups are given in the following classified list.

ILLINOIS SPECIES NOT FOUND IN THE ILLINOIS BASIN

WESTERN (2):
Hybognathus nubila
Flat-headed chub

SOUTHERN (10):
Harelipped sucker
Pigmy sunfish
Round sunfish
Eupomotis heros
Hadropterus ouachitæ
H. evides
Crystallaria asprella
Etheostoma obeyense
E. squamiceps
Brindled stonecat

GREAT LAKES (5):
Whitefish
Lake herring
Lake trout
Cottus ricei
Uranidea kumlienii

NORTHERN (2):
Long-nosed sucker
Nine-spined stickleback

MAIN MISSISSIPPI (1):
White sturgeon

SUBTERRANEAN (1):
Chologaster papilliferus

RARE IN ILLINOIS (2):
Brook lamprey
Long-nosed dace

As the Illinois basin contains 128 of the 150 species taken by us in the state, it is evident that the other and smaller basins must differ from this negatively rather than positively. Being not only much smaller, but also much less complex than the Illinois district, and offering less variety of situations for fishes as homes and places of resort, they may lack many species which find a fit environment somewhere in the Illinois or its dependent waters, but can contain relatively few not found there as well.

Regarded from this standpoint, the Michigan district is farthest removed from the Illinois ichthyologically, and of its fifty-seven species nine (16 per cent.) are wanting in the Illinois basin. The Cairo

district differs much less, eight of its one hundred and one fishes
being without representation in our collections from the Illinois system.   Next follows the Wabash basin in Illinois, with ninety-five
species and a difference from the Illinois basin of 6.1 per cent.; the
Galena district, with forty-four species and a difference of 4.6 per
cent.; the Saline district, with fifty-five species, and a difference of
3.8 per cent.; and the Mississippi and its marginal area, with ninety-
seven species, 3.2 per cent. of which are wanting to the Illinois
streams and lakes.   The Kaskaskia and the Big Muddy, on the other
hand, which are scarcely more than extensions of the Illinois district
downward to the southern end of the state, contain virtually no fishes
not in the main district, the Kaskaskia but one out of sixty-nine (1.4
per cent.), and the Big Muddy none out of forty-two species.   The
Rock River district differs from the Illinois by only three species out
of ninety-two (3.2 per cent.).   These data are presented more compactly in the table following.

DIFFERENCES BETWEEN THE SMALLER DISTRICTS AND THE ILLINOIS BASIN

Districts	Species in district	Species not found in Illinois basin	Ratios of difference
Illinois	128	—	—
Michigan	57	9	.16
Cairo	101	8	.08
Wabash	95	6	.061
Galena	44	2	.046
Saline	55	2	.038
Mississippi	97	3	.032
Rock River	92	3	.032
Kaskaskia	69	1	.014
Big Muddy	42	0	.000

Five species were found in the Illinois system and not in any other—three of them minnows of the genus *Notropis* (*anogenus*, *phenacobius*, and *pilsbryi*), one of them a sunfish (*Lepomis euryorus*), and one of them a darter (*Hadropterus evermanni*). All of these species have been very rare in our collections, occurring only from one to three times each, and it was probable that they would be found, if at all, where the largest number of collections was made.

The Galena district is distinguished from the Illinois basin especially by the presence of a minnow and a darter (*Hybognathus nubila* and *Crystallaria asprella*), the latter southern in its main range, and the former western, not occurring, indeed, farther east than western Illinois. These two fishes appear in the Rock River basin also, together with another distinctively western darter (*Hadropterus evides*). In the Michigan district, besides the five lake fishes already referred to—the whitefish, the lake herring, the lake trout, and two cottoids or miller's thumbs, *Cottus ricei* and *Uranidea kumlienii*—are the brook lamprey, the long-nosed sucker, the Great Lake catfish, and one of the sticklebacks (*Pygosteus pungitius*). All but the lamprey (which is rare in Illinois) are northern species not taken by us in the Illinois valley. The Mississippi district is distinguished from the Illinois by the presence of the rare white sturgeon (*Parascaphirhynchus albus*), hitherto taken only in the Mississippi itself, and by a southern darter and a western minnow already referred to. In the Kaskaskia district we find another southern darter (*Etheostoma squamiceps*). The six fishes of the Wabash district not found in the Illinois or its tributaries, are all southern species. The Big Muddy list contains no species not found in the Illinois basin; and the Saline River district contains two southern darters (*Etheostoma squamiceps* and *E. obeyense*). And, finally, among the eight species by which the Cairo district differs from the Illinois are three southern and two western species, a cave-fish, and two species of general distribution but rare in Illinois (*Lampetra wilderi* and *Rhinichthys cataractæ*).

Thus, of the twenty-three Illinois fishes not found by us in the waters of the Illinois basin, eight are distinctively southern, six are purely northern, if we include in this number the Great Lake fishes, four are western, one is an extremely local cave-fish, and four are so rare in Illinois that their appearance in any waters is a matter of unusual chance. The limitation upon the range of these imperfectly distributed species is thus climatic and general, and not geographic or local.   This state lies on the extreme borders of their proper territory, and they are not found more commonly in our waters because climatic and other general conditions most favorable to their maintenance, here reach the vanishing point.

LISTS OF SPECIES DISTINGUISHING DIFFERENT DISTRICTS FROM THE ILLINOIS BASIN

GALENA DISTRICT (2):
  Hybognathus nubila (Western)
  Crystallaria asprella (Southern)

ROCK RIVER DISTRICT (3):
  Hybognathus nubila (Western)
  Hadropterus evides (Western)
  Crystallaria asprella (Southern)

MICHIGAN DISTRICT (9):
  Brook lamprey (rare)
  Long-nosed sucker (Northern)
  Whitefish (Great Lakes)
  Lake herring (Great Lakes)
  Lake trout (Great Lakes)
  Great Lake catfish (Northern)
  Nine-spined stickleback (Northern)
  Cottus ricei (Great Lakes)
  Uranidea kumlienii (Great Lakes)

MISSISSIPPI STRIP (3):
  White sturgeon (rare; Mississippi only)
  Hybognathus nubila (Western)
  Crystallaria asprella (Southern)

KASKASKIA RIVER DISTRICT (1):
  Etheostoma squamiceps (Southern)

WABASH DISTRICT (6):
  Harelipped sucker (rare; Southern)
  Pigmy sunfish (Southern)
  Eupomotis heros (Southern)
  Hadropterus ouachitæ (Southern)
  Crystallaria asprella (Southern)
  Etheostoma squamiceps (Southern)

SALINE RIVER DISTRICT (2):
  Etheostoma obeyense (Southern)
  E. squamiceps (Southern)

CAIRO DISTRICT (8):
  Brook lamprey
  Hybognathus nubila (Western)
  Long-nosed dace (rare in Illinois)
  Flat-headed chub (Western)
  Chologaster papilliferus (subterranean)
  Pigmy sunfish (Southern)
  Eupomotis heros (Southern)
  Etheostoma squamiceps (Southern)

RELATIONS OF EACH DISTRICT TO ALL THE OTHERS

In the foregoing discussions and analyses the fishes of the various districts have been compared with those of the largest and most central district as a type; but a fuller and more accurate idea of the composition of the fish population of Illinois and of its relations in the various hydrographic divisions of the state may be obtained by a comparison of the species of each of our ten districts successively with those of all the others.   This may be done in an exact and uniform manner by determining for each pair of districts the ratio which the number of species common to the pair bears to the whole number of species occurring within the area of both the districts taken together as one.   In the Galena district, for example, there are 44 species recorded, and in the Saline River basin there are 55, a total of 99; but as 26 of these species have been found in both these districts, this number has been taken twice in the above addition, and the number of species found by us in the entire area of these two districts is consequently 73. The ichthyological affinity of these two areas is evidently to be measured by the ratio which the number of species common to both bears to the whole number of species found in either or both the areas—in this case, the ratio of 26 to 73, or 36 per cent. That is, 36 per cent. of the fishes found in either of these two districts have been found by us in both of them.

A similar analysis of the data for each of the forty-five pairs which it is possible to make up from our ten hydrographic districts, yields the material for the following table of common species and of ratios of affiliation. This table shows, in the lower left-hand part,

NUMBER OF SPECIES COMMON TO EACH PAIR OF DISTRICTS, AND RATIOS
OF SUCH COMMON NUMBERS TO THE WHOLE NUMBER
OF SPECIES IN EACH PAIR

Districts	1. Galena	2. Rock River	3. Illinois R.	4. Michigan	5. Mississippi	6. Kaskaskia	7. Wabash	8. Big Muddy	9. Saline	10. Cairo	Averages
1. Galena.............	....	45	32	20	41	40	38	28	36	37	.352
2. Rock River.........	42	....	68	35	69	59	63	40	47	62	.542
3. Illinois River	42	89	....	35	72	53	66	33	41	68	.52
4. Michigan...........	17	39	48	....	34	25	29	22	23	32	.283
5. Mississippi..........	41	77	94	39	....	54	61	34	42	66	.525
6. Kaskaskia	32	60	68	25	58	....	66	52	63	53	.517
7. Wabash............	38	72	89	34	73	66	....	41	53	63	.534
8. Big Muddy..........	19	38	42	18	35	38	40	....	70	39	.398
9. Saline River........	26	47	53	21	45	48	52	40	....	49	.471
10. Cairo..............	39	74	93	38	79	59	76	40	51	....	.521
Total species...........	44	92	128	57	97	69	95	42	55	101	
Number of collections.....	13	73	1115	20	57	41	103	10	18	95	

the number of species common to each pair of districts, and in the upper right-hand part the ratios which these numbers bear to the number of species occurring in each pair of districts taken as one. The number of species common to any two districts will be found in the lower left-hand part of the table, where the column for one district intersects with the line for the other, and the ratio of affiliation for the same pair of districts will be found in the opposite part of the table at the intersection of the line for the first with the column for the second. A simple inspection of the figures in the latter part shows at once which districts are most alike and which are most unlike in respect to their fish inhabitants. Thus, the Rock and Illinois basins and the Mississippi are the most closely related, according to these data, with affiliation ratios of 68–72 per cent. and an average of 70; and the Michigan, Galena, and Big Muddy districts are the least alike, with ratios of 20–28 per cent. and an average of 23. The two highest single ratios of ichthyological affiliation are those of the Illinois and Mississippi rivers (.72) and of the Big Muddy and Saline (.70).

The data of this table may be generalized by bringing into comparison the *average* of the ratios of affiliation for each district with those for all the rest, as shown in the column of figures farthest to the right. If the ten districts are arranged in the order of the size of their average ratios, they readily fall into two groups, the first of six districts, with relatively high ratios, and the second of four, with relatively low ratios. The first group comprises the basins of the larger rivers—the Mississippi, the Rock, the Illinois, the Kaskaskia, the Wabash, and the Ohio, each with its more or less complex system of tributaries. The average ratio for this group is 52.7 per cent. The second group is made up of small, widely separated districts, containing only small streams and lakes, except that one of them includes a little of the shallow southwestern border of Lake Michigan. In this group are the northwestern driftless area, the Saline River and its tributaries, the Big Muddy district, and the Michigan district, with an average affiliation ratio of 37.6.

If we average separately, for these groups, the ratios of each district to all the other districts of its group, we obtain for the first and higher group a ratio of mutual affiliation of 63 per cent., and for the lower group a similar ratio of 33 per cent. It is thus made clear that the districts most typical of our Illinois fauna are the first six

above mentioned, while those most individual and peculiar—least
closely affiliated among themselves and each with all the others—
are the Michigan, the Galena, the Saline, and the Big Muddy districts, excepting only the relation of the two last mentioned, which,
as already said, is unusually close.

THE FISHES OF NORTHERN, CENTRAL, AND SOUTHERN ILLINOIS

If mere difference in latitude, involving a climatic difference
within a range of five and a half degrees, limits the distribution
of any of our fishes, the fact should appear upon a comparison of
the species list of the northern, central, and southern sections of the
state, although due caution must, of course, be exercised that
other and more local causes are not confused with climatic ones.
The division of the state here adopted, is shown on Map I. of the
accompanying set.

The fishes of these three divisions number 119 species for
northern, 123 for central, and 119 for southern Illinois, respectively. Fourteen species have been found by us only in the northern
division, 9 only in the southern, and 5 only in the central, and 89 species are found in all three sections. Twelve species occur in both
northern and central Illinois, but not in southern, 17 in both southern and central Illinois, but not in northern, and 4 in both the northern and southern divisions of the state, but not in the central.

FISHES OF LIMITED DISTRIBUTION IN ILLINOIS

Illinois Distribution	General Distribution
Species Peculiar to Northern Illinois	
Whitefish	Great Lakes
Lake herring	" "
Lake trout	" "
Long-nosed sucker	Northern
Notropis anogenus	"
N. phenacobius	
N. pilsbryi	Southern
Great Lake catfish	Northern
Muskallunge	"
Brook stickleback	"
Nine-spined stickleback	"
Hadropterus evides	Rather general
Cottus ricei	Great Lakes
Uranidea kumlienii	" "
Species Peculiar to Southern Illinois	
Harelipped sucker	Southern
Long-nosed dace	General; rare in Illinois
Flat-headed chub	Western
Chologaster papilliferus	Local; cave
Pigmy sunfish	Southern
Round sunfish	"
Eupomotis heros	"
Hadropterus ouachitæ	"
Etheostoma obeyense	"

FISHES OF LIMITED DISTRIBUTION IN ILLINOIS—*concluded*

Illinois Distribution	General Distribution
Species in Northern and Central Illinois, but not in Southern	
Lake carp	Northern
Notropis cayuga	General
N. rubrifrons	"
Hybopsis hyostomus	"
Stonecat	Northern and southwestern
Pike	Northern
Menona top-minnow	"
Trout-perch	"
Lepomis ischyrus	
Sauger	General
Yellow perch	Northern
Burbot	Great Lakes
Species in Southern and Central Illinois, but not in Northern	
Paddle-fish	General
Shovel-nosed sturgeon	"
Alligator-gar	Southern
Mooneye	Northern
Black-horse	General
Ericymba buccata	"
Silver chub	"
Blue cat	Southern
Ictalurus anguilla	"
Freckled stonecat	"
Brindled stonecat	General

FISHES OF LIMITED DISTRIBUTION IN ILLINOIS—*concluded*

Illinois Distribution	General Distribution
Viviparous top-minnow	Southern
Lepomis symmetricus	"
Cottogaster shumardi	General
Green-sided darter	"
Etheostoma squamiceps	Southern

An examination of the general distribution of the species of these sectional lists of Illinois fishes shows, as was to have been expected, that the distinctively northern Illinois fishes are chiefly northern in their outside range, and that those of southern Illinois are mainly southern. Thus, of the 14 especially northern Illinois fishes, 11 are northerly in their general distribution and 1 is southerly; while of the 9 distinctively southern Illinois species, 6 are southerly in their general range, 1 is western, and 1 is a cave-fish local to Illinois. The species found in the northern and central sections of the state and not in the southern are varied in their distribution, 6 of them ranging northward from Illinois, and 4 of them in all directions, while 1 has been thus far found in Illinois only. The central and southern fishes, on the other hand, comprise 7 southern species, 1 of northern and 8 of general range, and 1 whose distribution is not recorded. Including only species whose general area shows that their restricted occurrence in Illinois is a feature of their geographical distribution at large, and excluding fishes special to the Great Lakes, we have twenty-six species whose distribution in this state seems limited by conditions connected with differences in latitude merely—twelve of these species essentially northern and fourteen of them southern.

ESPECIALLY NORTHERN SPECIES IN
ILLINOIS (16):

Whitefish
Lake herring
Lake trout
Long-nosed sucker
Lake carp
Notropis anogenus
Great Lake catfish
Mooneye
Pike
Muskallunge
Menona top-minnow
Brook stickleback
Nine-spined stickleback
Trout-perch
Cottus ricei
Uranidea kumlienii

ESPECIALLY SOUTHERN SPECIES IN
ILLINOIS (14):

Alligator-gar
Blue cat
Ictalurus anguilla
Freckled stonecat
Harelipped sucker
Notropis pilsbryi
Viviparous top-minnow
Pigmy sunfish
Round sunfish
Lepomis symmetricus
Eupomotis heros
Hadropterus ouachitæ
Etheostoma obeyense
E. squamiceps

USE OF LOCALITY MAPS

In the foregoing discussion of the sectional distribution of Illinois fishes no account has been taken of differences in the frequency of the occurrence of the species in the different sections in which they have been found, a single occurrence in southern Illinois, for example, counting for as much as fifty such occurrences in the northern part of the state. That highly interesting and important peculiarities of distribution are concealed by this gross method of comparison is made evident by an examination of the maps of the distribution of our collections of the various species accompanying this report, where the data are presented in a way to show, not the number of collections, it is true, in which each species was represented, but the number and distribution of localities from which the species has been obtained. From such a study of these maps it appears that the northern half or two thirds of this state is more favorable to a considerable number of species than the southern part, since these species have been taken there in a much larger number of localities; and also that a small group of species of wide general distribution has been found by us with surprising frequency in the Wabash drainage in this state as compared with that of adjacent districts.

The preference of certain species for the northern part of Illinois over the southern is clearly illustrated by the distribution maps of the following fifteen species: *Noturus flavus, Carpiodes thompsoni, Notropis cayuga, N. hudsonius, N. rubrifrons, Hybopsis dissimilis, H. kentuckiensis, Fundulus diaphanus, Percopsis guttatus, Eupomotis gibbosus, Stizostedion canadense, Perca flavescens, Etheostoma zonale, Roccus chrysops,* and *Morone interrupta.* With few and slight exceptions, all the species of this varied list, representing eight families and twelve genera, are so definitely limited to the northern half of this state that one gets the impression, as he examines these maps in succession, that some invisible barrier to their southward dispersal exists in the neighborhood of the Sangamon River.

PECULIARITIES OF DISTRIBUTION IN THE LOWER ILLINOISAN GLACIATION

That the distribution of these more northerly species is not limited by the watersheds is shown by the fact that they range across the state indifferently into all the stream systems of northern Illinois. It is not until we compare with our distribution maps a map of the surface geology of the state (Map III.) that we find a plausible explanation of a part, at least, of this peculiar distribution, for all but one of the species above mentioned are wholly excluded from the area of this glaciation, and this excepted species (*Hybopsis dissimilis*) appears in but one locality within the lower glaciation, and that a short distance within its border, on the upper Kaskaskia.

Especially significant in this relation are several cases in which species of this list range southward in the eastern part of the state upon the upper tributaries of the Kaskaskia and the Embarras, for in so doing they simply follow southward the course of the Shelbyville moraine which forms the boundary between the Wisconsin and the lower Illinoisan glaciations in east-central Illinois. The maps for *Noturus flavus, Hybopsis dissimilis, H. kentuckiensis,* and *Stizostedion canadense* are examples.

That this coincidence of distribution and surface geology points to a true explanation is further shown by the maps for twenty-two other species which range more definitely to the southward than the foregoing twelve, but which nevertheless avoid the southern glaciation more or less completely and to an unmistakable degree. For example, 19 of our 94 collection localities for the hogsucker (*Catostomus nigricans*) lie below the Springfield parallel, but only three of them are in the lower Illinoisan glaciation, and these are barely within its borders. Of our thirty localities for the short-headed redhorse (*Moxostoma breviceps*) only two are in this glaciation, and these are near its boundaries on the Embarras and the Kaskaskia. The very abundant minnow *Campostoma anomalum* was taken by us from one hundred and sixty localities, thirty-one of which are south of the Sangamon and eight of them from the non-glaciated area of the Cairo district, but only one of the entire number is within the lower glaciation, and that is on the upper Kaskaskia, just across the limiting moraine. The map for *Notropis cornutus* shows one hundred and sixty-one localities from which collections of this species were made, ninety of them below the Sangamon and twenty-nine in the Cairo district, but only three are in the southern glaciation. Other species testifying to the same effect will be found in the following list of fishes absent from this characteristic southern Illinois district.

ILLINOIS FISHES RARE OR WANTING IN THE LOWER ILLINOISAN GLACIATION

Short-nosed gar	*N. rubrifrons*
Common bullhead	Spotted shiner
Stonecat	Storer's chub
Lake carp	River chub
Quillback carp	Pike
Common sucker	Menona top-minnow
Hogsucker	Trout-perch
Short-headed red-horse	Pumpkinseed
Stone-roller	Small-mouthed black bass
Red-bellied dace	Sauger
Notropis cayuga	Yellow perch
N. heterodon	Banded darter
Straw-colored minnow	Rainbow darter
Notropis gilberti	Fan-tailed darter
Spot-tailed minnow	White bass
Common shiner	Yellow bass
Notropis jejunus	Miller's thumb

FISHES TOLERANT OF THE LOWER ILLINOISAN GLACIATION

Dogfish	Silver chub
Channel-cat	Grass pike
Yellow bullhead	Common top-minnow
Black bullhead	Viviparous top-minnow
Mud-cat	Pirate-perch
Tadpole cat	White crappie
Brindled stonecat	Round sunfish
Chub-sucker	Warmouth
Striped sucker	Green sunfish
Silvery minnow	Long-eared sunfish
Blunt-nosed minnow	Orange-spotted sunfish
Opsopœodus emiliæ	Large-mouthed black bass
Golden shiner	Black-sided darter
Bullhead minnow	*Boleosoma camurum*
Silverfin	Sand darter
Shiner	*Etheostoma jessiæ*
Blackfin	*Boleichthys fusiformis*
Ericymba buccata	

Among the ninety-eight Illinois species for which distribution maps have been prepared, thirty-four belong clearly to this group of fishes which seem to avoid the conditions common to the flat gray lands of the southern part of the state. Thirty-five species, on the other hand, are distributed over this glaciation in a way to indicate a tolerance of its conditions if not an indifference to them, the data concerning the remaining twenty-nine species being ambiguous or indecisive in this respect.

Two facts concerning the soil and waters of the lower Illinoisan glaciation may be held to account, at least in part, for the failure of certain species of fishes to thrive in its streams. Compared with the other regions of the state, this oldest of our glaciation areas has developed its drainage system to a point such that the rainfall runs off rapidly in a large number of small streams, leaving no marshes or ponds to hold back the waters during periods of dry weather. It is a level country whose streams fill up quickly and run down rapidly, the smaller ones drying up completely during the midsummer drought, which is here more marked than farther north. These variable and temporary creeks are, of course, less favorable to the maintenance of a varied and permanent fish population than the waters of the earlier Illinoisan or the Wisconsin areas.

As a further consequence of its geological antiquity, involving degenerative chemical changes and a long-continued leaching, the soil of this lower glaciation has become an extremely fine-grained, light-colored clay which, when compact, sheds water almost completely, but which washes into the streams as a fine detritus that remains persistently in suspension and renders the waters very turbid for a long time after a rain. Standing pools, indeed, never become even approximately clear. So persistent is this turbidity, due to very finely divided matter in suspension, that the chemists of the Water Survey find it almost impossible to free the water wholly from suspended solids even by repeated filtration. Furthermore, this soil has a definitely acid reaction, to which is due a notable physical difference between the soils of this area and those of the later glaciations west and north of it. A surplus of lime in a soil coagulates or granulates it, causing its ultimate particles to cohere in larger granules, while in an acid soil this effect is entirely wanting. This lack of granulation in a very finely divided soil increases, of course, the permanent muddiness of its waters as compared with those of the other areas in which lime in the soil renders it alkaline.

The acidity of this southern soil seems not to be of a kind or amount to affect the surface waters sensibly and directly, since the water samples from this region analyzed by the State Water Survey show a soft water, slightly alkaline, and chemically unobjectionable as a medium for fishes.

CLASSIFICATION AND USE OF ECOLOGICAL DATA

That these conditions are a part, at least, of the cause of the phenomenal distribution of southern Illinios fishes may be shown by a comparison of our ecological data for the fishes of the two lists—one composed of those adapted to the conditions of the lower Illinoisan glaciation and the other of those avoiding them. In the organization of the data of our collections of Illinois fishes, those concerning the character of the water body in which collections were made were classified in a way to show the number of collections of each species taken from each class of situation. By reducing these numbers to ratios of frequency of occurrence, we have a means of exhibiting the preference of species with respect to the situations in which each occurs. *Pimephales notatus*, for example, was found twenty times over a muddy bottom to thirty-four over a bottom of mud and sand, and to forty-six over a bottom of rock and sand. *Aphredoderus sayanus*, on the other hand, was found sixty-two times on a muddy bottom to nineteen times in each of the other situations.

By tabulating data of this description separately for each of the two lists of species referred to—thirty-four species in the one list and thirty-five in the other—and averaging the ratios for each group separately, significant evidence was obtained of the factors which affect the distribution of these fishes.

The species which distribute themselves freely over southern Illinois are those which are generally tolerant of turbid waters, as shown by the fact that 32 per cent. of all our collections of this group came from muddy streams and ponds, 34 per cent. from situations where the bottom was composed largely of rock and sand, and 24 per cent. from a bottom of sand and mud. The species avoiding the central area of southern Illinois, on the other hand, are, as a rule, intolerant of muddy waters, only 10 per cent. of all our data-bearing collections of this group coming from such situations, while 61 per cent..of them were from bottoms of rock and sand, and 29 per cent. from those of sand and mud. It is consequently clear that the suspended detritus of the streams of southern Illinois and the clay and mud of which their banks and bottoms are commonly composed, are an important part, at least, of the cause of the smaller variety of fishes in these waters; and these conditions trace back through the character of the soil to the geological history of the central part of southern Illinois.

FISHES OF THE OHIO AND OF THE MISSISSIPPI DRAINAGE

A comparison and classification of our distribution maps from another point of view enables us further to distinguish two rather definite groups of species coincident in great measure, but not wholly so, with the two groups which we have found in an opposite relation to the lower Illinoisan glaciation. No less than 27 of our species have either an exclusive or at least a strongly preponderant distribution in the Mississippi drainage in the western and northern parts of the state, while 8 species, on the other hand, are very definitely preponderant in the Ohio drainage in the southern and eastern parts. Nineteen of the 27 species of the first list are also on the list of species excluded from the region of the lower Illinoisan glaciation, while 6 of the 8 species of the second list are also on that of species distributed freely through this southern Illinois district. We have evidence here of another influence strongly affecting distribution, coincident in part with that already discussed, but independent of it also in part, the two causes, or sets of causes, operating together to determine the actual range of most of the species of limited distribution in this state.

The impression produced by an examination of the two sets of maps for the fishes above mentioned, is that of a small group of species, on the one hand, which enter the state from the south and east by way of the Wabash and the smaller tributaries of the Ohio, and, on the other hand, of a much larger group, most of which have entered the state from the west and north, making their way to its interior mainly by the Illinois and the Rock, but sometimes by the Kaskaskia and the Big Muddy also. Species of the Ohio group sometimes seem to spread into the headwaters of adjacent streams, especially into the branches of the Kaskaskia where these come nearest to the Embarras, and into those of the Big Vermilion of the Illinois which are nearest to the Little Vermilion of the Wabash. Some species, however, remain carefully within the tributaries of the Wabash system.

It seems possible that this appearance of an approach to the state and entrance upon its territory from opposite directions is not altogether deceptive, and that the annual movements of the fishes of the state, up the streams at the time of the spring floods, downwards with the recession of the waters, and still farther downwards, for many species, into deeper water in the winter, may take these two contingents of our fish population in opposite directions, from and towards local centers of population for the species, situated on opposite sides of the state. Whether and where such local centers of population actually exist, is a question which can not be answered definitely for lack of numerical or statistical data in the faunal lists and other literature of geographical distribution for the surrounding states. If they exist, the Wabash fishes would constitute one such system, and those of the Mississippi and its tributaries, another.

If we may speculate still further upon this subject, we may perhaps surmise that a general critical analysis of the fish population of the larger area of which Illinois forms the central part, would enable us to distinguish fairly well-defined districts, each with its characteristic assemblage of prevalent species, so associated and ecologically related as to form a balanced assemblage of species, all so adjusted to each other and so advantageously placed in their environment as to constitute a closed system, which the characteristic species of adjacent areas can not enter, or in which they can not permanently remain.

DISTRIBUTION CHIEFLY IN THE OHIO DRAINAGE

Brindled stonecat	Pirate-perch
Green-sided darter	*Notropis illecebrosus*
Boleichthys fusiformis	*Ericymba buccata*
Chub-sucker	Long-eared sunfish

DISTRIBUTION CHIEFLY IN THE MISSISSIPPI DRAINAGE

Short-nosed gar	White bass
Stonecat	Yellow bass
Lake carp	Common bullhead
Notropis cayuga	Short-headed red-horse

Spot-tailed minnow
Notropis rubrifrons
Spotted shiner
Pike
Menona top-minnow
Trout-perch
Pumpkinseed
Sauger
Yellow perch
Banded darter

Red-bellied dace
Notropis gilberti
Long-nosed gar
Dogfish
Mongrel buffalo
Black-head minnow
Hybognathus nubila
Redfin
Rock bass

BOUNDARY BETWEEN NORTHERN AND SOUTHERN SPECIES

Recurring next to the distinction made on another page between northern and southern fishes whose areas extend into Illinois but not beyond, and comparing the distribution of these groups within the state, as given on Map CIII., we see that northern and southern species meet and mingle in the western part of the state from Meredosia to Pekin on the Illinois, and from Quincy to Dallas City on the Mississippi, but that in eastern Illinois they are separated by a wide interval extending from Cook county to the mouth of the Embarras, in which interval we have never taken any representative of either group.

The distinctively southern species, although most abundant south of the line 28° 30″, nevertheless go up the Wabash to the Embarras, up the Kaskaskia to Shelby county, up the Mississippi to Henderson county, and up the Illinois to Pekin, also following the branches of the Sangamon to Logan county. The northern species, on the other hand, although most abundant above 40° 20″, come down the Illinois to Meredosia, and down the Mississippi to Quincy.

The boundary between the northern and southern species thus appears as a broad belt some fifty miles in width, extending two thirds of the way across the state just above its center, but widening to a distance of one hundred and seventy-five miles on the eastern boundary.

GENERAL FEATURES OF ECOLOGICAL DISTRIBUTION

In addition to the general distribution of Illinois fishes over the North American continent, their general or partial distribution within the state, and the unevenness of their distribution over the different divisions of the state, hydrographic, climatic, and geological, there are also recognizable differences and inequalities of distribution corresponding to the size of the water bodies in which the species are found, to the nature of the bottom and the consequent clearness and purity of the waters, and to the existence and rate of current or flow in the waters inhabited by them. In this class of divisions, geological distribution merges into ecological relation, the distribution of species being no longer by geological areas, but by ecological situations. In this sense two species may occupy precisely the same territory without ever coming into any effective contact with each other, because they are differently related to certain features of their environment.

As an explanation of the more general facts of distribution requires an analysis and interpretation of continental, terrestrial, and even cosmic agencies affecting it, so an understanding of what we may call the ecological distribution of a species requires a corresponding analysis of the ecological features of the region. Such an analysis can here be carried but a little way, since the ecological data borne by our collections are only of a very general type; but such as they are, they may, if used with discretion, add definiteness and detail and some degree of statistical precision to our knowledge of this part of the subject.

My statistics of associate occurrence exhibit in the most interesting manner the frequent tendency of closely allied species inhabiting the same territory to avoid each other's company and thus to evade competition with one another by the choice of different haunts and situations within the area of their common habitation. In consequence of this tendency, we sometimes find widely unlike species more closely and commonly associated in our collections than like, the ecological repulsion of each for its similars bringing dissimilars together into more or less definite associate groups. The sunfishes proper, for example—that is, the *Centrarchidæ* exclusive of the black bass—although a homogeneous group of species as to form and external structure, are a diverse assemblage as to ecological relationships. If we compare the proportionate frequency with which the closely similar species of the genus *Lepomis* have been taken together in our collections—in the same haul of the net, or from the same situation at the same time—with the frequency of associate occurrence of the widely dissimilar species of the other genera of the family, we find that the unlike species have been taken together much more frequently than the like—in a ratio of $1\frac{1}{2}$ to 1,—that the species of *Lepomis* have, indeed, been taken in company with species of other genera considerably more frequently than with each other. The sunfishes, consequently, are not an associate group, but tend to disperse themselves over a large variety of ecological situations, those least like each other being most likely to meet on common ground where their unlike capacities enable them to live together in a non-competitive way. Other striking examples of this reaction might be pointed out in the suckers, the minnows, the catfishes (especially the bullheads), and the top-minnows.

Ninety-seven of our species have been collected in large enough numbers, and from a sufficient variety of locations, to give us data for comparison with reference to the general character and size of the water bodies which they prefer; 62 species furnish available data concerning the bottom or substratum of these water bodies; and 49 species, data concerning current and rate of flow. The numbers of collections for the various species covered by these figures vary greatly from a minimum of 10 collections of a species to a maximum of 376. Unfortunately, the larger and more important fishes are commonly represented by the smaller numbers of collections, and statements made concerning these are less likely to be found fairly accurate and generally correct than are those concerning the smaller fishes, represented by larger numbers of collections.

One available set of our data may best be presented in tabular form, for such use as the student may wish to make of them; and to this table we add, as an illustration of its use, only a few statements concerning the more conspicuous ecological groups of our Illinois fishes.

By assorting the species according to the size of the ratios of frequency of occurrence for each class of situations distinguished in this table, we may separate those strongly preferring the given situation from those apparently avoiding it. In this way we learn that the species occurring in our collections with disproportionate frequency in the larger rivers of the state are the mud-cat (*Leptops olivaris*), one of the river carp (*carpio*), the toothed herring (*Hiodon tergisus*), and the sheepshead (*Aplodinotus*), among the larger fishes; and a small darter (*Cottogaster shumardi*), the trout-perch (*Percopsis guttatus*), and a minnow (*Hybopsis dissimilis*) among the smaller fishes.

The principal larger fishes of the smaller rivers make a much longer list, comprising the hogsucker, two of the native carp (*velifer* and *difformis*), a species of red-horse (*aureolum*), the rock bass, and the small-mouthed black bass; and the principal smaller species are six darters (*Etheostoma zonale, Hadropterus phoxocephalus, H. aspro, Diplesion blennioides, Etheostoma cœruleum*, and *Ammocrypta pellucida*), a stonecat (*Noturus flavus*), and *Hybopsis kentuckiensis*, and four other minnows, all of the genus *Notropis* (*rubrifrons, gilberti, blennius,* and *cornutus*)—their ratios running from 70 per cent. for *rubrifrons* to 41 per cent. for *cornutus*.

The species of our list which have from 50 to 100 per cent. of their representatives in creeks, as illustrated by our collections, include three sunfishes (the green sunfish, the round sunfish, and the long-eared sunfish), three suckers (the common sucker, the chubsucker, and the striped sucker), four darters, ten minnows, and the brindled stonecat.

The larger species found most abundantly in lakes, ponds, and other stagnant waters were the common bullhead, the buffaloes, the yellow perch, the white bass, the yellow bass, the large-mouthed black bass, and five sunfishes (both crappies, the warmouth, the pumpkinseed, and the bluegill); and the smaller kinds were the smallest of our fishes (*Microperca punctulata*), another darter (*Boleichthys fusiformis*), two minnows (*Notropis cayuga* and *N. heterodon*), the mud-minnow, and a killifish (*Fundulus dispar*).

Turning next to the 62 species for which our data of preference or avoidance of a muddy bottom are available, we find 7 species whose ratios of frequency of occurrence in such situations range from 43 to 88 per cent., and which may consequently be called limophagous fishes. These are the warmouth sunfish, the black and the yellow bullheads, the pirate-perch, a single darter (*Boleosoma camurum*), and two minnows, the golden shiner and the common shiner (*Notropis cornutus.*)

It is interesting to find, by an examination of our maps, that all these 7 species are freely distributed over the lower Illinoisan glaciation of the southern part of the state, where, as we have already shown, only fishes indifferent to a peculiarly persistent turbidity of the water are likely to occur.

By selecting from this same list of 62 species those with the lowest ratios of frequency over a muddy bottom, we get 13 species (with ratios of 4 to 10 per cent.) which evidently avoid such situations; and these, again, are without exception so distributed that the area of the lower Illinoisan glaciation is almost never entered by them. These are one of the native carp (*velifer*), a species of red-horse (*aureolum*), the small-mouthed black bass, two darters (*Hadropterus phoxocephalus* and *Etheostoma cœruleum*), five minnows (*Campostoma anomalum, Notropis heterodon, Ericymba buccata, Hybopsis kentuckiensis,* and *Notropis blennius*), two stonecats, and the little brook silverside (*Labidesthes*).

A more precise statement and a fuller discussion of the ecological relations of our fishes, including statistics of companionship for the various species, as shown by the frequency of their joint occurrence in collections, must be left for later contributions.

Attention may be profitably called, in conclusion, to the economic significance of the details of distribution of the various species as influenced both by geographical and ecological conditions, since a proper understanding and application of these facts will prevent wasteful efforts to introduce species where they do not belong and can not thrive. Indeed, the more detailed our knowledge of favorable, and even optimum, conditions for the different species, and the more exact, also, our acquaintance with the relations of each species of fish to its companion species in any associate assemblage, the more intelligent, and hence the more successful, in the long run, will be our efforts to extend the range and multiply the numbers of the more useful species and to lessen the numbers of those especially injurious.

ECOLOGICAL TABLE
ALL ILLINOIS SPECIES WITH AT LEAST TEN AVAILABLE RECORDS EACH*

Jordan and Evermann Nos.	Species	Water (97 species)					Current (49 species)				Bottom (62 species)			
		Available collections	Larger rivers	Smaller rivers	Creeks	Lakes, ponds, etc.	Available collections	Swift to moderate	Sluggish to stagnant	Variable	Available collections	Mud	Rock and sand	Mud and sand
151	Long-nosed gar.......	35	25	19	7	22	....	....	....	....	....	....	....	....
152	Short-nosed gar.......	57	28	24	4	25	....	....	....	....	....	....	....	....
155	Dogfish..............	37	18	7	6	30	....	....	....	....	....	....	....	....
207	Channel-cat..........	171	20	32	27	8	31	68	19	13	75	21	44	35
215	Yellow bullhead......	122	7	6	37	23	14	36	43	21	35	43	34	23
217	Common bullhead....	48	15	5	4	44	....	....	....	....	....	....	....	....
218	Black bullhead........	244	8	21	37	26	38	37	53	10	56	54	46	....
221	Mud-cat.............	30	53	21	5	8	....	....	....	....	....	....	....	....
222	Stonecat.............	41	10	53	34	....	15	60	13	26	24	8	58	34
223	Tadpole cat..........	193	17	5	23	41	21	48	43	9	45	29	27	44
231	Brindled stonecat.....	30	3	36	60	....	....	....	....	....	13	8	62	30
261	Red-mouth buffalo......	39	13	....	9	48	....	....	....	....	....	....	....	....
262	Mongrel buffalo.......	19	17	....	7	45	....	....	....	....	....	....	....	....
264	Small-mouth buffalo..	52	14	12	4	49	....	....	....	....	....	....	....	....
265	River carp...........	15	47	....	8	10	....	....	....	....	....	....	....	....
266	Blunt-nosed carp.....	102	9	42	30	12	16	50	25	25	47	21	36	43
268	Quillback carp.......	70	10	50	19	5	19	47	32	21	28	4	60	36

*The figures of this table, except those in the columns for available collections, are ratios of frequency of the species in our collections, computed with due reference to the comparative numbers of collections of all kinds made in each situation.

ECOLOGICAL TABLE—*continued*

ALL ILLINOIS SPECIES WITH AT LEAST TEN AVAILABLE RECORDS EACH

Jordan and Evermann Nos.	Species	Water (97 species)					Current (49 species)				Bottom (62 species)			
		Available collections	Larger rivers	Smaller rivers	Creeks	Lakes, ponds, etc.	Available collections	Swift to moderate	Sluggish to stagnant	Variable	Available collections	Mud	Rock and sand	Mud and sand
289	Common sucker	132	3	19	71	1	49	39	47	14	79	13	44	43
294	Hogsucker	•99	4	63	25	4	71	20	63	17	59	....	54	46
302a	Chub-sucker	131	9	12	57	14	23	52	48	....	57	32	39	29
303	Striped sucker	46	2	31	53	3	....	....	....	....	19	26	32	42
305	White-nosed sucker	18	7	44	20	6	....	....	....	....	....	....	....	....
314	Common red-horse	143	9	32	40	4	47	57	28	15	65	6	55	39
319	Short-headed red-horse	55	13	25	15	22	....	....	....	....	14	14	43	43
328	Stone-roller	195	3	37	55	1	65	63	23	14	105	7	57	36
334	Red-bellied dace	23	10	....	71	....	....	....	....	....	....	....	....	....
340	Silvery minnow	183	12	36	32	7	30	47	40	13	67	33	40	27
349	Black-head minnow	95	14	30	48	4	12	50	42	8	44	25	41	34
350	Blunt-nosed minnow	376	5	34	43	12	108	50	34	16	202	20	46	34
355	Horned dace	151	4	28	63	2	42	48	36	16	81	17	47	36
391	*Opsopœodus emiliæ*	40	13	6	36	32	....	....	....	....	....	....	....	....
394	Golden shiner	303	12	17	29	32	28	32	57	11	82	44	29	27
398	Bullhead minnow	187	17	31	28	7	36	67	17	16	62	11	44	45
405	*Notropis cayuga*	29	....	13	26	57	13	54	38	8	15	....	27	73
406	*N. heterodon*	92	19	1	19	60	....	....	....	....	14	7	22	71
408	Straw-colored minnow	185	7	44	37	3	63	49	26	25	103	10	50	40

ECOLOGICAL TABLE—*continued*
ALL ILLINOIS SPECIES WITH AT LEAST TEN AVAILABLE RECORDS EACH

Jordan and Evermann Nos.	Species	Water (97 species)					Current (49 species)				Bottom (62 species)			
		Available collections	Larger rivers	Smaller rivers	Creeks	Lakes, ponds, etc.	Available collections	Swift to moderate	Sluggish to stagnant	Variable	Available collections	Mud	Rock and sand	Mud and sand
420	*Notropis gilberti*.......	30	2	49	43	2	....	....	....	....	18	11	45	44
426	*N. illecebrosus*........	11	....	....	100	....								
428	Spot-tailed minnow...	147	28	5	2	39	....	....	....	....	10	20	80	....
432	Redfin..............	163	24	32	20	14	13	46	38	16	55	27	40	33
448	Silverfin.............	268	6	39	40	4	65	54	26	20	126	13	56	31
456	Common shiner.......	178	2	41	50	4	76	45	36	19	102	44	48	8
476	*Notropis jejunus*......	51	27	19	13	11	....	....	....	....	12	25	67	8
485	Shiner..............	206	20	36	15	11	23	57	30	13	48	21	64	14
489	*Notropis rubrifrons*....	13	4	70	26	....	11	45	18	36	11	....	82	18
498a	Blackfin.............	208	3	32	65	....	69	41	45	14	109	17	43	40
499	*Ericymba buccata*.....	74	1	18	81	....	14	43	29	28	38	8	63	29
501	Sucker-mouthed minnow...............	159	5	36	53	1	53	53	24	23	92	15	51	34
528	Spotted shiner........	11	50	27	22	....								
533	Silver chub..........	41	5	29	66	....	13	77	15	7	20	30	55	15
534	Storer's chub........	28	21	32	11	10	....	....	....	....				
536	River chub..........	129	4	41	51	1	55	53	24	23	74	8	43	49
674	Toothed herring......	10	46	....	....	16	....	....	....	....				
677	Gizzard-shad........	105	17	32	7	20	....	....	....	....	22	23	55	22
919	Mud-minnow........	34	7	24	8	49	....	....	....	....				

ECOLOGICAL TABLE—*continued*

ALL ILLINOIS SPECIES WITH AT LEAST TEN AVAILABLE RECORDS EACH

Jordan and Evermann Nos.	Species	Water (97 species)					Current (49 species)				Bottom (62 species)			
		Available collections	Larger rivers	Smaller rivers	Creeks	Lakes, ponds, etc.	Available collections	Swift to moderate	Sluggish to stagnant	Variable	Available collections	Mud	Rock and sand	Mud and sand
922	Grass pike	111	7	16	34	30	14	36	57	7	29	38	21	41
939	Menona top-minnow	17	....	....	42	49	....	....	....	....	....	....	....	....
966	Striped top-minnow	83	11	....	3	72	....	....	....	....	....	....	....	....
967	Common top-minnow	208	6	25	49	12	34	41	50	9	81	32	42	26
1000	Viviparous top-minnow	17	12	21	32	12	....	....	....	....	....	....	....	....
1145	Trout-perch	15	52	....	4	19	....	....	....	....	....	....	....	....
1147	Pirate-perch	100	18	5	42	21	14	21	72	7	37	62	19	19
1177	Brook silverside	120	13	28	13	36	16	31	44	25	21	10	62	28
1381	White crappie	166	15	19	17	34	14	64	29	7	43	35	49	16
1382	Black crappie	179	17	16	10	42	....	....	....	....	28	25	50	25
1383	Round sunfish	11	....	....	69	30	....	....	....	....	....	....	....	....
1385	Rock bass	48	7	49	24	13	20	55	15	30	27	....	48	52
1387	Warmouth	122	12	17	12	45	....	....	....	....	17	88	12	....
1391	Green sunfish	313	7	25	52	11	80	39	45	16	156	28	41	31
1397	*Lepomis miniatus*	23	10	....	11	41	....	....	....	....	....	....	....	....
1399	Long-eared sunfish	112	2	12	76	4	17	41	47	12	41	37	63	....
1400	Orange-spotted sunfish	174	12	25	34	20	21	38	38	24	60	30	35	35
1403	Bluegill	214	16	10	7	54	....	....	....	....	24	25	58	17
1408	Pumpkinseed	85	6	17	4	56	....	....	....	....	....	....	....	....

ECOLOGICAL TABLE—*continued*

ALL ILLINOIS SPECIES WITH AT LEAST TEN AVAILABLE RECORDS EACH

Jordan and Evermann Nos.	Species	Water (97 species)					Current (49 species)				Bottom (62 species)			
		Available collections	Larger rivers	Smaller rivers	Creeks	Lakes, ponds, etc.	Available collections	Swift to moderate	Sluggish to stagnant	Variable	Available collections	Mud	Rock and sand	Mud and sand
1409	Small-mouthed black bass..............	100	6	43	23	19	40	55	18	27	50	6	68	26
1410	Large-mouthed black bass..............	211	8	20	17	40	19	58	26	16	48	19	54	27
1413	Pike-perch...........	36	16	10	8	33	....	....	....	....	....	....	....	....
1414	Sauger..............	16	36	....	4	25	....	....	....	....	....	....	....	....
1415	Yellow perch.........	83	20	7	3	51	....	....	....	....	....	....	....	....
1417	Log-perch............	60	10	38	27	19	14	93	7	....	20	....	100	....
1418	*Hadropterus phoxocephalus*..............	85	7	57	27	3	32	87	13	....	48	6	94	....
1421	Black-sided darter....	159	6	42	47	1	49	70	30	....	76	16	84	....
1436	*Cottogaster shumardi*...	16	55	....	4	18	....	....	....	....	....	....	....	....
1443	Green-sided darter....	24	....	46	53	....	....	....	....	....	....	....	....	....
1446	Johnny darter........	234	3	25	53	16	71	68	32	....	126	11	89	....
1448	*Boleosoma camurum*...	107	9	23	42	17	17	41	59	....	39	60	40	....
1450	Sand darter..........	19	13	47	39	....	....	....	....	....	....	....	....	....
1461	Banded darter........	32	3	74	23	....	18	89	11	....	19	11	89	....
1474	*Etheostoma jessiæ*.....	158	20	19	16	24	12	83	17	....	31	23	67	....
1477	Rainbow darter.......	80	3	44	45	1	29	83	17	....	37	8	92	....
1489	*Etheostoma squamiceps*	10	....	35	64	....	....	....	....	....	....	....	....	....
1490	Fan-tailed darter.....	30	9	....	87	4	....	....	....	....	11	....	100	....

ECOLOGICAL TABLE—*concluded*

ALL ILLINOIS SPECIES WITH AT LEAST TEN AVAILABLE RECORDS EACH

Jordan and Evermann Nos.	Species	Water (97 species)					Current (49 species)				Bottom (62 species)			
		Available collections	Larger rivers	Smaller rivers	Creeks	Lakes, ponds, etc.	Available collections	Swift to moderate	Sluggish to stagnant	Variable	Available collections	Mud	Rock and sand	Mud and sand
1494	*Boleichthys fusiformis*..	56	1	12	24	62	.....	.....	.....	.....	21	33	67	....
1497	Least darter.........	12	....	....	4	95	.....	.....	.....	.....				
1529	White bass...........	56	28	....	8	46	.....	.....	.....	.....				
1531	Yellow bass.........	100	20	4	....	52	.....	.....	.....	.....				
1871	Sheepshead...........	57	29	16	1	27	.....	.....	.....	.....				

GENERAL SUMMARY

The principal conclusions of this article may be thus summarized:

1. The 150 native species of Illinois fishes here recognized, are so distributed within and without the state as to indicate an unequal commingling of the faunæ of the surrounding territories, southeastern species preponderating over southwestern, northeastern over northwestern, eastern over western, and southern over northern.

2. The Illinois basin may be taken as typical, in its fish population, of the ichthyology of the whole state—occupying, as it does, a central position, including more than half the area of the state, and containing a great variety of waters and situations fit for the habitation of fishes, and more than four fifths of the species found anywhere in Illinois. The more important fishes of the state not known from this basin are a few distinctively northern species, most of which are peculiar to the Great Lakes, and a few southern species which do not range as far north, in this state as the mouth of the Illinois. The

remainder are very rare in our territory, most of them coming from the west and south, and they are extremely insignificant elements of our fish fauna.

3. If the ten stream systems of the state be brought into comparison one with another, it appears that the six larger areas, containing the largest streams and presenting the greatest variety of situations, are much more closely affiliated ichthyologically than are the four smaller areas. The least closely affiliated with each other and with all the rest are the Michigan district of northeastern Illinois and the Big Muddy basin in the southwest. The closest relations are those between the Illinois, the Rock, and the Mississippi.

4. In the absence, in Illinois, of geographical barriers to the dispersal of fishes, the causes influencing their distribution are climatic, geologic, and ecological. As Illinois extends through 5.5° of latitude, differences of climate between the northern and the southern sections of the state are sufficient to affect, in considerable measure, the distribution of its plant and animal species—differences which, in its ichthyology, express themselves in the presence in northern Illinois, but not in southern, of 17 species of general northward range; and in southern Illinois, but not in northern, of 14 species of general southward range. These two groups of species meet and mingle in the great north and south rivers of the western half of the state, in an area of common occupation about fifty miles in width, from the latitude of Springfield northward; while on the eastern boundary of the state, occupied by small streams of various direction, these groups are separated by an interval of about a hundred and seventy-five miles over which no representative of either group has been taken.

5. Geological limitations to the dispersal of fishes are illustrated by peculiarities of distribution in southern Illinois as related to the area of the lower Illinoisan glaciation, which 34 species evidently avoid while 35 other species enter upon it freely and inhabit it successfully. A comparison of the ecological relations of these two groups of species as represented by our collection records, shows that they are strongly distinguished by the repugnance of the first group, and the indifference of the second, to waters with a muddy bottom, collections of the first group having been made from such situations in an average ratio more than three times as great as that for the second. The waters of this region, on the other hand, are re-

markably and persistently turbid, never clearing themselves spontaneously. This is owing in part to the extremely fine division of the soil, and in part to its generally acid character and the consequent lack of "granulation," or cohesion of its ultimate particles in granules, such as occurs in the alkaline soils of the other geological areas of the state. The surface waters of the district are soft and slightly alkaline, but contain much silica, and much solid matter in suspension which it is extremely difficult to remove completely by any ordinary filtering or precipitation process. The inference is plain that it is to this condition of the waters—due to the geological history of the soil of this region—that the unequal distribution of these fishes is largely to be attributed.

6. In consequence of another clearly recognizable inequality of distribution, partly coincident with the two preceding and partly independent of them, two additional groups may be distinguished; one of 8 species, distributed in this state mainly through the Ohio and Wabash drainage, and the other of 27 species, distributed through the Mississippi and its more northerly tributaries. The general distribution throughout the country at large of each of these two groups of species is quite varied, and offers no hint of a reason for these differences in Illinois. Two hypothetical explanations are suggested— the first presupposing different centers of population outside the state, from and towards which these species move, into and out of Illinois streams, with the spring rise, summer recession, and winter cooling of the waters, one of these centers to the west and north, and one to the east and south; and the second presupposing an organization of the fish population into more or less distinct communities of mutually well-adjusted species, each community so adapted to its environment that members of adjacent communities can not successfully intrude upon its territory.

7. An analysis of our statistical data of ecological distribution gives us many instances of a marked difference in preference of situation between nearly related species inhabiting the same area, the effect of which is to break the force of a competition between these species such as would prevail if they were similarly distributed ecologically as well as geographically. Closely related species are, as a consequence, often found much less frequently associated in their common territory than either is with widely unlike species of the same geographical range. Exceptions to this rule are found

where similar species occupy adjacent areas of distribution which merely overlap by their borders.

8. A table of the broader ecological relations of 97 species of Illinois fishes is made the basis of a few general statements, but that subject as a whole is reserved for more detailed treatment elsewhere.

List of Maps

The general map of the distribution of collections (Map IV.) shows, by the location of the red spots, all the localities from which collections of fishes have been made by us in the work of the Natural History Survey. The distribution maps for the various species indicate in the same way all the localities from which representatives of the species have been taken. *For an accurate idea of the significance of these species maps, each should be compared with Map IV.*

The following numbered list of the counties of the state corresponds to the figures on these maps.

1. Jo Daviess	35. Hancock	69. Madison
2. Stephenson	36. McDonough	70. Bond
3. Winnebago	37. Fulton	71. Fayette
4. Boone	38. Mason	72. Effingham
5. McHenry	39. Tazewell	73. Jasper
6. Lake	40. McLean	74. Crawford
7. Cook	41. Vermilion	75. Lawrence
8. Du Page	42. Champaign	76. Richland
9. Kane	43. Piatt	77. Clay
10. DeKalb	44. Dewitt	78. Marion
11. Ogle	45. Logan	79. Clinton
12. Lee	46. Menard	80. St. Clair
13. Carroll	47. Cass	81. Monroe
14. Whiteside	48. Schuyler	82. Randolph
15. Rock Island	49. Brown	83. Washington
16. Mercer	50. Adams	84. Perry
17. Henry	51. Pike	85. Jefferson
18. Bureau	52. Scott	86. Wayne
19. Putnam	53. Morgan	87. Edwards
20. La Salle	54. Sangamon	88. Wabash
21. Kendall	55. Christian	89. White
22. Grundy	56. Macon	90. Hamilton
23. Will	57. Moultrie	91. Franklin
24. Kankakee	58. Douglas	92. Jackson
25. Iroquois	59. Edgar	93. Williamson
26. Ford	60. Clark	94. Saline
27. Livingston	61. Coles	95. Gallatin
28. Marshall	62. Cumberland	96. Hardin
29. Woodford	63. Shelby	97. Pope
30. Stark	64. Montgomery	98. Johnson
31. Peoria	65. Macoupin	99. Union
32. Knox	66. Greene	100. Alexander
33. Warren	67. Calhoun	101. Pulaski
34. Henderson	68. Jersey	102. Massac

I. & II.   The three sections of Illinois, northern, central, and southern, and the ten stream systems of the state: I., the Galena District, II., the Rock River System, III., the Illinois River System, IV., the Lake Michigan drainage, V., the Mississippi River drainage, VI., the Kaskaskia River System, VII., the Wabash System, VIII., the Big Muddy River System, IX., the Saline River System, and X., the Cairo District.

III.   Glacial geology of Illinois.

IV.   Localities from which collections were made.

V.—CII.   Distribution of species.

CIII.   Northern species (●) and southern species (▲) in Illinois.

V.	Lepisosteus osseus	LIV.	A. nebulosus
VI.	L. platostomus	LV.	A. melas
VII.	Amia calva	LVI.	Leptops olivaris
VIII.	Dorosoma cepedianum	LVII.	Noturus flavus
IX.	Ictiobus cyprinella	LVIII.	Schilbeodes gyrinus
X.	I. urus	LIX.	S. miurus
XI.	I. bubalus	LX.	Umbra limi
XII.	Carpiodes carpio	LXI.	Esox vermiculatus
XIII.	C. difformis	LXII.	E. lucius
XIV.	C. velifer	LXIII.	Fundulus diaphanus
XV.	C. thompsoni		menona
XVI.	Erimyzon sucetta ob-	LXIV.	F. dispar
	longus	LXV.	F. notatus
XVII.	Minytrema melanops	LXVI.	Gambusia affinis
XVIII.	Catostomus commersonii	LXVII.	Percopsis guttatus
XIX.	C. nigricans	LXVIII.	Labidesthes sicculus
XX.	Moxostoma anisurum	LXIX.	Aphredoderus sayanus
XXI.	M. aureolum	LXX.	Pomoxis annularis
XXII.	M. breviceps	LXXI.	P. sparoides
XXIII.	Campostoma anomalum	LXXII.	Centrarchus macropterus
XXIV.	Chrosomus erythrogaster	LXXIII.	Ambloplites rupestris
XXV.	Hybognathus nuchalis	LXXIV.	Chænobryttus gulosus
XXVI.	H. nubila	LXXV.	Lepomis miniatus
XXVII.	Pimephales promelas	LXXVI.	L. megalotis
XXVIII.	P. notatus	LXXVII.	L. humilis
XXIX.	Semotilus atromaculatus	LXXVIII.	L. pallidus
XXX.	Opsopœodus emiliæ	LXXIX.	Eupomotis gibbosus
XXXI.	Abramis crysoleucas	LXXX.	Micropterus dolomieu
XXXII.	Cliola vigilax	LXXXI.	M. salmoides
XXXIII.	Notropis cayuga	LXXXII.	Stizostedion vitreum
XXXIV.	N. heterodon	LXXXIII.	S. canadense griseum
XXXV.	N. blennius	LXXXIV.	Perca flavescens
XXXVI.	N. gilberti	LXXXV.	Percina caprodes
XXXVII.	N. illecebrosus	LXXXVI.	Hadropterus phoxo-
XXXVIII.	N. hudsonius		cephalus
XXXIX.	N. lutrensis	LXXXVII.	H. aspro
XL.	N. whipplii	LXXXVIII.	Cottogaster shumardi
XLI.	N. cornutus	LXXXIX.	Diplesion blennioides
XLII.	N. jejunus	XC.	Boleosoma nigrum
XLIII.	N. atherinoides	XCI.	B. camurum
XLIV.	N. rubrifrons	XCII.	Ammocrypta pellucida
XLV.	N. umbratilis atripes	XCIII.	Etheostoma zonale
XLVI.	Ericymba buccata	XCIV.	E. jessiæ
XLVII.	Phenacobius mirabilis	XCV.	E. cœruleum
XLVIII.	Hybopsis dissimilis	XCVI.	E. squamiceps
XLIX.	H. amblops	XCVII.	E. flabellare
L.	H. storerianus	XCVIII.	Boleichthys fusiformis
LI.	H. kentuckiensis	XCIX.	Microperca punctulata
LII.	Ictalurus punctatus	C.	Roccus chrysops
LIII.	Ameiurus natalis	CI.	Morone interrupta
		CII.	Aplodinotus grunniens

I & II
Stream systems
and
Sections of
Illinois

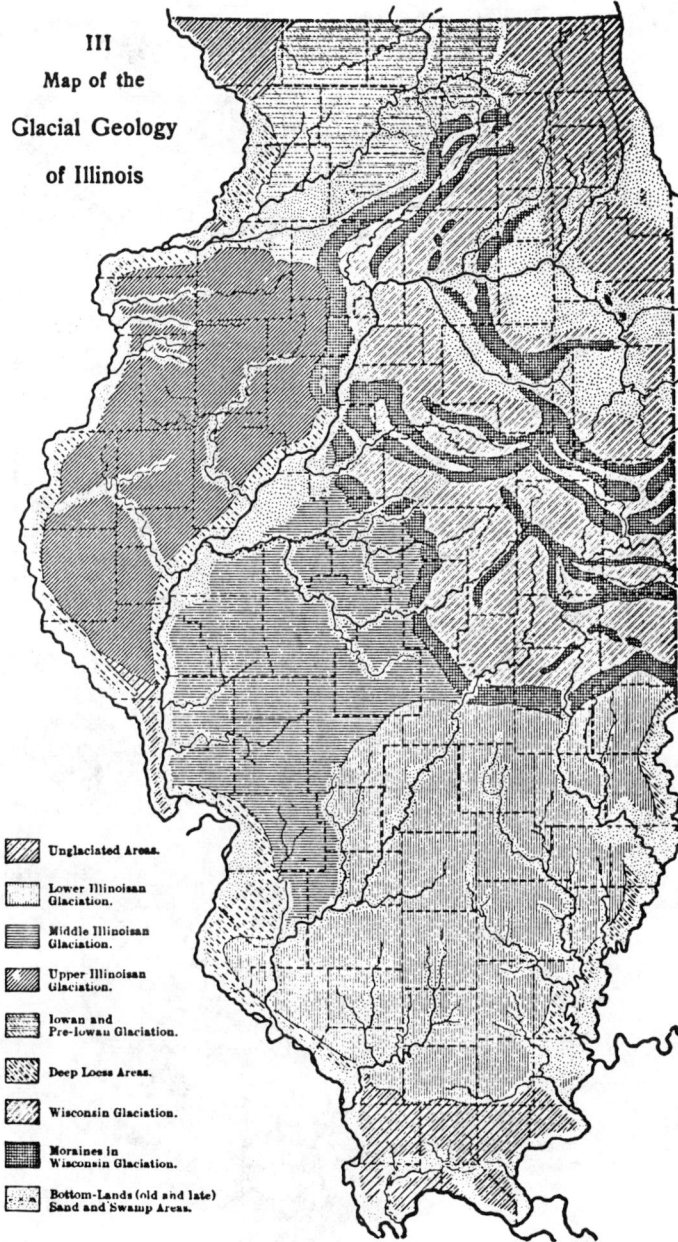

III

Map of the

Glacial Geology

of Illinois

Unglaciated Areas.

Lower Illinoisan
Glaciation.

Middle Illinoisan
Glaciation.

Upper Illinoisan
Glaciation.

Iowan and
Pre-Iowan Glaciation.

Deep Loess Areas.

Wisconsin Glaciation.

Moraines in
Wisconsin Glaciation.

Bottom-Lands (old and late)
Sand and Swamp Areas.

IV
Water Courses
and
Distribution
of
Collections

- - - - - Illinois and Michigan Canal
- - - - - Illinois and Mississippi Canal
- - - - - Drainage Canal
County Seat

HISTORY OF ECOLOGY
An Arno Press Collection

Abbe, Cleveland. **A First Report on the Relations Between Climates and Crops.** 1905

Adams, Charles C. **Guide to the Study of Animal Ecology.** 1913

American Plant Ecology, 1897-1917. 1977

Browne, Charles A[lbert]. **A Source Book of Agricultural Chemistry.** 1944

Buffon, [Georges-Louis Leclerc]. **Selections from Natural History, General and Particular, 1780-1785.** Two volumes. 1977

Chapman, Royal N. **Animal Ecology.** 1931

Clements, Frederic E[dward], John E. Weaver and Herbert C. Hanson. **Plant Competition.** 1929

Clements, Frederic Edward. **Research Methods in Ecology.** 1905

Conard, Henry S. **The Background of Plant Ecology.** 1951

Derham, W[illiam]. **Physico-Theology.** 1716

Drude, Oscar. **Handbuch der Pflanzengeographie.** 1890

Early Marine Ecology. 1977

Ecological Investigations of Stephen Alfred Forbes. 1977

Ecological Phytogeography in the Nineteenth Century. 1977

Ecological Studies on Insect Parasitism. 1977

Espinas, Alfred [Victor]. **Des Sociétés Animales.** 1878

Fernow, B[ernhard] E., M. W. Harrington, Cleveland Abbe and George E. Curtis. **Forest Influences.** 1893

Forbes, Edw[ard] and Robert Godwin-Austen. **The Natural History of the European Seas.** 1859

Forbush, Edward H[owe] and Charles H. Fernald. **The Gypsy Moth.** 1896

Forel, F[rançois] A[lphonse]. **La Faune Profonde Des Lacs Suisses.** 1884

Forel, F[rançois] A[lphonse]. **Handbuch der Seenkunde.** 1901

Henfrey, Arthur. **The Vegetation of Europe, Its Conditions and Causes.** 1852

Herrick, Francis Hobart. **Natural History of the American Lobster.** 1911

History of American Ecology. 1977

Howard, L[eland] O[ssian] and W[illiam] F. Fiske. **The Importation into the United States of the Parasites of the Gipsy Moth and the Brown-Tail Moth.** 1911

Humboldt, Al[exander von] and A[imé] Bonpland. **Essai sur la Géographie des Plantes.** 1807

Johnstone, James. **Conditions of Life in the Sea.** 1908

Judd, Sylvester D. **Birds of a Maryland Farm.** 1902

Kofoid, C[harles] A. **The Plankton of the Illinois River, 1894-1899.** 1903

Leeuwenhoek, Antony van. **The Select Works of Antony van Leeuwenhoek.** 1798-99/1807

Limnology in Wisconsin. 1977

Linnaeus, Carl. **Miscellaneous Tracts Relating to Natural History, Husbandry and Physick.** 1762

Linnaeus, Carl. **Select Dissertations from the Amoenitates Academicae.** 1781

Meyen, F[ranz] J[ulius] F. **Outlines of the Geography of Plants.** 1846

Mills, Harlow B. **A Century of Biological Research.** 1958

Müller, Hermann. **The Fertilisation of Flowers.** 1883

Murray, John. **Selections from** *Report on the Scientific Results of the Voyage of H.M.S. Challenger During the Years 1872-76.* 1895

Murray, John and Laurence Pullar. **Bathymetrical Survey of the Scottish Fresh-Water Lochs.** Volume one. 1910

Packard, A[lpheus] S. **The Cave Fauna of North America.** 1888

Pearl, Raymond. **The Biology of Population Growth.** 1925

Phytopathological Classics of the Eighteenth Century. 1977

Phytopathological Classics of the Nineteenth Century. 1977

Pound, Roscoe and Frederic E. Clements. **The Phytogeography of Nebraska.** 1900

Raunkiaer, Christen. **The Life Forms of Plants and Statistical Plant Geography.** 1934

Ray, John. **The Wisdom of God Manifested in the Works of the Creation.** 1717

Réaumur, René Antoine Ferchault de. **The Natural History of Ants.** 1926

Semper, Karl. **Animal Life As Affected by the Natural Conditions of Existence.** 1881

Shelford, Victor E. **Animal Communities in Temperate America.** 1937

Warming Eug[enius]. **Oecology of Plants.** 1909

Watson, Hewett Cottrell. **Selections from** *Cybele Britannica.* 1847/1859

Whetzel, Herbert Hice. **An Outline of the History of Phytopathology.** 1918

Whittaker, Robert H. **Classification of Natural Communities.** 1962